Investigating Chemistry

Investigating Chemistry

A Forensic Science Perspective

Matthew E. Johll
Illinois Valley Community College

W. H. Freeman and Company • New York

Publisher: Craig Bleyer
Executive Acquisitions Editor: Clancy Marshall
Developmental Editor: Donald Gecewicz
Market Development Manager: Kirsten Watrud
Marketing Manager: Anthony Palmiotto
Editorial Assistant: Carrie Wright
Design Manager: Diana Blume
Project Editor: Vivien Weiss
Illustration Coordinator: Susan Timmins
Illustrations: Network Graphics
Photo Editor: Patricia Marx
Photo Researchers: Julie Tesser, Elyse Rieder, Tracey Thompkins
Production Manager: Julia DeRosa
Media Editor: Victoria Anderson
Associate Editor: Amy Thorne
Composition: Matrix
Printing and Binding: RR Donnelley

COVER IMAGE: Gary S. Chapman/The Image Bank/Getty Images

Library of Congress Control Number: 2006921262

ISBN: 0-7167-6433-4
EAN: 9780716764335

First printing

W. H. Freeman and Company
41 Madison Avenue
New York, NY 10010
www.whfreeman.com

Contents

Preface

Investigating Chemistry: A Forensic Science Perspective is the first textbook of its kind: a chemistry textbook for liberal arts students with forensic science as its overarching theme. I chose this theme because of its demonstrated appeal to a wide audience and its ability to provide students a captivating context for learning the fundamental concepts of chemistry.

Society's fascination with criminal investigations is reflected in the popularity of a long line of literary whodunits, radio detective series, Hollywood murder mysteries, and hit television shows that focus on crime solving and police work. A few examples are Edgar Allan Poe's 1841 mystery *The Murders in the Rue Morgue,* the Sherlock Holmes stories of Arthur Conan Doyle, Agatha Christie's mystery novels, and the adventures of unforgettable crime solvers like Nancy Drew, Dick Tracy, Perry Mason, and Columbo. The application of science to criminal investigations plays an important part in many stories and is particularly emphasized in the current television series *CSI: Crime Scene Investigation,* which intrigues millions. In my teaching I have found that this interest in forensic science provides an excellent opportunity for engaging students in the study of chemistry.

This book is written specifically for liberal arts or nonscience majors. Students do not need a background in analytical chemistry to understand the forensic science—the textbook is written with the assumption that students will have had little, if any, previous experience with chemistry. The basic principles typically covered in this course are carefully explained.

Chemistry Through the Lens of Forensic Science

In teaching a chemistry course for liberal arts students, I have come up against several challenges that are well known to science instructors. The first is the reluctant audience of nonscience majors, who bring with them fears of chemistry and, perhaps, bad memories of past experiences with science courses that have not gone well for them. To draw students in, and to illustrate chemical principles vividly, I use crime-scene case studies, Sherlock Holmes stories, and true accounts of drug deals, murders, and thefts. This material is clearly of interest to students: The number of forensic science majors is growing, as are the ratings of television shows like *CSI: Crime Scene Investigation*. The public's interest in watching various series about police work—whether reality TV or more traditional scripted programs—shows no sign of lessening. Mysteries by such writers as Janet Evanovich, Carl Hiaasen, and Dan Brown top the bestseller lists, as do investigative accounts of white-collar crime.

The second challenge is the wide range of math abilities that students bring to the course. Many students struggle with even the most basic algebra.

I wrote a textbook that features flexible quantitative coverage, so that teachers can tailor the course to the needs and abilities of their students.

Because most of my students are taking the course to satisfy a science requirement, I endeavored to communicate to them the important chemical principles effectively. I selected a theme—forensic science—that has broad appeal but does not sacrifice the more quantitative aspects of chemistry.

Because this book covers the traditional material of a chemistry course, faculty can use it confidently without a formal forensic science background. In fact, almost all crime laboratory scientists have science backgrounds that include a traditional degree in either chemistry or biology.

The forensics theme gives instructors of chemistry an opportunity to attract more liberal arts students, and it provides those students with a new perspective on chemistry. One reviewer calls the chapters he has seen "a guaranteed draw," and, indeed, I hope you'll find that to be the case.

Goals of This Book

My goals in writing *Investigating Chemistry* are to communicate to liberal arts students the same excitement and enthusiasm that I have for chemistry, to explain chemical principles in a clear and accessible way, and to draw on the many intriguing examples afforded by forensic science to illustrate chemical concepts.

Why is studying criminal investigations and forensics such a good way to learn about the fundamental concepts of chemistry?

The investigation of a crime entails using procedures of the scientific method and very often involves chemistry. A criminal investigation also brings together more than one application of chemistry. Studying a criminal investigation means students must consider the evidence found at a crime site to be a substance whose composition must be determined, or a chemical compound that underwent chemical reactions or phase changes. Students need to consider quantitative measurements, pollution and poisons, and even how to trace and identify compounds using some fairly sophisticated equipment.

To me, this is what chemistry is all about, and I want to show students that these big ideas—ideas that have changed our way of thinking forever—are accessible to them, too.

Organization

The organizational scheme of *Investigating Chemistry* follows a pattern typical of many chemistry textbooks but sets the standard chemical principles within the context of forensic science. Every chapter begins with a true case study and ends with its resolution—after students are introduced to the chemistry principles that figure prominently in the case.

To illustrate science concepts as they are introduced, many additional forensic chemistry examples are contained within the chapters as well. Quantitative principles are illustrated with worked examples showing step-by-step solutions to problems throughout the chapter. As explained in the following review of chapter content, chemical concepts are systematically developed.

✦ **Chapter 1—Introduction to Forensic Chemistry**—begins with a case study that involves the disappearance of a U.S. Drug Enforcement Administration agent in Mexico. But the resolution of the story remains a mystery until the end of the chapter. The text leads students into basic principles first: matter and its forms; elements, compounds, and mixtures; the periodic table and its uses; names and symbols of the elements; and chemical formulas. These topics form the basis for understanding the nature of evidence gathered at crime scenes, and examples of such evidence are given throughout the chapter.

The importance of careful observation to forensic science—and to all areas of science—is also discussed. An evidence analysis section focuses on thin-layer chromatography as an analytical method to identify substances. Finally, the resolution of the case study concludes the chapter, revealing how careful observation of evidence plus a basic knowledge of matter provided key clues to the discovery of a massive cover-up surrounding the disappearance and death of the DEA agent.

✦ **Chapter 2—Evidence Collection and Preservation**—defines physical and chemical changes and how they relate to the preservation of forensic evidence. The chapter includes mass, weight, and units of measurement; the mathematics of unit conversions; errors and estimates in laboratory measurements; significant figures in calculations; accuracy and precision of experimental results; density and calculations using density measurements. A section on how to measure glass and soil evidence using physical properties helps students see the practical applications of these principles. The case study shows how analysis of soil sample evidence helped secure the conviction of mass murderer Thomas "Tommy Karate" Pitera, a member of a New York "crime family."

✦ **Chapter 3—Atomic Clues**—introduces atomic theory, beginning with ideas of the ancient Greek philosophers; touching on the contributions of Gassendi, Lavoisier, and Proust; exploring Dalton's atomic theory; and discussing our current understanding of atomic structure. Topics include subatomic particles, isotopes, atomic mass, orbitals, and electron configurations. The chapter explains the experimental evidence behind these discoveries, including J. J. Thompson's cathode-ray-tube experiments, Rutherford's alpha-particle scattering experiments, and emission spectra of the elements. A discussion of light energy includes concepts of wavelength, frequency, and the speed of light, as well as simple calculations involving these quantities. An explanation of how gunshot residue is analyzed demonstrates how the principles introduced in the chapter apply to real situations. The chapter case study hinges on analysis of elements in the cremated remains of the human body.

✦ **Chapter 4—Chemical Evidence**—begins with a discussion of regions of the periodic table and the terminology used to discuss the elements. The text then proceeds systematically through ionic and covalent compounds, rules for writing chemical formulas and naming compounds, the basics of chemical reactions, balancing equations, the mole concept, and calculations involving moles and stoichiometry. Worked examples throughout the chapter show step-by-step solutions to problems. Types of reactions—including precipitation, combustion, neutralization, and reduction-oxidation—are defined and examples are provided. A section on spectrophotometry describes how this widely used technique is applied to qualitative and quantitative identification of substances.

Chapters 5 and 6 cover a range of topics related to solutions and their properties. Considerable attention is given to these subjects because of their importance in chemistry and in forensic science—where evidence or the tests used to analyze evidence often involve solutions.

✦ **Chapter 5—Properties of Solutions I: Aqueous Solutions—** focuses on the process of dissolution, factors affecting the rates at which various substances dissolve, and properties of electrolyte and nonelectrolyte solutions. The concept of solubility is then explored further through solubility rules, discussion of saturated and unsaturated solutions, and use of net ionic equations to portray precipitation reactions. Molarity is introduced as a method for expressing the concentration of solutions, and examples of calculations involving molarity are given. Additional topics include acid and base solutions, the pH scale, and buffer solutions.

✦ **Chapter 6—Properties of Solutions II: Intermolecular Forces and Colligative Properties—**begins with surface tension as an example of how the properties of a liquid are related to intermolecular forces and, on a practical level in forensic science, how an understanding of surface tension informs blood spatter analysis. A discussion of the various types of intermolecular forces present in solution develops concepts needed to understand boiling point elevation, freezing point depression, and osmosis. A basic knowledge of intermolecular forces is also important to material covered in several of the succeeding chapters. High performance liquid chromatography is featured as a technique for evidence analysis.

✦ **Chapter 7—Drug Chemistry—**provides an introduction to organic chemistry. Drugs afford many fascinating examples of organic compounds and are a high-interest topic. The chapter begins by describing classes of organic compounds including alkanes, alkenes, alkynes, ethers, ketones, esters, amines, alcohols, aldehydes, and carboxylic acids. The text continues with the structural formulas for organic compounds, rules for naming of compounds, and examples for each class of compound. Branched isomers and cyclic compounds are also discussed. The use of infrared spectroscopy for drug analysis provides an interesting example of a common method for identifying unknown compounds.

✦ **Chapter 8—Chemistry of Addiction—**focuses on molecular geometry, a concept essential to understanding how drugs act in the human body and how analytical methods such as immunoassay work for drug identification. The chapter begins by exploring the nature of covalent bonds, a topic introduced briefly in Chapter 4. Concepts include the Lewis theory of bonding, how to write Lewis structures, valence shell electron pair repulsion (VSEPR) theory, and how the geometry of electron pairs in a molecule determines the structural arrangement of its atoms. Building on these concepts, the chapter shows how the complex three-dimensional shapes of drug molecules influence the mechanisms by which drugs affect the human body—for example, how drug molecules interact with neurons to produce a high.

✦ **Chapter 9—Arson Investigation—**starts off with the chemistry of fire and combustion reactions. Concepts of heat, temperature, and thermal equilibrium are explained, as are heat capacity, phase changes, the first law of thermodynamics, and calorimetry. An introduction to the process of petroleum refinement provides background information needed to understand various types of accelerants used in arson cases. The technique of gas chromatography illustrates an analytical method that exploits differences in the boiling points and polarities of compounds to separate the components of a mixture such as one collected from a scene where arson is suspected.

✦ **Chapter 10—Chemistry of Explosions—**begins with descriptive information about explosives and the chemistry of explosion reactions. Because explosions produce large quantities of gaseous products, the kinetic-molecular theory of gases is introduced in this chapter to explain how gases behave. The gas laws formulated by Boyle, Charles, Gay-Lussac, and Avogadro are explained, as is the ideal gas law. Examples of calculations involving pressure, volume, temperature, and moles of gas are given. Additional topics include Dalton's law of partial pressures, vapor pressure of solid explosives, and how molecules of an explosive can be detected in an air sample taken from the vicinity of the compound.

✦ **Chapter 11—Estimating the Time of Death—**introduces chemical kinetics and the collision theory of reactions. The chapter discusses how the rate of a reaction is affected by temperature, catalysts, and concentration of reactants. Zero-order and first-order reactions are explained. Forensic examples include blood detection by a reaction of luminol with hydrogen peroxide to produce a luminescent compound, and the oxidation of alcohol in the human body as an example of a zero-order reaction. The half-life of a reactant is shown to be a useful concept in determining the time it takes for a drug to be eliminated from the body. The chapter case study demonstrates the challenges of estimating time of death in a complex murder investigation.

✦ **Chapter 12—Dirty Bombs and Nuclear Terrorism—**presents concepts relating to radioactivity. The chapter proceeds from the discovery of natural radioactivity in the late nineteenth century to current medical and military applications of nuclear isotopes, the operation of nuclear power plants, and the potential hazards of terrorism through the use of dirty bombs. Other topics include radiation types (alpha and beta particles, gamma rays) and hazards, the half-lives of radioactive isotopes, and nuclear transmutations.

✦ **Chapter 13—Poisons—**leads into the subject of chemical equilibrium by describing the process by which poisons that have entered the bloodstream harm the human body. The ability of a toxic compound to pass out of the blood into cells and to disrupt cell function is largely determined by whether or not it will bind with albumin in the blood plasma. After explaining that the determination of the relative amounts of bound and unbound toxic substance requires an understanding of chemical equilibrium, the chapter proceeds through discussions of dynamic equilibrium, the equilibrium constant, Le Chatelier's principle, equilibrium systems involved in respiration, and solubility equilibrium.

✦ **Chapter 14—Identification of Victims: DNA Analysis**—begins with a discussion of the unprecedented challenges involved in identifying the victims who perished in the September 11, 2001, World Trade Center attack. Because ordinary means of identification such as fingerprint analysis or dental evidence could not be used for the majority of victims, scientists had to use DNA analysis technology to its fullest capacity. Before describing the basics of DNA analysis, the chapter introduces lipids, carbohydrates, and proteins. Examples are given for each type of compound, with particular attention paid to examples important in forensic identification, and chemical structures are discussed. Concepts involving protein structure are highlighted as essential background for understanding the nature of DNA and DNA analysis.

Features

Chapter-Opening Case Studies follow real-life crime dramas in which forensic chemistry was used to solve the case. In each chapter, a single case study involves students in the forensic process while demonstrating essential chemistry concepts.

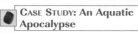

CASE STUDY: An Aquatic Apocalypse

The White River runs through central Indiana and provides drinking water and fishing and boating recreation for many communities, including the capital city of Indianapolis. In Anderson, Indiana, treated wastewater from both residential and industrial sources is discharged into the river. The Anderson Publicly Owned Treatment Works (Anderson POTW), a wastewater treatment plant, is responsible for ensuring that the water discharged into the White River meets all state and national regulations for quality. To monitor the quality of the discharge, workers at the facility take samples at reg-

An estimated 5 million fish weighing over 187 tons were killed between December 1999 and January 2000 along a 50-mile stretch of the White River that runs through central Indiana between the cities of Anderson and Indianapolis. (Photodisc Green/Getty Images)

Learning Objectives start each section and draw the students' attention to the key ideas they will encounter.

Visuals include vivid photos of forensic investigations and evidence that will intrigue students and lead them into the discussion of chemical concepts. At the same time, the text does not neglect the tried-and-true figures and photos that illustrate chemical concepts so memorably. Chapters include photos of reactions and of equipment, molecular models, tables, and graphs.

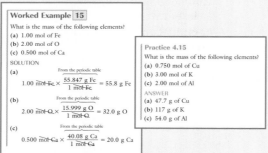

Worked Example 15

What is the mass of the following elements?
(a) 1.00 mol of Fe
(b) 2.00 mol of O
(c) 0.500 mol of Ca

SOLUTION

(a) From the periodic table

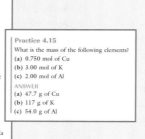

$$1.00\ \text{mol Fe} \times \frac{55.847\ \text{g Fe}}{1\ \text{mol Fe}} = 55.8\ \text{g Fe}$$

(b) From the periodic table

$$2.00\ \text{mol O} \times \frac{15.999\ \text{g O}}{1\ \text{mol O}} = 32.0\ \text{g O}$$

(c) From the periodic table

$$0.500\ \text{mol Ca} \times \frac{40.08\ \text{g Ca}}{1\ \text{mol Ca}} = 20.0\ \text{g Ca}$$

Practice 4.15

What is the mass of the following elements?
(a) 0.750 mol of Cu
(b) 3.00 mol of K
(c) 2.00 mol of Al

ANSWER
(a) 47.7 g of Cu
(b) 117 g of K
(c) 54.0 g of Al

Worked Examples Paired with Practice Problems give students a helpful roadmap for solving problems. Worked examples show step-by-step solutions, including the "simple" steps that are sometimes left out of texts on the assumption that all college students can do simple algebra. A practice problem follows each worked example, immediately reinforcing what is taught.

Mathematics Sections that go into greater depth on the quantitative aspects of subject matter are specially designated so that instructors can tailor the coverage of mathematics to their course syllabi.

4.10 Mathematics of Chemical Reactions: Limiting Reactants and Theoretical Yields

In many laboratory experiments, it is usual for at least one of the reactants to be present in an excess amount—beyond the amount that could be completely consumed by the other reactants. An alternate perspective is that one of the reactants will be completely consumed before the others are used up, at which point the reaction will come to a stop. The reactant that is completely consumed is called the **limiting reactant** (or limiting reagent). It is critical to know which reactant is the limiting reactant because the maximum amount of product that can be theoretically produced, the **theoretical yield**, is dictated by the limiting reagent. Firefighters use this same idea when approaching a fire fight. The goal of the firefighter is to prevent either further fuel or further oxygen from reaching the flames. When firefighters are successful, the fire will stop once the existing fuel or oxygen is consumed.

Limiting reagent calculations are modifications of the stoichiometry problems that were introduced earlier in this chapter. The essence of the limiting reagent calculation is to start with a mass of a reactant (it doesn't

Learning Objective
Apply stoichiometry calculations in solving limiting reactant problems.

Evidence Analysis | Thin-Layer Chromatography

To make a mixture, components can simply be combined physically. But separating a mixture into its components after they are mixed is not always easy. Separation requires an understanding of the properties and behavior of each component.

One of the main methods of separating mixtures in the forensic science laboratory is called **chromatography.** Chromatography was originally developed in the early 1900s to separate the colored pigments in flowers. The name *chromatography* literally translates as "color writing" because the individual colors separate from the mixture. Chromatography exploits the fact that different compounds are attracted to or repelled by other compounds to varying degrees, and the extent of attraction or repulsion is always the same for any two given compounds. We will talk about several forms of chromatography throughout this book, as it is one of the main methods used for identifying drugs, explosives, narcotics, and fibers.

If you place a drop of ink or a mixture of food coloring on a strip of filter paper and then dip the strip into a liquid such as water or alcohol, allowing the liquid to move upward over the spot, you will see chromatography in action. Most inks consist of a mixture of colored compounds, as does a combination of food coloring. The compounds in the mixture have different levels of attraction for the paper (called the *stationary phase*) and for the liquid (called the *mobile phase*) passing over them. Chromatography allows the different compounds to be separated, so that you can see different colored spots at different distances from the starting point.

The simplest form of chromatography described above is called *paper chromatography*. In most laboratories, however, a variation of this technique, called **thin-layer chromatography (TLC)**, is used. As with paper chromatography, there are always two parts to a TLC system, a mobile phase and a stationary phase. In TLC, the mobile phase is a liquid such as water; the stationary phase is a thin solid coating (such as silica gel) on a glass plate. The mixture to be separated is added onto the bottom of the stationary phase. The mobile phase is then allowed to move across the stationary phase. Each compound in the mixture is attracted to the liquid mobile phase and to the stationary solid phase to a different degree than are all of the other compounds in the mixture. If a compound is attracted only to the mobile phase, it will move as fast as the mobile phase. If a compound is attracted only to the stationary phase, it will not move at all. In most cases, a compound will be attracted somewhat to both the mobile phase and the stationary phase and it will move slowly across the stationary phase.

The key to chromatography is that each compound ends up moving at a speed slightly different from all other compounds; this separates each compound in the mixture. We can identify the compounds present in a mixture by determining if they move at the same speed as a known compound. The TLC plate shown in Figure 1.5 illustrates how a sample containing illegal drugs and narcotics can have each component identified. TLC is also used commonly for analyzing inks from kidnapping notes and for determining the types of explosives used in terrorist attacks.

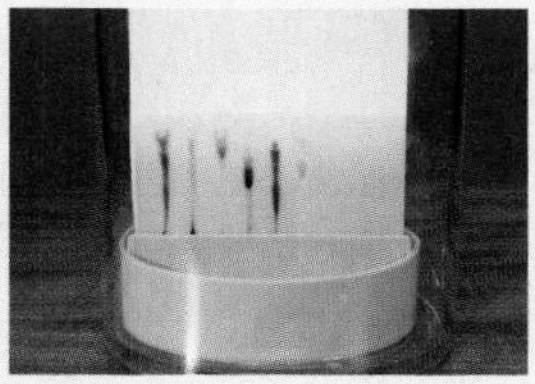

Water-soluble markers are often mixtures of several dyes that produce the desired colors. (Photo Researchers, Inc.)

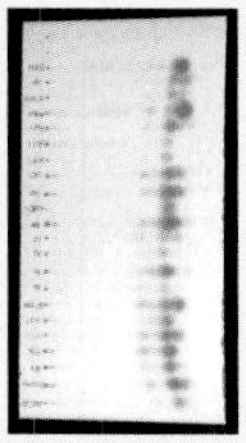

Figure 1.5 Thin-layer chromatography separates the compounds found in illegal drugs. Common drug samples are analyzed alongside the evidence. If a spot appears on both samples, a tentative identification is made. Further experiments provide a positive identification. (Missouri State Highway Patrol)

Evidence Analysis Boxes introduce laboratory techniques, instrumentation, and methods for studying evidence through chemistry.

End-of-Chapter Summaries in bulleted format help students to review the major concepts introduced in each chapter. A **Key Terms** list, arranged sequentially, accompanies the chapter summary.

Continuing the Investigation lists at the end of the chapter provide references for further research.

CONTINUING THE INVESTIGATION Additional Readings, Resources, and References

Baumann, E., and O'Brien, J. *Murder Next Door: How Police Tracked Down 18 Brutal Killers,* Chicago: Bonus Books, 1991.

Doyle, A. C. *Sherlock Holmes: The Complete Novels and Stories,* vol. 1, New York: Random House, 2003.

Lubasch, A. H. "Reputed Mobster Guilty in Six Narcotics Murders; Death Penalty Possible Under Federal Law," *New York Times,* New York, June 26, 1992, p. B3

Missouri State Highway Patrol Forensic Laboratory. *Forensic Evidence Handbook,* Jefferson City: Missouri State Highway Patrol, 2003.

Murray, R. C., and Tedrow, J. C. F. *Forensic Geology,* Englewood Cliffs: Prentice Hall, 1998.

For more information about the properties of glass and ceramics: www.mindrum.com/tech.html

End-of-Chapter Questions and Problems include paired exercises in which each odd-numbered question or problem is followed by an even-numbered exercise of the same type. **Forensic Chemistry Problems** provide students an opportunity to apply basic chemistry concepts to problems in crime investigation. **Case Study Problems** examine other crimes and situations that relate to the chapter content and show students how to investigate chemical concepts that come up in criminal and legal cases.

Answers to odd-numbered exercises are supplied at the end of the book. The *Student Solutions Manual* provides step-by-step solutions to odd-numbered problems.

Media/Supplements

Companion Web site at www.whfreeman.com/johll includes:

- **Chapter Outlines** to highlight key concepts and topics for every chapter
- **Self Quizzes** to help students study, review, and prepare for exams. Instructors can access results through an online database or can have results e-mailed directly to their account.
- **Animations/Videos**
- **Web Links** to provide various Web resources that tie into the concepts covered in the text
- **Chemistry in Action** commercial video clips, courtesy of Films for the Humanities and Sciences, to depict how chemistry is used in everyday life

- **Flashcards,** or key words and definitions from the textbook, set up in an interactive flashcard format
- **Interactive Periodic Table**
- **Learning Tutorials,** based on *CSI*-like crime scenarios, to probe students to think about the forensic science involved in this liberal arts chemistry course

Test Bank, 0-7167-7483-6 (printed) or 0-7167-7482-8 (computerized), by Mark Benvenuto, University of Detroit–Mercy
The test bank contains 100 multiple-choice and short-answer questions per chapter. The easy-to-use CD-ROM includes Windows and Mac versions on a single disc, in a format that lets instructors add, edit, and resequence questions to suit their needs.

Enhanced Instructor's Resource CD-ROM, 0-7167-7481-X
To help create lecture presentations, Web sites, and other resources, instructors can use this CD-ROM to **search** and **export** all the following resources by key term or by chapter:

- All text images
- Animations, Videos, Flashcards, and more
- PowerPoint files (lecture slides)
- Test Bank files

Overhead Transparencies, 0-7167-7484-4
Includes key illustrations, figures, and tables from the text.

***Investigating Chemistry Laboratory Manual*, 0-7167-7485-2, by David Collins, Colorado State University**
Written to specifically accompany Johll's *Investigating Chemistry,* this manual contains a wide variety of innovative experiments covering the basic topics of introductory chemistry and forensic science. Detailed instructions allow students to record their observations and reach conclusions while reinforcing key concepts from the text.

***Student Solutions Manual*, 0-7167-7486-0, by Jason Powell, Ferrum College**
This solutions manual contains step-by-step solutions and explanations to the odd-numbered questions and problems that appear at the end of each chapter. It is designed to help students understand the material better and avoid common mistakes.

Acknowledgments

First and foremost, I would like to thank my wife, Sally, for her encouragement, support, and understanding during the last several years. I could not have accomplished the writing of this book without her. Thank you to my sons, Benjamin and Alex, too, for pulling me away from my computer and reprioritizing my deadlines with playtime. I would also like to thank my parents, Greg and Barb, for making their children the focus and priority of their lives. Our collective successes are rooted in the foundation you gave us. My thanks also to Diane, Deb, Marty, and Mike for all the help, counseling, and encouragement a brother could ever need. Sally and I will always be grateful for the help you have given us over the years.

Finally, I would like to thank the editorial, marketing, and production team at W. H. Freeman and Company Publishers for their efforts in making this unique textbook a reality. Namely, Craig Bleyer, Publisher; Clancy Marshall, Executive Acquisitions Editor; Donald Gecewicz, outside Developmental Editor; Mary Ann Ryan, Assistant Developmental Editor; Kirsten Watrud, Market Development Manager; Anthony Palmiotto, Marketing Manager; Torie Anderson, Media Editor; Amy Thorne, Associate Editor; Carrie Wright, Editorial Assistant; Diana Blume, Design Manager; Vivien Weiss, Project Editor; Patricia Marx, Photo Editor; Julia DeRosa, Production Manager; and the rest of the Freeman team.

We are grateful to the following professors who reviewed the manuscript and offered helpful suggestions for improvement:

DeeDee A. Allen, *Shaw University*
Georgia Arbuckle, *Rutgers University*
Bruce Baldwin, *Spring Arbor College*
Theodore C. Baldwin, *Olympic College*
Harshavardhan Bapat, *University of Illinois–Springfield*
John Barbaro, *Rocky Mountain College*
Holly D. Bendorf, *Lycoming College*
Mark Benvenuto, *University of Detroit–Mercy*
Carol A. Bessel, *Villanova University*
Lea Blau, *Yeshiva University*
David L. Boatright, *University of West Georgia*
Henry C. Brenner, *New York University*
Justin Briggle, *East Texas Baptist University*
Patricia Brletic, *Washington and Jefferson College*
Aaron Brown, *Los Angeles City College*
Sherry Brown, *York College of Pennsylvania*
Heather A. Bullen, *Northern Kentucky University*
Bruce Burnham, *Rider University*
Andrew Burns, *Kent State University*
Francis Burns, *Grand Valley State University*
Annina Carter, *Adirondack Community College*
Ralph G. Christensen, *North Central Michigan College*
David C. Collins, *Colorado State University–Pueblo*
Jeannie T. B. Collins, *University of Southern Indiana*
Charles Cornett, *University of Wisconsin–Platteville*
Dagmar Cronn, *Oakland University*
Lauren G. Cross, *Wor-Wic Community College*
Mark S. Cubberley, *Alma College*
Dwane Davis, *Forsyth Technical Community College*
Maria A. Dean, *Coe College*
Michael De Rosa, *Pennsylvania State University–Delaware County*
Cielito DeRamos King, *Bridgewater State College*
Joyce Easter, *Virginia Wesleyan College*
Bret Findley, *Saint Michael's College*
Barbara E. Flowers, *Seton Hill University*
G. Craig Flowers, *Bluefield College*
Allison Flynn, *Mesa State College*
Donna Friedman, *St. Louis Community College*
Allan A. Gahr, *Gordon College*
Carolyn Gerdes, *Georgian Court College*
Luther D. Giddings, *Salt Lake Community College*
Emma W. Goldman, *University of Richmond*
Albert Gotch, *Mount Union College*
Cliff Gottlieb, *Shasta College*
Sapna Gupta, *Park University*
Christopher Hamaker, *Illinois State University*
Pete R. Hauck, *Stetson University*
Eric Helms, *SUNY Geneseo*
Carl E. Heltzel, *Transylvania University*
Fred Hilgeman, *Southwestern University*
Deborah Hokien, *Marywood University*
Philip Hunter, *Tacoma Community College*
Mark Jackson, *Florida Atlantic University*
Bret J. S. Johnson, *College of St. Scholastica*
Carol Jones, *Central Connecticut State University*
Booker Juma, *Fayetteville State University*
Sandor Kadar, *Salve Regina University*
Laya Kesner, *University of Utah*
Margaret G. Kimble, *Purdue University at Fort Wayne*
James L. Klino, *SUNY Cobleskill College*
Dennis Kraichely, *Cabrini College*
Bette Kreuz, *Michigan State University*
Peter Krieger, *Palm Beach Community College*
Anthony F. Lagalante, *Villanova University*
Richard Langley, *Stephen F. Austin State University*
Anna Larsen, *Ithaca College*
Joseph Laurino, *University of Tampa*
William T. Lavell, *Camden County College*
Doris Lewis, *Suffolk University*
George A. Lorzeno, *Eastern University*
Gary Lyon, *Governors State University*
Vicki MacMurdo, *Anoka-Ramsey Community College*

Marcy Marino, *Niagara University*
Ronald C. Marks, *North Greenville College*
Scott Mason, *Mount Union College*
Lawrence J. Mavis, *St. Clair County Community College*
Provi M. Mayo, *South Dakota State University*
Garrett McGowan, *Alfred University*
Roger McLaughlin, *Brock University*
Kathleen McNamara, *San Diego State University*
Nancy J. Mullins, *Florida Community College at Jacksonville*
Tom Munson, *Concordia University*
Andrew Napper, *Shawnee State University*
Mary Bethé Neely, *University of Colorado–Colorado Springs*
Karen Nordell, *Lawrence University*
Jung Oh, *Kansas State University*
MaryKay Orgill, *University of Nevada, Las Vegas*
Maria Pacheco, *Buffalo State College*
Shallee T. Page, *University of Maine*
Kim L. Pamplin, *Abilene Christian University*
John Parks, *Thompson Rivers University*
Gita Perkins, *Estrella Mountain Community College*
Lon A. Porter, Jr., *Wabash College*
Jason D. Powell, *Ferrum College*
Walda Powell, *Meredith College*
Lawrence Quarino, *Cedar Crest College*
Douglas Raynie, *South Dakota State University*
Darryl K. Reach, *University of Arkansas at Little Rock*
Mike Rennekamp, *Columbus State Community College*
Shashi Rishi, *Greenville Technical College*
Kresimir Rupnick, *Louisiana State University, Baton Rouge*
Sue Salem, *Washburn University*
Nicholas Schlotter, *Hamline University*
James Schreck, *University of Northern Colorado*
Bradley Sieve, *Northern Kentucky University*
Samuella B. Sigmann, *Appalachian State University*
Mary Sisak, *Slippery Rock University of Pennsylvania*
Charles A. Smith, *Our Lady of the Lake University*
Sharon Sowa, *Indiana University–Purdue*
Charlie Stinson, Jr., *Talladega College*
Paris Svoronos, *Queensborough Community College*
Soraya Svoronos, *Queensborough Community College*
E. Shane Talbott, *Somerset Community College*
Ronald Thompson, *Daley College of Chicago*
Cynthia Tidwell, *University of Montevallo*
Bruce Toder, *University of Rochester*
John Toedt, *Eastern Connecticut State University*
Michael van Aelstyn, *Sam Houston State University*
John B. Vincent, *University of Alabama*
Keith Vitense, *Cameron University*
Maria Vogt, *Bloomfield College*
Linda Waldman, *Cerritos College*
Kenneth Weed, *Oral Roberts University*
Thomas J. Wiese, *Fort Hays State University*
Patrice E. Williams-Gordon, *Northern Caribbean University*
Elva Wohlers, *Bentley College*
Catherine Woytowicz, *George Washington University*
Jim Zubricky, *Bowling Green State University*

We would also like to thank the following professors and their students for class testing *Investigating Chemistry: A Forensic Science Perspective:*

Bruce Burnham, *Rider University*
Andrew Craft, *University of Hartford*
Mark S. Cubberley, *Alma College*
Guillermo Muhlmann, *Capital Community College*
Maria Pacheco, *SUNY Buffalo*
Ann Paterson, *Williams Baptist College*
Sally Welch, *Marygrove College*

I would like to dedicate this book to the following people who were more than my teachers—you became my advisors, mentors, and role models. Each of you gave me the tools to succeed and then challenged me to do better. Whether you knew it or not, you have each had an impact on my life and changed who I am for the better.
I will always be thankful.

Carol Manning
Dennis Pratt
Ron Schultz
Jerry Sherwin
Coach Hahn
Minda Fortney
Hal Fenrick
Bob Hansen
Jim Hamilton
Dennis Johnson

CHAPTER 1

Introduction to Forensic Chemistry

Enrique "Kiki" Camarena (1948–1985), U.S. DEA agent. (AP Photo)

Case Study: Whose Side Are They On?

On February 7, 1985, an agent of the U.S. Drug Enforcement Administration (DEA) by the name of Enrique "Kiki" Camarena was kidnapped in broad daylight on a street in Guadalajara, Mexico. Witnesses observed several assailants forcing him into a car and provided descriptions to police. Camarena had a knack for developing informants in the major illegal drug operations and had been putting a dent into the profits of several very

powerful drug dealers. Unfortunately for Camarena, his success made a target of him. His disappearance was later shown to be part of a campaign by drug traffickers against supposed U.S. DEA agents.

The immediate response of the U.S. government was to put strong pressure on their Mexican counterparts to find Camarena. The intensity of the pressure puzzled Mexican officials because six other suspected DEA agents had disappeared from Guadalajara before Camarena, and the U.S. government had not made such unyielding demands. Also, the disappearance of Mexican agents was not uncommon. What the Mexican government didn't realize was that only Camarena was actually a DEA agent. The other Americans were mistakenly captured and killed by drug traffickers.

The pressure from the White House, State Department, and DEA continued to increase until March 5, 1985, when Mexican officials recovered two bodies after a police raid at a small-scale drug operation run by the Bravo family in Michoacán. The bodies were those of Enrique Camarena and a pilot who was often employed by the DEA. Because of the location of the bodies, it appeared that the Bravo family had murdered Camarena and the pilot. However, the police raid had killed all members of the Bravo family, leaving no witnesses to interview about the death of Kiki Camarena. Immediately, the Mexican government sent word to Washington about the recovery of the bodies, and the media descended on the site, broadcasting pictures of the ranch where the bodies were recovered.

Ron Rawalt spent his career as an FBI forensic geologist, making observations and drawing conclusions by comparing mineral specimens and linking them to each other or to a specific site. When he saw the images of the Bravo ranch and the two recovered bodies broadcast on television and in the newspapers, he instantly suspected that something was amiss in the story being told by the Mexican government. In fact, he suspected a cover-up. Agent Rawalt contacted embassy officials in Mexico City with a simple request for two soil samples, one from the soil clinging to the body of Enrique Camarena and a second sample from the Bravo ranch.

When Agent Rawalt was asked by the State Department how he could make such a fantastic claim, he informed them that it was actually quite simple . . .

1.1 Welcome

Learning Objective

Explain how forensic science can be used to learn chemistry.

Welcome to *Investigating Chemistry: A Forensic Science Perspective!* The book you are about to use takes a novel approach to teaching chemistry by focusing on **forensic science**. The word *forensic* comes from the Latin word *forensis*, meaning "of the forum." The Roman forum was a place where public debates and trials were held. Today forensic science deals with the application of science to legal matters and, in particular, to crime solving. Chemistry plays a vital role in forensic science, and the goal of this book is to teach chemistry by highlighting how it is used to solve crimes.

You might wonder how you can learn introductory chemistry by studying forensic science. Forensic science is an interdisciplinary field that grew out of the need to apply knowledge from multiple sciences—biology, geology, physics, psychology, and especially chemistry—to analyze evidence from crime scenes. A major focus of the modern forensic science crime laboratory is the chemistry of evidence, and the same principles and laws that are taught in a traditional chemistry course apply to the evidence. The essential role chemistry plays in forensic science allows us to draw upon many interesting cases and examples to introduce chemistry principles. We hope you find this new multidisciplinary approach to teaching chemistry engaging and interesting.

1.2 Chemistry, Crime, and the Global Society

Learning Objective

Gain an appreciation for the global role of forensic science.

When the topics of forensic science and crime laboratories are discussed, people tend to think in local terms: One person commits a crime against another within a given jurisdiction, and evidence from the scene is gathered and analyzed. However, crime can have much broader dimensions, often becoming national or international in scope. Consider the ever-present danger of terrorism. Terrorists have traditionally used explosives to inflict damage on targets. The risk of adding to such explosives radioactive waste from poorly secured nuclear plants presents an even greater danger and is a source of concern worldwide. The illegal hunting, harvesting, and exporting of endangered species for use in clothing, traditional health remedies, and jewelry can lead toward extinction of entire species. Forensic science is brought to bear on crime at all levels, from thefts at a local convenience store to the disappearance of radioactive material from a nuclear plant thousands of miles away.

Bears are illegally hunted globally for their gall bladders, which are used in traditional Asian medicines and sold illegally. (Sayapin Vladimir/ITAR-TASS/Corbis)

Forensic science is not simply a way of determining what happened at a crime scene. It is a way of thinking and approaching a problem in a scientific manner that in the end will provide a scientifically sound explanation of past events. We hope that by the end of this course you will know more about chemistry and forensic science, and their importance to solving crimes in the global society.

1.3 Physical Evidence: Matter and Its Forms

Learning Objective

Describe the three states of matter and distinguish elements from compounds.

> Wherever he steps, whatever he touches, whatever he leaves—even unconsciously—will serve as a silent witness against him. Not only his fingerprints and his shoeprints, but also his hair, the fibers from his clothes, the glass he breaks, the tool mark he leaves, the paint he scratches, the blood or semen he deposits or collects—all of these and more bear mute witness against him. This is evidence that does not forget. It is not confused by the excitement of the moment. It is not absent because human witnesses are. It is factual evidence. Physical evidence cannot be wrong; it cannot perjure itself; it cannot be wholly absent. Only in its interpretation can there be error. Only human failure to find, study, and understand it can diminish its value.
>
> —Paul L. Kirk, *Crime Investigation*

When crime scene investigators arrive on the scene of a crime, the site will usually have been secured from contamination by the first officers to arrive at the location. The site will hold evidence of the criminal's physical presence at the scene, and some evidence from the scene will have been carried away by the criminal. The evidence might consist of carpet fibers, pet hair, blood, or gunshot residues. All of this evidence is a form of matter. In broader terms, **matter** is the physical material of the universe. Analysis of evidence requires an understanding of the properties of matter and the states in which it can exist.

Officers cordon off the crime scene to preserve it. (Reuters/Corbis)

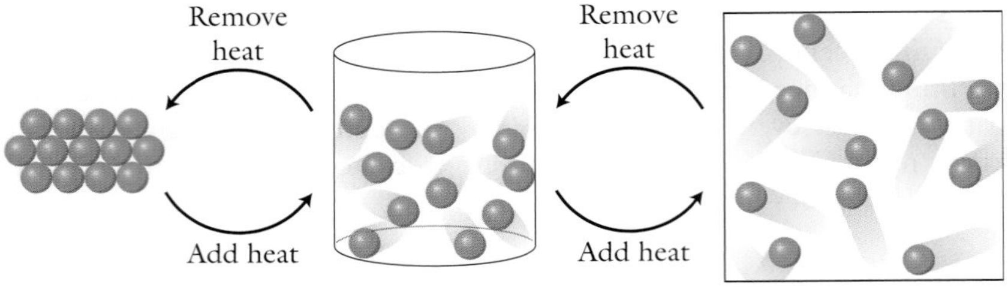

Figure 1.1 The physical states of matter are solid, liquid, and gas. The addition or removal of energy changes matter from one state to another. The volume, shape, and compressibility of matter depend on the physical state.

States of Matter

The three primary states of matter are the solid, liquid, and gaseous states, as illustrated in Figure 1.1. A **solid** is characterized by an orderly arrangement of the particles that compose it. The particles—which can be atoms, molecules, or ions—are in close proximity and are held together by forces that keep them in fixed positions. Because of this, solids do not flow as liquids or gases do. Solids are very difficult to compress because the particles that make them up are already closely packed. The volume and shape of a solid are constant because of the fixed position of its particles and their inability to flow. When a solid is heated, the particles gain energy and the solid becomes a liquid.

Liquids are characterized by particles that are farther apart than those of a solid, but are still fairly close to one another. The greater distance between particles in a liquid allows them to move around freely, a property that is evidenced by the ability of liquids to flow. Liquids cannot easily be compressed because the particles remain in contact even as they move about. The volume of a liquid is constant, but because the liquid can flow, its shape changes to fit the container that holds it. If a liquid is cooled it will form a solid, but if heat is added it will form a gas.

Gases are characterized by very large distances between particles that are moving at high speeds—approximately 300 meters per second! Gases can be compressed because of the large distance between particles, and gases readily flow because the particles are in continuous motion. Gases will always fill the volume and shape of the container in which they are placed.

Pure Substances—Elements and Compounds

The evidence recovered at a crime scene may be either a pure substance or a mixture. A **pure substance** is any form of matter that has a uniform composition and cannot be separated by physical methods such as filtration or evaporation into more than one component.

Elements and compounds are two subclasses of pure substances. An **element** is the simplest form of a pure substance. The smallest unit of an element that retains all of the properties of that element is called the **atom**. Each element has unique atoms that are unlike the atoms of any other element. One common example of an element is aluminum, as found in aluminum foil. A less familiar example is sodium, which is often found in

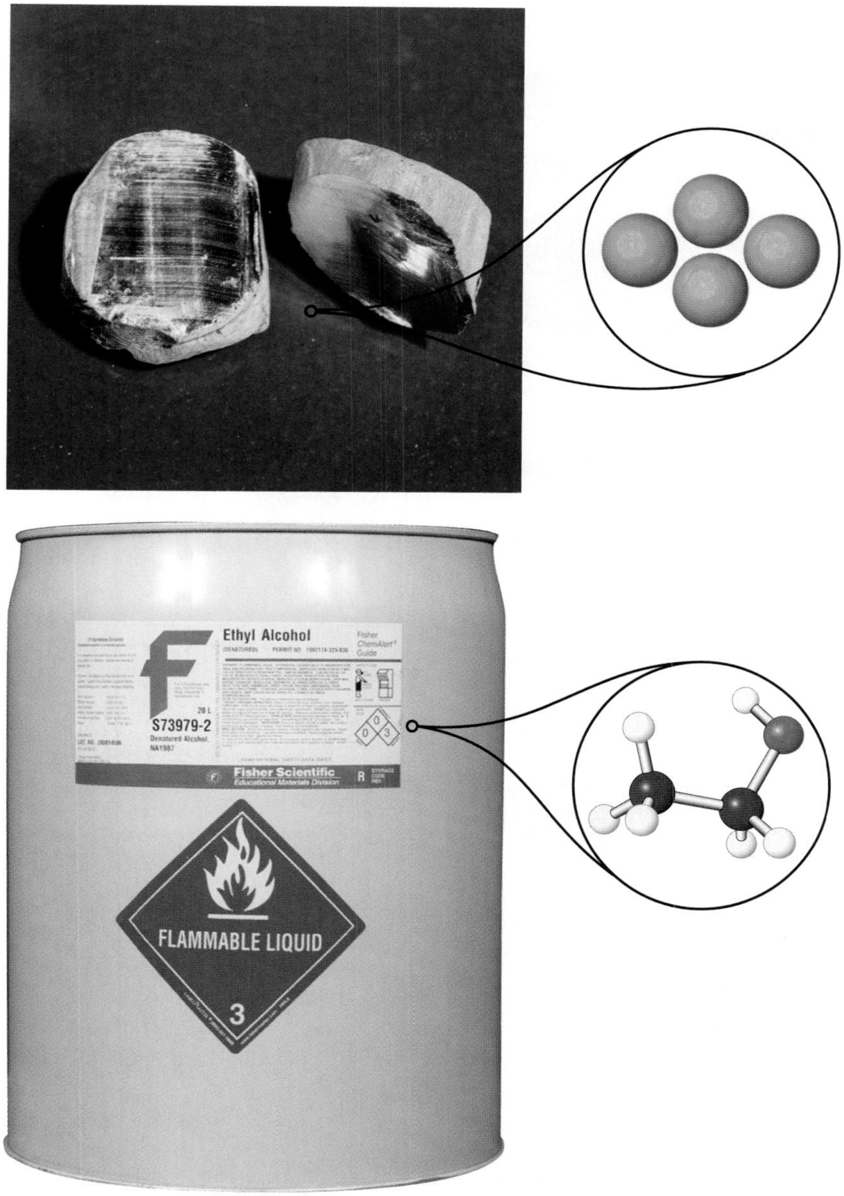

Pure substances can be found as pure elements (such as sodium, top) or as pure compounds (such as ethanol, bottom). (Top photo: Science Photo Library/Photo Researchers, Inc.; bottom photo: Tom Schultz)

clandestine drug labs where methamphetamine is manufactured. Sodium is very dangerous because it reacts violently with water—even water vapor in the air—and must be stored under a layer of oil to keep it from reacting. Gold is an element known since ancient times and has been valued for use in jewelry and coins from ancient to modern times. Our fascination with gold is reflected by Hollywood movies that come out almost yearly involving the theft of large amounts of the precious metal.

It is rare to find substances in elemental form at a crime scene or anywhere else because most elements react with other substances to

Figure 1.2 Homogeneous mixtures commonly found at crime scenes. (a) The steel in a metal truck bumper is an alloy of iron, carbon, and a few other trace elements. (b) Lighter fluid is a mixture of several flammable compounds such as butane, isobutane, and propane. (c) Lead bullets contain trace impurities such as bismuth and antimony. (d) Wine is a complex mixture that contains many compounds; the major components are water, ethanol, sugar, and colored molecules called tannins that give wine its color. (a: istockphoto; b: Elyse Rieder; c: Catherine dée Auvil/istockphoto; d: Medioimages/Getty)

form **compounds**, the second class of pure substances. A compound is a substance that is made up of two or more elements chemically bonded together. There are two broad classes of compounds, **molecular compounds** and **ionic compounds**. A **molecule** is the simplest unit of a molecular compound while a **formula unit** is the simplest unit of an ionic compound. The properties of molecular and ionic compounds will be discussed in detail in future chapters, but for now it is important to know that the elements that make up either type of compound cannot be separated by physical methods but only by a chemical reaction.

Examples of compounds that are often recovered at clandestine drug labs are sodium chloride, in which sodium is bonded to chlorine, or ethyl ether, in which carbon, hydrogen, and oxygen are bonded in a chemical combination. Other compounds that are sometimes prominent in criminal cases are poisonous substances such as compounds of arsenic and mercury. Historically, arsenic compounds were used as rat poisons and pesticides and the commercial availability of arsenic compounds led to their widespread use by those with murderous intent.

Cocaine is extracted from the leaves of the coca plant. (Science Photo Library/Photo Researchers, Inc.)

Mixtures

Much of the physical evidence recovered at a crime scene is in the form of a mixture. A *mixture* is two or more pure substances that are physically combined but not chemically bonded together. Mixtures also have two subclasses. A **homogeneous mixture** is one in which the substances that compose it are so evenly distributed that a sample from any one part of the mixture will be chemically identical to a sample from any other part. A common example of a homogeneous mixture is a solution of sugar in water. The "date rape" drug Rohypnol has proven to be especially dangerous because it can be dissolved in a person's drink to form a homogeneous solution without giving any visual evidence that the drink has been altered. The manufacturer has recently changed the formulation of Rohypnol, a prescription sedative; it is now a slow-dissolve tablet that releases a blue dye to help prevent the misuse of the drug.

Figure 1.2 shows several additional examples of homogeneous solutions commonly found at crime scenes. Note that solids can be homogeneous mixtures; alloys are mixtures of several metallic elements. In the case of a hit-and-run incident, traces of alloy components from a truck bumper may be used to include or exclude suspected vehicles.

Gases can form homogeneous mixtures, the most common example being the air we breathe. Air is a homogeneous mixture of 78% nitrogen, 21% oxygen, and 1% carbon dioxide, helium, argon, and a few other gases.

A **heterogeneous mixture** is one in which the composition varies from one region of a sample to another. For example, although cocaine itself is a pure substance with the formula $C_{17}H_{21}NO_4$, samples of cocaine sold on the illegal drug market are actually heterogeneous mixtures. The samples usually contain impurities from extracting and processing the cocaine from the coca leaves. The impurities with the cocaine are sometimes analyzed and their components identified to determine whether two samples are likely to have come from the same source. It is also common for drug dealers to boost profits by mixing cocaine with other compounds that act as fillers to increase the sample size. Forensic drug chemists have found all sorts of compounds used as fillers—including sugar, starch, baking soda, talcum powder, and even Tang® breakfast drink powder!

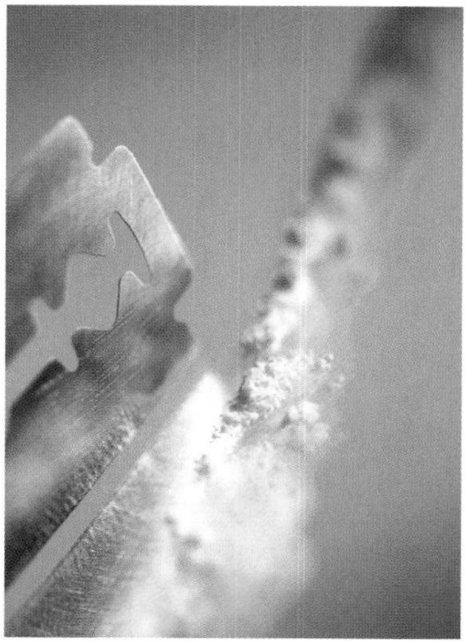

Cocaine is diluted with fillers to increase the profit margins of drug dealers. (Getty Images)

It takes great care to recover evidence from a heterogeneous mixture because of the variation in the sample. If too small a sample is taken, not all of the components in the mixture may be included. Some examples of heterogeneous mixtures found at crime scenes are shown in Figure 1.3.

Mixtures can be separated into their components by physical means such as evaporation or filtering. For example, if you wanted to separate salt water into salt and water, simply boil the water off, trap the vapor, and then condense the water vapor back to liquid water in a separate container by cooling it. Boiling works in this case because the solid salt, which is dissolved in the water, is not altered by heating water to its boiling point.

A summary of the many forms of matter can be found in Figure 1.4.

Figure 1.3 Heterogeneous mixtures from crime scenes. (a) Arson debris contains partially burnt carpeting, wood, plastics, and other building materials. (b) Soil contains pebbles, sand, silt, and clay of different minerals and decaying vegetable matter. (c) Evidence recovered from natural waters will contain microorganisms, silt, algae, and aquatic invertebrates. (d) Various fibers, hairs, and dust are recovered from a suspect's car. (a: Steve Crise/Corbis; b: Getty Images; c: Jim Vecchi/Corbis; d: M. Angelo/Corbis)

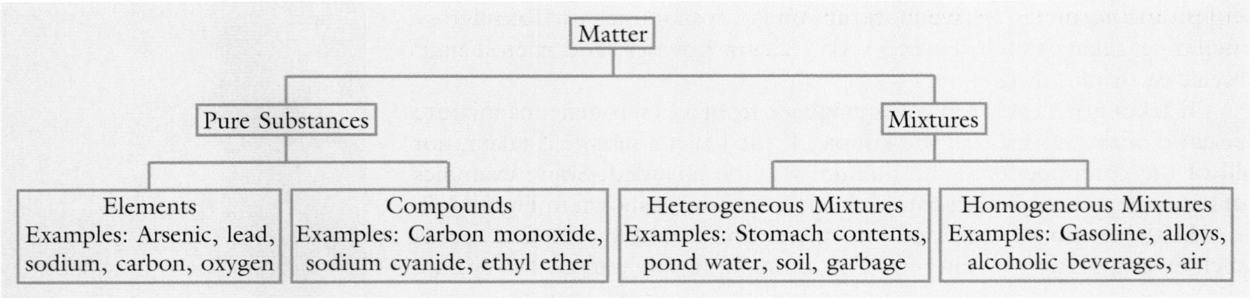

Figure 1.4 All matter, the physical material of the universe, can be classified as either a pure substance or a mixture. If found as a pure substance, it must be in the form of a pure element or a pure compound. If found as a mixture, it will either be a heterogeneous mixture with differing chemical composition throughout or a chemically uniform homogeneous mixture.

Worked Example 1

Label each of the containers below as either a pure substance or a mixture.

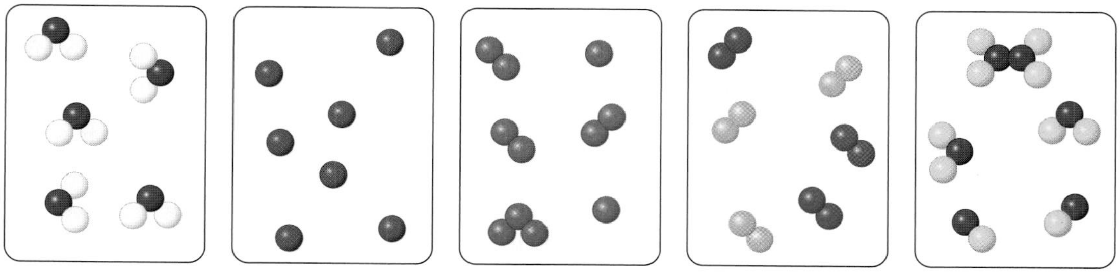

A B C D E

SOLUTION

A: Pure substance, B: Pure substance, C: Mixture, D: Mixture, E: Mixture

Practice 1.1

Label each container from Worked Example 1 as containing either elements, compounds, or both.

ANSWER

A: Compounds, B: Elements, C: Both, D: Compounds, E: Compounds

1.4 The Periodic Table

Learning Objective

Trace the development of the periodic table.

One of the most widely used tools in science is the periodic table, copies of which are displayed on the inside covers of nearly all chemistry textbooks (including this one) and the walls of laboratories and science classrooms throughout the world. The **periodic table** shows all the known chemical elements, arranged in a specific pattern. The importance of the table stems from the fact that it places an extensive amount of information at our fingertips. For example, it allows us to predict chemical reactions, the formulas of many compounds, relative sizes of atoms and molecules, shapes of molecules, and whether compounds will dissolve in water or oil.

The periodic table is so crucial to scientists that an international organization, the International Union of Pure and Applied Chemistry (IUPAC), maintains and updates any changes or additions to the periodic table. Although scientists long ago discovered all of the naturally occurring elements, scientists now create new elements that need to be characterized and named.

Development of the Periodic Table

One of the best examples to illustrate the power of the periodic table is the story of its creation in 1869 by Dmitri Mendeleev, a Russian chemist. The necessity of developing some kind of organizing scheme for chemi-

Evidence Analysis | Thin-Layer Chromatography

To make a mixture, components can simply be combined physically. But separating a mixture into its components after they are mixed is not always easy. Separation requires an understanding of the properties and behavior of each component.

One of the main methods of separating mixtures in the forensic science laboratory is called **chromatography.** Chromatography was originally developed in the early 1900s to separate the colored pigments in flowers. The name *chromatography* literally translates as "color writing" because the individual colors separate from the mixture. Chromatography exploits the fact that different compounds are attracted to or repelled by other compounds to varying degrees, and the extent of attraction or repulsion is always the same for any two given compounds. We will talk about several forms of chromatography throughout this book, as it is one of the main methods used for identifying drugs, explosives, narcotics, and fibers.

If you place a drop of ink or a mixture of food coloring on a strip of filter paper and then dip the strip into a liquid such as water or alcohol, allowing the liquid to move upward over the spot, you will see chromatography in action. Most inks consist of a mixture of colored compounds, as does a combination of food coloring. The compounds in the mixture have different levels of attraction for the paper (called the *stationary phase*) and for the liquid (called the *mobile phase*) passing over them. Chromatography allows the different compounds to be separated, so that you can see different colored spots at different distances from the starting point.

The simplest form of chromatography described above is called *paper chromatography.* In most laboratories, however, a variation of this tech-

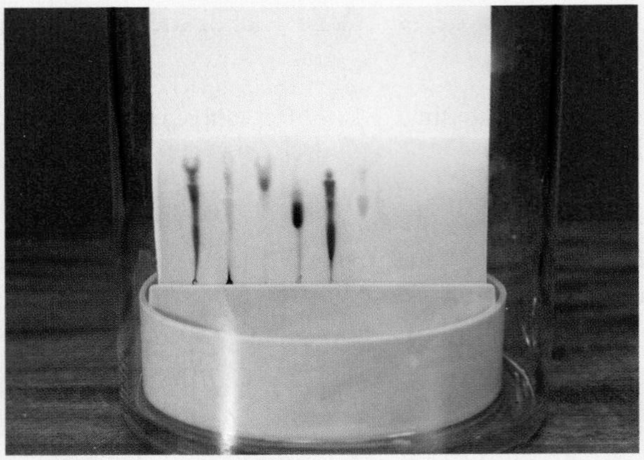

Water-soluble markers are often mixtures of several dyes that produce the desired colors. (Photo Researchers, Inc.)

Dmitri Mendeleev (1837–1907), creator of the modern periodic table. (Getty Images)

cal elements and their properties had become evident around Mendeleev's time because of the success of chemists in discovering new elements. Some elements such as gold, silver, and lead had been known from ancient times, but most elements were unknown until the eighteenth and nineteenth centuries. (Remember that most elements exist in compounds rather than in pure form in nature.) Over 20 new elements were isolated in the 1700s and, by the time Mendeleev started working on the periodic table in 1869, the number had grown by another 30, for a total of 63 known elements.

Mendeleev noticed that some elements had very similar reactions and properties. He also noticed that as the mass of the atoms increased, these properties seemed to repeat periodically. Mendeleev organized the elements according to their repeating, periodic properties, grouping into columns elements with similar properties. When the known elements were arranged this way, there were several gaps in the periodic table. Mendeleev boldly stated that the gaps were due to elements that existed but had not yet been discovered.

The first hole in his table is where we now find element number 31, gallium (Ga). Mendeleev called this *eka-aluminum*, which means "simi-

nique, called **thin-layer chromatography (TLC)**, is used. As with paper chromatography, there are always two parts to a TLC system, a mobile phase and a stationary phase. In TLC, the mobile phase is a liquid such as water; the stationary phase is a thin solid coating (such as silica gel) on a glass plate. The mixture to be separated is added onto the bottom of the stationary phase. The mobile phase is then allowed to move across the stationary phase. Each compound in the mixture is attracted to the liquid mobile phase and to the stationary solid phase to a different degree than are all of the other compounds in the mixture. If a compound is attracted only to the mobile phase, it will move as fast as the mobile phase. If a compound is attracted only to the stationary phase, it will not move at all. In most cases, a compound will be attracted somewhat to both the mobile phase and the stationary phase and it will move slowly across the stationary phase.

The key to chromatography is that each compound ends up moving at a speed slightly different from all other compounds; this separates each compound in the mixture. We can identify the compounds present in a mixture by determining if they move at the same speed as a known compound. The TLC plate shown in Figure 1.5 illustrates how a sample containing illegal drugs and narcotics can have each component identified. TLC is also used commonly for analyzing inks from kidnapping notes and for determining the types of explosives used in terrorist attacks.

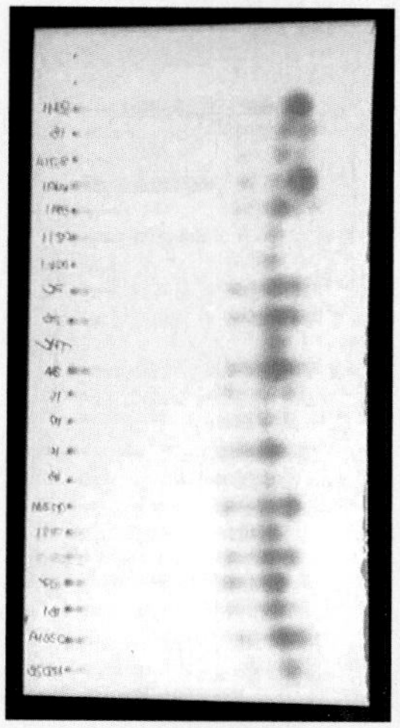

Figure 1.5 Thin-layer chromatography separates the compounds found in illegal drugs. Common drug samples are analyzed alongside the evidence. If a spot appears on both samples, a tentative identification is made. Further experiments provide a positive identification. (Missouri State Highway Patrol)

lar to aluminum." The second missing element was number 32, which today is called germanium (Ge), but Mendeleev called the undiscovered element *eka-silicon*. Table 1.1 illustrates the power of the periodic table as Mendeleev made predictions about each of the undiscovered elements. In the following chapters, we will discuss atomic mass, density, and

Table 1.1 Mendeleev's Predictions and Actual Values

Property	Eka-aluminum Predictions	Actual Properties of Gallium	Eka-silicon Predictions	Actual Properties of Germanium
Atomic mass (amu)	About 68	69.7	About 72	72.6
Melting point (°C)	Low	29.8	—	—
Density (g/cm^3)	5.9	5.94	5.5	5.47
Oxide formula	X_2O_3	Ga_2O_3	XO_2	GeO_2
Chloride formula	XCl_3	$GaCl_3$	XCl_4	$GeCl_4$
Discovered (year)		1886		1875

formulas, but for now we focus on how closely Mendeleev was able to predict the values for the two undiscovered elements. Both gallium and germanium were discovered and isolated a few years after Mendeleev's predictions.

1.5 Learning the Language of Chemistry

Learning Objective

Interpret and use atomic symbols, the periodic table, and chemical formulas.

Throughout this book, we will be learning how to use the periodic table to determine the chemical formulas of compounds, what the various numbers represent, and how to predict both physical and chemical properties of the elements. However, it is important for you to start learning the names and symbols of the elements now so that the language of chemistry will be familiar to you when the periodic table is explored in more depth. The periodic table lists all of the elements by their symbols. The corresponding name and spelling of each element can be found in Table 1.2.

Most symbols of the elements are formed from the first letter of the element's name, and in many cases, either the second or the third letter of the name follows the first letter. For example, the symbol for helium is He, and for manganese is Mn. However, you will notice that some elements have very unusual symbols that do not derive from their contemporary names. Au is the symbol for gold and Na is the symbol for sodium. These symbols come from the Latin names of the elements: *aurum* for gold and *natrium* for sodium.

1	2	3	4	5	6	7	8	9	10	11	12	13	14	15	16	17	18
1 H 1.0079																	2 He 4.003
3 Li 6.941	4 Be 9.012											5 B 10.811	6 C 12.011	7 N 14.007	8 O 15.999	9 F 18.998	10 Ne 20.180
11 Na 22.990	12 Mg 24.305											13 Al 26.982	14 Si 28.086	15 P 30.974	16 S 32.065	17 Cl 35.453	18 Ar 39.948
19 K 39.098	20 Ca 40.078	21 Sc 44.956	22 Ti 47.867	23 V 50.942	24 Cr 51.996	25 Mn 54.938	26 Fe 55.845	27 Co 58.933	28 Ni 58.693	29 Cu 63.546	30 Zn 65.409	31 Ga 69.723	32 Ge 72.64	33 As 74.922	34 Se 78.96	35 Br 79.904	36 Kr 83.798
37 Rb 85.468	38 Sr 87.62	39 Y 88.906	40 Zr 91.224	41 Nb 92.906	42 Mo 95.94	43 Tc (98)	44 Ru 101.07	45 Rh 102.906	46 Pd 106.42	47 Ag 107.868	48 Cd 112.411	49 In 114.818	50 Sn 118.710	51 Sb 121.760	52 Te 127.60	53 I 126.905	54 Xe 131.293
55 Cs 132.905	56 Ba 137.327	57 La 138.906	72 Hf 178.49	73 Ta 180.948	74 W 183.84	75 Re 186.207	76 Os 190.23	77 Ir 192.217	78 Pt 195.078	79 Au 196.967	80 Hg 200.59	81 Tl 204.383	82 Pb 207.2	83 Bi 208.980	84 Po (209)	85 At (210)	86 Rn (222)
87 Fr (223)	88 Ra (226)	89 Ac (227)	104 Rf (261)	105 Db (262)	106 Sg (266)	107 Bh (264)	108 Hs (277)	109 Mt (268)	110 Ds (271)	111 Rg (272)							

58 Ce 140.116	59 Pr 140.908	60 Nd 144.24	61 Pm (145)	62 Sm 150.36	63 Eu 151.964	64 Gd 157.25	65 Tb 158.925	66 Dy 162.500	67 Ho 164.930	68 Er 167.259	69 Tm 168.934	70 Yb 173.04	71 Lu 174.967
90 Th 232.038	91 Pa 231.036	92 U 238.029	93 Np (237)	94 Pu (244)	95 Am (243)	96 Cm (247)	97 Bk (247)	98 Cf (251)	99 Es (252)	100 Fm (257)	101 Md (258)	102 No (259)	103 Lr (262)

The periodic table of elements.

Table 1.2 Atomic Symbols and Names

Ac	Actinium	Gd	Gadolinium	Po	Polonium
Ag	Silver	Ge	Germanium	Pr	Praseodymium
Al	Aluminum	H	Hydrogen	Pt	Platinum
Am	Americium	He	Helium	Pu	Plutonium
Ar	Argon	Hf	Hafnium	Ra	Radium
As	Arsenic	Hg	Mercury	Rb	Rubidium
At	Astatine	Ho	Holmium	Re	Rhenium
Au	Gold	Hs	Hassium	Rf	Rutherfordium
B	Boron	I	Iodine	Rg	Roentgenium
Ba	Barium	In	Indium	Rh	Rhodium
Be	Beryllium	Ir	Iridium	Rn	Radon
Bh	Bohrium	K	Potassium	Ru	Ruthenium
Bi	Bismuth	Kr	Krypton	S	Sulfur
Bk	Berkelium	La	Lanthanum	Sb	Antimony
Br	Bromine	Li	Lithium	Sc	Scandium
C	Carbon	Lr	Lawrencium	Se	Selenium
Ca	Calcium	Lu	Lutetium	Sg	Seaborgium
Cd	Cadmium	Md	Mendelevium	Si	Silicon
Ce	Cerium	Mg	Magnesium	Sm	Samarium
Cf	Californium	Mn	Manganese	Sm	Tin
Cl	Chlorine	Mo	Molybdenum	Sr	Strontium
Cm	Curium	Mt	Meitnerium	Ta	Tantalum
Co	Cobalt	N	Nitrogen	Tb	Terbium
Cr	Chromium	Na	Sodium	Tc	Technetium
Cs	Cesium	Nb	Niobium	Te	Tellurium
Cu	Copper	Nd	Neodymium	Th	Thorium
Db	Dubnium	Ne	Neon	Ti	Titanium
Ds	Darmstadtium	Ni	Nickel	Tl	Thallium
Dy	Dysprosium	No	Nobelium	Tm	Thulium
Er	Erbium	Np	Neptunium	U	Uranium
Es	Einsteinium	O	Oxygen	V	Vanadium
Eu	Europium	Os	Osmium	W	Tungsten
F	Fluorine	P	Phosphorus	Xe	Xenon
Fe	Iron	Pa	Protactinium	Y	Yttrium
Fm	Fermium	Pb	Lead	Yb	Ytterbium
Fr	Francium	Pd	Palladium	Zn	Zinc
Ga	Gallium	Pm	Promethium	Zr	Zirconium

Worked Example 2

What are the correct names for the following elements?

(a) F **(b)** Be **(c)** Ni

SOLUTION

The elements listed in this example are some of those for which names are commonly misspelled by beginning chemistry students.

(a) Fluorine, commonly misspelled as ~~flourine~~.

(b) Beryllium, commonly misspelled as ~~berryllium~~.

(c) Nickel, commonly misspelled as ~~nickle~~.

Practice 1.2

What are the correct names for the following elements that have their atomic symbols derived from their Latin names?

(a) Ag **(b)** Na **(c)** Sn

ANSWER

(a) Silver **(b)** Sodium **(c)** Tin

Worked Example 3

What are the atomic symbols for the elements listed below?

(a) Magnesium and manganese

(b) Boron, beryllium, barium, and bromine

(c) Potassium and phosphorus

SOLUTION

The elements listed above commonly have their atomic symbols confused.

(a) Mg and Mn

(b) B, Be, Ba, and Br

(c) K and P

Practice 1.3

What are the correct atomic symbols for the following elements, which have their atomic symbols derived from their Latin names?

(a) Lead **(b)** Gold **(c)** Mercury

ANSWER

(a) Pb **(b)** Au **(c)** Hg

Regions of the Periodic Table

The periodic table can be divided into many subregions. It is often necessary to know if an element is a metal, nonmetal, or metalloid because that will influence the way a compound formula is written. Many elements are classified as metals and dominate the middle and left sides of the periodic table, as shown in Figure 1.6. All **metals** share the following properties: They conduct electricity and heat; they are solids at room temperature except for mercury, which is a liquid; and the **melting point**, the temperature at which most metals melt to form a liquid, is generally very high.

The nonmetals occupy the upper right corner of the periodic table. **Nonmetals** do not conduct electricity or heat very well and can be found as a solid, such as carbon, a liquid, such as bromine, or a gas, such as helium. In general, nonmetals have melting points that are much lower than metals.

There is a third class of elements known as metalloids. The **metalloids** have properties that are between the two extremes shown by metals and nonmetals. The location of metals, metalloids, and nonmetals on the periodic table is shown in Figure 1.6.

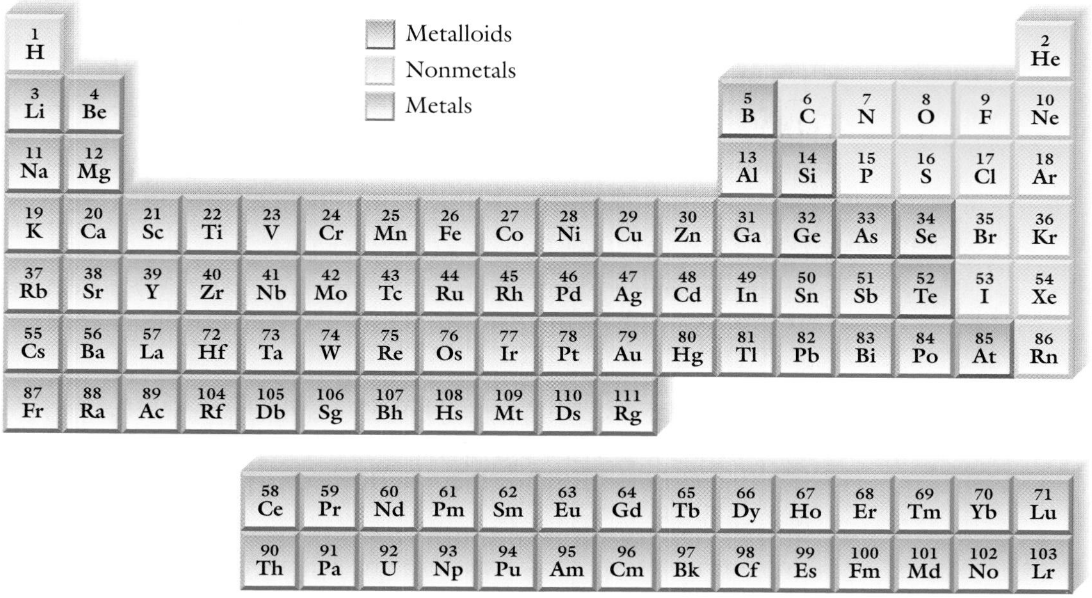

Figure 1.6 Regions of the periodic table: metals, metalloids, and nonmetals.

Worked Example 4

List the names and atomic symbols for five metals that could be found in a student dormitory. Include what object is made from the metal.

SOLUTION

Answers may vary, but some possibilities include:

Gold, Au: jewelry
Aluminum, Al: soda can
Copper, Cu: power cord wires
Iron, Fe: steel bed frames
Titanium, Ti: lightweight bicycle parts

Practice 1.4

List the names and atomic symbols for two nonmetals that could be found in a student dormitory. Include what object is made from the nonmetal.

ANSWER

Answers may vary, but some examples include:

Carbon, C: pencil graphite
Neon, Ne: advertising sign
Oxygen, O: air (as O_2)

Chemical Formulas

Formulas are a chemist's shorthand for showing the elements in a compound and the numbers of the atoms of each element that make up the compound. Using a chemical formula instead of writing out the name saves time and simplifies the writing process, provided that you know the symbols of the elements. Chemical formulas make it easier to communicate, in an exact manner, the composition of a substance. Notice that each **subscript** in the formulas refers only to the element directly preceding it. Metals and metalloids are listed before nonmetals when writing formulas.

Worked Example 5

Arsenic poisoning commonly results from using As_2O_3. What elements are present and how many atoms of each are in the compound?

SOLUTION

The two elements in the compound are arsenic (As) and oxygen (O). The subscript number 2 refers to two atoms of arsenic (As), and the subscript number 3 refers to three atoms of oxygen (O).

Practice 1.5

How many atoms of each element are present in (a) As_2O_5 and (b) $AsCl_3$, which are two additional arsenic compounds that are poisonous?

ANSWER

(a) Two arsenic atoms and five oxygen atoms

(b) One arsenic atom and three chlorine atoms

Worked Example 6

Ludwig van Beethoven is believed to have died of lead poisoning. Foul play is not suspected because the dangers of lead and lead compounds were unknown at the time, and he could have been exposed accidentally. Write the formula for the compound lead sulfate that has one lead atom, one sulfur atom, and four oxygen atoms.

SOLUTION

When a compound contains only one atom of an element, we do not use a subscript (the subscript "1" is understood). Therefore, the formula has only a subscript number 4 after the oxygen atom: $PbSO_4$.

Practice 1.6

The compound mercury(II) chloride is extremely poisonous and has one mercury atom and two chlorine atoms. Write the correct formula for the compound.

ANSWER

$HgCl_2$

1.6 The Most Important Skill of a Forensic Scientist: Observation

Learning Objective

Describe a critical skill scientists must learn.

The single most important skill a forensic scientist can have is the ability to make accurate observations, whether at a crime scene or in the laboratory. The value of forensic evidence often lies in finding trace components that indisputably link one environment to another. Many great discoveries have been made in the laboratory by an observant scientist who noticed what others had missed. Scientists are trained in making observations early in their careers by spending countless hours in the laboratory.

The conclusion of the case study that was introduced at the beginning of this chapter will demonstrate how one forensic scientist saw a picture of a crime scene and was observant enough to expose a cover-up behind the murder of a DEA agent.

1.7 Case Study Finale: Whose Side Are They On?

What was FBI Special Agent Rawalt's reason for not believing the Mexican government? The color of the soil from the bodies did not match the color of the soil from the Bravo ranch. This mismatch suggested that the bodies could not have been buried at the ranch. Samples of soil were soon smuggled out of Mexico and into the hands of Rawalt.

The soil sample from Camarena's body was a much darker color than the soil from the Bravo ranch, which is what Rawalt had initially observed. However, once Rawalt had the sample, he discovered that the soil was saturated with fatty tissue from the body. This fatty tissue had to be removed so that he could make a clear comparison with the soil from the Bravo ranch. Rawalt placed the sample into a device that rid the sample of fatty tissue by converting all of the carbon and hydrogen atoms in the fatty tissue to carbon dioxide and water. When the Camarena soil was cleaned, it turned out to be much lighter in color than the Bravo ranch soil—just the opposite of what had originally triggered Rawalt's suspicion. Under Rawalt's microscope, the Bravo soil was determined to contain very dark greenish-gray globular obsidian, whereas the Camarena soil was a tan-white rhyolite ash that could not have come from the Bravo ranch. Now that there was evidence that Enrique Camarena had not been originally buried at the Bravo ranch, investigators focused on the important task of finding the actual burial location.

The soil from Camarena's body held a very important clue to its geographic origin. Less than 1% of the sample contained a mixture of three minerals that would help trace the original location of the burial site. The first mineral was bixbyite, which has a very dark black color, and the second was the opalized and clear forms of cristobalite. The final component of the soil was a pink glass of such an unusual color that Special Agent Rawalt could only compare it with something you might find in a candy jar. Although both the soil sample from the body and the one from the ranch are a kind produced by volcanoes, the Camarena soil was of a type that would be found on a mountain slope, not the lower elevation of the Bravo ranch.

Rawalt's next step was to search the scientific literature and publications for soil surveys done in the Mexican mountains. As luck would have it, he came across a geologist at the Smithsonian Institution who had done work on a mountain in Jalisco State Park in Mexico. The geologist said that the park is the only place where the type of soil on the body could be found. Rawalt filed his report and was told he would be going immediately to Mexico to locate the original burial site. Within hours he was on a captured drug runner's plane flying under the Mexican radar with the paperwork of a DEA

The body of Kiki Camarena is repatriated to the United States under military escort. Soil evidence taken from his body would prove critical to unraveling who was responsible for his murder. (Lenny Ignelzi/AP Photo)

agent, the deception being necessary because FBI agents are not allowed to enter Mexico. After spending some time searching the mountain in Jalisco State Park, he was able to locate the burial site.

Proof that Camarena's body had been buried in Jalisco cast serious doubt on the story of the Mexican police that he had been killed by the Bravo family. In the aftermath of the extensive investigations that followed, it became clear that certain Mexican officials had conspired with a major drug family to produce the body of Enrique Camarena in order to pacify the U.S. government. Ultimately, 22 Mexican officials were convicted of conspiracy, and members of the drug family were tried and found guilty of the abduction and murder of Enrique Camarena.

The careful observation of one scientist had led to the unraveling of a story designed to cover up the identity of Camarena's actual murderers.

The Secret to Success

The secret to success in a chemistry course is very simple: You must practice solving problems long before an exam is ever given. Success in any aspect of life comes only after repeated practice. Professional athletes paid millions of dollars can be seen running the same practice drills as high school and college athletes. Why? Practice makes perfect! For the same reason, musicians will spend 90% of their time practicing an arrangement for a very few limited public performances. The end of each chapter of this book contains many problems for you to practice so that you will be able to succeed at exam time.

You will notice that throughout the book we have used examples with a forensic science theme. However, the end-of-chapter problems start with many examples of problems without a forensic theme. This was done so that you have the chance to practice a wide variety of problems and master the required skills and calculations. There are additional problems at the end of the problem section that return to the forensic science theme for you to practice your newly mastered skills. Odd-numbered problems have their corresponding answers provided at the end of the book so that you can check your answers. If you feel you need additional resources, you can purchase the *Student Solutions Manual*, which has the answers to all odd-numbered problems worked out in full, step-by-step detail. Even-numbered problems provide you with more practice and mirror the odd-numbered problems, but answers are not provided.

CHAPTER SUMMARY

- Chemical principles and the chemistry of evidence are an important part of forensic science and form the basis for much of the work in a modern forensic crime laboratory.
- Matter is the physical material of the universe and can exist in the solid, liquid, or gas states. Matter consists of either pure substances (elements and compounds) or mixtures.
- Elements are characterized by having atoms that are identical. Elements are represented by symbols that are usually derived from the first few letters of their modern or familiar name, although some are derived from Latin names.
- The periodic table is divided into the metals, nonmetals, and metalloids. All elements belonging to a class have similar properties. Metals have the ability to conduct electricity whereas nonmetals do not conduct electricity.
- Compounds are made up of two or more atoms of different elements bonded together. Chemical formulas indicate each element in the compound and the number of each type of atom.
- Mixtures can be classified as heterogeneous or homogeneous. A heterogeneous mixture is one that has a different chemical makeup from one region of a sample to another, whereas a homogeneous mixture is uniform throughout. Mixtures can be separated into their components by physical methods such as filtering, evaporation, or thin-layer chromatography.

KEY TERMS

forensic science, p. 3
matter, p. 5
solid, p. 6
liquid, p. 6
gas, p. 6
pure substance, p. 6
element, p. 6
atom, p. 6
compound, p. 8
molecular compound, p. 8
ionic compound, p. 8
molecule, p. 8
formula unit, p. 8
homogeneous mixture, p. 9
heterogeneous mixture, p. 9
periodic table, p. 11
chromatography, p. 12
thin-layer chromatography, p. 13
metals, p. 16
melting point, p. 16
nonmetals, p. 16
metalloids, p. 16
formula, p. 17
subscript, p. 18

CONTINUING THE INVESTIGATION Additional Readings, Resources, and References

"Death of a Narc," *Time*, November 7, 1988, p. 132.

Kirk, P. L. *Crime Investigation*, John Wiley & Sons: New York, 1974, pp. 1–8.

McPhee, J. "Death of an Agent," *The New Yorker*, 1996, vol. 71, no. 46, pp. 60–69.

Plummer, C. M. "The Forgery Murders," *Chemistry Matters*, December 1995, pp. 8–11.

For an interactive Web site about the case in Problem 37: www.crimelibrary.com/criminal_mind/forensics/mormon_forgeries

REVIEW QUESTIONS AND PROBLEMS

Questions

1. What distinct areas of science are important to forensic science?
2. What is the difference between compounds and elements?
3. Explain how you can distinguish between a heterogeneous mixture and a homogeneous mixture.
4. Illustrate the difference between a heterogeneous and homogeneous mixture by drawing two containers, each containing "●", "▲", and "■" particles.
5. How could you determine whether a sample is a pure substance or a homogeneous mixture?
6. Give an example of a homogeneous mixture not listed in the textbook. What could you

add to the example you provided to make it a heterogeneous mixture?

7. Explain how a mixture can be separated into its pure components.
8. Why is the periodic table such a valuable tool?
9. Why is the ability to make observations important to a scientist?
10. What is the organization that is responsible for maintaining the periodic table?
11. Why are some atomic symbols not based on the contemporary name of the element?
12. Sketch the periodic table and indicate where metals, metalloids, and nonmetals are located.
13. Why are elements rarely found as pure substances?

Problems

14. Identify each of the following substances as either a pure substance or a mixture.
(a) Gasoline
(b) Air
(c) Water
(d) Steel
15. Identify each of the following substances as either a pure substance or a mixture.
(a) Ethanol
(b) Soda
(c) Soil
(d) Brass
16. Identify each of the following substances as either an element or a compound.
(a) Silicon
(b) Carbon dioxide
(c) Arsenic
(d) Water
17. Identify each of the following substances as either an element or a compound.
(a) Sugar
(b) Carbon
(c) Lead
(d) Rust
18. What are the names of the elements represented by the atomic symbols listed below?
(a) Li
(b) Cl
(c) C
(d) Al
19. What are the names of the elements represented by the following atomic symbols?
(a) H
(b) N
(c) O
(d) S
20. What are the names of the elements represented by the atomic symbols listed below?
(a) Cr
(b) Mg
(c) Cu
(d) Kr
21. What are the names of the elements represented by the atomic symbols listed below?
(a) As
(b) Ne
(c) Ca
(d) Pt
22. What are the atomic symbols of the elements listed below?
(a) Helium
(b) Manganese
(c) Potassium
(d) Chlorine
23. What are the atomic symbols of the elements listed below?
(a) Magnesium
(b) Iron
(c) Silver
(d) Mercury
24. What are the atomic symbols of the elements listed below?
(a) Tin
(b) Aluminum
(c) Boron
(d) Sulfur
25. What are the atomic symbols of the elements listed below?
(a) Iodine
(b) Silicon
(c) Chromium
(d) Titanium
26. Examine each of the elements and symbols listed below and correct any spelling errors or incorrect atomic symbols. If there are no errors, indicate by writing *correct.*
(a) Cadmium, Ca
(b) Potasium, K
(c) Chlorine, Cl
(d) Zinc, Zi

27. Examine each of the elements and symbols listed below and correct any spelling errors or incorrect atomic symbols. If there are no errors, indicate by writing *correct.*
(a) Sulfer, S
(b) Aluminium, Al
(c) Manganese, Ma
(d) Boron, B

28. Identify each of the following elements as a metal, metalloid, or nonmetal.
(a) Silicon
(b) Phosphorus
(c) Potassium
(d) Antimony

29. Identify each of the following elements as a metal, metalloid, or nonmetal.
(a) Calcium
(b) Gallium
(c) Iodine
(d) Arsenic

30. Identify each of the following elements as a metal, metalloid, or nonmetal.
(a) Ba
(b) Co
(c) Te
(d) N

31. Identify each of the following elements as a metal, metalloid, or nonmetal.
(a) Ag
(b) H
(c) F
(d) Al

32. Which formula below corresponds to the compound containing one magnesium atom, one sulfur atom, and four oxygen atoms?
(a) MnSO4
(b) $MgSO^4$
(c) $MgSuO_4$
(d) $MgSO_4$

33. Which formula below corresponds to the compound containing three sodium atoms, one phosphorus atom, and four oxygen atoms?
(a) So_3PO_4
(b) Na_3PO_4
(c) Na^3PO^4
(d) So3PO4

34. Identify each of the following substances as either a heterogeneous or homogeneous mixture.
(a) Soil
(b) Air
(c) Diesel fuel
(d) Concrete

35. Identify each of the following substances as either a heterogeneous or homogeneous mixture.
(a) River water
(b) Brass alloy
(c) Coffee
(d) Whiskey

36. Write the formula for the mineral hematite, which is made of two iron atoms and three oxygen atoms.

Forensic Chemistry Problem

37. A document discovered in 1985, reportedly one of the very first documents printed on a printing press in the North American colonies in 1640, was being sold for $1.5 million. The Oath of a Freeman was a loyalty oath all citizens of the Massachusetts Bay Colony had to take. The age of the paper and ink were verified by carbon-14 dating, but an observant scientist determined it was a fraud. Printing presses of that period were based on arranging each letter individually on the page, inking the surface, and pressing the paper against it. All letters and individual characters were on blocks of exactly the same dimension. By looking at the document, can you determine why it is a forgery?

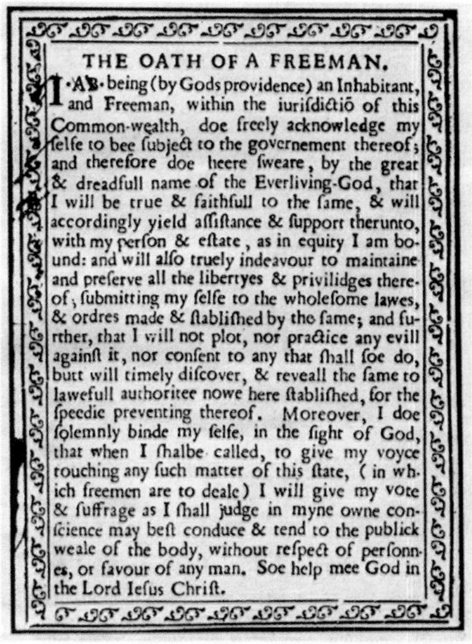
THE OATH OF A FREEMAN.

I ·AB· being (by Gods providence) an Inhabitant,
and Freeman, within the iuriſdictiõ of this
Common-wealth, doe freely acknowledge my
ſelfe to bee ſubject to the governement thereof;
and therefore doe heere ſweare, by the great
& dreadfull name of the Everliving-God, that
I will be true & faithfull to the ſame, & will
accordingly yield aſſiſtance & ſupport therunto,
with my perſon & eſtate, as in equity I am bo-
und: and will alſo truely indeavour to maintaine
and preſerve all the libertyes & privilidges thereof, ſubmitting my ſelfe to the wholeſome lawes,
& ordres made & ſtabliſhed by the ſame; and further, that I will not plot, nor practice any evill
againſt it, nor conſent to any that ſhall ſoe do,
butt will timely diſcover, & reveall the ſame to
lawefull authoritee nowe here ſtabliſhed, for the
ſpeedie preventing thereof. Moreover, I doe
ſolemnly binde my ſelfe, in the ſight of God,
that when I ſhalbe called, to give my voyce
touching any ſuch matter of this ſtate, (in which freemen are to deale) I will give my vote
& ſuffrage as I ſhall judge in myne owne con-
ſcience may beſt conduce & tend to the publick
weale of the body, without reſpect of perſonn-
es, or favour of any man. Soe help mee God in
the Lord Ieſus Chriſt.

(Utah Lighthouse Ministry)

CHAPTER 2

Evidence Collection and Preservation

Federal prosecutors believe Thomas Pitera was involved in nearly 30 murders while an active member of the Bonanno crime family, including the murder of an FBI informant against John Gotti (pictured above), the head of the Gambino family. *(Time Life Pictures/Getty Images)*

Case Study: Grave Evidence

In 1985 an assistant United States Attorney released the information that Wilfred "Willie Boy" Johnson had been serving as an FBI informant against John Gotti, the head of the Gambino crime family. This controversial strategy was used in an attempt to force Johnson to testify against Gotti and enter into the federal witness protection program. When this occurred, Johnson immediately feared for his life—and rightly so. He had long served as an informant but had always had the understanding that he would never testify against anyone in the family because betraying the family was a death sentence. He also knew that he didn't have to fear just the Gambino crew, but any member of the New York crime families and others looking to enhance their reputations by killing a perceived traitor.

The only chance Johnson had to escape the wrath of the families was to refuse to testify. By refusing to cooperate with the U.S. attorney, he was able to get a short reprieve from the Gambino family since the federal case against Gotti failed to secure a conviction. However, he had crossed too many individuals, and ultimately, a contract was placed on his life.

Several members of the Bonanno crime family eventually carried out the murder in 1988. Two members of the family, Vincent "Kojak" Giattino and Thomas "Tommy Karate" Pitera, took the contract. As they saw it, doing a favor for Gotti, the "boss of bosses," might be of benefit if they should ever need a favor in return. Both would stand trial for the murder of Johnson, but only Giattino was convicted. Pitera was acquitted and seemed to have escaped the justice system to the dismay of the FBI and federal prosecutors. It appeared that Pitera would return to his former life—that is, until a small sample of soil would betray Tommy Karate as a mass murderer. What information does a bit of soil on a shovel hold that might persuade a jury to convict someone and sentence him to life in prison?

Applying the scientific method, we begin with . . .

2.1 Preserving Evidence: Reactions, Properties, and Changes

Learning Objective

Distinguish chemical changes and properties from physical changes and properties.

One of the requirements in the collection of physical evidence from a crime scene is to preserve the evidence in its original state. Specially trained police officers usually do the evidence collection, although in some police departments forensic laboratory personnel come to the crime scene. The evidence to be collected is first photographed in place and then collected for processing at a crime laboratory. Each item is packaged separately to prevent cross contamination.

Chemical Change and Chemical Properties

When matter undergoes a chemical reaction and forms a new substance, we refer to this process as a **chemical change**. In some instances, it is important to prevent chemical change from taking place in order to preserve evidence at crime scenes. For example, bloody clothes collected at a crime scene are dried out to prevent the blood from decomposing, a process that involves chemical changes. In other instances, chemical changes are used in a beneficial way for the collection of evidence, as in the development of fingerprints. Latent fingerprints are invisible to the

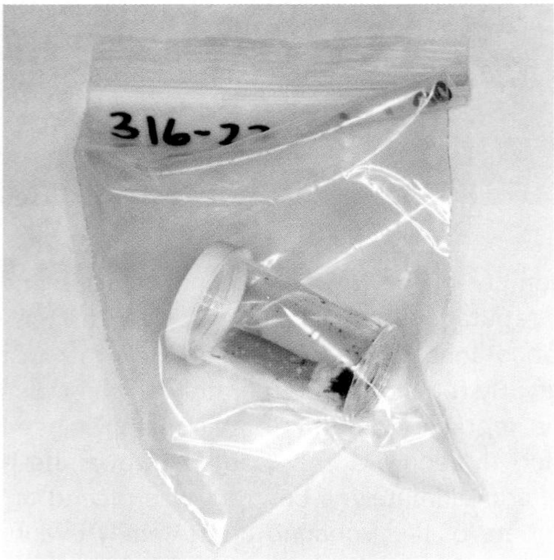

Evidence is collected into clean containers, sealed, and labeled with information specific to the crime scene. If clothing is wet from blood or other liquids, it is collected in a temporary plastic container and repackaged after being air-dried in a controlled atmosphere. (Dan Chavkin/Getty Images)

naked eye, but when treated with a variety of chemicals, images of the fingerprints become visible.

Chemical properties are used to describe the potential chemical reactions a substance can undergo. Examples of chemical properties include flammability (the potential of a substance to react rapidly with oxygen and burst into flames, releasing light and heat), rusting (the ability of iron to react with oxygen and form iron oxide—rust), and explosiveness (the tendency of compounds such as nitroglycerin to decompose violently, producing a mixture of gases that expand rapidly).

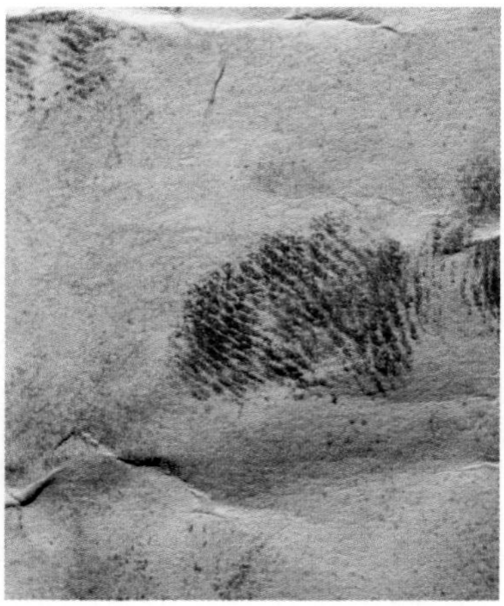

Amino acids in the sweat from your fingers react with a compound called ninhydrin to produce a visible fingerprint image. (Mauro Fermariello/Science Photo Library)

It is important to understand the chemical properties of the evidence being gathered so it can be collected in proper containers. Corrosive compounds such as battery acid are collected in either glass or plastic bottles. If battery acid were placed in a metal container, it would react with the metal, corroding it. Crime scene investigators use proper storage containers, refrigerate and freeze evidence, and separate incompatible types of evidence to preserve the evidence in its original state.

Physical Change and Physical Properties

A **physical change** occurs when matter is transformed in a way that does not alter its chemical identity, such as when matter changes between the solid, liquid, or gaseous states. One example of a physical change in evidence collection involves a sample collected from the suspected area of a fire's origin in an arson investigation. The area will often contain some of the flammable liquid used to start the fire. The liquid sample is collected into an airtight metal canister, but over time, the liquid evaporates into the air space inside the canister. However, the flammable vapor is still the same chemical compound; it has merely changed from the liquid state to the gaseous state.

Smokeless gunpowder, represented by the simplified formula $C_6H_6O_5(NO_2)_3$, undergoes a chemical reaction and produces gaseous compounds such as carbon dioxide (CO_2), water vapor (H_2O), nitrogen gas (N_2), hydrogen gas (H_2), and carbon monoxide (CO), which propel the bullet down the barrel. The carbon monoxide and hydrogen gas further react (ignite) with oxygen gas from the atmosphere to produce the muzzle flash. (Corbis)

Water is being vaporized by the heat of the fire to form steam. However, the chemical makeup of water remains the same, H_2O. (Michael Salas/Getty Images)

Physical properties can be measured without altering the chemical identity of a substance. Some common physical properties are the color, melting point, boiling point, odor, refractive index, hardness, texture, solubility, electrical properties, and density of a substance. When laboratory personnel are presented with an unknown compound, they often begin their analysis by determining the melting point. Although the melting point does not specifically identify the compound, it allows all compounds with different melting points to be eliminated. The **refractive index** is a measure of how much light is bent when it passes through a transparent substance such as glass. Density, a property that we will discuss in depth later in this chapter, is a very useful tool in determining the identity of evidence—such as glass fragments from a window smashed by an intruder or from a hit-and-run crime scene.

2.2 Critical Thinking and the Crime Scene: The Scientific Method

Learning Objective

Think like a scientist to address a problem.

Investigators start to piece together the sequence of events that may have occurred at a crime scene as soon as they arrive and begin gathering information and evidence. This is the beginning of a process during which the investigators may periodically rethink and revise their conclusions of what happened based on new information or new evidence. Ultimately, the investigators want to follow the trail of evidence so that they can prove, beyond a reasonable doubt, what happened and who was responsible for the crime. In pursuing this goal, investigators make use of the **scientific method**, a systematic method for problem solving, described as follows.

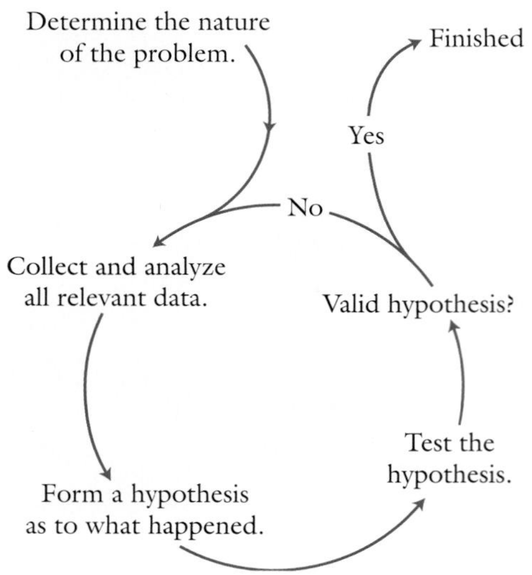

The scientific method has been simplified into the basic components illustrated in the circular process shown here. If a hypothesis is proven invalid, that information can be used in the formation of a new hypothesis. Great care must be taken in designing experiments that serve to test the hypothesis and not merely confirm a set of preconceived ideas.

The Scientific Method

1. Determine the nature of the problem. (Who committed a crime and how?)
2. Collect and analyze all relevant data. (Consider all physical evidence, witness statements, alibis, etc.)
3. Form an educated guess, called a **hypothesis**, as to what happened. (The butler did it in the library with a candlestick.)
4. Test the hypothesis. (Test alibis, reconstruct the crime scene, consider timelines, etc.)
5. If your hypothesis holds up to the testing, you are finished. If not, go back to step two.

A common misconception is that only laboratory scientists use the scientific method when, in truth, many people, from mechanics to medical doctors, commonly use it. You can probably even think of a problem that you have solved in your own life using the scientific method.

Worked Example 1

Discuss the steps a doctor might use to determine what is wrong with a patient who describes a shooting pain that developed in her arm after she fell down a flight of stairs.

SOLUTION

1. Is the arm broken or merely bruised? Hypothesis to test: The arm may be broken.

Detectives apply the scientific method at a crime scene to determine what items to photograph and what to collect as evidence. (Corbis)

2. Ask the patient to describe how she landed. Feel the arm and apply slight pressure to it.
3. Consider the patient's reaction to pressure, the details of the accident description, and how the arm physically feels.
4. If the patient's reaction is not indicative of an arm break, send her home with a pain reliever. If any doubt remains, order an X-ray to confirm a break.
5. Evaluate the X-ray and prescribe appropriate treatment.
6. If the patient returns within days still complaining of pain, repeat the procedure or refer her to a specialist.

Practice 2.1

The alternator in a vehicle supplies electricity to the car when the engine is running and recharges the car battery. If the alternator is malfunctioning, the car will drain the battery. Describe how a mechanic would use the scientific method to determine whether the problem is the alternator or the battery.

ANSWER

Answers vary individually. Have a classmate read your answer to see if it seems reasonable.

Evidence Collection Guidelines for Missouri State Highway Patrol

Liquids: 1 ounce

Drugs: Small seizure, entire sample
Large seizure, representative sample

Blood: 2 vials, 10 mL each

Urine: 50 mL

Stomach contents: 50 mL

Hair: 50 head, 25 pubic

Clothing: Dried, packaged separately

Learning Objective

Distinguish mass from weight and explain the importance of units of measurement.

What Constitutes Relevant Data?

One of the earliest decisions the crime scene investigators have to make is which items present should be considered evidence and which are not vital to the investigation. It would be extremely costly and time-consuming if the entire crime scene were tagged as evidence and analyzed. Even if it were possible to analyze everything, just sifting through the useless information would take detectives a tremendous amount of time. Therefore, investigators must use their experience and judgment to determine what material is critical evidence and should be collected for analysis at a crime laboratory. The police can also secure crime scenes for several days to prevent anyone from entering the site. This allows investigators to come back if there is a need to search for more evidence or if a new hypothesis is formed in evaluating the case.

2.3 Physical Evidence Collection: Mass, Weight, and Units

Because the evidence collected at a crime scene may need to be analyzed using several laboratory methods, it is important for investigators to know how large a sample to collect. If an insufficient sample is collected, the laboratory might not be able to completely analyze the evidence. On the other hand, forensic crime laboratories are often backlogged with evidence to analyze, which can create problems for storage if needlessly large samples are collected from crime scenes. Therefore, most crime laboratories publish guidelines for sample size.

The **mass** of a sample is a measure of how much matter is contained in the sample. The unit commonly used for reporting mass is the gram. In the laboratory, mass is measured on a balance, which compares the amount on the pan to a known mass and displays the value. Accuracy in determining the mass of a sample is important for many types of evidence

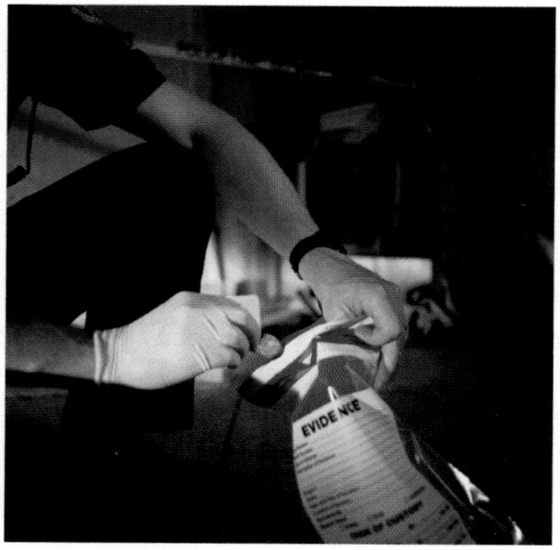

Collecting sufficient evidence at a crime scene is critical so that the forensic crime lab can run multiple tests on the sample without consuming the entire sample of evidence. (Brand X Pictures/Alamy)

but particularly when a drug arrest is made. The charge that will be filed—"intent to deliver" versus "intent for personal use"—is defined in terms of the mass of drugs recovered.

The mass of a sample is not to be confused with its weight. **Weight** is a measure of how strongly gravity is pulling on matter. If an object is moved to a location where the pull of gravity is lower, the weight of the object decreases. An astronaut on the moon weighs only 1/6 as much as on earth because the moon's gravity is 1/6 of the earth's. However, the astronaut's mass remains the same. The mass of an object does not depend on gravity and, therefore, does not change from one location to another.

Weight is commonly reported in units of pounds and is measured with a scale. A scale has a spring that compresses as the force of gravity acts on an object placed on the scale. A dial attached to the spring indicates the weight.

It should be pointed out that in the laboratory we use *only* the metric system for measurements. However, since many other agencies, from police departments to engineering firms, use the older English measurement units, such as feet and pounds, a scientist should be able to convert from English units to metric units. Appendix A lists the metric-English equivalents for converting between the two sets of units. Table 2.1 summarizes the units used in a laboratory to take measurements.

Table 2.1 Units of Measurement

Measurement	Units	Abbreviation
Mass	grams	g
Volume	liters	L
Distance	meters	m
Time	seconds	s

The standard units are used to measure experimental variables to facilitate communication between scientists.

Units of measurement are often modified to reflect the size of the sample being measured. When police seize a major shipment of drugs, they commonly report how many "kilos" were seized. The term *kilo* is a slang expression for kilogram (kg), which is equal to 1000 grams. It is easier to communicate the size of a large sample by saying "100 kilos" rather than "100,000 grams." On the other hand, when police officers search a vehicle during a routine traffic stop and come across a small quantity of illegal narcotics, they measure how many grams were seized rather than kilos. It is simpler to describe the mass of a small sample as 5 grams rather than 0.005 kilograms. Other units commonly modified with prefixes are milliliters and centimeters. The prefix modifiers often used in the laboratory are summarized in Table 2.2.

The **unit** of a measurement contains critical information about what system of measurement is used and whether the base unit is modified with a prefix. Therefore, it is always important to state the unit along with the number. If a police officer were to write that an amount of cocaine seized in an arrest was "10," the information would be useless. It is very important to know whether the officer meant 10 grams, 10 kilograms, or 10 milligrams.

Table 2.2 Common Prefix Modifiers

Prefix	Abbreviation	Multiplier
mega	M	1,000,000
kilo	k	1,000
deci	d	0.1
centi	c	0.01
milli	m	0.001
micro	μ	0.000001

Prefix modifiers are added to units in order to convey large or small quantities in a clear and concise manner.

2.4 Mathematics of Unit Conversions

Learning Objective

Use conversion factors properly in solving mathematical problems.

Solving problems in forensic science and in chemistry often requires converting from one unit of measurement to another by means of a **conversion factor.** A conversion factor is simply an equality by which a quantity is multiplied to convert the original units of the quantity to the new units that are desired. To use a conversion factor, divide by the units you wish to cancel and multiply by the units you want in the final answer. Because you will be doing conversions throughout this course, it is advisable for you to master this skill now.

Worked Example 2

How many meters are in 245 mm?

SOLUTION

The equality 1 mm = 0.001 m provides the information needed for a conversion factor between millimeters and meters.

$$245\ \cancel{\text{mm}} \times \frac{0.001\ \text{m}}{1\ \cancel{\text{mm}}} = 0.245\ \text{m}$$

Notice that when you divide millimeters by millimeters, the units cancel out, leaving the meter as the remaining unit. You can use the same conversion factor for converting meters to millimeters.

Practice 2.2

How many millimeters are in 5.00 cm?

ANSWER

50.0 mm

Worked Example 3

How many millimeters are in 1.055 m?

SOLUTION

$$1.055\ \cancel{\text{m}} \times \frac{1\ \text{mm}}{0.001\ \cancel{\text{m}}} = 1055\ \text{mm}$$

The equality needed to solve this is 1 mm = 0.001 m. But how do we know whether to use 1 mm/0.001 m or 0.001 m/1 mm as the conversion factor? In this case, the question asks us to find *millimeters*. Therefore, we want *meters* to cancel out in the calculation, and we use 1 mm/0.001 m as the conversion factor.

Practice 2.3

How many kilometers are in 575 meters?

ANSWER

0.575 km

It is often necessary to use multiple conversion factors to get from the given information to the desired units. For example, how many centimeters are there in 2.000 feet? You probably do not know the conversions for centimeters to feet. However, you know that 1 foot = 12 inches and from Appendix A you can find that 2.54 centimeters = 1 inch.

Worked Example 4

How many centimeters are in 1.50 ft.?

SOLUTION

$$1.50 \text{ ft.} \times \frac{12 \text{ in.}}{1 \text{ ft.}} \times \frac{2.54 \text{ cm}}{1 \text{ in.}} = 45.7 \text{ cm}$$

Practice 2.4

How many centimeters are in 0.750 yd.?

ANSWER

68.6 cm

2.5 Errors and Estimates in Laboratory Measurements: Significant Figures

Learning Objective

Identify significant figures in measurements and know how to determine them.

In making measurements for scientific purposes, we intentionally include one number that is an estimate in every measurement. It seems odd, but by including one digit that is an estimate (which means it may contain error), the measurement is actually more accurate than if we used only the digits that are exactly known. Consider the bloody footprint shown in Figure 2.1. How would you report the width of the partial print? It is greater than 11 cm but it is definitely less than 12 cm. An estimate of 11.5 cm would be reasonable. If you thought 11.4 or 11.6, those estimates would also be acceptable. Because the last decimal place is an estimate, it will vary from one person's observation to another's. The rule in the laboratory is that we keep all digits that are known exactly, plus one digit that is an estimate and contains some error. Collectively these digits are called **significant figures**.

Worked Example 5

A buret is a piece of scientific glassware used to accurately measure out solutions during experiments. What is the volume reading on the buret

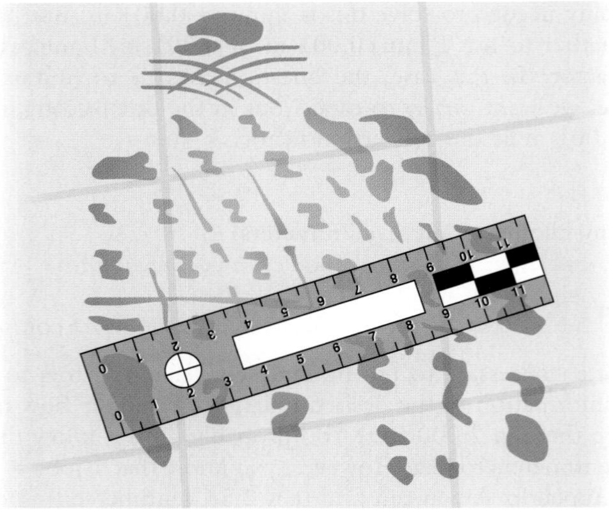

Figure 2.1 When making measurements, it is necessary to include an estimated digit in the answer to obtain the most accurate value. When taking a photograph of evidence, a ruler is always included for later reference. It is critical that the camera be at a 90-degree angle to the ruler. Otherwise, the readings will be distorted.

below? (*Hint:* The solution forms a concave shape, called a *meniscus.* Read the volume at the lowest point of the meniscus.)

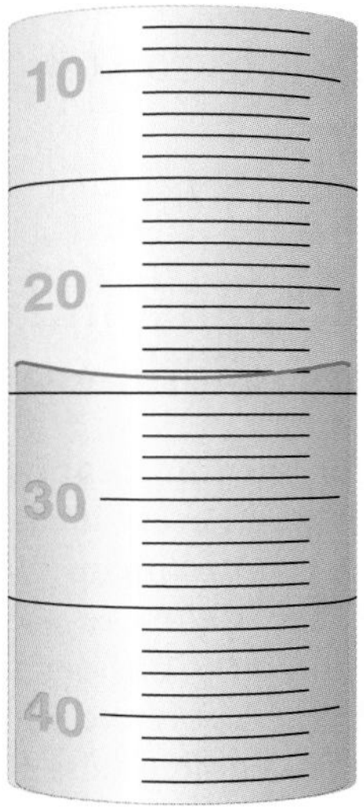

SOLUTION

The volume is more than 24.0 mL but less than 25.0 mL, so an estimate of 24.2 mL would be reasonable.

Practice 2.5

Using a metric ruler, measure the height of a can of soda in units of centimeters.

ANSWER

12.2 cm

It is often necessary to use measured quantities in calculations. When this is done, it is necessary to know how many significant figures are in each number involved in the calculation, as the answer must reflect the proper number of significant figures. There are guidelines that are helpful in doing this.

Zero and Nonzero Numbers in Measurements

The first rule for determining significant figures is that all nonzero numbers in a measurement are significant. For example, the number 459.61 cm contains five significant figures.

The rules for determining significant figures get more complex when there is a zero in the number. The rules for zeros are summarized below.

Zero Rules

1. Zeros located between nonzero numbers are significant.
   Example: 101 has three significant figures.
2. Zeros at the beginning of a number containing a decimal point are *not* significant.
   Example: 0.015 has two significant figures.
3. Zeros at the end of a number containing a decimal point are significant.
   Example: 25.20 has four significant figures.
4. Zeros at the end of a number not containing a decimal point are ambiguous.
   Example: 100 contains at least one significant figure but could contain up to three significant figures, depending on whether the estimate is in the tens or the ones place.

How should the number 100 be written to indicate the correct number of significant figures? The solution is to use **scientific notation** because only those zeros that are considered significant are included in the number, as shown below. There are some older methods, such as underlining the significant zeros or placing a decimal at the end to indicate significant digits. These antiquated methods are *not* used in modern laboratories and are not considered proper methods for indicating significant figures.

Scientific Notation and Significant Figures

100 written with 1 significant digit is 1×10^2

100 written with 2 significant digits is 1.0×10^2

100 written with 3 significant digits is 1.00×10^2

The procedure for writing numbers in scientific notation is:

1. Count the number of places the decimal point is moved to have one nonzero digit to the left of the decimal place.
2. The number of places you moved the decimal is the number used as the exponent.

   (a) If you moved the decimal to the right, you make the exponent a negative value.

   (b) If you moved the decimal to the left, you make the exponent a positive value.

Worked Example 6

Write the number 12,000 with two significant figures.

SOLUTION

1. Count the number of places the decimal point is shifted to the left until you have one nonzero digit to the left of the decimal:

$$\underset{\leftarrow\ \leftarrow\leftarrow\leftarrow}{12{,}000}$$

2. Write the number in scientific notation with the exponent equal to the number of shifts. The exponent is positive because you shifted to the left: 1.2×10^4.

Practice 2.6

Write the number 100,000 using four significant figures.

ANSWER

1.000×10^5

Worked Example 7

Write the number 0.000120 in scientific notation. *Note:* This number contains three significant digits with no ambiguity. However, we commonly write very small or very large numbers in scientific notation form.

SOLUTION

1. Count the number of places the decimal point is shifted to the right until you have one nonzero digit to the left of the decimal:

$$\underset{\rightarrow\rightarrow\rightarrow\rightarrow}{0.000120}$$

2. Write the number in scientific notation with the exponent equal to the number of shifts. The exponent is negative because the decimal was shifted to the right: 1.20×10^{-4}.

Practice 2.7

How would you write the number 0.03210 using scientific notation?

ANSWER

3.210×10^{-2}

2.6 Mathematics of Significant Figure Calculations

Learning Objective

Explain the rules for determining the number of significant figures required in answers to calculations.

Some measurements used in a calculation may have more significant figures than others. In such a case, how do we determine the number of significant figures the answer should have? The last digit, which contains the estimate in a measurement, affects any number by which it is multiplied, divided, added to, or subtracted from. The rules for counting significant figures in mathematical operations are listed below. The digit containing the estimate is colored in blue, and any number it affects is also colored blue to help you visualize why the rules apply as they do.

Rules for Mathematical Operations of Significant Figures

1. Addition and Subtraction: The answer can have only as many decimal places as the number with the fewest digits after the decimal place.
2. Multiplication and Division: The answer can have only as many significant figures as the number with the fewest significant figures.

When writing a number with the correct number of significant figures from a mathematical problem, it will often need to be rounded up or down according to the rules below.

Rules of Rounding

1. If the number being dropped is greater than 5, increase the last saved digit by 1.
2. If the number being dropped is less than 5, leave the last saved digit as it is.
3. If the number being dropped is exactly 5, flip a coin.*

*This rule changes from one source to another, as there is not a standard procedure for this situation. A truly random coin flip prevents a bias from being introduced.

Worked Example 8

What is the answer to 145.056 + 7.01 + 22.0261?

SOLUTION

$$\begin{array}{r l} 145.056 & \\ 7.01 & \leftarrow \text{fewest digits after decimal place} \\ +\ 22.0261 & \\ \hline 174.0921 & \end{array}$$

The answer is 174.09 since there are two decimal places in 7.01.

Practice 2.8

What is the answer to 61.83 − 59.241?

ANSWER

2.58

Worked Example 9

What is the answer to: 10.31 × 2.5?

SOLUTION

$$\begin{array}{r l} 10.31 & \\ \times\ \ 2.5 & \leftarrow \text{fewest significant digits} \\ \hline 5155 & \\ +\ 2062 & \\ \hline 25.775 & \end{array}$$

The answer is 26 since there are two significant digits in 2.5.

Practice 2.9

What is the answer to 0.00101/0.10?

ANSWER

0.010

There are some problems that require multiplication or division and addition or subtraction in the same problem. For these mixed operation problems, always follow the rules for order of operations giving priority to any mathematical steps in parentheses first, followed by multiplication/division, and finally addition/subtraction. When a problem includes multiple steps, never round a number until the calculation is completely finished. To determine the correct number of significant digits in a number, examine each step in the order completed and use the rules appropriately for each step.

Worked Example 10

What is the answer to $\dfrac{(14.01 - 1.025)}{0.0120}$?

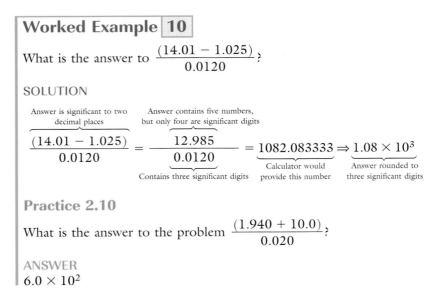

SOLUTION

$$\frac{\overbrace{(14.01 - 1.025)}^{\text{Answer is significant to two decimal places}}}{0.0120} = \frac{\overbrace{12.985}^{\text{Answer contains five numbers, but only four are significant digits}}}{\underbrace{0.0120}_{\text{Contains three significant digits}}} = \underbrace{1082.083333}_{\text{Calculator would provide this number}} \Rightarrow \underbrace{1.08 \times 10^3}_{\text{Answer rounded to three significant digits}}$$

Practice 2.10

What is the answer to the problem $\dfrac{(1.940 + 10.0)}{0.020}$?

ANSWER

6.0×10^2

2.7 Experimental Results: Accuracy and Precision

Learning Objective

Distinguish between accuracy and precision in experimental results.

The forensic scientists who analyze evidence are often called to court to explain the results of their analysis. There are two aspects of the results generated by the scientists that are critical for both the prosecution and defense to understand.

Accuracy

The first aspect of importance is how close the experimental results are to the true or real value for the quantity being measured. This represents the **accuracy** of the results. For example, if a cocaine sample were 50.0% pure and the lab reported back a value of 50.1%, the analysis would be considered accurate because the experimental value was close to the true

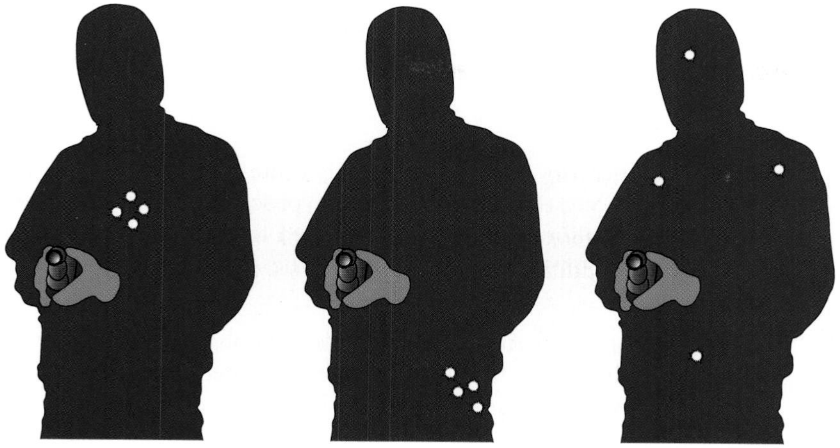

Figure 2.2 Law enforcement officers must train using human-shaped targets because the visual difference between a traditional bull's-eye target and human target can affect both their accuracy and precision.

value. If the lab had reported a value of 45%, the results would be inaccurate. How is the true or real value of a sample determined? Many times it is not possible to know the exact value. However, there are analytical methods and advanced statistics that enable scientists to predict how close the results are to the real or true value.

Precision

Another important aspect of the experimental results is the **precision** of a measurement. Precision refers to how reproducible a measurement is if the same sample is measured multiple times. From the cocaine example, if the scientists reported that the cocaine sample was analyzed three times and the results were 45.0%, 44.9%, and 45.2%, those results are numerically close and the measurement would be considered precise. However, because the measurements are not close to the true value of 50%, they are said to be inaccurate. For reported results to be considered valid, measurements must be both precise and accurate. This would be the case if the cocaine analysis gave results of 49.9%, 50.0%, and 50.2%. The concepts of accuracy and precision are also illustrated in Figure 2.2.

Worked Example 11

Label each of the sets of data below as accurate, precise, inaccurate, or imprecise, as appropriate. In each case, the true value of the measurement is 13.25.

(a) 12.22, 14.21, 13.24

(b) 15.24, 15.21, 15.28

(c) 13.24, 13.21, 13.23

SOLUTION

(a) Scattered data = imprecise. Not close to true value = inaccurate.

(b) Close data = precise. Not close to true value = inaccurate.

(c) Close data = precise. Close to true value = accurate.

Practice 2.11

When forensic chemists testify at trial, it is quite common for the defense attorney to question them about the procedures, methods, and chemicals used to calibrate the instrument used for the analysis. Explain how this line of inquiry is used to question the accuracy of the results.

ANSWER

The calibration of an instrument ensures that the instrument is providing accurate results. If the instrument were calibrated incorrectly, then all of the results would be inaccurate.

2.8 How to Analyze Evidence: Density Measurements

Learning Objective

Use density as another physical property to investigate chemical evidence.

Earlier in this chapter we introduced some physical properties of matter—such as melting point, boiling point, and density—that can aid in the identification of unknown substances collected as evidence. Physical properties of the evidence are compared with physical properties of a series of known materials. This comparison allows investigators to eliminate from the list of possible substances any material that does not match. A tentative identification of the evidence can be made if its physical properties match those of a known substance. More extensive experiments need to be done to positively identify the material, but the job is much easier if the possibilities have been greatly narrowed.

One of the physical properties that can be used to identify evidence is density. **Density** is defined as the ratio of the mass of an object to its volume. Density is often confused with weight or mass. This is understandable because objects that have a high density, such as cement or iron, are those that are often described as "heavy." Objects that have a low density, such as foam or feathers, are typically ones that we consider "light." We can therefore think of density as the "heaviness" of a material, but density should not be confused with either the mass or the weight of an object. Glass evidence lends itself to analysis by the examination of density, as there are many different types of glass, each with different densities. Listed in Table 2.3 are some physical properties of various types of glass.

Worked Example 12

If the density of a glass sample is determined to be 2.26 ± 0.02 g/mL, what are the possible types of glass the sample could be? The $\pm$ sign signifies that any glass that has a density within 0.02 units of 2.26 g/mL would be a possibility.

SOLUTION

The range of acceptable answers is 2.24 to 2.28 g/mL. The following glass samples all fall within this range: borosilicate, alkali barium borosilicate, soda borosilicate, and alkali strontium.

Practice 2.12

If the refractive index for the same glass sample is determined to be 1.48 ± 0.01, what types of glass could the sample be?

ANSWER

Borosilicate, alkali barium borosilicate, or soda borosilicate

Table 2.3 Physical Properties of Glass

Type	Softening Point[1] (°C)	Density (g/mL)	Refractive Index[2]
Alkali barium	646	2.64	1.511
Alkali barium (optical)	647	2.60	1.512
Alkali barium borosilicate	712	2.27	1.484
Alkali borosilicate	718	2.29	1.486
Alkali strontium	688	2.26	1.519
Alkali zinc borosilicate	720	2.57	1.523
Borosilicate	720	2.28	1.490
Baria alumina borosilicate	844	2.76	1.530
Barium-alumina borosilicate	847	2.96	1.545
Borosilicate	821	2.23	1.473
Lanthanum barium	759	3.98	1.678
Lead borosilicate	447	5.46	1.860
Lead zinc borosilicate	370	3.80	—
Lithia potash borosilicate	—	2.13	1.469
Potash borosilicate	820	2.16	1.465
Potash soda lead	630	3.05	1.560
96% Silica	1530	2.18	1.458
96% Silica (porous)	1530	1.50	—
Silica (99.9% fused)	1585	2.20	1.459
Soda borosilicate	808	2.27	1.476
Soda alumina borosilicate	705	2.17	1.468
Soda-lime	696	2.47	1.510

[1]The softening point is the temperature at which heated glass starts to deform under its own weight.
[2]The refractive index of all samples is measured at a wavelength of 589.3 nm.

2.9 Mathematics of Density Measurements

Take a look at the equations on the next page. Equation 1 can be used to determine the density of an object, provided that the mass and volume of the object are known. Furthermore, it is possible to rearrange equation 1 so that either mass or volume can be calculated from the data. This is shown in equations 2 and 3.

Learning Objective

Reinforce the use of density measurements and illustrate their use in the forensic laboratory.

$$\text{Density} = \frac{\text{mass}}{\text{volume}} \text{ or } D = \frac{m}{V} \tag{1}$$

$$\text{Initial equation: } D = \frac{m}{\underbrace{V}_{\text{Solve}}} \Rightarrow \text{Goal is to solve for } V \tag{2}$$

Step 1: Multiply both sides by $V \Rightarrow D \times V = \frac{m}{\not{V}} \times \not{V}$

Step 2: Divide both sides by $D \Rightarrow \frac{\not{D} \times V}{\not{D}} = \frac{m}{D}$

Step 3: Final equation: $V = \frac{m}{D}$

$$\text{Initial equation: } D = \frac{\overbrace{m}^{\text{Solve}}}{V} \Rightarrow \text{Goal is to solve for } m \tag{3}$$

Step 1: Multiply both sides by $V \Rightarrow D \times V = \frac{m}{\not{V}} \times \not{V}$

Step 2: Final equation: $D \times V = m$ or $m = D \times V$

The units of density are most commonly g/mL, g/cc, or g/cm^3 ("cc" is an older abbreviation for a cubic centimeter, cm^3). Because 1 mL = 1 cc = 1 cm^3, all of the previously listed density units are interchangeable. You will occasionally hear the term *cc* instead of *milliliter* on a TV medical drama, and *cc* is also commonly used in engineering and material science literature. Therefore, you should be aware of its usage and meaning.

Worked Example 13

A shattered glass jar was found at a burglary scene and was determined to be made of soda-lime glass. A glass fragment recovered from a suspect's home had a mass of 4.652 g. When the glass fragment was placed into a graduated cylinder with 20.00 mL of water, the level of the water rose to 21.53 mL. Does the glass fragment link the suspect to the crime scene? Consult Table 2.3.

SOLUTION

$$D = \frac{m}{V} = \frac{4.652 \text{ g}}{(21.53 - 20.00)} = \frac{4.652 \text{ g}}{1.53 \text{ mL}} = 3.04 \text{ g/mL}$$

No, the glass samples do not match. Soda-lime glass has a density of 2.47 g/mL. The fragment recovered is most likely potash soda lead glass, which has a density of 3.05 g/mL.

Practice 2.13

Determine the identity of a 12.471-g sample of glass that displaces 2.33 mL of water.

ANSWER

$D = 5.35$ g/mL, which is closest to lead borosilicate glass.

Worked Example 14

The typical density range for urine samples recovered from an autopsy is 1.002 to 1.028 g/cm^3. Density values higher than this range can indicate the victim suffered from vomiting, excessive sweating, or dehydration shortly before death. Density values lower than this range can indicate that the victim was diabetic or had acute renal failure due to exposure to certain metals or solvents. Calculate the acceptable mass range that a 25.00-mL urine sample could have and still be considered normal.

SOLUTION

Two calculations are necessary. The first is based on the low density value of 1.002 g/cm^3 and the second is based on the high density value of 1.028 g/cm^3.

$$\text{Low density: } m = D \times V \Rightarrow m = 1.002 \frac{\text{g}}{\cancel{\text{cm}^3}} \times \overbrace{25.00\ \cancel{\text{cm}^3}}^{1\text{ mL} = 1\text{ cm}^3} = 25.05\text{ g}$$

$$\text{High density: } m = D \times V \Rightarrow m = 1.028 \frac{\text{g}}{\text{cm}^3} \times 25.00\ \cancel{\text{cm}^3} = 25.70\text{ g}$$

Practice 2.14

Calculate the mass of 75.00 mL of urine with a density of 1.018 g/cm^3.

ANSWER

76.35 g

Worked Example 15

Cerebral edema is swelling of the brain tissue caused by the absorption of water and by increased water content in the brain cavity, which can be caused by a blunt force trauma. The average human brain is 1.30×10^3 g and the density of the normal human brain tissue is 1.05 g/mL. Calculate the initial volume the brain tissue occupies and then the volume it would occupy after absorption of water decreases the brain tissue density to 1.01 g/mL.

SOLUTION

$$\text{Initial: } V = \frac{m}{D} \Rightarrow V = \frac{1.30 \times 10^3\ \cancel{\text{g}}}{1.05\text{ g/mL}} = 1.24 \times 10^3\text{ mL}$$

$$\text{Final: } V = \frac{m}{D} \Rightarrow V = \frac{1.30 \times 10^3\ \cancel{\text{g}}}{1.01\ \cancel{\text{g}}\text{/mL}} = 1.29 \times 10^3\text{ mL}$$

Practice 2.15

The density of PEK-1 plastic explosive is 1.45 g/cc. A single cartridge of PEK-1 has a mass of 1.00×10^2 g. What is the volume of the PEK-1 cartridge?

ANSWER

69.0 cc

2.10 How to Analyze Glass and Soil Evidence: Using Physical Properties

Learning Objective

Show how a physical property such as density is used to analyze evidence.

> [Holmes] Tells at a glance different soils from each other. After walks has shown me splashes upon his trousers, and told me by their colour and consistence in what part of London he had received them.
>
> —*A Study in Scarlet,* Sir Arthur Conan Doyle

Glass fragments and soil samples are two types of evidence that have historically been analyzed by measuring their physical properties, especially density. Today there are more modern instrumental methods for analyzing the exact elemental makeup of evidence, and we will discuss these methods in the next chapter. However, density is still used today for an initial evaluation of glass and soil.

Glass Evidence

There are many types of glass and glasslike materials such as Plexiglas, which is used for car windows and headlights. When glass material is collected at a crime scene, density can be measured to help identify the material and to see if it matches evidence collected from a suspect. The density measurement is often made by the **sink-float method**, in which the glass fragment is placed into a solution of known density. If the glass has a higher density than the liquid, it sinks; if it has a lower density than the liquid, it floats. If the glass has the same density as the liquid, it will stay suspended in the solution.

Soil Evidence

Soil evidence is analyzed in a slightly different manner. Soil is a heterogeneous mixture containing a variety of components such as minerals, dust, organic materials, pollen, clay, pebbles, and so on. Each component has a different density, and ideally, the components are a unique mixture found only around the crime scene area. If this is true and matching soil is recovered from the suspect, an evidence trail has been established. Remember that the soil evidence does not prove guilt, merely that the person was present at that location. More investigative work is needed to prove whether a suspect was, in fact, at the location at the time of the crime and committed the criminal act.

The method used historically to analyze soil is called the **density gradient method**. A tall cylinder contains a solution that has a high density at the bottom of the cylinder and a low density near the top of the cylinder. When the soil sample is introduced to the top of the cylinder, the particles begin to sink down through the lighter density liquid until they reach a place where the density of the liquid matches the density of the particles. The particles remain suspended at the level where their density and the density of the liquid are equal. The result is that the soil forms bands throughout the cylinder, as illustrated in Figure 2.3. When soil evidence is obtained from other sources, such as a suspect's shovel, matching the density bands allows investigators to make a preliminary assumption as to whether the new soil sample came from the same area as a

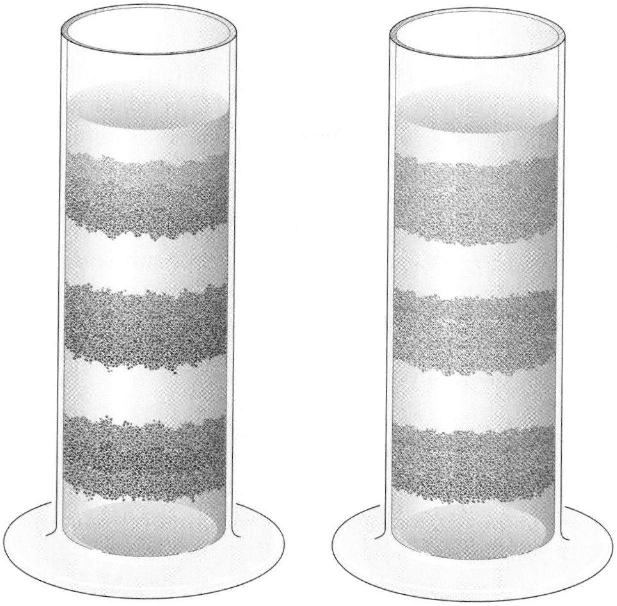

Figure 2.3 Soil samples separate to form bands of minerals by the various minerals sinking until they reach a liquid region of matching density.

previous one, such as a crime scene. Matching the soil density bands works best when the soil components significantly change from one area to another, making the soil near the crime scene unique. One problem with using a gradient tube is that many different minerals have nearly identical density values and a unique band structure might not be obtained even though the samples are different.

Modern forensic geologists will use a host of physical properties to determine whether the soil sample from a crime scene matches one obtained from a suspect. The first step in the analysis of soil is to compare the color of the two soil samples and determine if they match. The human eye is one of the best methods for detecting even slight differences in colors and is an accepted method for determining if soils match. We have already seen in the Camarena case in Chapter 1 how differences in soil color provided a significant clue. In order for soil colors to be compared properly, it is important that both soil samples have the same moisture content, since moisture will change the color of the soil. Moisture level is commonly adjusted by heating the samples to dryness or, less often, by adding excess moisture to both samples. Soil taken from the surface of a dead body may also have to be cleared of any fatty deposits, as became apparent in the Camarena case. If the colors of the soil samples match, the forensic geologist will next compare the textures of the soil samples.

> Observation tells me that you have a little reddish mould adhering to your instep. Just opposite the Wigmore Street Office they have taken up the pavement and thrown up some earth, which lies in such a way that it is difficult to avoid treading in it in entering. The earth is of this peculiar reddish tint which is found, as far as I know, nowhere else in the neighborhood.
>
> —*The Sign of Four,* Sir Arthur Conan Doyle

The soil texture depends on how much sand, silt, and clay are present; the texture can change dramatically if the ratio of these components changes. The terms *sand*, *silt*, and *clay* refer only to the physical sizes of the particles, not to their chemical identities. Two sand samples can have a different chemical makeup. However, if two soil samples come from a common point of origin, the ratio of sand to silt to clay should be constant. The particles of soil also demonstrate another physical property used to characterize the soil—the shape of the particles. The three shape descriptions a forensic geologist uses are angular, sub-rounded, and well-rounded. Figure 2.4 illustrates physical properties used to analyze soils.

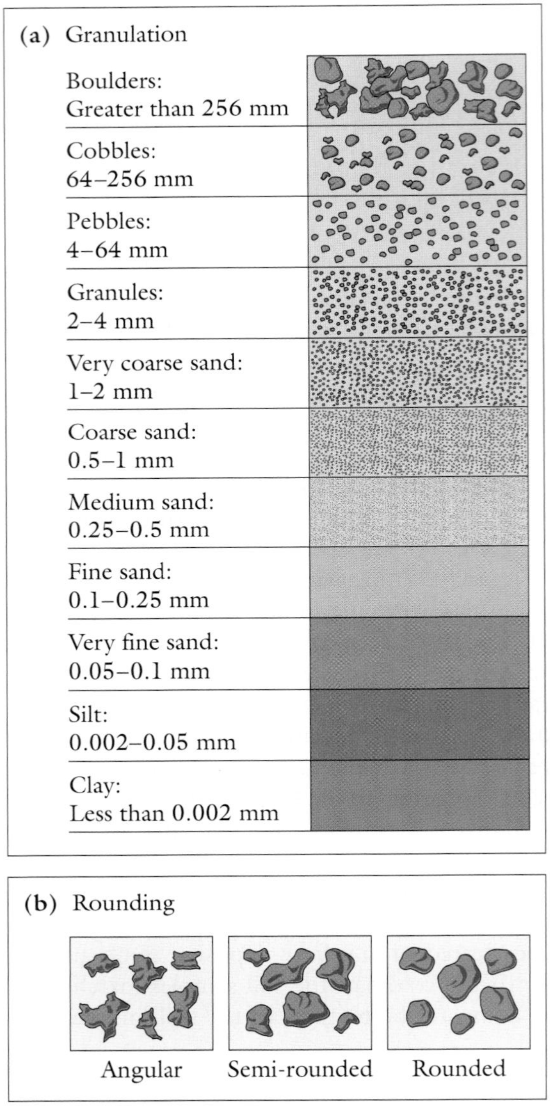

Figure 2.4 The size and shape of soil components dictate the overall texture and appearance of soil samples. The size of the particle, not the chemical makeup of the particle, determines whether a soil particle is classified as sand, silt, or clay. (Adapted from Jones, N. W., and Jones, C. E., *Laboratory Manual for Physical Geology,* 4th ed., 2003)

The final step forensic geologists can perform is analyzing the mineral composition of the soil. Each mineral is a unique chemical compound. For example, the mineral quartz has the chemical formula SiO_2 and the mineral calcite has the formula $CaCO_3$. Mineral identification can be based on physical appearance under a microscope because the shape and color of a mineral can be quite distinctive. Further identification is also possible using advanced laboratory instrumentation. Figure 2.5 illustrates the distinctive shapes and appearances of minerals.

The analysis of physical properties of soils is an important tool in the hands of law enforcement because it can provide a physical link between a suspect and a specific location. It was just such an analysis that led to the downfall of one member of a New York crime family. Now back to the Tommy Karate Pitera case study . . .

(a) Silicon dioxide, SiO_2 (quartz)

(b) Calcium carbonate, $CaCO_3$ (calcite)

(c) Aluminum hydroxide, $Al(OH)_3$ (gibbsite)

Figure 2.5 Geologists give unique names for most minerals, whereas chemists tend to refer to each mineral by the chemical name, from which the formula can be determined. (a: Geoff Thompkinson/Photo Researchers, Inc.; b: Mark A. Schneider/Photo Researchers, Inc.; c: The Fine Mineral Company)

2.11 Case Study Finale: Grave Evidence

Tommy Karate Pitera had evaded the justice system for the murder of Willie Boy Johnson, but his luck was running out. An associate of Pitera's who had assisted him on several of his murderous drug deals confessed his role to police and agreed to testify against Pitera in exchange for a lighter sentence and protection. The U.S. attorney wanted more evidence than just the word of a co-conspirator to the crime. Wire taps had recorded incriminating statements made by Pitera to his associates, but the prosecutors wanted to be able to physically link Pitera to the crimes. Soil evidence would provide that positive link from Pitera to a mass grave site containing suitcases filled with dismembered body parts that were discovered outside of the William R. Davis Wildlife Sanctuary located on Staten Island, New York.

The suitcases contained the remains of seven people who, it was later determined, had the misfortune of crossing Pitera. Pitera was in charge of the drug dealing aspects of the Bonanno crime family, and his main work entailed stealing drug shipments and money from other drug dealers. If the drug dealers protested the loss of their money and drugs, they were murdered. Pitera's other victims were those whom he suspected of being informants or who had insulted him.

The detectives at the mass grave site decided to take soil samples at the location of the graves, hoping that the soil at the grave site would provide a color, texture, and composition that could be linked to Pitera. The value of such evidence lies in its uniqueness. If the soil all over Staten Island were uniform, there would be no way to link a suspect to the grave site rather than to any other location on the island. In order to prove that the soil around the graves was unlike soil found elsewhere on Staten Island, detectives took samples not only at the site but also at various distances from the grave location to show that the soil changes in composition as one goes away from the site.

The soil samples from Staten Island were sent to Special Agent Bruce W. Hall, supervisor of the forensic mineralogy unit at FBI headquarters. He was also given one other sample: the soil that was stuck in a shovel obtained from a home on East 12th Street in the Gravesend section of Brooklyn, the home of Pitera. Shovels have a small loop on top for the foot to apply pressure, as shown in the picture, opposite. This area works as an excellent soil sampler since it takes a sample and then, because it is filled, does not allow future soil to mix with the sample.

Hall was able to examine the color, texture, and composition of the soil found in the shovel and link that shovel to the burial site. When defense attorneys argued that the shovel had been used for gardening and the soil could have come from other locations, Hall had samples taken from the alibi locations and was able to show that the soils were vastly different from the soil on the shovel. Did this mean the shovel was not used at the alibi locations? Not at all—it simply meant that no soil from those sites was found on the shovel, but soil from the burial site was.

Tommy Karate Pitera had escaped conviction for the murder of Willie Boy Johnson, but his conviction for the murders of the victims whose bodies were found at the mass grave site on Staten Island resulted in a life

The rolled-over steel flange used to press a shovel into the soil takes an excellent horizontal soil sample. (Corbis)

sentence in prison. A sample of soil stuck to a shovel and a scientist who could show how the physical properties of that soil connected it to a burial site contributed to getting a mass murderer off the streets.

CHAPTER SUMMARY

- Physical and chemical properties of matter are important in collection and analysis of evidence. Physical changes do not alter the chemical identity of a substance, whereas chemical changes result in the formation of a new substance.
- When investigating matter and the changes that it undergoes, scientists use a systematic, logical approach to problem solving called the scientific method. The scientific method consists of collecting data, making a hypothesis, and testing the hypothesis.
- The reproducibility of an experimental measurement is an indication of the precision of the measurement. The accuracy of measurements depends on how close the results are to the true value. Both precision and accuracy are important in the interpretation of experimental results.
- There is always error in a measurement because numbers resulting from measurements consist of all known digits plus one digit that contains an estimate. These digits are collectively called the significant figures of a number.
- Density, the ratio of mass to volume of a substance, is a physical property that can aid in the identification of glasses and soils. Color and texture of soils are also used for identification, as are the distinctive shapes of mineral crystals contained in the soil.

KEY TERMS

chemical change, p. 25
chemical properties, p. 26
physical change, p. 26
physical properties, p. 27
refractive index, p. 27
scientific method, p. 27
hypothesis, p. 28
mass, p. 30
weight, p. 31
units, p. 31
conversion factors, p. 32
significant figures, p. 33
scientific notation, p. 35
accuracy, p. 38
precision, p. 39
density, p. 40
sink-float method, p. 44
density gradient method, p. 44

CONTINUING THE INVESTIGATION Additional Readings, Resources, and References

Baumann, E., and O'Brien, J. *Murder Next Door: How Police Tracked Down 18 Brutal Killers,* Chicago: Bonus Books, 1991.

Doyle, A. C. *Sherlock Holmes: The Complete Novels and Stories,* vol. 1, New York: Random House, 2003.

Lubasch, A. H. "Reputed Mobster Guilty in Six Narcotics Murders; Death Penalty Possible Under Federal Law," *New York Times,* New York, June 26, 1992, p. B3

Missouri State Highway Patrol Forensic Laboratory. *Forensic Evidence Handbook,* Jefferson City: Missouri State Highway Patrol, 2003.

Murray, R. C., and Tedrow, J. C. F. *Forensic Geology,* Englewood Cliffs: Prentice Hall, 1998.

For more information about the properties of glass and ceramics: www.mindrum.com/tech.html

REVIEW QUESTIONS AND PROBLEMS

Questions

1. What is the key difference between chemical changes and physical changes?

2. Give an example of a chemical and physical change not listed in the textbook.

3. Why are physical properties useful in identifying unknown substances?

4. What are some examples of chemical properties?

5. Explain how the scientific method is used to solve problems.

6. Why do many different professions use the scientific method?

7. What is the difference between mass and weight?

8. Why are units important in a measurement?

9. What are the standard units used for measuring mass? Distance?

10. What are the two other common units that are interchangeable with the milliliter (mL)?

11. What are the common prefix modifiers and their abbreviations for the metric system?

12. Explain how conversion factors are used to change units on a number.

13. What is the conversion factor between centimeters and inches?

14. Why is there always error involved in making a measurement?

15. How can estimating the last digit on a measurement provide you with a more accurate value?

16. Draw a sketch of a dartboard and illustrate, using "X" marks, the concepts of accuracy and precision.

Problems

17. Identify each of the following as either a chemical change or a physical change.
(a) Evaporation of gasoline
(b) Toasting a marshmallow
(c) Filtering a pond water sample
(d) Burning documents

18. Identify each of the following as either a chemical change or a physical change.
(a) Formation of clouds and rain
(b) Freezing biological samples for storage
(c) Baking bread dough
(d) Detonation of TNT

19. Identify each of the following as either a chemical property or a physical property.
(a) Hardness (c) Flammability
(b) Corrosiveness (d) Color

20. Identify each of the following as either a chemical property or a physical property.
(a) Refractive index (c) Explosiveness
(b) Boiling point (d) Inertness

21. Which of the following step(s) is *not* part of the scientific method?
(a) Collect relevant data
(b) Evaluate data
(c) Create a hypothesis
(d) Eliminate conflicting data

22. Which of the following steps is *not* part of the scientific method?
(a) Test the hypothesis
(b) Validate the original hypothesis
(c) Repeat the process, if necessary
(d) Determine the nature of a problem

23. Identify whether each of the following statements relates to mass or weight.
(a) Depends on gravity
(b) Measured in grams
(c) Measures the amount of matter
(d) Measured with a scale

24. Identify each of the following statements as part of the definition for mass or weight.
(a) Measured in pounds
(b) Measures gravity pulling on matter
(c) Measured with a balance
(d) Independent of gravity

25. Write the abbreviation and multiplier for the following prefix modifiers.
(a) micro (b) kilo (c) centi

26. Write the abbreviation and multiplier for the following prefix modifiers.
(a) mega (b) deci (c) milli

27. Write the prefix for multiplying by each of the numbers below.
(a) 0.001 (b) 1000 (c) 0.1

28. Write the prefix for multiplying by each of the numbers below.
(a) 0.01 (b) 1,000,000 (c) 0.000001

29. Convert each of the following numbers to the units indicated below.
(a) 0.001 g = _______ mg
(b) 5755 mL = _______ L
(c) 0.59 m = _______ dm
(d) 750 cm = _______ m

30. Convert each of the following quantities to the units indicated below.
(a) 1×10^6 g = _______ Mg
(b) 1.56 m = _______ cm
(c) 9.5×10^{-4} s = _______ ms
(d) 1.75 g = _______ mg

31. Convert each of the following quantities to the units indicated below.
(a) 144 in. = _______ ft.
(b) 85.0 ft. = _______ yd.
(c) 1.04 m = _______ mm
(d) 6.35 in. = _______ cm

32. Convert each of the following quantities to the units indicated below.
(a) 34.0 ft. = _______ cm
(b) 238 cm = _______ yd.
(c) 71.7 in. = _______ yd.
(d) 2.98 yd. = _______ m

33. How many significant figures are in each of the numbers below?
(a) 1011
(b) 0.00193
(c) 0.08050
(d) 1000

34. How many significant figures are in each of the numbers below?
(a) 1.00×10^4
(b) 5029
(c) 0.00890
(d) 1.001

35. Write the following numbers in scientific notation.
(a) 1000
(b) 0.00001
(c) 0.043
(d) 5,120,000

36. Write the following numbers in scientific notation.
(a) 2500
(b) 85
(c) 0.00329
(d) 0.0000000477

37. Write the following numbers in regular decimal notation.
(a) 3.32×10^3
(b) 4.110×10^{-2}
(c) 9.00×10^5
(d) 6.617×10^{-3}

38. Write the following numbers in regular decimal notation.
(a) 7×10^5
(b) 4.401×10^{-1}
(c) 9.0×10^1
(d) 2.2279×10^{-6}

39. Record the value of the buret readings below with the proper number of significant figures.

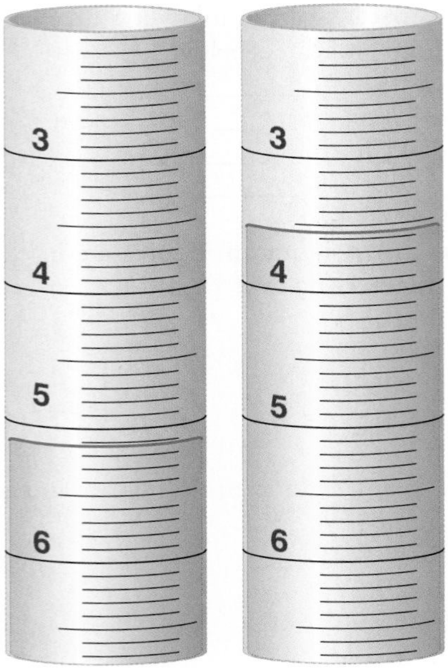

(a) (b)

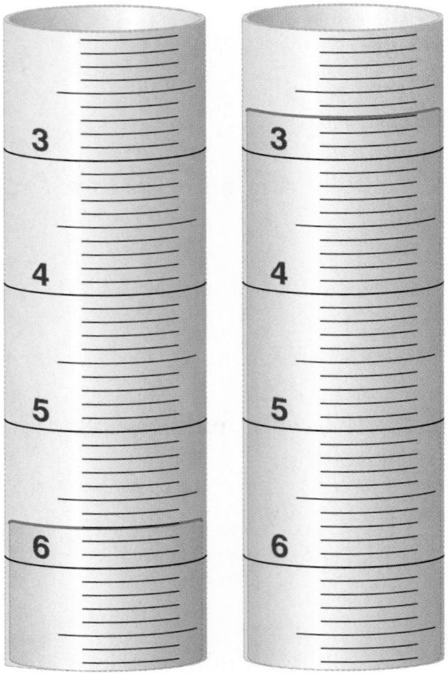

(c) (d)

40. Record with the proper number of significant figures the value of the temperatures shown in the following thermometers.

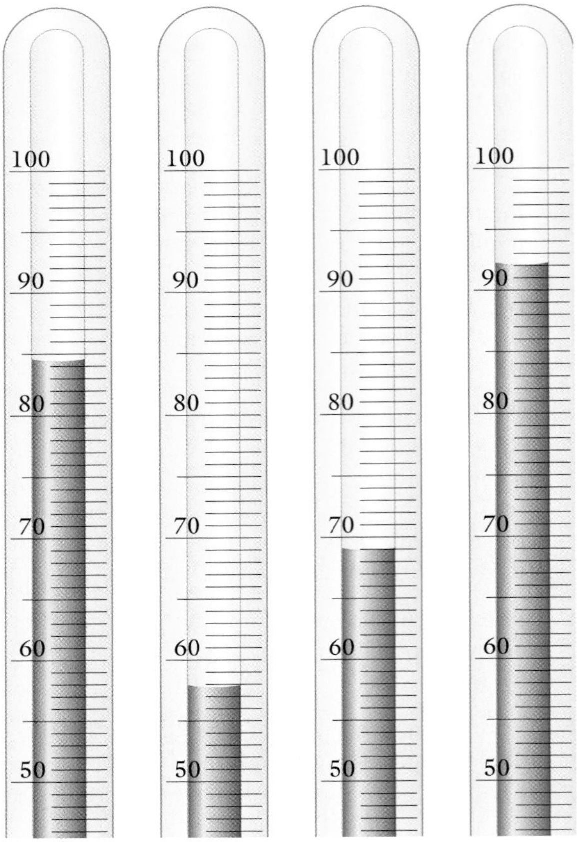

(a) (b) (c) (d)

41. Which of the following numbers contains three significant figures? If a number does not contain three significant figures, alter the number so it will contain exactly three significant figures.
(a) 1.0×10^{-3}
(b) 0.02
(c) 0.00215
(d) 100

42. Which of the following numbers contains three significant figures? If a number does not contain three significant figures, alter the number so it will contain exactly three significant figures.
(a) 100,200
(b) 0.010
(c) 0.1000
(d) 1.010×10^{-11}

43. Rewrite the following numbers in scientific notation. Round the answer to contain the number of significant figures indicated in parentheses.

(a) 1000 (2)
(b) 0.00010024 (3)
(c) 1,584,982 (4)
(d) 8832 (2)

44. Rewrite the following numbers in scientific notation. Round the answer to contain the number of significant figures indicated in parentheses.
(a) 0.000038620 (5)
(b) 102,883 (5)
(c) 5,837,930 (4)
(d) 0.0010025 (4)

45. Which set of numbers below represents measurements that have good precision and good accuracy? The true value of the measurement is 4.75.
(a) 3.64, 3.82, 3.74, 3.98
(b) 4.71, 3.98, 5.78, 3.03
(c) 4.79, 4.68, 4.81, 4.83
(d) 4.32, 4.93, 4.05, 4.11

46. Which set of numbers below represents measurements that have good precision but poor accuracy? The true value of the measurement is 18.44.
(a) 17.81, 17.10, 19.99, 18.43
(b) 18.46, 18.39, 18.52, 18.48
(c) 18.10, 19.21, 17.44, 17.99
(d) 17.35, 17.41, 17.29, 17.38

47. Write the answer to the mathematical problems below with the correct number of significant figures.
(a) $101.34 - 92.1 - 1.793 =$ _______
(b) $345.3 + 12.12 + 16.10 =$ _______
(c) $14.5 + 12.34 - 8.991 =$ _______
(d) $33.9 - 15.60 + 12 =$ _______

48. Write the answer to the mathematical problems below with the correct number of significant figures.
(a) $45.5 + 0.0023 + 17 =$ _______
(b) $34.4 - 7.92 - 0.0731 =$ _______
(c) $56 - 17.98 + 0.02 =$ _______
(d) $1.45 + 101 - 12.02 =$ _______

49. Write the answer to the mathematical problems below with the correct number of significant figures.
(a) $12.2 \div 3.4 \div 0.0127 =$ _______
(b) $14.9 \div 12.29 \times 0.020 =$ _______
(c) $3.0 \times 2.34 \times 329 =$ _______
(d) $76.3 \div 875.023 \times 31.1 =$ _______

50. Write the answer to the mathematical problems below with the correct number of significant figures.
(a) $642 \div 32.90 \div 100.0 =$ _______
(b) $47 \times 23.3 \times 10.1 =$ _______
(c) $82.901 \div 26.8 \times 3.33 =$ _______
(d) $3967 \times 0.022 \div 9.09 =$ _______

51. Write the answer to the mathematical problems below with the correct number of significant figures.
(a) $(19.83 \times 2.3) + 4.100 =$ _______
(b) $14.3 - 2.3 \div 0.2 =$ _______
(c) $0.020 \times 211.2 - 40.0 =$ _______
(d) $12.11 \times (2.8 - 13.3) =$ _______

52. Write the answer to the mathematical problems below with the correct number of significant figures.
(a) $76.3 - 23.345 \div 16.0 =$ _______
(b) $8.240 \times 37.2 - 119.00 =$ _______
(c) $(1.003 \times 23.0) + 173.90 =$ _______
(d) $56.2 \div 2.300 + 9 =$ _______

53. A glass sample has a softening point in the range of 718°C to 723°C. List the possible glass types (consult Table 2.3). What are the possibilities if the density is determined to be 2.6 g/cm^3?

54. A glass sample has a softening point in the range of 819°C to 833°C. List the possible glass types (consult Table 2.3). What are the possibilities if the refractive index is determined to be 2.2?

55. Calculate the density of an object given the masses and volumes below.
(a) Mass = 14.45 g, volume = 10.0 cc
(b) Mass = 12.2 g, volume = 3.43 mL
(c) Mass = 9.02 g, volume = 6.23 cm^3
(d) Mass = 7.02 g, volume = 8.29 mL

56. Calculate the density of an object given the masses and volumes below.
(a) Mass = 12.82 g, volume = 13.28 cc
(b) Mass = 2.34 g, volume = 1.11 mL
(c) Mass = 9.23 g, volume = 6.67 cm^3
(d) Mass = 4.73 g, volume = 5.72 mL

57. Calculate the volume of an object given the densities and masses below.
(a) Density = 0.982 g/mL, mass = 14.45 g
(b) Density = 3.231 g/mL, mass = 10.0 g
(c) Density = 1.34 g/cc, mass = 4.71 g
(d) Density = 2.90 g/cm^3, mass = 11.67 g

58. Calculate the volume of an object given the densities and masses below.
(a) Density = 0.864 g/mL, mass = 44.99 g
(b) Density = 2.77 g/cc, mass = 21.4 g
(c) Density = 11.0 g/mL, mass = 5.76 g
(d) Density = 8.76 g/cm^3, mass = 2.003 g

59. Calculate the mass of an object given the densities and volumes below.
(a) Density = 0.935 g/mL, volume = 23.30 mL
(b) Density = 1.45 g/cc, volume = 12.22 cc
(c) Density = 13.6 g/mL, volume = 9.32 mL
(d) Density = 2.25 g/cm^3, volume = 5.60 cm^3

60. Calculate the mass of an object given the densities and volumes below.
(a) Density = 3.91 g/mL, volume = 9.44 mL
(b) Density = 0.791 g/mL, volume = 10.9 mL
(c) Density = 2.34 g/cc, volume = 8.45 cc
(d) Density = 7.44 g/cm^3, volume = 11.08 cm^3

Forensic Chemistry Problems

61. Describe how the scientific method was used in the soil analysis done by Special Agent Bruce Hall in the Tommy Karate case study.

62. Soil analysis is a comparative science, which means that two samples are compared to determine whether, in fact, they come from the same source. Describe what physical properties are used to compare soils. If a sample from a suspect matches that from a crime scene, what else must be proven to show that the suspect was at the crime scene location?

63. Why would it be more difficult to compare soil recovered from the blade of a shovel at a crime scene than soil from the rounded-over flange of the shovel?

64. Most states are adopting as the legal limit for driving under the influence of alcohol a blood alcohol content (BAC) of 0.08 g/dL. Many clinical and forensic laboratories will report BAC in units of mg/mL, g/L, or mg/dL. Convert 0.08 g/dL to each of the alternate units for BAC.

65. Diazodinitrophenol (DDNP) is an explosive with a density of 1.63 g/mL. What volume (cm^3) would a 1.00-pound sample of DDNP occupy? (1 lb = 453.5 g)

66. One complication of examining a cadaver for BAC is that the decomposition process can produce alcohol. Postmortem alcohol production is usually less than 70 mg/dL but can be as high as 100 mg/dL. Would it be possible to mistakenly report that a sober victim had been legally drunk at the time of death? Assume that the legal BAC level is 0.08 g/dL.

67. Alcohol is eliminated from the human body at an approximate rate of 15 mg/(dL · hr). What is the rate of alcohol elimination in units of g/(dL · min)?

68. A piece of glass with a mass of 7.89 g was recovered from a crime scene. When submerged in a graduated cylinder, the volume of water in the cylinder rose from 24.00 mL to 27.54 mL. Identify from Table 2.3 the type of glass found at the crime scene.

69. If a 17.84-g sample of alkali barium glass is submerged into a cylinder containing 30.00 mL, what is the final level of the water in the cylinder?

70. Does a chemical or physical change force the bullet out of a gun? Is the smoke coming out the end of this gun a homogeneous or heterogeneous mixture? Explain your answer.

(Index Stock Photos/Fotosearch)

71. If a suspected arson sample sent to the laboratory comes back negative for petroleum-based accelerants, does that prove that the fire was not deliberately set? If the sample comes back positive, does that automatically mean the fire was deliberately set? Explain how an investigator might use this information (whether affirmative or negative laboratory results) in determining the chain of events leading up to the fire.

72. Discuss whether or not Sherlock Holmes is applying the scientific method correctly based on the following quotes:
 (a) "We approached the case, you remember, with an absolutely blank mind, which is always an advantage. We had formed no theories. We were simply there to observe and to draw inferences from our observations."
 —*The Adventure of the Cardboard Box*
 (b) "It is a capital mistake to theorize before you have all the evidence. It biases the judgment." —*A Study in Scarlet*
 (c) "How often have I said to you that when you have eliminated the impossible, whatever remains, however improbable, must be the truth?" —*The Sign of Four*

Case Study Problems

73. A Molotov cocktail is a fuel-filled glass bottle that has a cloth rag acting as both a plug and a fuse. When thrown at a target, the glass breaks and releases the fuel, which is immediately ignited by the fuse. An arsonist used such a device to burn down a property in Dallas, Texas. The suspect was a professional arsonist from the Kansas City, Kansas, region. The only evidence from the crime scene was a shard of glass. How could a forensic scientist link the shard of glass found in Texas to a suspect from Kansas? No DNA or fingerprints were found on the glass. Use the scientific method and remember from the Tommy Pitera example that it is often as important to exclude possibilities as it is to find matching samples.

74. An eyewitness positively identified an individual as the person seen fleeing the scene of a homicide. There were two blood types found at the scene, that of the victim and another blood type matching the suspect. The suspect had no known connections to the victim and claimed he had spent the evening at home watching TV by himself. The suspect was convicted and sentenced to life in prison. Fifteen years later, DNA analysis of evidence from the crime scene proved the convicted man innocent. Based on the limited information within this problem, was it reasonable to convict the suspect? Was the scientific method properly applied in the original case and subsequent acquittal?

CHAPTER 3

Atomic Clues

Investigators search for human remains on the property of Ray Marsh in Noble, Georgia, in February 2002. *(Tammi Chappell/Reuters/Corbis)*

Case Study: To Burn or Not to Burn

The sight was worse than a scene in a Stephen King novel or a Hollywood horror movie. The body of an elderly man lay decomposing under the hot glare of the sun, shaded only by the junked remains of a 1976

Buick Century. Another corpse, most likely that of a teenage boy, was half submerged in a stagnant puddle. The skeleton of a young woman protruded from the grass and weeds on a hillside. A middle-aged man's body in an unusually well-preserved state slumped in the corner of a locked work shed. But these horrors were just the tip of the iceberg: Investigators were about to find 334 bodies in all in this grisly landscape.

Not a single one of the deceased had been reported missing by their families. How could so many deaths have occurred without someone raising an alarm? It gradually became clear on investigation that the dead were people whose bodies had been sent to a crematorium after death. Their families believed that the remains of their loved ones rested in the funeral urns returned to them from the crematorium.

This case garnered national attention in 2002 when a family-owned crematorium in Noble, Georgia, was found to have been accepting bodies for cremation for several years, charging families for cremation services, and then dumping the bodies anywhere on the property. What, then, was in the urns that supposedly held the ashes of the deceased?

This case study focuses on the content of one of the urns about which a family asked a simple question: "Does this urn contain Chigger, or is he somewhere in the acre of rotting remains yet to be identified?" But before this question could be answered, the family would have to know if the cremated remains, called cremains, in the urn they had been given were even human, let alone those of their loved one. The family obtained legal advice, and their attorney turned to Dr. Bill Bass, one of the leading experts in the world on death and decomposition.

This case, however, presented Bass with a new problem. Bass specializes in working with skeletal remains and decomposed bodies, not piles of ashes. He needed the advice of a chemist, and he turned to Dr. Al Hazari of the University of Tennessee in Knoxville. The problem for Hazari was to find a method for identifying the contents of the urn as either ashes from a human body or merely filler material of some kind, such as cement or wood ashes. DNA evidence could not be used to answer this question because the fires of a crematorium destroy all traces of DNA. In fact, cremation converts all carbon in the body to carbon dioxide gas, and the ashes consist of substances that do not burn or escape as gases.

Hazari solved the problem by considering known facts about what is left when a human body is cremated. Then, to determine what was present in the remains reported to be those of Chigger, he used methods of analysis that depend on knowledge of atomic structure and the behavior of electrons when energy is added to an atom.

First, let's investigate atoms and their structures . . .

3.1 Origins of the Atomic Theory: Ancient Greek Philosophers

Learning Objective

Distinguish the two early schools of thought on the nature of the atom.

The very presence or absence of certain atoms in an evidence sample can mean the release of an innocent suspect or the conviction of a criminal. The ability to determine the elemental makeup of evidence is based on our knowledge of atomic structure and how atoms behave. The gradual elucidation of the nature and structure of atoms constitutes one of the greatest detective stories in science. It took over 2300 years to solve the case! In fact, the case was closed only about seventy years ago. Moreover, the first mug-shot of an atom was taken only in 1983 when IBM researchers invented an instrument called the scanning tunneling microscope (STM), which was capable of sensing the presence of an individual atom and creating an image. Figure 3.1 shows an STM image of an oval of iron atoms arranged on a copper surface. The STM image is not a traditional picture; it is a computer-generated image depicting the forces between the electrons in an atom and the STM instrument.

The mystery began in the year 440 B.C. in ancient Greece. Two philosophers, Leucippus and his student Democritus, started a revolution of thought by claiming that matter was made of small, hard, indivisible particles they called *atoms*. According to their theory, atoms came in various sizes, shapes, and weights, were in constant motion, and combined to make up all the various forms of matter. The observable properties of matter could be directly related to the types of atoms it contained. Wherever atoms did not exist, there was a void or vacuum.

This theory of Leucippus and Democritus set the stage for a confrontation with the philosopher Aristotle, who believed in a continuous model of matter. Aristotle taught that matter could be infinitely divided, which meant that no indivisible particles (atoms) existed. He convinced the majority of his colleagues that the atomic view of matter was illogical.

Democritus (460 B.C.–370 B.C.) (Nimatallah/Art Resource, NY)

Aristotle (384 B.C.–322 B.C.) (Bettman/Corbis)

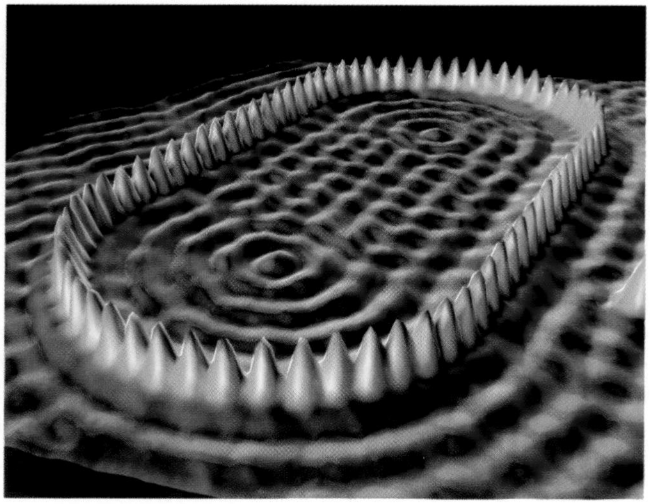

Figure 3.1 A ring of iron atoms on a copper surface imaged by an STM. The colors in the image are computer generated. (Crommie, Lutz & Eigler/IBM Corporation)

The argument for the existence of atoms went into the cold case file for the next 1900 years. In fact, teachings contrary to Aristotle, including the atomic theory, were against the law in some parts of Europe! To make matters worse, the atomic theory was associated with atheistic beliefs.

3.2 Foundations of a Modern Atomic Theory: Gassendi, Lavoisier, and Proust

Learning Objective

Understand how the political and scientific groundwork for a modern atomic theory was established.

In the 1600s, Pierre Gassendi, a priest, philosopher, and scientist, reopened the atomic debate by publishing a work that defended the atomic theory and argued that the principles of atomic theory were not contradictory to Christian beliefs. He modified the original principles of Leucippus and Democritus to state that all atoms were actually created by God and that their motion was a gift of God. Current atomic theory is neutral on the existence of a supreme being, but Gassendi's work was important because it allowed people to engage in open debate about atomic theory without fear of retribution.

Pierre Gassendi, a seventeenth-century priest, philosopher, and scientist. (Science Photo Library/Photo Researchers, Inc.)

Even though the atomic theory could now be debated and discussed in academic settings, there were no significant developments in the case until the late 1700s. Experiments that laid the groundwork for a better understanding of matter and chemical change were carried out around that time and became necessary precursors to the further development of atomic theory.

Antoine Lavoisier, who is considered the founder of modern chemistry, made a discovery in 1785 that was to become known as the law of conservation of mass. The **law of conservation of mass** states that mass is neither created nor destroyed in a chemical reaction but merely changes form. For many centuries scientists had pondered what happened to matter during chemical reactions. Perhaps even our earliest ancestors who mastered the art of fire wondered what happened to the wood that was burning in their campfires. The fire seemed to consume the wood, since the only tangible evidence left from the fire was a very small amount of ash.

The key to unraveling this mystery was the awareness that matter undergoing a chemical change (such as burning wood) in an open environment only appears to lose mass because one or more of the products of the reaction escapes into the surroundings as a gas. Lavoisier conducted experiments in closed systems in which the substances undergoing a transformation could not exchange matter with the surroundings, as shown in Figure 3.2.

Lavoisier heated mercury(II) oxide in a sealed system to form elemental mercury, which is a silver metallic liquid, and colorless oxygen gas, as shown in Figure 3.3. Taking careful quantitative measurements, he found that the mass of the mercury(II) oxide before heating equaled the combined masses of the liquid mercury and oxygen gas that formed after heating. Thus there was no overall change in mass as long as the reaction was done in a sealed system.

The ashes left by countless campfires could finally be explained as the solid remains of a combustion reaction that also produced carbon dioxide

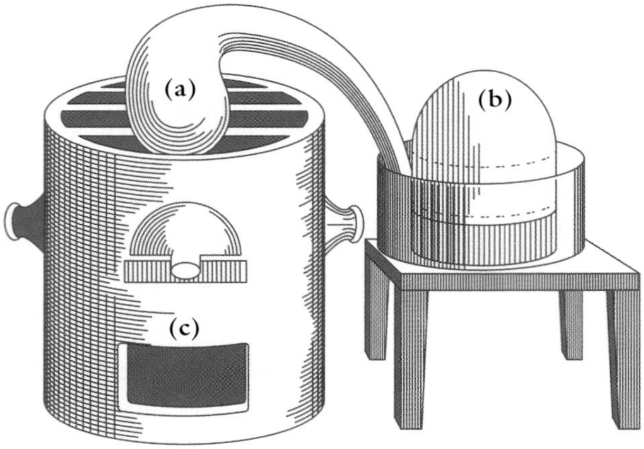

Figure 3.2 Lavoisier's reaction vessel (a) could be heated by the furnace (c) and any gases that formed during the reaction would be collected in the jar (b).

gas and water vapor. The carbon dioxide and water vapor escape into the surrounding air.

Worked Example 1

Lavoisier studied the production of water from hydrogen and oxygen gases. If he produced 36.0 g of water starting with 4.0 g of hydrogen, how many grams of oxygen must have reacted with the hydrogen?

SOLUTION According to the law of conservation of mass, the total mass of reactants must equal the total mass of products:

$$\text{grams hydrogen} + \text{grams oxygen} = \text{grams water}$$
$$4.0\ \text{g} + \text{g oxygen} = 36.0\ \text{g}$$
$$\text{g oxygen} = 36.0\ \text{g} - 4.0\ \text{g} = 32\ \text{g oxygen}$$

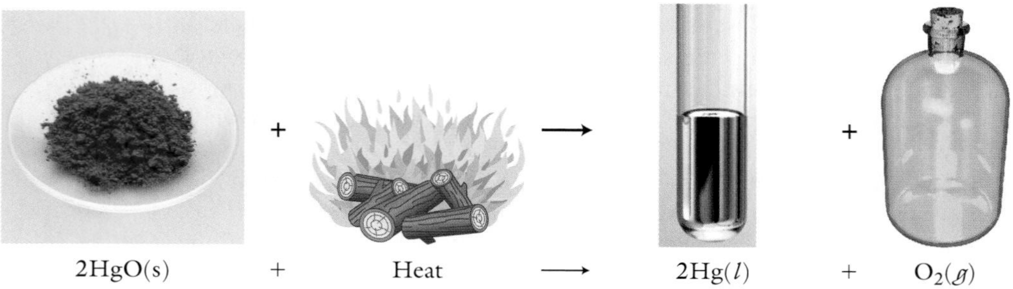

Figure 3.3 In Lavoisier's reaction, the heating of the mercury(II) oxide caused the solid compound to decompose into its constituent elements, mercury and oxygen gas. The reaction was a visually dramatic illustration of the law of conservation of mass: A red solid was converted by heat into a silver liquid and a colorless gas. (Mercury oxide: Charles D. Winters/Photo Researchers, Inc.; liquid mercury: Richard Mengus/Fundamental Photographs)

Practice 3.1

Alcohol in our digestive system reacts with oxygen gas to form carbon dioxide and water. How many grams of carbon dioxide are released if a 10.00-g alcohol sample reacts with 20.85 g of oxygen gas and produces 11.77 g of water?

ANSWER

19.08 g CO_2

Another series of important experiments, carried out by Joseph Louis Proust in 1797, led to a principle that became known as the law of definite proportions (also called the law of constant composition). The **law of definite proportions** states that a compound is always made up of the same relative masses of the elements that compose it. At the time Proust was investigating compounds, some scientists believed that a compound could contain any ratio of its elements and still be the same substance. Their argument was that only the type of elements in the compound mattered, not necessarily the amounts of each element. Proust was the first to show that analysis of a carefully prepared and purified compound always showed the same mass ratio of elements—even if different methods were used to prepare the compound—from one experiment to another.

The key to Proust's work was that he did it very carefully so that he was able to obtain both accurate and precise results. The work by earlier scientists had not been carried out as carefully as Proust's, and therefore earlier scientists based their conclusions on inaccurate data.

3.3 Dalton's Atomic Theory

Learning Objective

Show how Dalton's atomic theory could explain and predict the behavior of matter.

John Dalton is credited with giving the world the first modern atomic theory. The word *theory* in the sciences is the term for the best current explanation of a phenomenon. It is *not* synonymous with an opinion and does *not* change as a result of political or personal agendas. A **theory** is accepted by the scientific community until such time that new data contradict the theory.

Dalton's passion was studying meteorology and weather patterns, and this led him to study how gas mixtures behave. As a result of his studies, he developed an atomic theory that explained the behavior of gases and the previous findings of Lavoisier and Proust. Dalton's atomic theory of matter, published in 1803, had four basic tenets:

Dalton's Atomic Theory

1. All matter is made up of tiny, indivisible particles called atoms.
2. Atoms cannot be created, destroyed, or transformed into other atoms in a chemical reaction.
3. All atoms of a given element are identical.
4. Atoms combine in simple, whole-number ratios to form compounds.

With **Dalton's atomic theory,** the law of conservation of mass could now be explained. When a chemical reaction occurs, the atoms of each

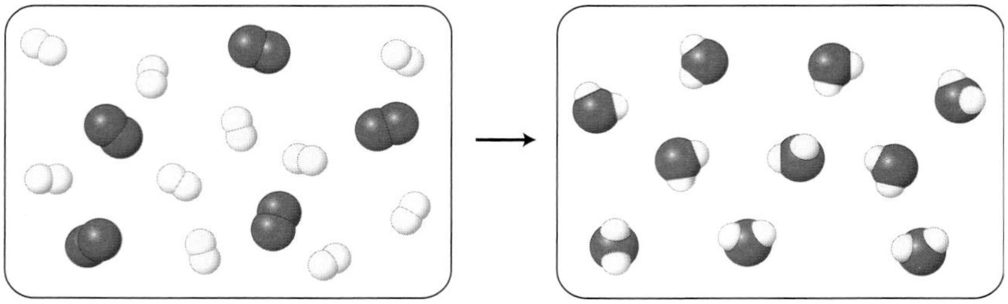

Hydrogen ( ) reacts with oxygen ( ) to form water ( )

Figure 3.4 The law of conservation of mass explains a chemical reaction as the rearrangement of atoms. The number of atoms before or after the reaction is constant (10 oxygen and 20 hydrogen); the arrangement of chemical bonds changes.

compound are rearranged to form new compounds, as shown in Figure 3.4, but the number of atoms remains the same. Thus no mass is lost or gained in the process.

Dalton's atomic theory could also explain the law of definite proportions. According to the atomic theory, compounds are made up of atoms in simple ratios. Pure water always has two hydrogen atoms to every oxygen atom, regardless of where the water comes from or how it is created. This constant atomic ratio provides the reason the mass ratio of elements in the compound is also constant.

Finally, Dalton's atomic theory predicted the law of multiple proportions formulated by Dalton himself. The **law of multiple proportions** states that any time two or more elements combine in different ratios, different compounds are formed. For example, laughing gas, which has the chemical formula N_2O, is quite different from the gas NO_2, a by-product of explosions and burning diesel fuel. Figure 3.5 shows various compounds of nitrogen and oxygen.

In spite of the success of Dalton's theory in explaining the behavior of matter and the widespread support the theory received, it took almost another hundred years before the atomic theory of matter was completely accepted. In fact, there were still a few major scientists who disbelieved the atomic theory into the early 1900s. It would take more than logical reasoning and circumstantial evidence to get a unanimous decision by a jury of scientists.

Nitrogen ( )
Oxygen ( )

Figure 3.5 Nitrogen atoms can combine with oxygen atoms in a variety of simple ratios. Each combination represents a unique compound, illustrating the law of multiple proportions.

Worked Example 2

Label each of the following drawings as illustrating the law of conservation of mass (LCM), the law of multiple proportions (LMP), the law of definite proportions (LDP), or more than one law, if applicable.

(a)

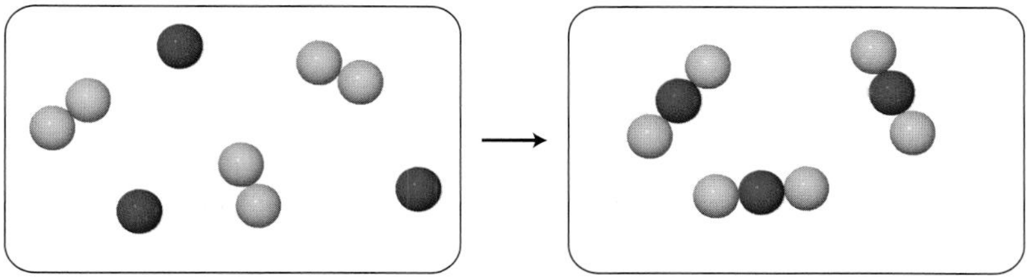

(b)

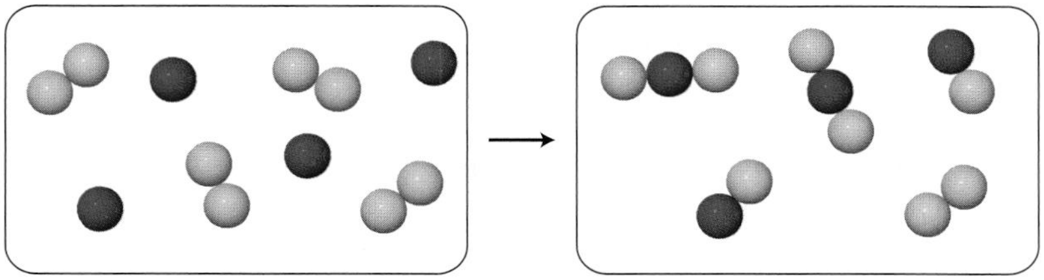

(c)

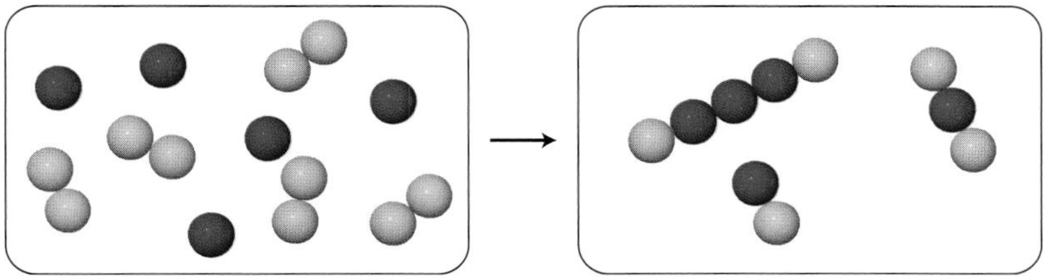

SOLUTION

(a) LCM, LDP. The reaction illustrates the law of conservation of mass because both reactants and products have an identical number of atoms. The law of definite proportions is also illustrated because a single compound has been created with a constant formula.

(b) LCM and LMP. The reaction illustrates the law of conservation of mass because both reactants and products have an identical number of atoms. The law of multiple proportions is also illustrated because several compounds have been created, each with a unique formula.

(c) LMP. The law of multiple proportions is illustrated because several compounds have been created, each with a unique formula.

Practice 3.2

Write the chemical formulas for all the nitrogen-oxygen compounds shown in Figure 3.5. Explain how the figure illustrates the law of multiple proportions.

ANSWER

NO, NO_2, N_2O, N_2O_3, N_2O_4, N_2O_5. The law of multiple proportions states that elements can combine in multiple whole-number ratios of their elements (1:1, 1:2, 2:1, and so forth), but each new ratio of atoms represents a new compound.

3.4 Atomic Structure: Subatomic Particles

Learning Objective

List the main subatomic particles and show how their discovery furthered our insight into of the nature of the atom.

Some of the most convincing evidence for the existence of atoms came from experiments that actually contradicted the first principle of Dalton's atomic theory. Experimental evidence began to show that atoms were not indivisible; in fact, there were three major components to the atom. These subatomic particles were the positively charged **proton,** the negatively charged **electron,** and the **neutron,** which had no electrical charge and was thus electrically neutral.

Experiments that led to the discovery of the electron depended on the invention in the 1850s of the cathode ray tube, the direct ancestor of the conventional television picture tube and computer monitor. A **cathode ray tube** is a sealed glass tube from which nearly all gases inside have been removed to create a vacuum. In the cathode ray tube are two metal electrodes called the *cathode* and the *anode* to which a large voltage is applied. When this process occurs, rays can be seen emanating from the cathode. The rays can be deflected by the poles of a magnet, moving away from the north-seeking (N) pole and toward the south-seeking (S) pole, as shown in Figure 3.6.

In 1897, the British physicist J. J. Thomson was able to measure the deflection of the cathode rays from magnetic and electric fields. Using the data from his deflection experiments, he calculated that the mass-to-charge ratio of the cathode ray particles was less than 1/1000 of the value calculated for any other known element. Thomson thus discovered that whatever made up the cathode rays was smaller than the smallest atom! Thomson correctly concluded that the cathode rays consisted of negatively-charged subatomic particles called *electrons.*

The discovery of the electron presented a problem for scientists. They knew that atoms are electrically neutral, yet now they had evidence that atoms also contain negative particles. A positive charge had to balance the negative charge, but where in the atom is it? One popular theory at the time was that the electron particles were randomly stuck into a ball of positive charge. This model was commonly called the *plum pudding model,* referring to the way the fruit is distributed throughout a plum pudding. A more familiar model today might be a chocolate chip muffin.

In 1909, Ernest Rutherford set out to investigate the plum pudding model, which he thought was correct, by taking an extremely thin layer of gold foil and exposing it to a radioactive element. **Radioactivity** is the spontaneous emission of particles from unstable elements. The alpha particle, which is emitted from some radioactive elements, has a positive electrical charge and a mass four times that of a hydrogen atom. In Ruther-

Plum pudding

Chocolate chip muffin

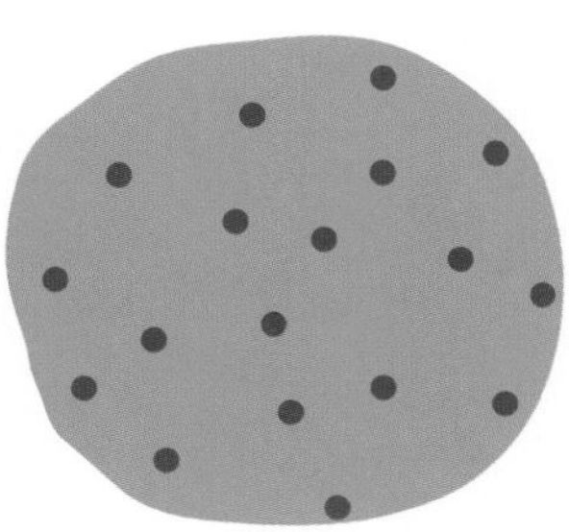

Plum pudding model

The plum pudding model consisted of a positively charged matrix in which the negative electrons are imbedded into the positive region. (Top: Foodfolio/Alamy; center: Index Stock)

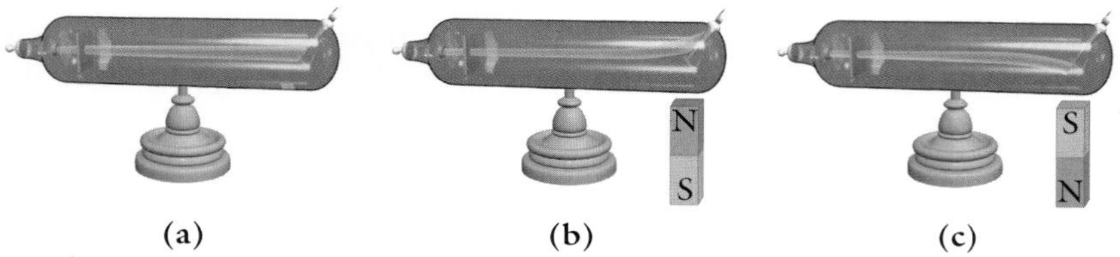

Figure 3.6 A beam of cathode rays is attracted to the south-seeking pole of a magnet and deflected by the north-seeking pole.

ford's experiment, the alpha particle emitter was aimed at the target placed inside a detector that would show a flash of light when an alpha particle struck it (Figure 3.7).

The three major experimental observations made during the gold foil scattering experiment, with the conclusions Rutherford made from each observation, are as follows.

Observation 1: The vast majority of alpha particles passed directly through the solid gold foil.

Conclusion 1: The atom must consist mostly of empty space for the alpha particles to go directly through.

Observation 2: Occasionally an alpha particle would veer from a straight-line path and hit the detector on the side.

Conclusion 2: (a) The positively charged alpha particle was pushed off course when it came close to a positive region of the atom because like charges repel. (b) The positive region had to be small because only a few alpha particles ever came close enough to be repelled.

Observation 3: Rarely, an alpha particle would bounce directly back toward the alpha particle source after striking the gold foil.

Conclusion 3: (a) In these instances the alpha particle made a direct hit on a very dense object that caused it to bounce back. (b) Since the direct hits were a rare event, the dense object must be very small and must be positively charged to repel the alpha particle.

The final result stunned Rutherford. He would later compare it with firing an artillery shell at a piece of paper and having it bounce back at you! The dense positive region Rutherford discovered in the atom is called the **nucleus.** Rutherford continued to explore the structure of the atom and would later discover that the positively charged hydrogen ion has the simplest nucleus consisting of only one positively charged subatomic particle called the *proton*. The model Rutherford put forward is illustrated in Figure 3.8.

The task of determining the structure and components of an atom was complete except for an explanation for the fact that protons made up

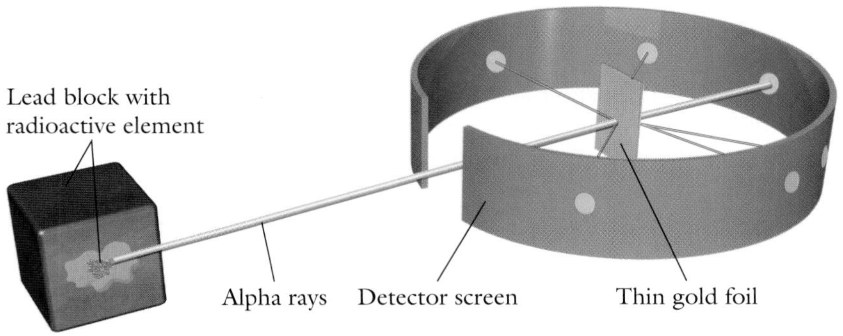

Figure 3.7 In Rutherford's gold foil scattering experiment, alpha rays from a radioactive element are directed out of a lead container toward a gold foil target surrounded by a detector screen that flashes at any location struck by an alpha particle.

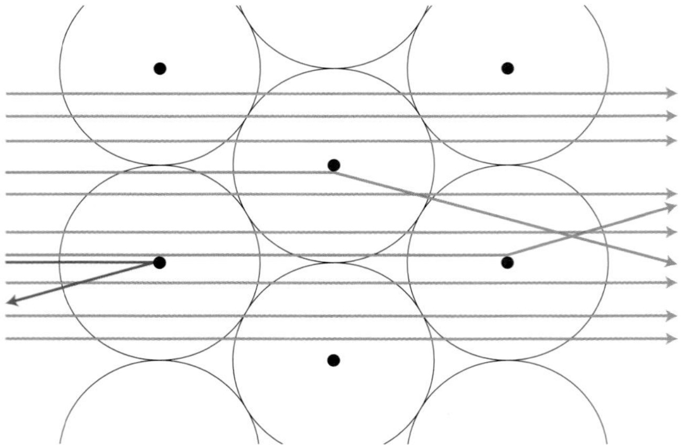

Figure 3.8 Rutherford's model of the atom consisted almost entirely of empty space in which the protons were located in a very small, dense nucleus. This model could account for the three experimental observations: (1) Most rays pass through (gray). (2) Some rays are deflected (orange). (3) Rarely, an alpha ray is returned (purple).

only a portion of the mass of most atoms. The difference in mass could not be explained by the presence of electrons because they were so much lower in mass that they contributed very little to the overall mass of the atom. The search for the missing mass continued until 1932, when James Chadwick correctly interpreted several nuclear experiments that were producing heavy neutral particles called *neutrons.* A neutron has a mass equal to that of a proton but no electronic charge.

Neutrons were later linked to the violation of another principle of Dalton's original atomic theory. Dalton had postulated that all atoms of an element are identical, but, in fact, some atoms of the same element have different numbers of neutrons than others. These atoms are called **isotopes.** Isotopes of an element behave and react chemically like all other atoms of that element, but they differ in their atomic masses.

Because the masses of single atoms and subatomic particles are so small, using units of grams to express mass does not make sense. Instead, mass is expressed in a unit called the **atomic mass unit (amu).** The amu is defined as 1/12 the mass of a carbon atom that contains 6 protons and 6 neutrons. The properties of the subatomic particles are summarized in Table 3.1. Note how small the mass of the electron is compared with the masses of the proton and the neutron. The electron's mass is only 1/1838 (or 0.0005486) the mass of a proton or neutron. For most applications, the mass of the proton and neutron can be rounded to 1 amu, and the mass of the electron to 0 amu.

Table 3.1 Subatomic Particles

Particle	Charge	Mass (amu)	Symbol
Electron	−1	0.0005486	e^-
Proton	+1	1.0073	p or p^+
Neutron	0	1.0087	n

3.5 Nature's Detectives: Isotopes

Learning Objective

Distinguish isotopes and describe how they can be used in scientific research.

Isotopes are used quite often for solving chemical and medical mysteries. For example, radioactive isotopes are used routinely at most hospitals to investigate whether a patient has properly functioning organs or to locate traces of cancer. Isotopes are also commonly used in the laboratory to investigate how each step of a chemical reaction occurs. A closer look at the periodic table and the nature of isotopes is warranted because of the important role isotopes play in science and medicine.

Isotopes differ only in the masses of their atoms and therefore have the same chemical symbol. How then do we represent the different isotopes of an element symbolically to distinguish one from another? The goal of any symbolic form used in chemistry is to clearly relay the relevant information in an abbreviated form.

For example, the element hydrogen has three isotopes, as shown in Table 3.2. The **mass number (A)** is the mass of the isotope measured in atomic mass units (amu). The mass of the isotope is the mass of the protons plus the mass of the neutrons; each particle has a mass of 1 amu. The mass of an electron is so small that it does not affect the mass of the atom to any important extent. It would take the mass of almost 2000 electrons to equal the mass of just one proton or neutron.

Table 3.2 Isotopes of Hydrogen

Isotope Name	Number of Neutrons	Number of Protons	Number of Electrons	Mass Number	Model Protons • Neutrons •
Protium	0	1	1	1	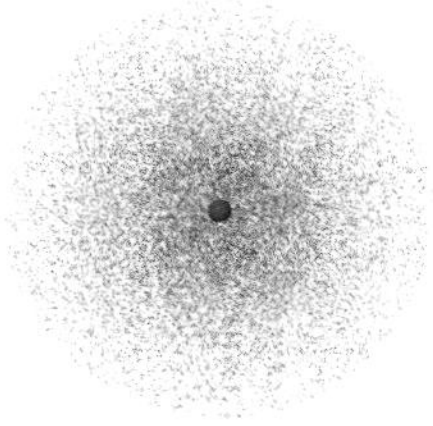
Deuterium	1	1	1	2	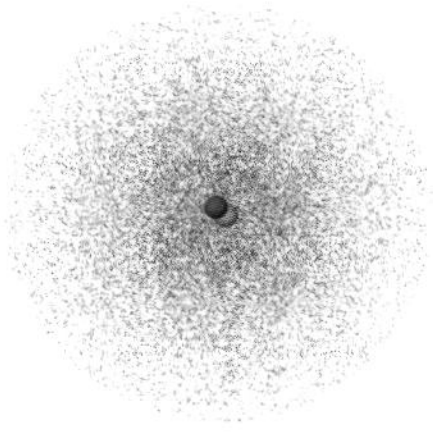
Tritium	2	1	1	3	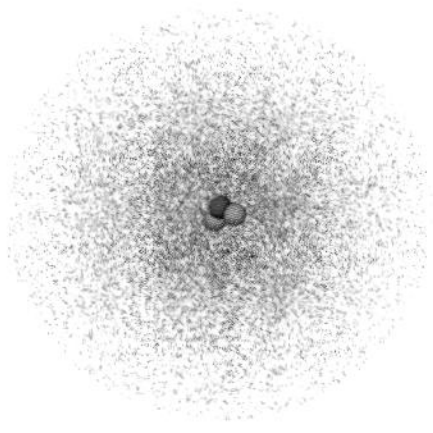

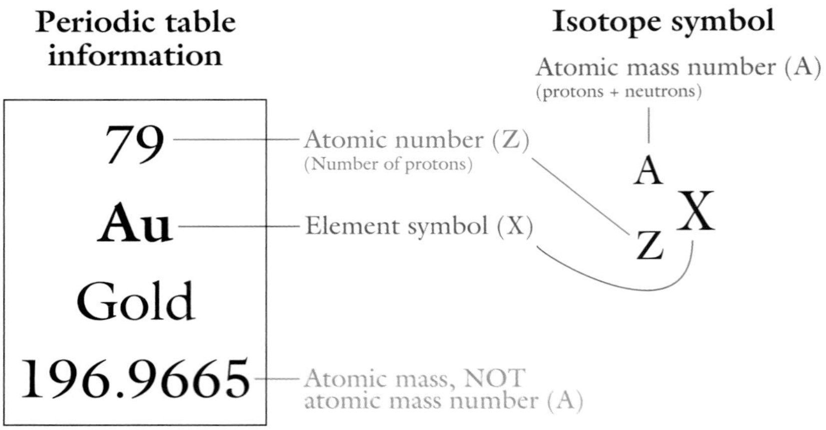

Figure 3.9 Relationship between the information in the periodic table and the information in the isotope symbol.

The number of protons in each isotope of an element is the same. In this example, each isotope of hydrogen has one proton. All atoms of an element, regardless of the number of neutrons, have the same number of protons. The periodic table is organized by ascending number of protons, shown in the upper center or upper right corner of the block for each element. The number of protons in an element is called the **atomic number (Z).** Figure 3.9 shows the pattern that is used to represent specific isotopes.

The isotopes of hydrogen can then be written $^{1}_{1}H$ for protium, $^{2}_{1}H$ for deuterium and $^{3}_{1}H$ for tritium. This system makes clear the distinction between isotopes without the need for memorization of names or formulas. One common modification of this method is the omission of the atomic number. The three isotopes of hydrogen could just as easily be written ^{1}H, ^{2}H, and ^{3}H. This shortcut is valid because all isotopes of hydrogen have the same atomic number, which can be determined from the periodic table.

Giving each isotope a distinct name would be impractical. Therefore, an isotope is simply given the name of the element and the atomic mass number (Z). For example, ^{3}H is written *hydrogen-3* and is properly pronounced "hydrogen three." This terminology may be familiar to you: The technique of carbon-14 dating of artifacts is commonly mentioned in the popular media in discussions of archeological discoveries.

Worked Example 3

Write the isotope symbol for each of the following isotopes:

(a) Carbon atom containing 6 neutrons

(b) Nitrogen atom containing 8 neutrons

(c) Uranium atom containing 143 neutrons

SOLUTION

(a) Carbon's atomic number is 6 (also the number of protons). The mass number is equal to protons + neutrons: 6 + 6 = 12. The isotope symbol is $^{12}_{6}C$.

(b) Nitrogen's atomic number is 7 (also the number of protons). The mass number is equal to protons + neutrons: 7 + 8 = 15. The isotope symbol is $^{15}_{7}N$.

(c) Uranium's atomic number is 92 (also the number of protons). The mass number is equal to protons + neutrons: 92 + 143 = 235. The isotope symbol is $^{235}_{92}U$.

Practice 3.3

Write the isotope symbols for the following:

(a) Atom containing 26 protons and 28 neutrons

(b) Atom containing 14 protons and 15 neutrons

(c) Atom containing 16 protons and 18 neutrons

ANSWER

(a) $^{54}_{26}Fe$ **(b)** $^{29}_{14}Si$ **(c)** $^{34}_{16}S$

Given the isotope symbol, it is possible to determine the number of protons, neutrons, and electrons of any given atom using simple mathematics. The mass number of an atom is the sum of protons and neutrons. Subtracting the number of protons (the atomic number) from the mass number yields the number of neutrons. The number of electrons present in an atom is always equal to the number of protons as long as the atom does not have a positive or negative charge.

Worked Example 4

Determine the number of protons, neutrons, and electrons found in each of the following isotopes:

(a) $^{22}_{10}Ne$ **(b)** $^{136}_{56}Ba$ **(c)** $^{106}_{46}Pd$

SOLUTION

(a) $^{22}_{10}Ne \Rightarrow {}^{p+n=22}_{p=10}Ne$
$p = 10$
$n = 22 - 10 = 12$
$e^- = p = 10$

(b) $^{136}_{56}Ba \Rightarrow {}^{p+n=136}_{p=56}Ba$
$p = 56$
$n = 136 - 56 = 80$
$e^- = p = 56$

(c) $^{106}_{46}Pd \Rightarrow {}^{p+n=106}_{p=46}Pd$
$p = 46$
$n = 106 - 46 = 60$
$e^- = p = 46$

Practice 3.4

Determine the number of protons, neutrons, and electrons in each of the following isotopes:

(a) Krypton-84 **(b)** Tin-117 **(c)** Zinc-66

ANSWER

(a) 36 p, 48 n, 36 e^- **(b)** 50 p, 67 n, 50 e^- **(c)** 30 p, 36 n, 30 e^-

Some elements, such as beryllium and fluorine, do not have any naturally occurring isotopes. Other elements, such as mercury and osmium, have seven naturally occurring, stable (nonradioactive) isotopes. When an element has isotopes, the proportion of each isotope is generally constant in natural samples. For example, 92.23% of silicon atoms are silicon-28, 4.67% of silicon atoms are silicon-29, and the remaining 3.10% are silicon-30. Distribution of isotopes and percentages is unique to each element.

The fact that the abundance of each isotope is constant in natural samples can be a useful tool in the fight against terrorism. For example, 98.90% of carbon atoms are ^{12}C, whereas 1.10% of carbon atoms exist as ^{13}C. An explosives manufacturer could replace the standard ^{12}C atoms with ^{13}C atoms at specific positions in the compound and make a code that could identify the company and the production batch from which it came. If the explosive were then used by a criminal or terrorist, a forensic scientist could determine where the explosive came from and investigators could try to link the explosive to a suspect.

This idea is not currently being used in the United States because of concerns about added costs for production and record keeping. Switzerland, however, has mandated since 1980 that all explosives produced or sold in their country be physically or chemically tagged. There is also a concern that the manufacturers may open themselves to potential lawsuits by those who might be harmed through misuse of their product, just as some gun manufacturers have had to face lawsuits from victims of shootings.

3.6 Atomic Mass: Isotopic Abundance and the Periodic Table

Learning Objective

Determine the atomic mass of an element with isotopes.

What is the mass of a single atom? If you're thinking not very much, you're right. A single atom of hydrogen has a mass of 1.67×10^{-24} g. For most purposes scientists do not use grams to communicate atomic masses. They use the atomic mass unit (amu), introduced earlier in this chapter. One amu is defined as 1/12 of the mass of a carbon-12 atom. The masses of the rest of the elements are calculated by the ratio of the mass of an element to that of carbon-12. For example, if the mass of an atom were twice that of carbon-12, the mass would be near 24 amu.

The existence of isotopes complicates the matter of reporting the atomic weights of the elements in the periodic table. How do we determine the atomic mass of an element with isotopes? To explore this problem, consider the isotopes of silver, ^{107}Ag and ^{109}Ag. The natural abundance of ^{107}Ag is 51.8% and that of ^{109}Ag is 48.2%. The atomic mass reported on the periodic table should reflect that a little more than 50% of all silver atoms have a mass of 107 amu, and a little less than 50% of all silver atoms have a mass of 109 amu. It should be no surprise that the atomic mass of silver is reported as 107.865 amu, just under 108 amu. No single atom of silver will ever have a mass of exactly 107.865 amu, but on average it is the correct value. This method of determining atomic mass by natural abundance is called a **weighted average.**

Worked Example 5

Oxygen has three isotopes: oxygen-16, oxygen-17, and oxygen-18. Which of these isotopes has the highest natural abundance?

SOLUTION The atomic mass of oxygen in the periodic table is very close to 16 amu (its actual value is 15.999 amu), which indicates that the vast majority of oxygen atoms exist as the oxygen-16 isotope.

Practice 3.5

Bromine has two isotopes: bromine-79 and bromine-81. Which of these isotopes has the highest natural abundance?

ANSWER

Bromine-79 has a slightly higher abundance.

3.7 Atomic Structure: Electrons and Emission Spectra

Learning Objective

Assess data from clues that the emission spectra of elements provide about the location of electrons within an atom.

To complete the understanding of atomic structure, scientists had to investigate where electrons are located in an atom. Do electrons orbit the nucleus the way planets orbit the sun, or do the electrons move around the nucleus in random paths? The clues needed to solve this atomic mystery came from experiments involving light and its relationship to the behavior of electrons in atoms.

When white light passes through a prism, all the colors of the rainbow appear in a **continuous spectrum,** as shown in Figure 3.10. It is extremely difficult to look at a continuous spectrum and be able to pick an exact point where one color starts and another stops. Instead, the colors gradually transition from one to another. A musical analogy of a continuous spectrum is the unbroken progression of sound made by a trombone as the trombone player pushes the slide all the way out and back in one smooth motion.

The opposite of a continuous spectrum is a **line spectrum,** the single lines of various colors that are clearly separated when certain types of light pass through a prism. A musical analogy to a line spectrum is a scale of individual notes played on an instrument.

Experimentally it had been shown that atoms produce light when placed in a flame or when electricity is passed through a gas vapor of the element. The colors of fireworks are an example of light produced by atoms in a flame, whereas neon light is an example of light that results from electricity flowing through a gas vapor. If the light emitted from excited atoms (such as those in a flame) passes through a prism, it appears as the thin bands of a line spectrum. What is especially interesting

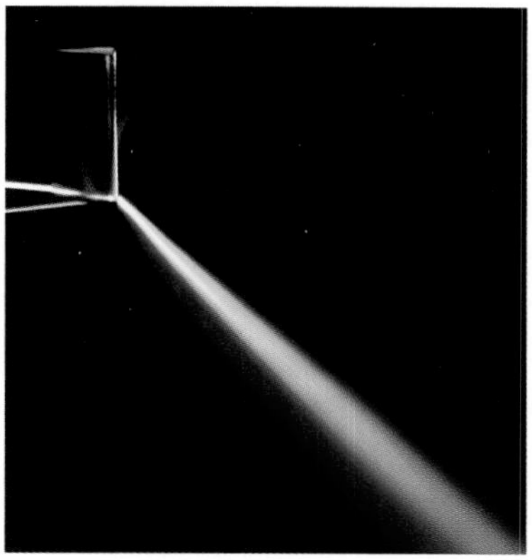

White light is being separated into its continuous spectrum by a prism. (Paul Silverman/Fundamental Photographs)

Line spectrum

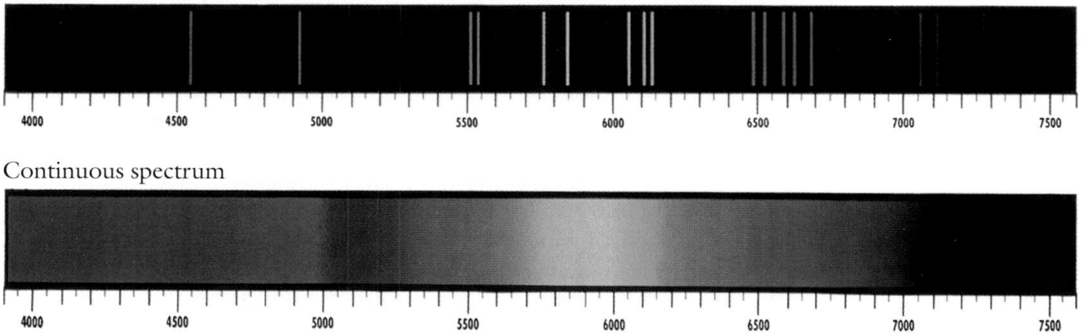

Figure 3.10 A continuous spectrum has no clear distinction between colors, whereas a line spectrum has clear separation between colors. (Wabash Instrument Corp./Fundamental Photographs)

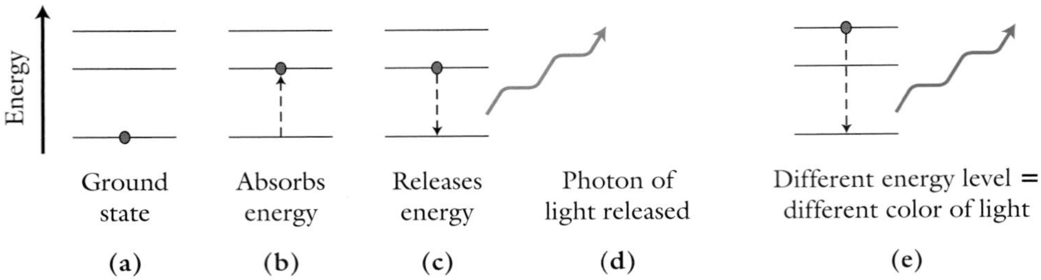

Figure 3.11 (a) The ground-state electron (b) absorbs energy and then (c) spontaneously releases excess gained energy as (d) a photon of light. If the electron were to absorb a different amount of energy corresponding to a higher excited state, the photon of light produced as the electron releases the excess energy would be of a different color (e).

is that every element produces a different line spectrum. The unique spectrum of each element helps forensic scientists detect the presence of trace elements in samples such as arsenic in hair, barium and antimony in gunshot residue, and trace elements in glass fragments.

The discovery of line spectra posed a scientific mystery: Why do they occur? The answer turned out to provide a strong clue to how electrons are arranged in the atom. When atoms are at room temperature, the electrons are located in the lowest possible energy level, called the **ground state** (Figure 3.11a). When an atom absorbs energy, the electrons are pushed into higher energy levels, called **excited states** (Figure 3.11b). The excited state of an atom is unstable, and electrons have a tendency to go back to the lower energy ground state. For the electron to go back to the ground state, the excess energy it absorbed must be released (Figure 3.11c–e).

The electron emits the excess energy in the form of a **photon**—a single "particle" or bundle of light. Multiple colors of light, corresponding to photons of different energies, are produced when electrons make the transition from higher energy levels back to the ground state. The energy of the photon, which equals the difference in energy between an excited state and the ground state of an electron, dictates the color.

The production of line spectra with a limited number of lines of color suggests that electrons in atoms are found in a few fixed energy levels. In the terminology of science, we say that the electrons are found in **quantized energy levels.** An analogy for quantized energy levels would be stair steps: The allowed states are a fixed distance apart, and it is impossible to pause partway between stair steps. If the energy levels of electrons were not quantized, electrons in excited atoms could emit photons of all energies and colors and would produce continuous spectra.

Each line in a spectrum is the result of an electron releasing energy from a higher quantized excited state to the ground state. Figure 3.12 shows the visible line spectra for several of the Group 1 elements. Many lines are produced outside the visible spectrum, such as in the infrared region and the ultraviolet region, which are not shown in the figure. Note how the complexity of the line spectra increases for elements with more electrons. The complex spectra result because a greater number of electrons are available for excitation, and there are more possible excited energy levels for the electrons to occupy.

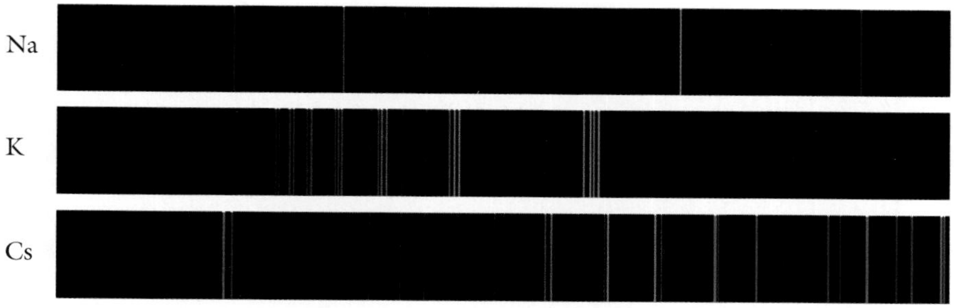

Figure 3.12 Line spectra of the elements in the first column of the periodic table. Each spectrum increases in complexity as the number of electrons increases. (Wabash Instrument Corp./Fundamental Photographs)

3.8 Mathematics of Light

Learning Objective

Examine and use the relevant mathematical equations to describe the behavior of light.

Light travels through space in the form of waves, much as waves travel across the oceans. Waves can be described mathematically by their **wavelength,** symbolized by the Greek letter lambda (λ). The wavelength is the distance from one wave peak to another, as illustrated in Figure 3.13. The unit used to measure wavelength for light in the visible spectrum is typically the nanometer (nm). The prefix *nano* represents the multiplier 1×10^{-9}, or one billionth of a meter.

Waves are also mathematically described by their frequency, symbolized by the Greek letter nu (ν). **Frequency** is the number of times the wave peak passes a point in space within a specific time period. It is measured in units called *hertz* (Hz), and 1 Hz represents a wave that passes a point once per second (1 Hz = 1 s^{-1}). Mathematically, s^{-1} and 1/s are equivalent.

Wavelength and frequency are related to the **speed of light** (c), as shown in equation 1. The speed of light in a vacuum is a constant value equal to 3.00×10^8 m/s. Equation 1 is commonly rearranged to solve for the wavelength, as shown in equation 2, or for the frequency, as shown in equation 3. One of the most common mistakes made in using equations 1–3 is failing to make sure that the units of distance are the same. The speed of light is measured in meters per second (m/s), whereas light from the visible spectrum is measured in nanometers. Just remember to use the appropriate conversion factors for units!

$$c = \lambda \times \nu \quad (1)$$

$$\frac{c}{\nu} = \frac{\lambda \times \cancel{\nu}}{\cancel{\nu}} \Rightarrow \lambda = \frac{c}{\nu} \quad (2)$$

$$\frac{c}{\lambda} = \frac{\cancel{\lambda} \times \nu}{\cancel{\lambda}} \Rightarrow \nu = \frac{c}{\lambda} \quad (3)$$

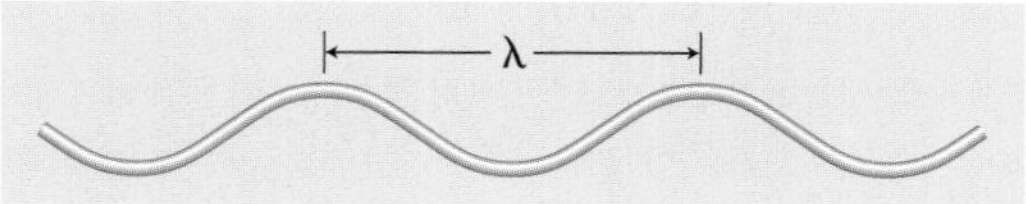

Figure 3.13 Light travels as a wave described by the wavelength (the distance between peaks) and the frequency of wave oscillations that occur in a given time period.

Worked Example 6

Napoleon Bonaparte was exiled to the island of St. Helena after his defeat at Waterloo. Soon after his exile, his health started to deteriorate. Historical documents show that his symptoms matched those of arsenic poisoning, although his illness had been diagnosed at the time as stomach cancer. Arsenic, when in an excited state, will emit light at wavelength 193.7 nm. In 2001, Napoleon's hair was analyzed and tested positive for arsenic. Calculate the frequency (Hz) of light that corresponds to the wavelength of arsenic.

SOLUTION Using equation 3,

$$\nu = \frac{c}{\lambda}$$

but first we must convert 193.7 nm to meters:

$$193.7\ \cancel{\text{nm}} \times \frac{1 \times 10^{-9}\ \text{m}}{1\ \cancel{\text{nm}}} = 1.937 \times 10^{-7}\ \text{m}$$

$$\nu = \frac{c}{\lambda} = \frac{3.00 \times 10^{8}\ \text{m/s}}{1.937 \times 10^{-7}\ \text{m}} = 1.549 \times 10^{15}\ 1/\text{s} = 1.55 \times 10^{15}\ \text{Hz}$$

Practice 3.6

The detection of lead from gunshot residue is accomplished by measuring emitted light that corresponds to a wavelength of 220.4 nm. What is the frequency (Hz) of this wavelength of light?

ANSWER
1.36×10^{15} Hz

Worked Example 7

Glass fragments recovered from a crime scene can be analyzed for trace elements to determine whether or not the fragments are consistent with the fragments recovered from a suspect. Determine the wavelength (nm) of light used for measuring cobalt given that the frequency of the light is 1.2976×10^{15} Hz.

SOLUTION Using equation 2,

$$\lambda = \frac{c}{\nu}$$

$$\lambda = \frac{c}{\nu} = \frac{3.00 \times 10^{8}\ \text{m/s}}{1.298 \times 10^{15}\ 1/\text{s}} = 2.3112 \times 10^{-7}\ \text{m}$$

$$2.3112 \times 10^{-7}\ \cancel{\text{m}} \times \frac{1\ \text{nm}}{1 \times 10^{-9}\ \cancel{\text{m}}} = 231.12\ \text{nm} \Rightarrow 231\ \text{nm}$$

Practice 3.7

Crystal is a distinctive form of glass that contains a high level of lead. The lead content can be used to determine whether two fragments are likely to have a common origin. Calculate the wavelength of light that lead atoms in an excited state will emit given that the frequency of light is 1.361×10^{15} Hz.

ANSWER

2.20×10^{2} nm

Recall that the color of each line present in a spectrum represents the transition from an excited-state energy level to a lower ground-state energy level. The difference in energy between these two levels can be calculated by the frequency of light according to equation 4, where E is energy with units of joules (J) and h is *Planck's constant,* which has a value of 6.626×10^{-34} J · s.

$$E = h \times \nu \tag{4}$$

$$\overbrace{E = h \times \nu \text{ and } \nu \times \frac{c}{\lambda}}^{\text{Combine (4) and (3)}} \Rightarrow E = \frac{h \times c}{\lambda} \tag{5}$$

Worked Example 8

Determine the energy emitted from an arsenic atom as it releases a photon of light with a wavelength of 193.7 nm.

SOLUTION

$$E = \frac{h \times c}{\lambda} = \frac{(6.626 \times 10^{-34}\ \text{J} \cdot \cancel{\text{s}}) \times (3.00 \times 10^{8}\ \cancel{\text{m}}/\cancel{\text{s}})}{193.7\ \cancel{\text{nm}} \times \dfrac{1 \times 10^{-9}\ \cancel{\text{m}}}{1\ \cancel{\text{nm}}}}$$

$$= \frac{1.9908 \times 10^{-25}\ \text{J}}{1.937 \times 10^{-7}} = 1.03 \times 10^{-18}\ \text{J}$$

Practice 3.8

Determine the energy emitted from a lead atom as it releases a photon of light with a wavelength of 220.4 nm.

ANSWER

9.02×10^{-19} J

Worked Example 9

Determine the energy emitted from a cobalt atom as it releases a photon of light with a frequency of 1.2976×10^{15} Hz.

SOLUTION

$$E = h \times \nu = (6.626 \times 10^{-34}\ \text{J} \cdot \cancel{\text{s}})(1.2976 \times 10^{15}\ 1/\cancel{\text{s}}) = 8.598 \times 10^{-19}\ \text{J}$$

Practice 3.9

Determine the energy emitted from a lead atom as it releases a photon of light with a frequency of 1.361×10^{15} Hz.

ANSWER

9.018×10^{-19} J

3.9 Atomic Structure: Electron Orbitals

Learning Objective

Describe the modern model of the atom and contrast it to earlier, obsolete models.

The quantized energy levels in which electrons exist led to speculation about where these levels were located around the nucleus. Early models showed electrons orbiting the nucleus as planets orbit the sun. In fact, we still draw a similar model when discussing the emission of photons from an excited energy level. However, you should recognize that what is fixed are the energy levels of electrons, *not* the locations of the electrons. Scientists often use models that are known to be oversimplified, but these tools help us understand a topic, and we also recognize that the model is not perfect. An analogy to police work would be the use of profiles to catch a serial murderer. The police know that even though the profiles are not 100% accurate, they provide helpful information for understanding the criminal.

One challenge that arises when we attempt to locate the exact position of an electron is that we cannot do so without disturbing the energy of that electron. If we attempt to measure the energy of an electron, the corresponding location is altered. Therefore, finding the exact location of electrons that have a specific, fixed energy level is impossible. This physical limitation to the measurement of electrons is summarized in the **Heisenberg uncertainty principle:**

> The more precisely the position is determined, the less precisely the momentum is known in this instant, and vice versa.
>
> —Werner Heisenberg, 1927

The energy of a particle is related to the momentum to which Heisenberg refers in his statement.

The best scientists can do is to predict the highest probability of the location of an electron of a given energy around the nucleus of an atom. The details of how these predictions are made are in an area of study known as quantum mechanics, which requires an extensive background in calculus, chemistry, and physics to grasp. However, a full understanding of quantum mechanics is not necessary to comprehend and use the results. Just as we do not need an extensive knowledge of computer chip technology to use a computer, we do not have to dwell on the details of quantum mechanics to gain some understanding of electrons in atoms.

Examine the crime map shown in Figure 3.14. Although police officers do not know the exact location of future crimes, they can examine a map that shows where crimes have been committed in the past and predict where the greatest probability of the occurrence of crimes will be. In Figure 3.14 the circled area shows the greatest density of purse snatchings, so it makes sense to expect this crime to have the highest probability in the future of occurring in the circled region.

Through the mathematics of quantum mechanics, density maps for electrons of a fixed energy level have been determined. The **orbital** of an electron is a three-dimensional region of space in which an electron has the highest probability of being located. In a similar fashion to the crime map example, an area of highest probability is defined and shown in Figure 3.15. Figure 3.15a shows the cross-sectional (two-dimensional) electron density map for locating the single electron of the hydrogen atom. It is apparent from the figure that the greatest probability of locating the

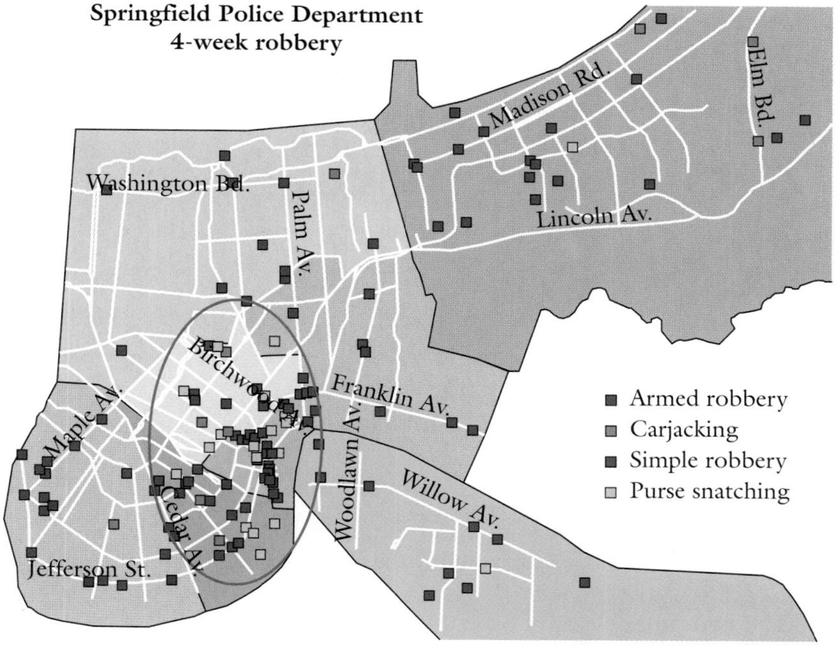

Figure 3.14 This crime map of Springfield shows a high density of purse snatchings occurring over four weeks in a roughly oval-shaped area. The probability of purse snatching occurring in the future is greatest in this region.

electron is in a region surrounding the center of the atom. The probability of finding the electron decreases at distances farther away from the nucleus. Figure 3.15b is the three-dimensional representation of the orbital, which has a spherical shape. A sphere-shaped orbital is known as an ***s* orbital.** (The *s* actually stands for "sharp," not "spherical," as one might predict. The term *sharp* was used because it describes the emission lines observed when the electrons in the atom are excited.)

The shape of the electron orbital changes as the energy level of an electron increases. Beyond the *s* orbital, there are three other orbitals, starting with the ***p* orbital,** shown as a probability density map in Figure 3.16a. The *p* orbital has a dumbbell shape: Electrons are located on either side of the nucleus in a teardrop-shaped lobe. One major difference between *p* and *s* orbitals is that there are three of the *p* orbitals, all having

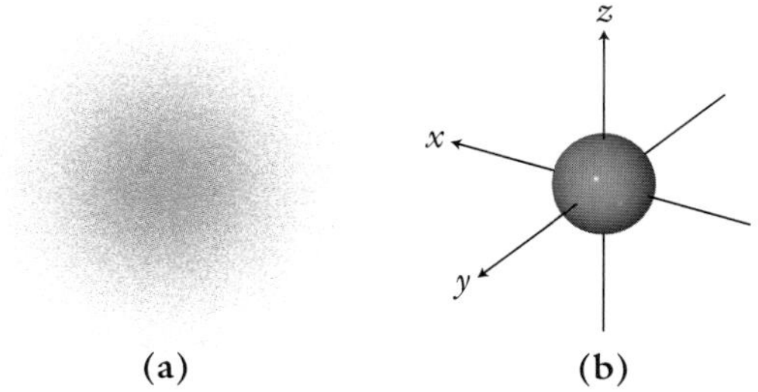

Figure 3.15 (a) Cross-sectional model and (b) spherical three-dimensional model of the *s* orbital.

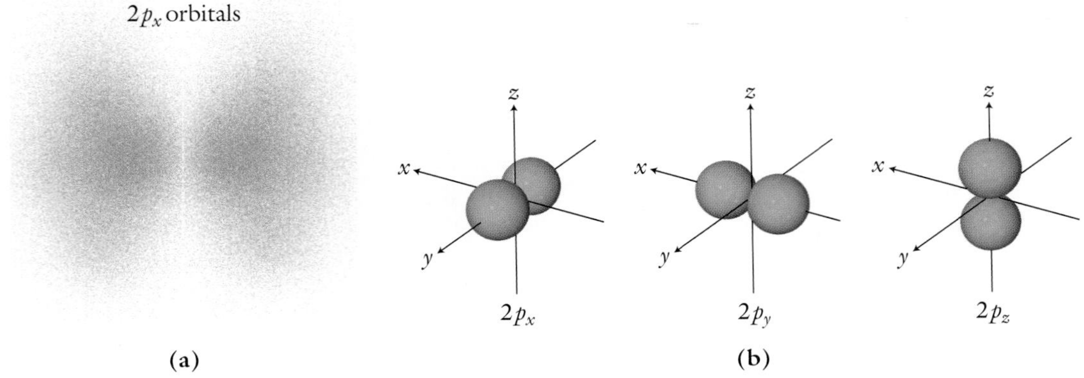

Figure 3.16 (a) Cross-sectional model and (b) the three dumbbell-shaped three-dimensional models of the *p* orbitals.

identical energies. As you can see in Figure 3.16b, one of the *p* orbitals is lined up on the *x* axis (p_x), one on the *y* axis (p_y), and the third on the *z* axis (p_z).

The two other electron orbital types, the ***d*** and ***f* orbitals,** are shown in Figure 3.17. The number and complexity of the orbitals increase dramatically, as shown in the figure, and are provided only for completeness; this book does not delve further into the orbital discussion.

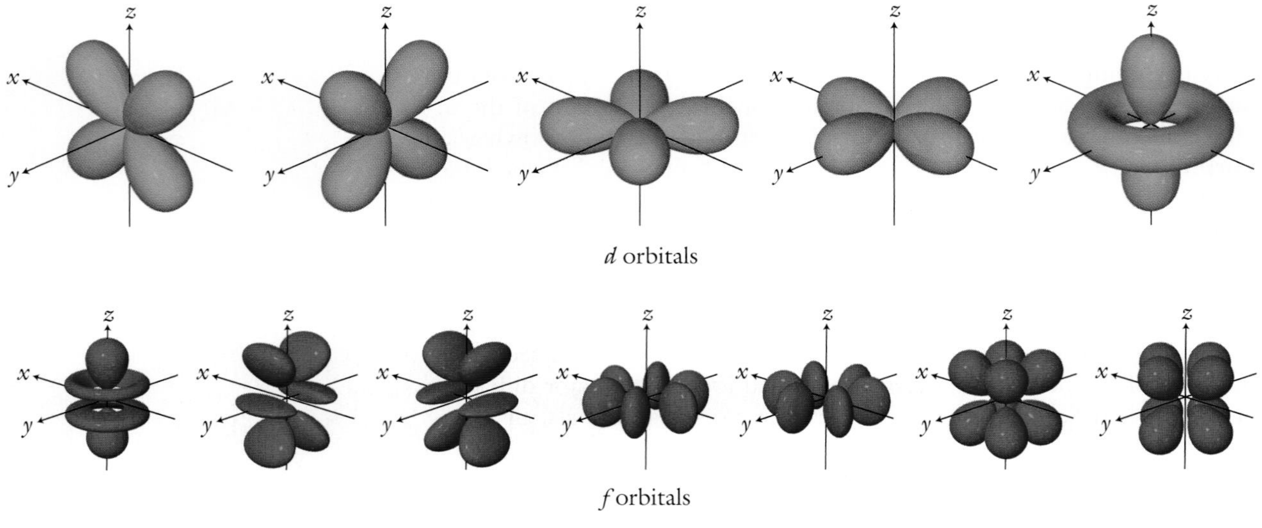

Figure 3.17 There are five separate *d* orbitals and seven separate *f* orbitals. The shapes of the orbitals in a set vary dramatically, but the energy of the orbitals in a set is equivalent.

3.10 Electron Configurations

Learning Objective

Write the electron configuration for an atom of an element.

Let's take a moment for a brief summary of what has been presented thus far about the behavior of electrons in atoms. The atomic clues that are used to solve crimes come from the transitions of electrons between a ground-state orbital and a vacant, excited-state orbital. When energy is absorbed by an electron, it moves from a lower energy level (usually the ground state) to a higher energy orbital (an excited state). As the electron then returns to the ground state, it emits a photon. The line emission spectra for atoms with fewer electrons (the lighter elements) are simpler than those of heavier elements. Heavier elements have a greater number of electrons and a larger variety of possible excited states.

For atoms with multiple electrons a question then arises: Exactly which electron is being excited to a higher excited state? To answer this question, a system is needed to identify each electron in an atom. This system is called the *electron configuration* of the element. The periodic table is a necessary tool for understanding electron configurations.

Two rules govern the distribution of electrons in orbitals:

Rules for Electron Orbitals

1. A single orbital can contain a maximum of two electrons.
2. When filling electrons in the *p*, *d*, and *f* subsets, each orbital of the subset gets a single electron before any orbital of the subset receives a second electron.

Electron configurations can be thought of in much the same way as house numbers in your street address. Each set of orbitals is like a city block. Figure 3.18 is a color-coded map of the orbital blocks. Remember that the periodic table is arranged by increasing number of protons. For each proton that is gained, so is one electron. The electrons go into the orbital designated by the color-coded blocks in the periodic table.

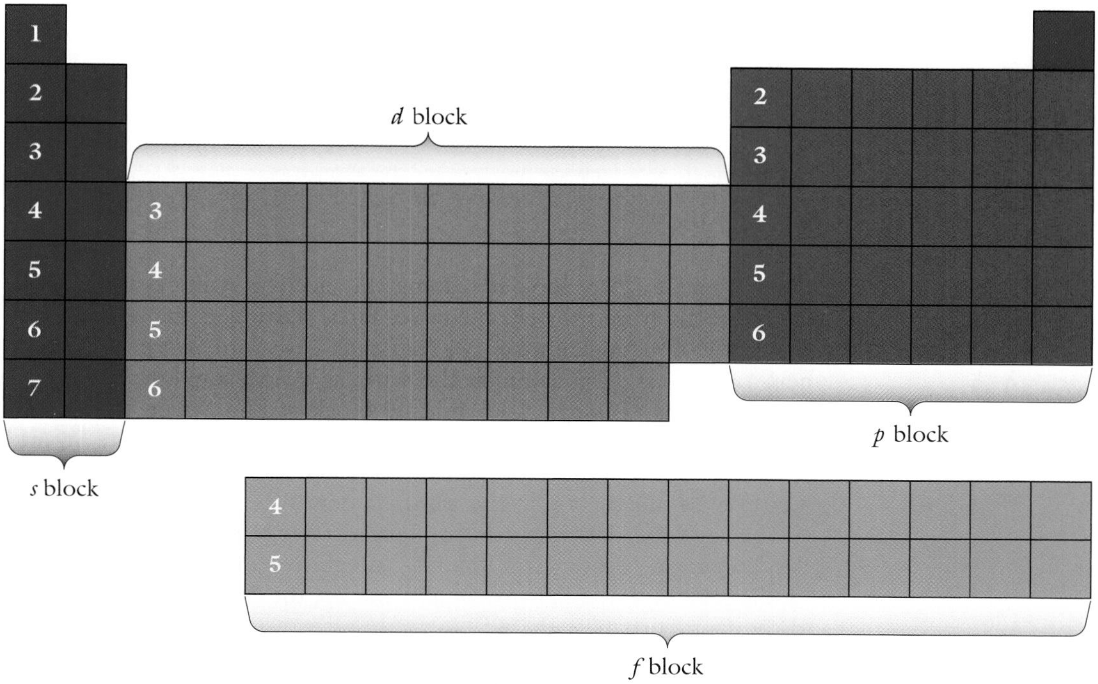

Figure 3.18 In examining the elements by increasing atomic number, it is apparent that electrons first occupy the 1*s* orbital. The second orbital to fill is the 2*s* followed by the 2*p*, and so on.

Let's examine which orbitals the electrons occupy as we increase the atomic number.

Element	Number of e^-	Orbital
H	1	First *s* orbital
He	2	Both e^- located in first *s* orbital
Li	3	2 e^- located in first *s* orbital, 1 e^- in second *s* orbital
Be	4	2 e^- located in first *s* orbital, 2 e^- in second *s* orbital

Writing the electron location in this fashion is cumbersome and time-consuming. Therefore, a shorthand method has been developed:

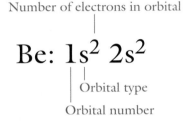

Number of electrons in orbital

$$\text{Be: } 1s^2\ 2s^2$$

Orbital type

Orbital number

Rewriting and expanding on our list of electron configurations:

Element	Number of e^-	Orbital
H	1	$1s^1$
He	2	$1s^2$
Li	3	$1s^22s^1$
Be	4	$1s^22s^2$
B	5	$1s^22s^22p^1$*
C	6	$1s^22s^22p^2$
N	7	$1s^22s^22p^3$

*Note that the first *p* orbital is in the second energy level.

Let's take a closer look at reading the electron configuration for nitrogen directly from the periodic table. First, starting at the beginning of the periodic table in Figure 3.18, read each row until you get to the element of interest. This is much the same as giving someone driving directions such as "Pass the 100 and 200 block of Elm Street, turn left on Cambridge Street, and go three houses down to 606 Cambridge Street." For nitrogen, we would say, "Pass the $1s^2$ and $2s^2$ block, and when you get to the $2p$ block, stop at the third element."

The key to reading electron configurations is to practice writing them until the pattern becomes ingrained. You should note on a copy of the periodic table that the first *p* orbital is the $2p$, the first *d* orbital is the $3d$, and the first *f* orbital is the $4f$.

Worked Example 10

Write the electron configurations for nitrogen through argon.

SOLUTION

N: $1s^2 2s^2 2p^3$
O: $1s^2 2s^2 2p^4$
F: $1s^2 2s^2 2p^5$
Ne: $1s^2 2s^2 2p^6$
Na: $1s^2 2s^2 2p^6 3s^1$
Mg: $1s^2 2s^2 2p^6 3s^2$
Al: $1s^2 2s^2 2p^6 3s^2 3p^1$
Si: $1s^2 2s^2 2p^6 3s^2 3p^2$
P: $1s^2 2s^2 2p^6 3s^2 3p^3$
S: $1s^2 2s^2 2p^6 3s^2 3p^4$
Cl: $1s^2 2s^2 2p^6 3s^2 3p^5$
Ar: $1s^2 2s^2 2p^6 3s^2 3p^6$

Practice 3.10

Write the full electron configurations for xenon, arsenic, and ruthenium.

ANSWER

Xe: $1s^2 2s^2 2p^6 3s^2 3p^6 4s^2 3d^{10} 4p^6 5s^2 4d^{10} 5p^6$
As: $1s^2 2s^2 2p^6 3s^2 3p^6 4s^2 3d^{10} 4p^3$
Ru: $1s^2 2s^2 2p^6 3s^2 3p^6 4s^2 3d^{10} 4p^6 5s^2 4d^6$

Writing the entire electron configuration for the heavier elements can become time-consuming and repetitive. Additionally, the electrons that are of interest to chemists are the outermost *s* and *p* orbital electrons. These electrons are collectively called the **valence shell electrons** and are of interest because chemical reactions involve the gaining, losing, and sharing of valence shell electrons.

The inner electrons that make up the beginning of the electron configuration are called the **core electrons.** Core electrons are not involved in chemical reactions except in rare cases. The separation of electrons into those that participate in reactions and those that do not leads to a method of abbreviating and shortening the electron configurations. When you provide directions to someone, you usually don't list every single house they will pass. Instead, you provide a reference point to start the direction such as, "Go down to the Y mart, turn left, go two blocks, and stop at the sixth house."

The reference point for electron configurations is the noble gas elements. Always start at the noble gas element that directly precedes the element of interest. For example, if we were to write the electron configuration of sodium, we would start at neon, the noble gas directly before sodium. The newly abbreviated electron configuration of sodium is $[Ne]3s^1$.

Worked Example 11

Write out the abbreviated electron configurations for iron, iodine, phosphorus, zinc, cadmium, and potassium.

SOLUTION

Fe: $[Ar]4s^2 3d^6$
I: $[Kr]5s^2 4d^{10} 5p^5$
P: $[Ne]3s^2 3p^3$
Zn: $[Ar]4s^2 3d^{10}$
Cd: $[Kr]5s^2 4d^{10}$
K: $[Ar]4s^1$

Practice 3.11

Which elements have the following electron configurations?

$[Ar]4s^23d^2$ $[Kr]5s^24d^{10}5p^1$ $[Ne]3s^23p^2$ $[Ar]4s^23d^{10}4p^5$

ANSWER

Titanium Indium Silicon Bromine

Table 3.3 lists the full and abbreviated electron configurations of the first 54 elements for your reference. An excellent use for this table is in practicing writing both the full and abbreviated electron configurations of all the elements: Check your work against the table. Several elements do not follow the rules for writing electron configurations; these elements are highlighted in red.

Table 3.3 Electron Configuration of the Elements Hydrogen through Xenon

	Full	Abbreviation		Full	Abbreviation
H	$1s^1$	—	Ni	$1s^22s^22p^63s^23p^64s^23d^8$	$[Ar]4s^23d^8$
He	$1s^2$	—	Cu	$1s^22s^22p^63s^23p^64s^1d^{10}$	$[Ar]4s^13d^{10}$
Li	$1s^22s^1$	$[He]2s^1$	Zn	$1s^22s^22p^63s^23p^64s^23d^{10}$	$[Ar]4s^23d^{10}$
Be	$1s^22s^2$	$[He]2s^2$	Ga	$1s^22s^22p^63s^23p^64s^23d^{10}4p^1$	$[Ar]4s^23d^{10}4p^1$
B	$1s^22s^22p^1$	$[He]2s^22p^1$	Ge	$1s^22s^22p^63s^23p^64s^23d^{10}4p^2$	$[Ar]4s^23d^{10}4p^2$
C	$1s^22s^22p^2$	$[He]2s^22p^2$	As	$1s^22s^22p^63s^23p^64s^23d^{10}4p^3$	$[Ar]4s^23d^{10}4p^3$
N	$1s^22s^22p^3$	$[He]2s^22p^3$	Se	$1s^22s^22p^63s^23p^64s^23d^{10}4p^4$	$[Ar]4s^23d^{10}4p^4$
O	$1s^22s^22p^4$	$[He]2s^22p^4$	Br	$1s^22s^22p^63s^23p^64s^23d^{10}4p^5$	$[Ar]4s^23d^{10}4p^5$
F	$1s^22s^22p^5$	$[He]2s^22p^5$	Kr	$1s^22s^22p^63s^23p^64s^23d^{10}4p^6$	$[Ar]4s^23d^{10}4p^6$
Ne	$1s^22s^22p^6$	$[He]2s^22p^6$	Rb	$1s^22s^22p^63s^23p^64s^23d^{10}4p^65s^1$	$[Kr]5s^1$
Na	$1s^22s^22p^63s^1$	$[Ne]3s^1$	Sr	$1s^22s^22p^63s^23p^64s^23d^{10}4p^65s^2$	$[Kr]5s^2$
Mg	$1s^22s^22p^63s^2$	$[Ne]3s^2$	Y	$1s^22s^22p^63s^23p^64s^23d^{10}4p^65s^24d^1$	$[Kr]5s^24d^1$
Al	$1s^22s^22p^63s^23p^1$	$[Ne]3s^23p^1$	Zr	$1s^22s^22p^63s^23p^64s^23d^{10}4p^65s^24d^2$	$[Kr]5s^24d^2$
Si	$1s^22s^22p^63s^23p^2$	$[Ne]3s^23p^2$	Nb	$1s^22s^22p^63s^23p^64s^23d^{10}4p^65s^24d^3$	$[Kr]5s^24d^3$
P	$1s^22s^22p^63s^23p^3$	$[Ne]3s^23p^3$	Mo	$1s^22s^22p^63s^23p^64s^23d^{10}4p^65s^24d^4$	$[Kr]5s^24d^4$
S	$1s^22s^22p^63s^23p^4$	$[Ne]3s^23p^4$	Tc	$1s^22s^22p^63s^23p^64s^23d^{10}4p^65s^24d^5$	$[Kr]5s^24d^5$
Cl	$1s^22s^22p^63s^23p^5$	$[Ne]3s^23p^5$	Ru	$1s^22s^22p^63s^23p^64s^23d^{10}4p^65s^1d^7$	$[Kr]5s^1d^7$
Ar	$1s^22s^22p^63s^23p^6$	$[Ne]3s^23p^6$	Rh	$1s^22s^22p^63s^23p^64s^23d^{10}4p^65s^1d^8$	$[Kr]5s^14d^8$
K	$1s^22s^22p^63s^23p^64s^1$	$[Ar]4s^1$	Pd	$1s^22s^22p^63s^23p^64s^23d^{10}4p^64d^{10}$	$[Kr]4d^{10}$
Ca	$1s^22s^22p^63s^23p^64s^2$	$[Ar]4s^2$	Ag	$1s^22s^22p^63s^23p^64s^23d^{10}4p^65s^1d^{10}$	$[Kr]5s^14d^{10}$
Sc	$1s^22s^22p^63s^23p^64s^23d^1$	$[Ar]4s^23d^1$	Cd	$1s^22s^22p^63s^23p^64s^23d^{10}4p^65s^24d^{10}$	$[Kr]5s^24d^{10}$
Ti	$1s^22s^22p^63s^23p^64s^23d^2$	$[Ar]4s^23d^2$	In	$1s^22s^22p^63s^23p^64s^23d^{10}4p^65s^24d^{10}5p^1$	$[Kr]5s^24d^{10}5p^1$
V	$1s^22s^22p^63s^23p^64s^23d^3$	$[Ar]4s^23d^3$	Sn	$1s^22s^22p^63s^23p^64s^23d^{10}4p^65s^24d^{10}5p^2$	$[Kr]5s^24d^{10}5p^2$
Cr	$1s^22s^22p^63s^23p^64s^13d^5$	$[Ar]4s^13d^5$	Sb	$1s^22s^22p^63s^23p^64s^23d^{10}4p^65s^24d^{10}5p^3$	$[Kr]5s^24d^{10}5p^3$
Mn	$1s^22s^22p^63s^23p^64s^23d^5$	$[Ar]4s^23d^5$	Te	$1s^22s^22p^63s^23p^64s^23d^{10}4p^65s^24d^{10}5p^4$	$[Kr]5s^24d^{10}5p^4$
Fe	$1s^22s^22p^63s^23p^64s^23d^6$	$[Ar]4s^23d^6$	I	$1s^22s^22p^63s^23p^64s^23d^{10}4p^65s^24d^{10}5p^5$	$[Kr]5s^24d^{10}5p^5$
Co	$1s^22s^22p^63s^23p^64s^23d^7$	$[Ar]4s^23d^7$	Xe	$1s^22s^22p^63s^23p^64s^23d^{10}4p^65s^24d^{10}5p^6$	$[Kr]5s^24d^{10}5p^6$

Evidence Analysis | Scanning Electron Microscopy

Figure 3.19 The smoke cloud created during the discharge of a firearm leaves a residue on any object in the close vicinity of the firearm. (BrandX Pictures)

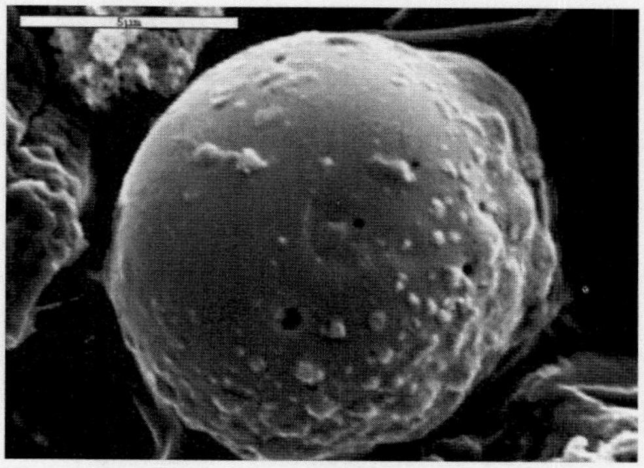

Figure 3.20 Scanning electron microscope image of gunshot residue. (International Association for Microanalysis)

Figure 3.19 shows the smoke cloud that surrounds the end of the barrel and the chamber of a pistol after firing. The cloud consists of gases formed during the explosive discharge of the gunpowder and small amounts of unburned gunpowder. The gunshot residue is deposited onto the hands and arms of the shooter. Gunshot residue is also deposited on the target if the target is close enough to the end of the barrel. Investigators use the amount and pattern of gunshot residue on a victim to determine an approximate distance to the shooter.

To determine whether or not a person has gunshot residue on his or her hands, detectives take several small metal cylinders that contain an adhesive coating and pat down the suspected shooter's hands. Gunshot residue, if present, sticks to the adhesive on the metal cylinder, which is then placed into a scanning electron microscope (SEM). A scanning electron microscope functions like a traditional microscope except that the SEM uses electrons bouncing off a surface to form an image. The traditional microscope depends on photons of light bouncing off a surface. There are several advantages to using electrons over photons; the main advantage is that electrons make it possible to obtain an image of much smaller particles.

Particles of gunpowder come in very distinctive shapes—cylinders, tubes, discs, spheres, and slabs—a portion of which remain unburned in gunshot residue. The SEM image in Figure 3.20 shows how gunshot residue recovered from a suspect's hand would appear.

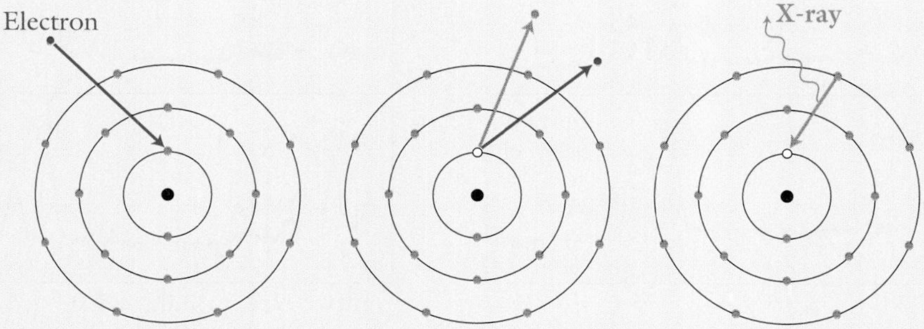

Figure 3.21 Basic principle of an energy-dispersive X-ray spectrometer (EDS).

(*continued*)

Evidence Analysis | Scanning Electron Microscopy (*continued*)

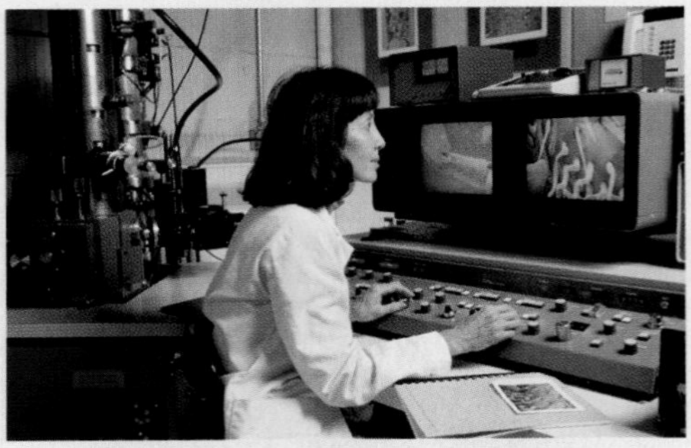

Typical SEM-EDS system. (Lawrence Migdale/Photo Researchers, Inc.)

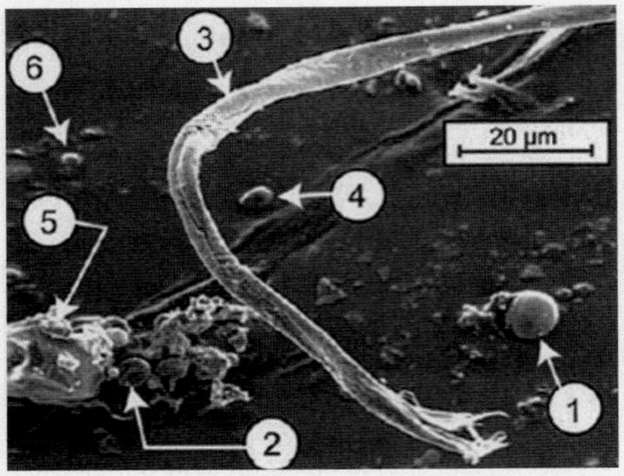

Figure 3.22 Six distinct particles collected from skin near the burn wound of the victim. (Kosanke, K. L.; Dujay, R. C.; Kosanke, B. *J. Forensic Sci,* May 2003, vol. 48, no. 3)

The shape of a particle is not sufficient evidence for the forensic scientist to conclude whether the particle is, in fact, gunshot residue. The scientist must also evaluate the chemical makeup of the particle, which may be only 1.0×10^{-5} m (10 μM) in size! The elemental analysis of the particles can be done simultaneously with the SEM imaging of the particle by an instrument called an energy dispersive X-ray spectrometer (EDS), which is attached to the SEM. The combined analytical system is abbreviated SEM-EDS.

To understand how the EDS system detects the presence of elements, Figure 3.21, showing a simplified orbital model of an atom, can be of help. (Although we know that electrons are not actually in orbits that resemble those of the planets, this simplified model is a convenient device for demonstrating how the EDS system works.) In part (a), an electron from the SEM strikes an atom. Because the electron is so small, it can penetrate the atom and collide with enough energy to force an inner core electron to leave the atom. Part (b) shows the atom with a vacancy in the inner core electrons. This condition is not stable; electrons always go to the lowest possible energy level. Part (c) shows an electron from one of the outer shells, which has a higher energy, dropping down to fill the vacancy in the lower energy orbital and releasing excess energy as a photon.

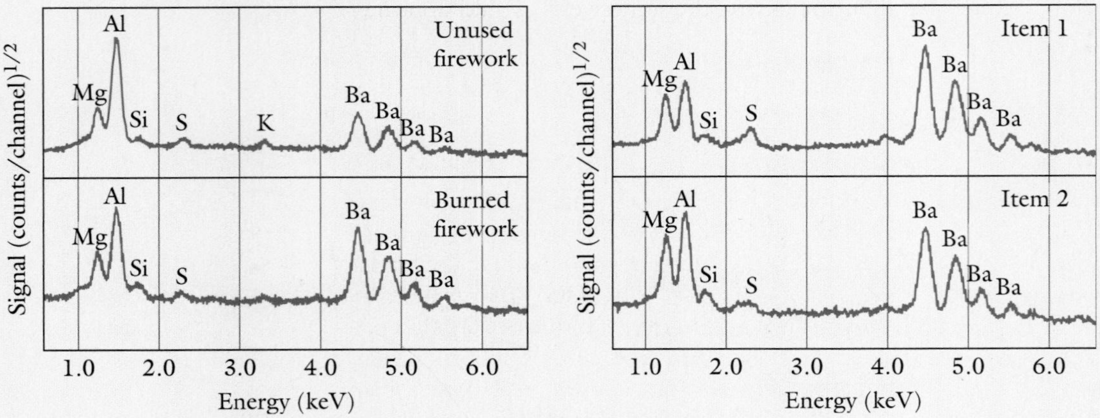

Figure 3.23 EDS spectra. Elemental analysis of the explosive particles in an unused firework and a carefully burned firework are compared with particles found on the victim and suspected to be from the firework.

The photons emitted from atoms in this process are X-rays. The process involves the same basic concepts studied earlier for the creation of line spectra from energetically excited electrons relaxing to a lower energy level. Just as the line spectra of excited electrons in atoms can fingerprint an element, each element detected in an EDS emits X-rays that have a unique energy for that element.

An SEM-EDS system can be used to investigate any material in which the combination of high magnification and elemental analysis would be useful to investigators. Consider an investigation into an accidental explosion of consumer fireworks. A person claimed to have been severely burned by the negligent use of fireworks by neighbors. The neighbors admitted to having launched fireworks but disputed the claim that they were responsible for causing the burns. If the victim's claim were true, residue from the explosives in the fireworks should be found in the vicinity of the burn. Figure 3.22 is an SEM image of the trace evidence residue obtained from skin near the burn.

The next step of the investigation was to determine if any of the particles were in fact from the firework itself. The forensic scientists obtained unused fireworks of the same type used on the evening in question and compared the elemental makeup of the particles shown in Figure 3.22 to the unused fireworks. Two recovered particles were determined to be residue from an explosive material and had the same elemental composition, as shown in Figure 3.23.

3.11 Case Study Finale: To Burn or Not to Burn

(Tammi Chappell/Reuters/Corbis)

Dr. Al Hazari contemplated the fate of Chigger's remains and the puzzle of how to determine if the ashes were human or if a filler material had been used to dupe the family. The investigation into the crematorium had led investigators to believe that a variety of materials, from regular wood ash to cement mix, were being used to fill family urns. Dr. Hazari came up with a simple method for solving the problem. He considered the known facts about the elemental content of a typical human body, shown in Figure 3.24. He noted that the human body contains only about 1 gram of silicon, and he knew that this element would have survived the heat of a crematorium oven. He calculated that if the contents of the urn were, in fact, human ashes, only a small fraction—less than 1%—would be silicon. If the cremains were in actuality cement or mortar mix, the sample would contain a much higher proportion of silicon.

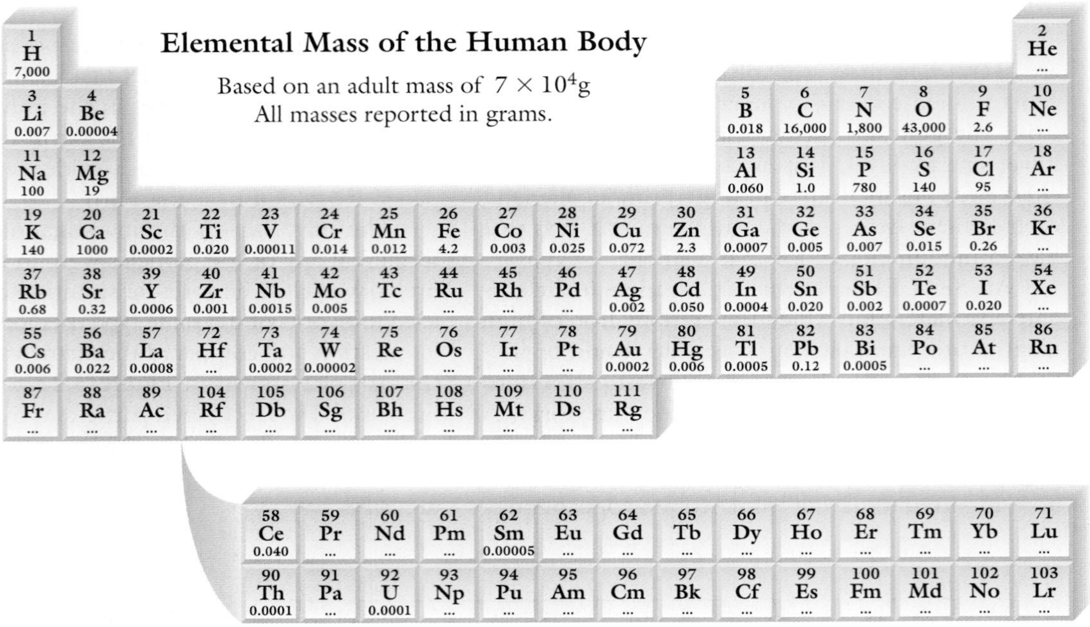

Figure 3.24 Periodic table of the human body.

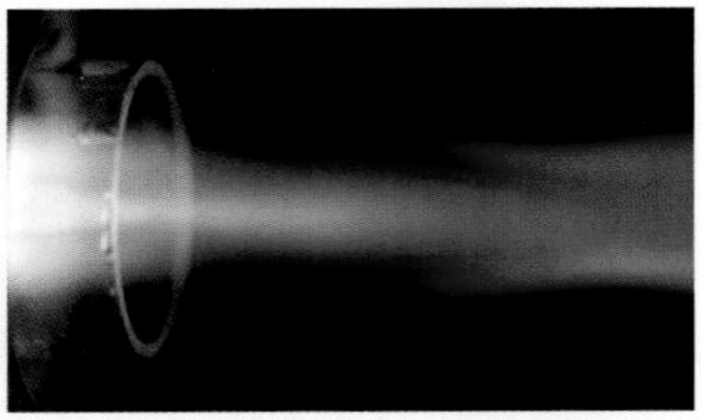

The temperature of the plasma in an ICP-OES can reach 10,000°C. (Courtesy of Dave Aeschliman)

The cremains were sent to a professional laboratory for analysis of silicon content. The instrument used to analyze the sample was an inductively coupled plasma-optical emission spectrometer, or ICP-OES. This instrument is commonly used to detect elemental concentrations as low as one part per billion.

When silicon atoms are heated to extremely high temperatures, the electrons reach a higher-energy excited state and emit light at a wavelength of 516 nm. An ICP-OES is capable of exciting the silicon atoms in the plasma and detecting the wavelength of light emitted by the excited atoms. The **plasma** of the ICP-OES is a gaseous region that is electrically conductive and can reach a temperature of 10,000°C. The temperature of a plasma is about three times hotter than the temperature of a flame. It causes more atoms to be excited and permits lower detection limits.

The results of the analysis came back as the family had feared: 19% silicon. The cremains were certainly not those of Chigger or any other human. The skeletal remains of Chigger were later identified among the 334 bodies by the use of DNA analysis. This time the family had Dr. Bill Bass provide an independent identification and also personally supervise the cremation of the remains.

CHAPTER SUMMARY

- Leucippus and Democritus first proposed an atomic theory of matter in ancient Greece. In the 1700s, work by Antoine Lavoisier and Joseph Proust provided sufficient experimental data for John Dalton to develop his atomic theory of matter.
- Experiments to prove the atomic theory produced results that seemed to contradict the indivisible nature of the atom. The experiments would lead to the discovery of the subatomic particles: electrons, protons, and neutrons.

- In Rutherford's model of the atom, most of the atom was empty space. The protons were in an extremely small region called the nucleus. Later experiments showed that the nucleus also contained neutrons.
- Isotopes are atoms of the same element that have a different number of neutrons. The isotopes undergo identical chemical reactions, which allow them to be used as medical and investigative indicators. The atomic weight of an element listed in the periodic table is actually a weighted average of all the isotopes, taking into consideration the natural abundance of each isotope.
- When energy is absorbed by an atom, the electrons can be promoted to higher quantized energy levels. The excited atoms then release photons of light as the electrons relax back to the ground state, forming a unique line spectrum that can be used to identify the element.
- Determining the exact location and the energy of an electron is impossible, according to the Heisenberg uncertainty principle, because measuring one variable inevitably alters the other. Modern atomic theory uses quantum mechanics to determine the region in space with the highest probability of containing an electron. There are four distinct orbital types, the *s*, *p*, *d*, and *f* orbitals. The *s* orbital is spherical in shape and the *p* orbital is shaped like a dumbbell. The *d* and *f* orbitals are more complex.
- Electron configurations provide a method for identifying each electron within an atom. The configuration lists each orbital in the atom and the number of electrons located in the orbitals.
- The SEM-EDS is a powerful investigative tool. It uses electrons scattered from the surface of an object to provide magnification beyond that available using a traditional light microscope, easily viewing objects that are only 10 μm in size. As the electrons strike the surface, core electrons are ejected. As a higher energy valence electron assumes the place of the missing core electron, it releases a photon of light corresponding to the X-ray region of the spectrum. The energy of the X-ray is unique to the element producing it and provides the chemical makeup of the particles being viewed.

KEY TERMS

CONTINUING THE INVESTIGATION Additional Readings, Resources, and References

Bass, B., and Jefferson, J. *Death's Acre,* Putnam, New York, 2003.

Beard, B. L., and Johnson, C. M. Strontium isotope composition of skeletal material can determine the birthplace and geographic mobility of humans and animals. *Journal of Forensic Science* 45(5), September 2003.

Emsley, J. *Nature's Building Blocks: An A–Z Guide to the Elements,* Oxford University Press, New York, 2001.

Kosanke, K. L., Dujay, R. C., and Kosanke, B. Characterization of pyrotechnic reaction residue particles by SEM/EDS. *Journal of Forensic Science* 48(3), May 2003.

REVIEW QUESTIONS AND PROBLEMS

Questions

1. Describe atoms according to the model of Leucippus and Democritus.
2. What did Leucippus and Democritus propose that existed where atoms did not?
3. How did Aristotle view the nature of matter?
4. Although Gassendi's work is not part of the modern atomic theory, why was it important at the time he published it?
5. What does the law of conservation of mass state?
6. How was the law of conservation of mass important to the development of modern atomic theory?
7. What did Lavoisier do differently from other scientists that led to his success?
8. What does the law of definite proportions state?
9. How was the law of definite proportions important to the development of modern atomic theory?
10. Why did Joseph Proust succeed with his experiments where others had failed?
11. How is the term *theory* used differently in the sciences and in popular usage?
12. Can a theory be disproved?
13. What are the four basic principles of Dalton's atomic theory?
14. Explain the law of conservation of mass in terms of Dalton's atomic theory.
15. Explain the law of definite proportions in terms of Dalton's atomic theory.
16. Explain how the law of multiple proportions came from Dalton's atomic theory.
17. Sketch a reaction between five O_2 molecules and five C atoms that follows the law of conservation of mass.
18. Using ● as a carbon atom and ○ as an oxygen atom, illustrate the concept of the law of definite proportions for CO_2.
19. Using ● as a carbon atom and ○ as an oxygen atom, illustrate the concept of the law of multiple proportions for CO and CO_2.
20. Describe a cathode ray tube and explain what the results of cathode ray tube experiments meant in terms of the properties of electrons.
21. What are the three subatomic particles discussed in this chapter?
22. What were the three major observations Rutherford made in his gold foil experiment?
23. What conclusions did Rutherford make about atomic structure from the gold foil experiment?
24. Sketch Rutherford's atomic model.
25. What is the main difference between isotopes of the same element?
26. Explain how isotopes are used to investigate chemical mysteries.
27. What type of light produces a continuous spectrum? A line spectrum?
28. Describe what happens to the electrons in an atom when they are excited by electricity or a flame.
29. Why is the line spectrum produced by excited cesium atoms more complex than the line spectrum produced by excited lithium atoms?
30. What is the Heisenberg uncertainty principle?
31. What happens to the energy of an electron when the location of the electron is determined?
32. Sketch and label the regions of the periodic table that correspond to the *s*, *p*, *d*, and *f* electron orbitals.
33. Sketch the shape of the *s* and *p* orbitals.
34. Describe how an X-ray photon is generated by a SEM-EDS.

Problems

35. Place the following scientists in chronological order of their contributions to atomic theory.
    (a) Lavoisier (c) Dalton
    (b) Democritus (d) J. J. Thomson
36. Place the following scientists in chronological order of their contributions to atomic theory.
    (a) Gassendi (c) Chadwick
    (b) Rutherford (d) Proust
37. Which of the following statements about the experiments that led to the law of conservation of mass is false?
    (a) The chemical reactions were studied in closed systems.
    (b) The chemical reactions neither gained nor lost mass.
    (c) The chemical reactions could exchange mass with the surroundings.
    (d) The chemical reactions could be heated externally from the surroundings.

38. Which of the following statements about the experiments that led to the law of definite proportions is false?
(a) A compound is made up of two or more elements.
(b) A compound produced in the laboratory is identical to the same compound found in nature.
(c) A compound is made up of differing relative ratios of the same elements.
(d) A compound must be properly purified and carefully analyzed to obtain accurate results.

39. What is the mass of the missing reactant or product in each of the following reactions given that the reaction goes to completion?
(a) 9.0 g water + 22.0 g carbon dioxide → _______ g carbonic acid
(b) _______ g sodium hydroxide + 20.0 g hydrofluoric acid → 18.0 g water + 42.0 g sodium fluoride
(c) 25.0 g calcium carbonate + 20.0 g sodium hydroxide → 26.5 g sodium carbonate + _______ g calcium hydroxide
(d) 4.0 g sodium hydroxide + _______ g carbon dioxide → 8.4 g sodium hydrogen carbonate

40. What is the mass of the missing reactant in the following reactions given that the reaction goes to completion?
(a) 27.0 g water + _______ g carbon dioxide → 93 g carbonic acid
(b) 16.0 g sodium hydroxide + 8.0 g hydrofluoric acid → _______ g water + 16.8 g sodium fluoride
(c) 0.40 g calcium carbonate + 0.32 g sodium hydroxide → _______ g sodium carbonate + 0.30 g calcium carbonate
(d) 2.65 g sodium hydroxide + 1.85 g carbon dioxide → _______ g sodium hydrogen carbonate

41. In Rutherford's gold foil experiment, the vast majority of alpha particles passed directly through the gold foil. This observation leads to which conclusion?
(a) The positive region of the atom has to be small.
(b) The majority of the atom must consist of empty space.
(c) The alpha particle makes a direct hit on the positive region.
(d) The positive region of the atom is very dense.

42. In Rutherford's gold foil experiment, occasionally the alpha particle veered from a straight-line path. This observation leads to which conclusion?
(a) The positive region of the atom has to be small.
(b) The majority of the atom must consist of empty space.
(c) The alpha particle makes a direct hit on the positive region.
(d) The positive region of the atom is very dense.

43. Fill in the missing information in the following table of subatomic particles.

Particle	Charge	Mass (amu)	Symbol
_______	−1	_______	_______
_______	_______	1	_______
_______	_______	_______	n

44. Fill in the missing information in the following table of subatomic particles.

Particle	Charge	Mass (amu)	Symbol
_______	_______	1.0073	_______
_______	0	_______	_______
_______	_______	0.0005486	_______

45. Using the periodic table on the inside front cover, fill in the missing information in the following table:

Protons	Neutrons	Electrons	Isotope Symbol
_______	_______	_______	$^{59}_{27}CO$
76	124	_______	_______
_______	_______	_______	$^{14}_{\_}C$
21	24	21	_______

46. Using the periodic table on the inside front cover, fill in the missing information in the following table.

Protons	Neutrons	Electrons	Isotope Symbol
_______	146	_______	$^{__}_{92}$_______
80	_______	80	$^{200}_{__}$_______
_______	_______	_______	$^{58}_{\_}Ni$
32	40	_______	_______

47. Write the isotope symbol for each of the following isotopes.
(a) Z = 12, neutrons = 13
(b) Z = 20, neutrons = 24
(c) Z = 33, neutrons = 42
(d) Z = 72, neutrons = 105

48. Write the isotope symbol for each of the following isotopes.
(a) Z = 18, neutrons = 22
(b) Z = 37, neutrons = 50
(c) Z = 50, neutrons = 70
(d) Z = 46, neutrons = 64

49. Which isotope of magnesium is most likely to be in the highest percentage?
(a) Mg-24
(b) Mg-25
(c) Mg-26

50. Which isotope of argon is most likely to be in the highest percentage?
(a) Ar-36
(b) Ar-38
(c) Ar-40

51. Which of the following statements about the production of line spectra is false?
(a) Electrons can absorb energy from being heated in a flame.
(b) Electrons are stable in excited states.
(c) Electrons release energy as they relax to the ground state.
(d) Photons are produced when electrons release energy.

52. Which of the following statements about the production of line spectra is false?
(a) The energy of the photon can have any value.
(b) The energy of the photon is related to the energy levels of the electron orbitals.
(c) Electrons can be excited by passing electricity through a gaseous vapor.
(d) Electrons always relax back to the ground state.

53. Calculate the frequency (Hz) of photons with the following wavelengths.
(a) 900 nm (c) 400 nm
(b) 650 nm (d) 330 nm

54. Calculate the frequency (Hz) of photons with the following wavelengths.
(a) 1075 nm (c) 770 nm
(b) 510 nm (d) 212 nm

55. Calculate the wavelength (nm) of photons with the following frequencies.
(a) 1.25×10^{15} Hz (c) 1.82×10^{15} Hz
(b) 1.33×10^{15} Hz (d) 2.20×10^{15} Hz

56. Calculate the wavelength (nm) of photons with the following frequencies.
(a) 1.00×10^{15} Hz (c) 1.39×10^{15} Hz
(b) 1.66×10^{15} Hz (d) 1.14×10^{15} Hz

57. Determine the energy of each photon in Problem 53.

58. Determine the energy of each photon in Problem 54.

59. Determine the energy of each photon in Problem 55.

60. Determine the energy of each photon in Problem 56.

61. Write the full electron configuration for an atom of each of the following elements.
(a) Pd (c) Br
(b) S (d) Cr

62. Write the full electron configuration for an atom of each of the following elements.
(a) Mg (c) K
(b) Ni (d) Al

63. Write the abbreviated electron configuration for an atom of each of the following elements.
(a) Cl (c) B
(b) Se (d) Cs

64. Write the abbreviated electron configuration for an atom of each of the following elements.
(a) P (c) Be
(b) Sr (d) Co

65. Which elements have the following electron configurations?
(a) $1s^22s^22p^63s^23p^64s^23d^{10}4p^65s^24d^4$
(b) $1s^22s^22p^63s^23p^1$
(c) $1s^22s^22p^63s^23p^64s^23d^2$
(d) $1s^22s^22p^63s^23p^64s^23d^7$

66. Which elements have the following electron configurations?
(a) $1s^22s^22p^63s^23p^64s^23d^{10}4p^65s^24d^{10}$
(b) $1s^22s^22p^63s^23p^64s^2$
(c) $1s^22s^22p^63s^23p^64s^23d^{10}4p^65s^1$
(d) $1s^22s^22p^4$

67. Which elements have the following electron configurations?
(a) $[Ne]3s^23p^2$
(b) $[Kr]5s^24d^{10}5p^5$
(c) $[Ne]3s^2$
(d) $[Kr]5s^24d^{10}5p^3$

68. Which elements have the following electron configurations?
(a) $[Ar]4s^23d^3$
(b) $[Ar]4s^23d^{10}4p^6$
(c) $[He]2s^22p^3$
(d) $[He]2s^1$

69. The electron configuration for an atom of each of the following elements is incorrect. Correct the errors.
(a) Y: $[Ar]5s^24d^1$
(b) Sc: $[Ar]4s^24d^1$
(c) Fe: $1s^22s^23p^64s^25p^66s^27d^6$
(d) F: $1s^22s^22p^6$

70. The electron configuration for an atom of each of the following elements is incorrect. Correct the errors.
(a) Ca: $1s^2s^2p^3s^3p^4s$
(b) P: $[Na]3s^23p^3$
(c) Ge: $[Ar]4s^24d^{10}4p^2$
(d) Sn: $1s^22s^22p^53s^23p^54s^23d^{10}4p^55s^24d^{10}5p^2$

Forensic Chemistry Problems

71. Strontium, a commonly found trace element in food, is incorporated into human bone structure because of its similarity in properties to calcium. Four stable isotopes of strontium are found in nature: ^{88}Sr (82.5%), ^{87}Sr (7.0%), ^{86}Sr (10.0%), and ^{84}Sr (0.5%). However, the actual distribution of the isotopes tends to vary greatly from one geographic location to another. Strontium isotope analysis can be used to identify ancient migration patterns as well as the origin of commingled human remains in battlefield burial sites. Determine the number of protons, neutrons, and electrons in each isotope of strontium.

72. Thallium compounds have historically been used as rat poisons and inevitably also by people with murderous intentions. The author Agatha Christie is credited, perhaps unjustly, with initiating a wave of thallium poisonings when she described thallium poisoning in the 1961 book *The Pale Horse*. The symptoms of thallium poisoning can mimic other disorders. Thallium has two stable isotopes, thallium-203 and thallium-205. Determine the number of protons, neutrons, and electrons in each isotope of thallium.

73. Wolfgang Amadeus Mozart died on December 5, 1791, at the age of 35 after battling an illness for three months. Some suspected that he was being poisoned by a rival, but it is now believed that he had been accidentally poisoned by an antimony compound prescribed by doctors. Years later, Antonio Salieri, a contemporary composer and rival, while suffering from senility, is said to have confessed to murdering Mozart, but he would not have had access to Mozart to commit such a crime. Write the full electron configuration of an atom of antimony. Antimony is determined by monitoring the light emitted from an ICP at 206.8 nm. Calculate the energy and frequency corresponding to this wavelength.

74. A suspect is apprehended in an assault case and is found to have smears of makeup on his clothing. The victim was known to have been wearing a distinctive pearl-white makeup. The pearl effect in cosmetics is from a bismuth compound, BiOCl. Bismuth can be detected by monitoring the light emitted from an ICP at 223.1 nm. Calculate the energy and frequency corresponding to this wavelength.

75. Using the Internet as a resource, determine what other elements are commonly found in gunshot residue particles when analyzed by SEM-EDS.

Case Study Problems

76. In the latter portion of the Evidence Analysis case study (about the victim burned by fireworks), why was it important to analyze the pyrotechnical mixture from both an unburned firework and one that was burned in the laboratory? Could a conclusion have been drawn from an analysis of only the evidence recovered from the burn? Use your knowledge of the scientific method to help answer this question.

77. In the Evidence Analysis case study, the evidence recovered from the burn victim matched the elemental profile of the explosives used by the neighbors. Does this fact prove the neighbors guilty? Given the facts, provide an alternative explanation of the data that would exonerate the neighbors, if true.

CHAPTER 4

Chemical Evidence

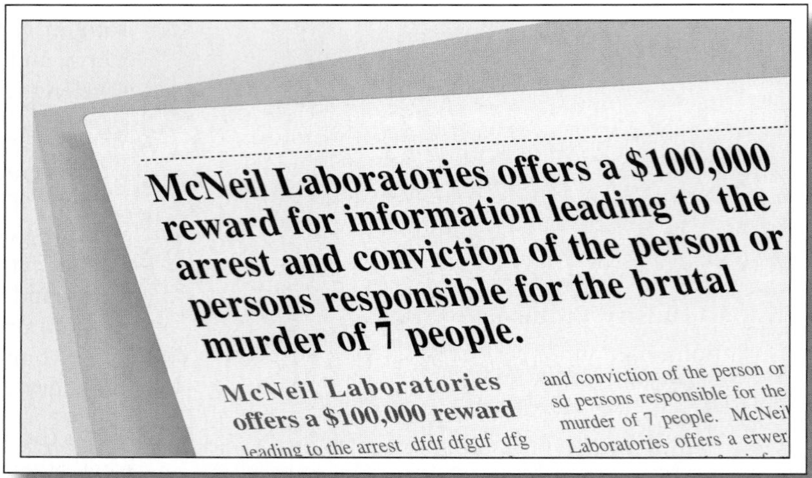

This reward has gone unclaimed for 24 years. The brutal murders gripped America in fear and changed forever the way companies do business.

CASE STUDY: A Killer Headache

In September of 1982, Adam Janus took the day off from his job as a postal supervisor to get a few things done around the house. He spent the morning running errands, picking up a neighbor's daughter from preschool, and getting flowers for his wife. Around lunchtime Adam began to experience some minor chest pains. He opened a new bottle of pain reliever and took a capsule. As Adam left the kitchen, he started feeling dizzy and nauseated, and his head began throbbing. Before he could get to a couch to lie down, he collapsed to the floor. His chest tightened and he felt as if something were ripping the oxygen out of his body. Within minutes he lay convulsing on the kitchen floor as his wife frantically called for an ambulance.

Adam was rushed to Northwest Community Hospital in the suburbs of Chicago, Illinois. He passed away at 3:15 P.M. despite the best efforts of Dr. Thomas Kim to revive him. Adam, at age 27, had mysteriously died leaving a grieving widow and two children. The utter shock of losing a family member so quickly and unexpectedly stunned the entire Janus family as they gathered at the hospital.

Adam's brother Stanley, two years younger, couldn't believe Adam was gone. The two brothers had always been close and looked out for each other. So many questions raced through Stanley's mind that he could hardly focus as he and his wife, Theresa, drove from the hospital to Adam's house where the Janus family agreed to gather to make funeral arrangements. The shock of losing his brother and the overwhelming stress gave Stanley a pounding headache.

When Stanley and Theresa arrived at Adam's house, the family was assembled around the kitchen table. Stanley spotted a bottle of pain reliever on the table, a bottle of Extra Strength Tylenol. He opened the bottle, took one, and offered the bottle to the rest of the family. The only other person who took one was his wife, Theresa.

The Janus family funeral for Adam, Stanley, and Theresa. *(Bruno Torres/Bettmann/Corbis)*

As Stanley walked around the kitchen he felt a bit dizzy, his stomach was turning over, and his headache was getting worse, if that was possible. His breathing quickened and he began to gasp for air. Within minutes he was on the floor convulsing. The Janus family went from a horrible tragedy into an unimaginable nightmare as they watched Stanley duplicate his brother's last moments. Paramedics were called to the house for the second time that day and arrived just before 6 P.M. to stabilize Stanley. In another grim twist of fate, Theresa fell to the floor in the kitchen as the paramedics worked on her husband. She, too, was falling victim to the same mysterious killer. The same doctor who had tried to save his brother pronounced Stanley dead at 8:15 that evening. Theresa survived for several days on life support before she succumbed.

In the early phase of the investigation, it became clear that something was poisoning the Janus family members, but what it was and how they were being exposed to it was unclear. It would take two off-duty firefighters from different firehouses, a doctor from a poison control center, and some quick laboratory work to determine how they died.

Who caused their deaths would remain a mystery, which, to this day, has never been solved . . .

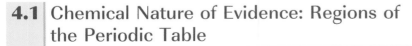

4.1 Chemical Nature of Evidence: Regions of the Periodic Table

Learning Objective

Describe the various regions of the periodic table and the terminology used to discuss the elements.

This chapter will extend the discussion of chemical compounds and formulas that began in Chapter 1. Here we will explore two broad classes

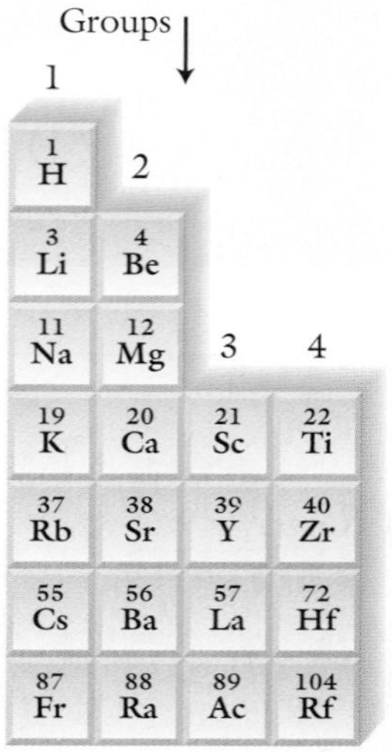

of compounds called *ionic* and *covalent* compounds, their formulas and names, and the kinds of reactions they undergo. But to understand these topics, more information on the periodic table is needed.

As we saw in Chapter 1, the periodic table is arranged so that elements with common properties and reactions are in the same column. Each of the columns is referred to as a **group** or **family**. Sodium and potassium, two elements that are in the same group, are both soft metals and form similar reaction products. Reactions 4.1 and 4.2 illustrate how sodium and potassium, respectively, undergo similar reactions with water to form hydrogen gas (H_2) and metal hydroxides (NaOH or KOH). Reactions 4.3 and 4.4 illustrate how magnesium and calcium, respectively, undergo similar reactions with water to form hydrogen gas and the resulting metal hydroxides ($Mg(OH)_2$ or $Ca(OH)_2$).

The elements within a group share many common traits, as illustrated previously with sodium and potassium. There are also differences between

$2Na + 2H_2O \rightarrow 2NaOH + H_2$
(Reaction 4.1)

$2K + 2H_2O \rightarrow 2KOH + H_2$
(Reaction 4.2)

$Mg + 2H_2O \rightarrow Mg(OH)_2 + H_2$
(Reaction 4.3)

$Ca + 2H_2O \rightarrow Ca(OH)_2 + H_2$
(Reaction 4.4)

(a, b: Charles D. Winters/Photo Researchers, Inc.; c: Martin F. Chillmaid/Science Photo Library/Photo Researchers, Inc.; d: Andrew Lambert Photography/Science Photo Library/Photo Researchers, Inc.)

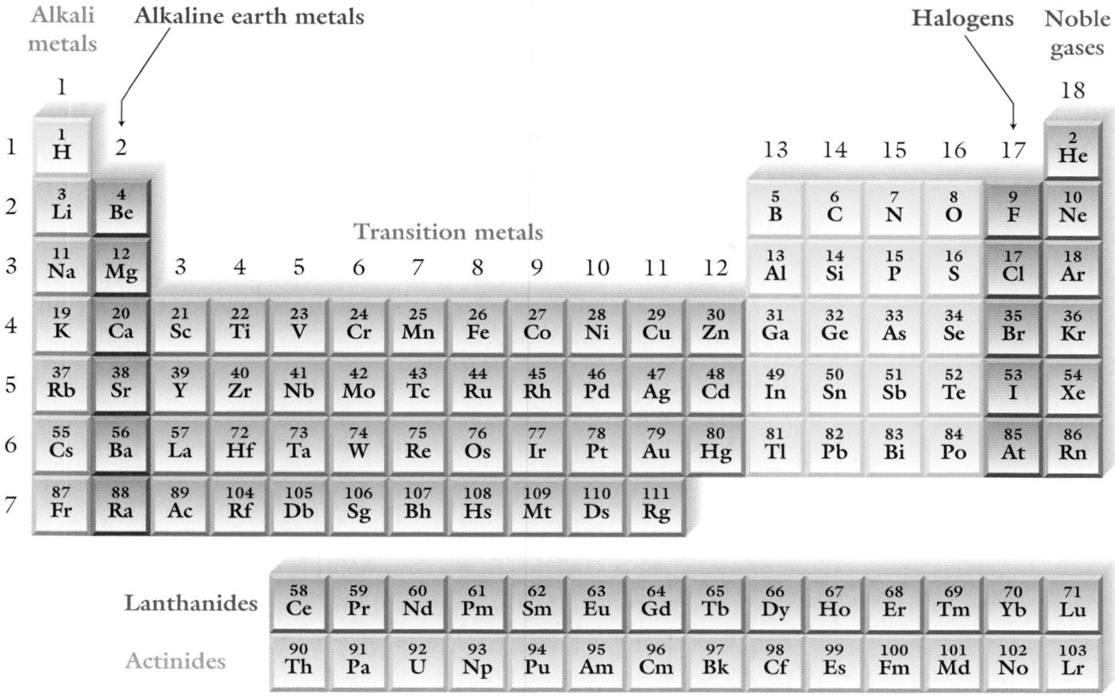

Figure 4.1 Regions of the periodic table.

members of a group such as the number of protons, neutrons, electrons, atomic mass, and isotopes of the elements.

There are three different methods for numbering the groups of the periodic table. This book uses the simplest version, which is also the method used by the International Union of Pure and Applied Chemistry, the body that governs changes to the periodic table. This method numbers the columns from 1 to 18, starting from the left and proceeding to the right, as shown in Figure 4.1

Another method for identifying elements is the use of group names. Except for hydrogen, all the elements of Group 1 of the periodic table (Li to Fr) are often referred to as the **alkali metals**. The elements of Group 2, Be to Ra, are called the **alkaline earth metals**. Not all groups are named individually. The elements of Groups 3 to 12 are collectively known as the **transition metals**. The elements of Groups 13 to 16 do not have a common name. Continuing across the periodic table, Group 17 elements are known as the **halogens,** and finally, Group 18 elements are designated the **noble gases**. Figure 4.1 shows a color-coded periodic table representing the members of each group with group names indicated.

The rows of the periodic table are referred to as **periods**. The periods are simply numbered from 1 to 7, as shown on the left side of Figure 4.1. It should be noted that the two unnumbered rows at the bottom of the table actually belong with the transition metals and are called the **inner transition metals**. The elements with atomic numbers 58–71 belong after lanthanum (La), which has atomic number 57. For this reason they are called the *lanthanides*. Similarly, the elements with atomic numbers 90–103 would fit after actinium (Ac, atomic number 89) and are called the *actinides*. However, inserting these elements would make the periodic table awkward to use, so they are set apart at the bottom.

Worked Example 1

Lead, arsenic, antimony, barium, and copper are common components of gunshot residue. Identify the group number, group name (if any), and period of each component.

SOLUTION Locate each element on the periodic table and then locate the column number at the top and period number at the side of the table. The common names can be identified from Figure 4.1.

Lead (Pb): Group 14, no common group name, Period 6
Arsenic (As): Group 15, no common group name, Period 4
Antimony (Sb): Group 15, no common group name, Period 5
Barium (Ba): Group 2, alkaline earth metals, Period 6
Copper (Cu): Group 11, transition metals, Period 4

Practice 4.1

The following elements are part of compounds commonly found as major components in fingerprint residue: Na, Ca, Fe, and Br. Identify the group number, common group name, and period number.

ANSWER
Sodium (Na): Group 1, alkali metal, Period 3
Calcium (Ca): Group 2, alkaline earth metal, Period 4
Iron (Fe): Group 8, transition metal, Period 4
Bromine (Br): Group 17, halogens, Period 4

Worked Example 2

Identify the following elements by their position on the periodic table.

(a) The element used to provide thermal shock resistance in glass is located in Group 13, Period 2.

(b) The element that is typically found in high concentrations in common window glass is located with the alkaline earth metals, Period 4.

(c) Crystal glass has added to it high levels of this potentially toxic element, which is located in Group 14, Period 6.

SOLUTION

(a) Count over 13 columns and down 2 rows on the periodic table to get to boron. Note: The periods are numbered on the left side of the periodic table. Boron is in Period 2, despite being the first element of Group 13.

(b) Alkaline earth metals are Group 2. The fourth row down is calcium.

(c) Counting over 14 columns and down 6 rows gives the element lead.

Practice 4.2

Identify the following elements, which are all found in clandestine drug labs.

(a) Group 13, Period 3 element, used to make screening smoke.

(b) Alkaline earth metal in Period 3, used for making infrared decoy flares.

(c) Group 14, Period 3 element, used as a priming agent in explosives.

ANSWER

(a) Aluminum

(b) Magnesium

(c) Silicon

4.2 Types of Compounds: Ionic Compounds

Learning Objective

Write the name and determine the formulas of ionic compounds.

In Chapter 1, a compound was defined as a pure substance that is composed of two or more elements bonded together. We will now consider compounds in more detail. There are two types of chemical bonds that form between atoms within a compound. Each bonding type imparts different physical and chemical properties to the compound. The two types of compounds must then be handled differently when being collected and analyzed as evidence. Each type also has its own distinct set of rules for naming compounds.

If the bond between two atoms forms by one atom transferring an electron (or electrons) to the other atom, the resulting compound is an **ionic compound**. When an atom loses or gains electrons during the formation of an ionic compound, it becomes electrically charged and is referred to as an **ion**. More specifically, the loss of an electron leaves an atom with a +1 charge (recall that electrons have a −1 charge), thus forming a positively charged ion called a **cation**. The gain of an electron gives an atom a −1 charge, forming a negatively charged ion called an **anion**. An **ionic bond** is the attractive force between the positively charged cation and negatively charged anion.

Ionic compounds are found as solid powders with a variety of colors. Pictured here (clockwise from top center): iron(III) sulfate, $Fe_2(SO_4)_3$; copper(II) sulfate, $CuSO_4$; copper(II) carbonate, $CuCO_3$; sodium chloride, NaCl; and iron(II) sulfate, $FeSO_4$. (Andrew Lambert Photography/Science Photo Library/Photo Researchers, Inc.)

Naming Ionic Compounds

Ionic compounds always contain at least one cation and one anion. The name of an ionic compound begins with the name of the cation followed by the name of the anion. The rule for naming a cation is quite simple: Write the name of the element followed by the word *ion*. The reaction below shows the formation of the sodium ion from elemental sodium. Notice that the charge of the sodium ion is indicated as a superscript following the atomic symbol, as shown below.

$$Na \rightarrow Na^+ + e^- \qquad \textbf{(Reaction 4.5)}$$

The rule for naming an anion is to change the ending of the element name with the suffix *-ide* and add the word *ion*. The next reaction shows the formation of the oxide ion from atomic oxygen. For any ion with a charge magnitude greater than 1, the number appearing before the positive (+) or negative (−) charge indicates the magnitude of the charge.

$$O + 2e^- \rightarrow O^{2-} \qquad \textbf{(Reaction 4.6)}$$

Worked Example 3

Name each of the following ions, which are major components in fingerprint residue:

(a) K^+

(b) Br^-

(c) Ca^{2+}

(d) I^-

SOLUTION

(a) Cations have the same name as the element but with *ion* added: potassium ion.

(b) Anions have the ending of the element name modified with *-ide*: bromide ion.

(c) Cations have the same name as the element but with *ion* added: calcium ion.

(d) Anions have the ending of the element name modified with *-ide*: iodide ion.

Practice 4.3

Name each of the following ions:

(a) Li^+: An ingredient in the original recipe for 7-Up that was removed in the 1950s when scientists discovered it had a psychological effect (it is used to treat manic-depressive disorders).

(b) F^-: If detected in urine, it is a sign of inhalant abuse of solvents containing fluorine.

(c) Cl^-: Low levels in postmortem urine analysis can indicate that the victim experienced severe vomiting or diarrhea before death; high levels could indicate use of certain prescription drugs called *corticosteroids* for asthma or allergies.

ANSWER
(a) Lithium ion
(b) Fluoride ion
(c) Chloride ion

Determining Formulas of Ionic Compounds

The periodic table provides assistance in determining whether an atom will form a cation or anion and, to some degree, what the charge of the ion will be. Metals tend to lose electrons and form cations, while nonmetals tend to gain electrons and form anions. (The metals and nonmetals can be distinguished by their positions on the periodic table in Figure 1.6.)

The charges of ions for some groups of elements follow a predictable pattern. For example, the alkali metals (Group 1) form cations that have a charge of +1. Cations formed by the alkaline earth metals (Group 2) have a +2 charge. The Group 16 elements typically form −2 anions, and the halogens (Group 17) form −1 anions. The noble gases do not form ions.

As for other elements with charges that tend to be consistent, silver in Group 11 always forms a +1 cation, zinc and cadmium in Group 12 form +2 cations, and aluminum in Group 13 forms a +3 cation. Finally, nitrogen and phosphorus in Group 15 tend to form −3 anions. It may be helpful for you to write the common charges of the ions on a copy of the periodic table, as shown in Figure 4.2, for reference purposes while

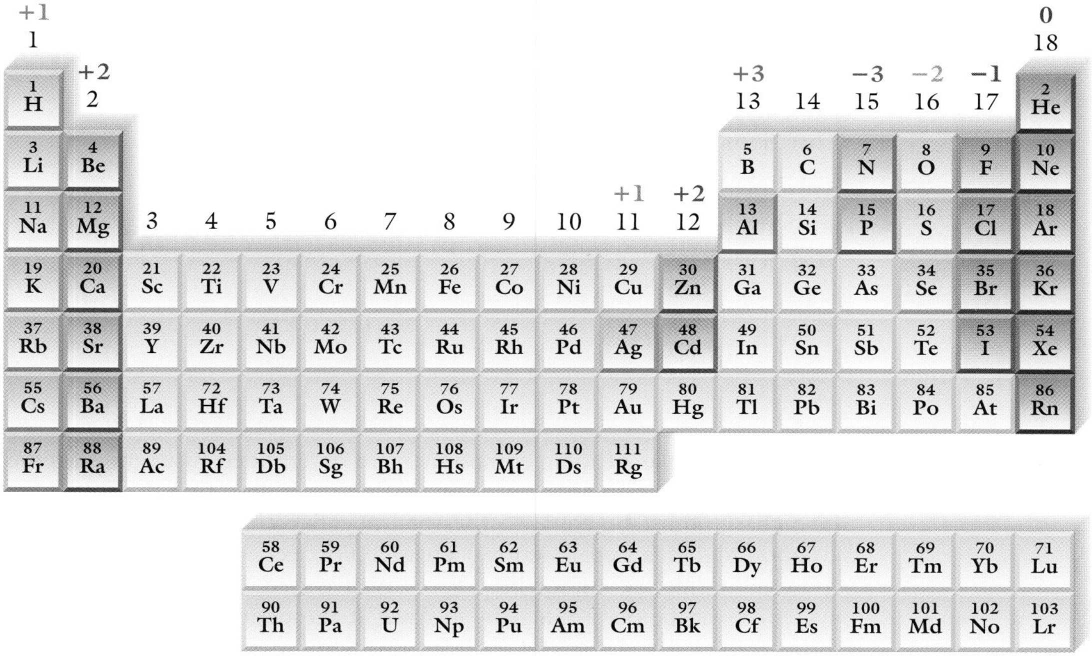

Figure 4.2 The charges of ions can be predicted based on the location of the elements on the periodic table. The noble gas elements do not form ions because of the stability of having filled the outermost *s* and *p* orbitals with eight electrons. The stability of having eight valence electrons is called the *octet rule* and explains why the ionic charges of the elements follow trends. All halogens tend to gain one electron, forming a −1 anion, to fill the outer *s* and *p* orbitals. For the alkali metals, it is easier to lose one electron than to gain seven electrons to reach the octet. The octet rule will be studied in greater depth in Chapter 8.

doing homework. The remaining elements not listed above can have several different charges and will be discussed later in the chapter.

Worked Example 4

Predict the charge on the following ions, which can be found in the analysis of glass fragments recovered from a crime scene.

(a) Barium ion

(b) Aluminum ion

(c) Lithium ion

(d) Oxide ion

SOLUTION

(a) Ba is a member of Group 2, alkaline earth metals, and forms ions with a +2 charge.

(b) Al is a member of Group 13 and forms ions with a +3 charge.

(c) Li is a member of Group 1, alkali metals, and forms ions with a +1 charge.

(d) O is a member of Group 16 and forms ions with a −2 charge.

Practice 4.4

Predict the charge on the following ions found in cerebrospinal fluid, which can serve as markers for determining time since death, although more accurate methods are generally available.

(a) Potassium ion

(b) Magnesium ion

(c) Sodium ion

(d) Chloride ion

ANSWER

(a) +1

(b) +2

(c) +1

(d) −1

The formulas of ionic compounds are determined by examining the charges of the cation and the anion and remembering that the total positive charge must be equal to the total negative charge so that the final compound has a zero (neutral) charge. For example, the formula for sodium fluoride, a common blood preservative used when collecting blood evidence, can be determined by looking at the periodic table and recalling the rules for charges of each group. Sodium forms the cation Na^+ and fluorine forms the anion F^-. Because the +1 and −1 charges sum to zero, the formula for sodium fluoride is NaF. Remember that both elements exist as charged ions within the compound, which is neutral overall. To determine the formula for calcium chloride, a common component in fingerprint residue, note that calcium belongs to the Group 2 elements, which form +2 cations, and that chlorine forms a −1 anion. The final formula must have a zero charge. Therefore, the formula must have two chloride ions for each calcium ion, forming $CaCl_2$.

To determine the formula for ionic compounds when the charges are not equal, it is necessary to find the lowest common multiple of the ionic charges. For example, Ca^{2+} and P^{3-} have a lowest common multiple of 6 ($2 \times 3 = 6$). Therefore, the formula would require three Ca^{2+} ions and two P^{3-} ions to form the neutral compound Ca_3P_2. If one carefully examines the magnitude of the ion charges, those values indicate the number of atoms of the opposite ion needed, as shown below for Ca_3P_2. Always verify that the formula produces a neutral (zero charge) compound and the least common multiple is used.

$$Ca^{2+}\ P^{3-} \rightarrow Ca_3P_2 \qquad \text{Because:} \quad \begin{array}{r} 3 \times Ca^{2+} = +6 \\ 2 \times P^{3-} = -6 \\ \hline 0 \end{array}$$

Worked Example 5

Predict the formulas for the following ionic compounds formed from the elements indicated below.

(a) A compound formed with K and Br that can detect blood in a stain by reacting with it to form distinctive crystals.

(b) A compound formed with Ca and O, called *quicklime*, mistakenly believed to accelerate the decomposition of human remains.

(c) A compound formed with Al and O and produced from the use of H-6 military grade explosives.

SOLUTION

(a) KBr: The +1 charge of K^+ and the −1 charge of Br^- balance to zero.

(b) CaO: The +2 charge of Ca^{2+} and the −2 charge of O^{2-} balance to zero.

(c) Al_2O_3: The +3 charge of Al^{3+} and the −2 charge of O^{2-} must balance to zero; this requires two $Al^{3+}$and three O^{2-}.

Practice 4.5

Predict the formula for the ionic compound formed from the elements indicated below.

(a) A compound formed with Al and Cl and used as a dehydrating agent in autopsied bodies.

(b) A compound formed with Ca and Cl and found in small amounts in H-6 military grade explosives.

(c) A compound formed with K and I that makes distinctive branching needlelike crystals when it reacts with the drug phencyclidine (PCP).

ANSWER

(a) $AlCl_3$

(b) $CaCl_2$

(c) KI

Drawing upon your personal experiences, you probably already know the rule for naming ionic compounds. What is the chemical name and formula for table salt? The answer is, of course, sodium chloride, or NaCl, which consists of the sodium ion and the chloride ion. Simply combine the names of the cation with the anion, omitting the word *ion* from each.

Worked Example 6

Name the compounds from Worked Example 5.

SOLUTION

(a) Potassium ion and bromide ion form the compound potassium bromide.

(b) Calcium ion and oxide ion form the compound calcium oxide.

(c) Aluminum ion and oxide ion form the compound aluminum oxide.

Practice 4.6

Name the compounds from Practice 4.5.

ANSWER

(a) Aluminum chloride

(b) Calcium chloride

(c) Potassium iodide

Cations with Multiple Charges and Polyatomic Ions

As shown in Figure 4.2, the majority of transition metals as well as the metals of Group 13 through Group 15 did not have a common charge. These elements will form cations; however, the charges of the ions cannot be easily predicted, and many of the elements can form ions with different charge magnitudes. Table 4.1 lists some of the common elements and the various charges their cations can have.

Different ions of the same element can have greatly different properties. For example, Cr^{6+} is a cancer-causing pollutant used commercially in the electroplating and steel refining industries; it has been the subject of several high-profile lawsuits. However, the Cr^{3+} ion is an essential nutrient in our diets. The names of compounds containing transition metal cations, or cations from the metals found in Groups 13 through 15 that

Table 4.1 Cations with Multiple Common Charges

Element	Common Ionic Forms
Iron	Fe^{2+}, Fe^{3+}
Chromium	Cr^{2+}, Cr^{3+}, Cr^{6+}
Cobalt	Co^{2+}, Co^{3+}
Copper	Cu^{1+}, Cu^{2+}

Pictured here are two chloride salts of copper: copper(II) chloride, $CuCl_2$ (bottom left); and copper(I) chloride, CuCl (top right). (Richard Megna/ Fundamental Photographs)

do not follow a pattern, should indicate which ion of the element is present in the compound.

The method for distinguishing the ions from each other is to place the charge of the ion in Roman numerals after naming the metal element. For example, Cr^{3+} is named chromium(III) ion, and the Cr^{6+} ion is named chromium(VI) ion. When naming a compound, the Roman numeral charge is included as part of the name. Thus, the name of $CrCl_3$ is chromium(III) chloride.

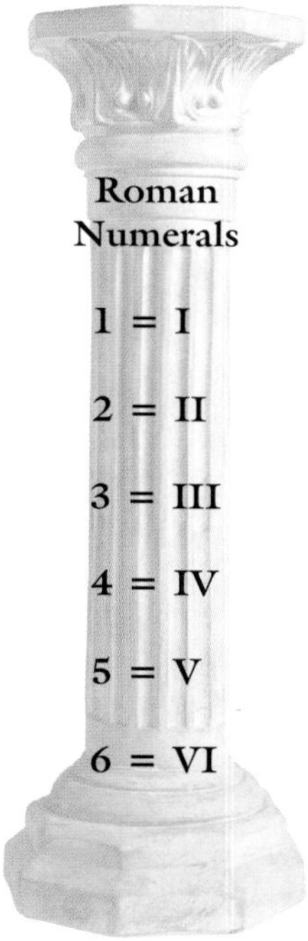

(Getty Images)

Worked Example 7

Predict the formula and name the compound formed from the following ions:

(a) Au^{2+} and Cl^- form a compound prescribed in the late 1800s to combat alcoholism.

(b) Cu^{2+} and Cl form an ingredient that is used to form crystals with the illegal drug GHB.

(c) Fe^{3+} and Cl^- form a compound used to restore serial numbers from stainless steel.

SOLUTION

(a) Two chloride ions are needed to balance the +2 charge on the gold(II) ion: $AuCl_2$, gold(II) chloride.

(b) Two chloride ions are needed to balance the +2 charge on the copper(II) ion: $CuCl_2$, copper(II) chloride.

(c) Three chloride ions are needed to balance the +3 charge on the iron(III) ion: $FeCl_3$, iron(III) chloride.

Practice 4.7

Predict the formula and name the compound formed from the following ions:

(a) Cu^{2+} and Cl^- form a compound used to restore serial numbers on cast iron and steel.

(b) Ti^{4+} and O^{2-} form a compound used as a pigment for white paint that wasn't used until the 1800s. One method of detecting forged artwork is to look for this compound on works of art supposedly from earlier periods.

(c) Hg^{2+} and Cl^- form a compound that produces rosettelike crystals when mixed with heroin.

ANSWER

(a) $CuCl_2$, copper(II) chloride

(b) TiO_2, titanium(IV) oxide

(c) $HgCl_2$, mercury(II) chloride

Thus far we have discussed only ions that are formed when one element gains or loses electrons. These are commonly referred to as **monatomic ions**, meaning ions formed from a single atom. There is another group of ions called **polyatomic ions**. These are groups of atoms bound together very tightly and acting as a single particle that has an electrical charge.

One of the most infamous polyatomic ions in forensic science is the cyanide ion, which has the formula CN^-. The cyanide ion consists of a carbon atom tightly bonded to a nitrogen atom in a unit that has an overall -1 charge. It is important to note that the carbon and nitrogen atoms stay bonded together under typical conditions in compounds containing the cyanide ion. Table 4.2 has a list of common polyatomic ions.

Table 4.2 Polyatomic Ions

Polyatomic Ion	Symbol
Ammonium ion	NH_4^+
Nitrate ion	NO_3^-
Hydroxide ion	OH^-
Acetate ion	$C_2H_3O_2^-$
Cyanide ion	CN^-
Permanganate ion	MnO_4^-
Chlorate ion	ClO_3^-
Carbonate ion	CO_3^{2-}
Sulfate ion	SO_4^{2-}
Chromate ion	CrO_4^{2-}
Phosphate ion	PO_4^{3-}

Worked Example 8

Predict the formula for the compounds consisting of the following ions:

(a) K^+ and NO_3^- form a compound that is an ingredient in pyrotechnic igniters.

(b) Fe^{2+} and NO_3^- form a compound used to develop fingerprints.

(c) Na^+ and CN^- form a substance that is a deadly poison.

SOLUTION

(a) KNO_3: The $+1$ and -1 ion charges balance to zero.

(b) $Fe(NO_3)_2$: The Fe^{2+} requires two NO_3^- to balance the charge to zero.

(c) NaCN: The +1 and −1 ion charges balance to zero.

Practice 4.8

Name the compounds shown in the solution to Worked Example 8.

ANSWER

(a) Potassium nitrate

(b) Iron(II) nitrate

(c) Sodium cyanide

Worked Example 9

Paint chip analysis is commonly performed with a SEM-EDS instrument for the following compounds. Predict their formulas.

(a) Calcium sulfate, a cheap whitening agent

(b) Iron(II) cyanide, a blue pigment

(c) Zinc phosphate, a corrosion-resistant white pigment

SOLUTION

(a) $CaSO_4$: Calcium always forms +2 ions and sulfate is SO_4^{2-}.

(b) $Fe(CN)_2$: Iron(II) is Fe^{2+} and cyanide is CN^-.

(c) $Zn_3(PO_4)_2$: Zn always forms +2 ions and phosphate is PO_4^{3-}. To achieve a formula with a neutral charge, a lowest common multiple of 6 is needed. This requires three Zn^{2+} ions and two PO_4^{3-} ions.

Practice 4.9

Determine the charge on each ion within the following compounds.

(a) $Fe(NO_3)_3$

(b) $CuCl_2$

(c) $Cd_3(PO_4)_2$

ANSWER

(a) Fe^{3+}, NO_3^-

(b) Cu^{2+}, Cl^-

(c) Cd^{2+}, PO_4^{3-}

Properties of Ionic Bonds

The strength of an ionic bond comes from the attraction between the opposite charges of the cation and the anion. The positive cation is attracted to the negative anion, just as the south pole of one magnet is attracted to the north pole of another. The ions are held together in fixed positions within a crystal structure. The crystal structure has a three-dimensional, repeating pattern called the **crystal lattice** in which the cations and anions are arranged. Shown in Figure 4.3 is the crystal lattice for KBr. Note that the crystal is made up of many potassium ions and many bromide ions. There is not actually a single KBr unit comparable to a molecule in

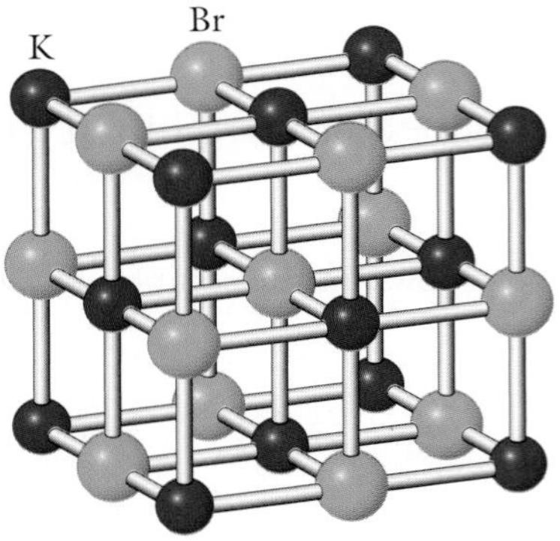

Figure 4.3 The lattice structure of KBr shows the three-dimensional repeating units of a crystal.

a covalent compound. (The term *formula unit* rather than *molecule* is used to describe the simplest ratio of the ions that make up an ionic compound.) Each ion is effectively held in place by the attractive forces of the other ions surrounding it on all sides. The crystal lattice imparts the properties of hardness and rigidity that ionic compounds share.

The **melting point** for an ionic compound is the temperature at which the ions have sufficient energy to escape the attractive forces holding them in the lattice structure. When the melting point is reached, the crystal lattice collapses and a liquid is formed in which the ions are able to move past one another. In general, the melting point of an ionic compound is very high and increases with the magnitude of the charge on the ions.

Worked Example 10

The melting point of CaO is 2572°C and that of KBr is 730°C. Explain why there is such a large difference in melting points.

SOLUTION

Because CaO has the higher melting point, the Ca^{2+} ions must be attracted more strongly to the O^{2-} ions than the K^{+} ions are attracted to the Br^{-} ions. This makes sense because the attractive force between a +2 charge and a −2 charge would be stronger than the force between a +1 charge and a −1 charge.

Practice 4.10

Place the following compounds in order from lowest melting point to highest melting point:

(a) AlP

(b) NaCl

(c) MgO

ANSWER

NaCl < MgO < AlP

4.3 Types of Compounds: Covalent Compounds

Learning Objective

Write the names and formulas of covalent compounds.

The second form of bonding in compounds occurs when two atoms share electrons. This sharing is referred to as a **covalent bond;** compounds that have this type of bonding are called *covalent compounds.* Covalent bonding occurs when the atoms from two nonmetal elements form a bond, such as H_2S, CO_2, and NH_3. The covalent compounds discussed in this chapter will be limited to binary (two-element) compounds. A much larger class of covalent compounds consists of the organic compounds, which are covered in depth in Chapter 7.

How exactly are electrons shared between two atoms? Recall that the electrons are located in orbitals that describe the area of highest probability of finding the electron. If atoms are to share electrons, then their orbitals must overlap with one another. In the reaction shown in Figure 4.4a, a covalent bond forms between the two hydrogen atoms by the overlap of *s* orbitals to form hydrogen gas (H_2). In a similar fashion, the reaction shown in Figure 4.4b illustrates the overlap of a *p* orbital of fluorine with an *s* orbital of hydrogen to form HF.

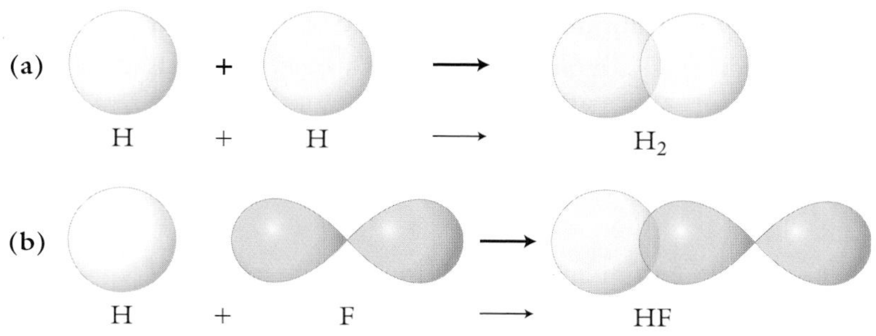

Figure 4.4 The formation of covalent bonds occurs through the overlap of electron orbitals from each of the atoms participating in the bond.

Names and Formulas of Covalent Compounds

Determining the name and formulas of covalent compounds differs greatly from ionic compounds, as one cannot use the periodic table to make predictions for covalent compounds. Instead, the number of atoms of each element in the covalent compound must be specified in the name. As an example, consider the covalent compound carbon dioxide, CO_2. The name of the first element (carbon) is given first, followed by the name of the second element (oxygen) with an *-ide* ending and a prefix that indicates the number of atoms of each element. The prefixes used are listed in Table 4.3. Thus, CO_2 is named *carbon dioxide.* It is not named *monocarbon dioxide* as it is customary to omit the prefix *mono-* if the first element in the formula has only one atom. The prefix identifying the number of oxygen atoms in the compound is needed because carbon can combine with oxygen to form carbon monoxide (CO) and the name must clearly distinguish one from the other.

If the name of an element starts with a vowel, omit the letters *a* or *o* from the end of the prefix for ease of pronunciation (as was shown above for carbon monoxide). However, do not omit the letter *i* in the prefixes *tri-* or *di-*, as in carbon dioxide.

Table 4.3 Prefixes for Covalent Compounds

Prefix	Number
mono-	1
di-	2
tri-	3
tetra-	4
penta-	5
hexa-	6
hepta-	7
octa-	8
nona-	9
deca-	10

Worked Example 11

Predict the formula for the following compounds:

(a) Carbon tetrachloride, a solvent abused as an inhalant

(b) Sulfur dioxide, a radioactive form of which can be used to detect fingerprints on fabrics and adhesives.

(c) Dihydrogen monosulfide, a compound produced during the bloating stage of bodily decomposition.

SOLUTION

(a) One carbon atom and four chlorine atoms form CCl_4.

(b) One sulfur atom and two oxygen atoms form SO_2.

(c) Two hydrogen atoms and one sulfur atom form H_2S.

Practice 4.11

Name the following compounds:

(a) PF_5

(b) SO_3

(c) SF_2

(d) ICl

ANSWER

(a) Phosphorus pentafluoride

(b) Sulfur trioxide

(c) Sulfur difluoride

(d) Iodine monochloride

Worked Example 12

In Section 3.3, the law of multiple proportions was illustrated using various compounds composed of nitrogen and oxygen. Name each of the following nitrogen/oxygen compounds.

(a) NO

(b) N_2O

(c) N_2O_3

SOLUTION

(a) *Mono-* represents one atom, but *mono-* is not used before the first element in a compound's name. Use *mono-* only in front of the second element. The name of NO is nitrogen monoxide.

(b) *Di-* represents two atoms, *mono-* represents one atom. The name of N_2O is dinitrogen monoxide.

(c) *Di-* represents two atoms, *tri-* represents three atoms. The name of N_2O_3 is dinitrogen trioxide.

Practice 4.12

Name each of the following nitrogen/oxygen compounds.

(a) NO_2

(b) N_2O_5

(c) N_2O_4

ANSWER

(a) Nitrogen dioxide

(b) Dinitrogen pentoxide

(c) Dinitrogen tetroxide

Properties of Covalent Compounds

The sharing of electrons that occurs in covalent compounds is not necessarily an equal sharing of electrons. The tendency of elements to pull electrons within a bond closer to themselves is a property called **electronegativity**. Because elements differ in their electronegativity, the shared electrons in a covalent bond tend to be found closer to the more electronegative element. The periodic table can be used to predict the relative electronegativity of the elements because electronegativity increases from bottom to top within a group and from left to right within a period. The most electronegative element is fluorine at the top of Group 17 on the upper right of the table. (The noble gas elements rarely form compounds and we do not concern ourselves with their electronegativity.) Chapter 6 will go into further detail on the effects of electronegativity.

When a covalent compound is in the solid state, the attractive forces between molecules are weak. The addition of only a small amount of energy is needed to melt the solid. For this reason, the melting point of most covalent compounds is typically low as compared with that of ionic compounds. For example, the covalent compound HF has a melting point of $-35°C$, whereas the ionic compound NaF has a melting point of $995°C$. A similar pattern exists with the boiling points of covalent and ionic compounds.

Summary of Rules for Naming Ionic and Covalent Compounds

It is quite common for beginning students to confuse the rules for naming ionic compounds with the rules for naming covalent compounds,

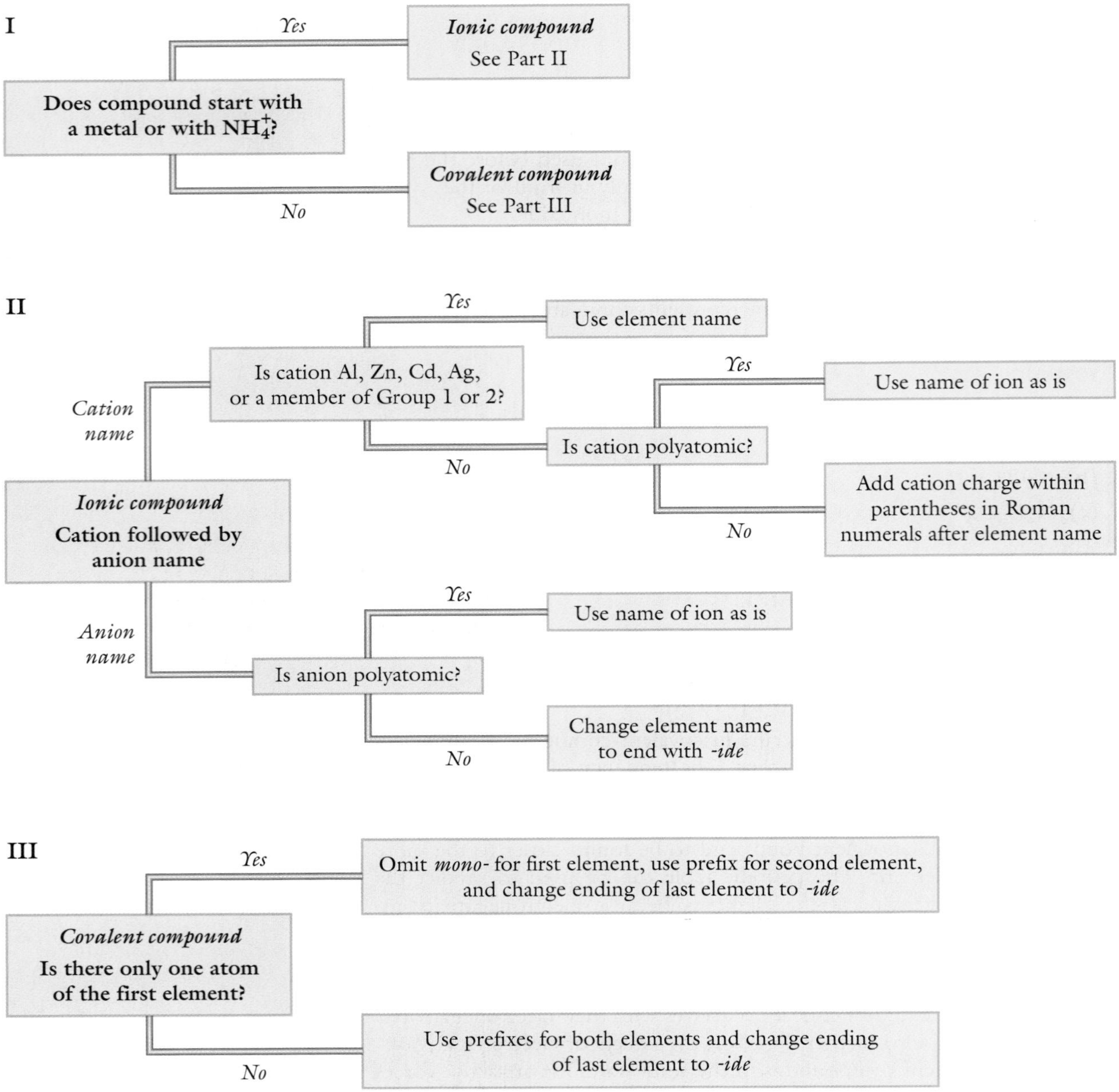

especially under the stress of an exam. The best way to prevent this is, of course, to practice! The second is to remember the rules and follow a systematic method for determining the correct name. Listed in the flowchart are some helpful questions to ask while naming compounds.

Learning Objective

Designate chemical compounds when common names are used.

4.4 Common Names and Diatomic Elements

Historically, many compounds have been given names that were not as systematically determined as those described in the previous sections. Despite the implementation of the standard rules for naming compounds,

Pictured here (clockwise from bottom left): water (dihydrogen monoxide, H_2O), alcohol (ethanol, C_2H_5OH), muriatic acid (hydrochloric acid, HCl), salt (sodium chloride, NaCl), and lye (sodium hydroxide, NaOH). (Richard Megna/Fundamental Photographs)

some of these older names are still used in industry and laboratories around the world. The best example is water, which has the formula H_2O. The systematic name for water is dihydrogen monoxide, which is not used because it would cause more confusion than clarity. Another name commonly used is *salt*, which most people interpret to mean *table salt* with the formula NaCl. However, the term **salt** scientifically refers to a large class of ionic compounds (all those that do not have H^+ or OH^- as one of the ions), not just NaCl. Sodium cyanide fits the definition of a salt and could be deadly if mistaken as table salt!

A well-known compound for which a nonscientific name continues to be widely used is alcohol. In common use, *alcohol* refers specifically to ethanol, the alcohol found in beer, wine, and hard liquors. However, to a chemist the term *alcohol* refers to a whole class of organic compounds that have an $-OH$ group covalently bonded to a carbon atom. The U.S. government taxes ethanol for human consumption very highly, but ethanol that is to be used for industrial purposes or scientific research is exempt from this tax. To make sure that the ethanol being sold for the latter purposes is not being sold on the black market for human consumption, the suppliers mix a small amount of methanol into the ethanol. Methanol is an alcohol that can cause a person to become desperately sick, can cause permanent blindness, and is potentially lethal. This mixture of ethanol and methanol is called **denatured alcohol.**

Seven elements in their natural states are found combined with another atom of the same element, forming diatomic (two-atom) molecules. For example, oxygen gas is known to exist as O_2, not simply O. Anyone referring to oxygen in ordinary conversation is usually speaking about molecular oxygen (O_2) and not atomic oxygen (O). The seven diatomic elements are hydrogen (H_2), nitrogen (N_2), oxygen (O_2), fluorine (F_2), chlorine (Cl_2), bromine (Br_2), and iodine (I_2). It can be helpful to remember that the elements themselves appear to form the number seven on the periodic table, as shown in Figure 4.5.

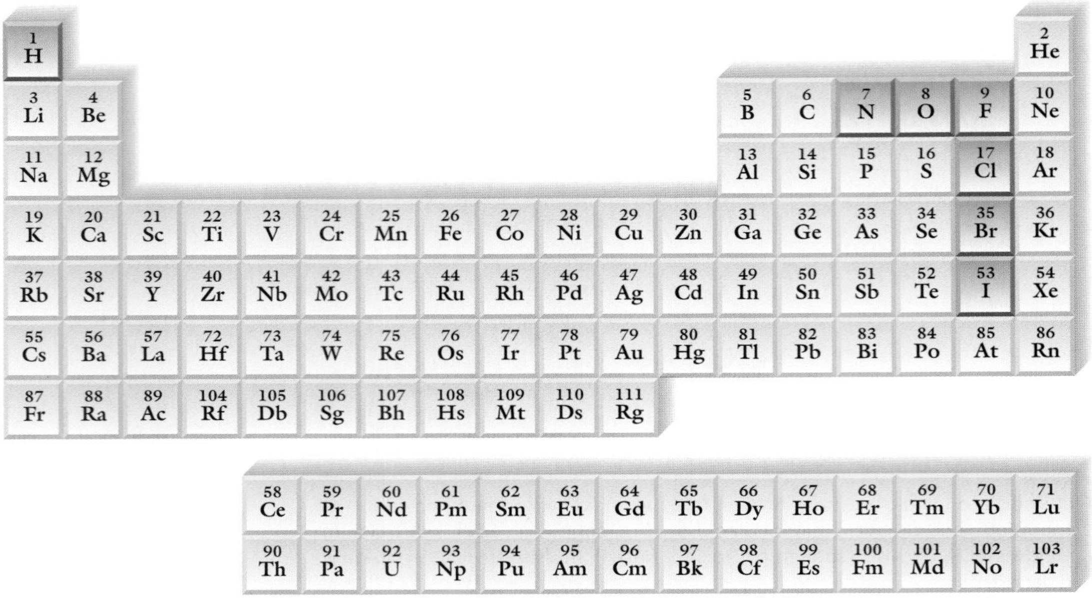
1 H
2 He
3 Li 4 Be 5 B 6 C 7 N 8 O 9 F 10 Ne
11 Na 12 Mg 13 Al 14 Si 15 P 16 S 17 Cl 18 Ar
19 K 20 Ca 21 Sc 22 Ti 23 V 24 Cr 25 Mn 26 Fe 27 Co 28 Ni 29 Cu 30 Zn 31 Ga 32 Ge 33 As 34 Se 35 Br 36 Kr
37 Rb 38 Sr 39 Y 40 Zr 41 Nb 42 Mo 43 Tc 44 Ru 45 Rh 46 Pd 47 Ag 48 Cd 49 In 50 Sn 51 Sb 52 Te 53 I 54 Xe
55 Cs 56 Ba 57 La 72 Hf 73 Ta 74 W 75 Re 76 Os 77 Ir 78 Pt 79 Au 80 Hg 81 Tl 82 Pb 83 Bi 84 Po 85 At 86 Rn
87 Fr 88 Ra 89 Ac 104 Rf 105 Db 106 Sg 107 Bh 108 Hs 109 Mt 110 Ds 111 Rg
58 Ce 59 Pr 60 Nd 61 Pm 62 Sm 63 Eu 64 Gd 65 Tb 66 Dy 67 Ho 68 Er 69 Tm 70 Yb 71 Lu
90 Th 91 Pa 92 U 93 Np 94 Pu 95 Am 96 Cm 97 Bk 98 Cf 99 Es 100 Fm 101 Md 102 No 103 Lr

Figure 4.5 The seven diatomic elements (shown in pink).

4.5 Basics of Chemical Reactions

Learning Objective

Employ the basic terminology of chemical reactions.

One of the key methods for identifying the type of evidence at a crime scene or in a laboratory is the use of chemical reactions. For example, when detectives are attempting to determine whether a white powder found on a kitchen table is an illegal narcotic or an innocent baking ingredient, they can conduct a field test for the presence of illegal drugs. A field test requires a small amount of the suspected drug, which is mixed with chemicals that are known to react with illegal drugs. The chemical reactions will produce colored compounds with illegal drugs but not with most other substances.

Officers can make a preliminary identification of illegal drugs by placing a small sample in a field test reaction vial. A positive reaction is indicated by the formation of a product whose color matches that shown on the vial—orange for amphetamines, red for cannabis (marijuana), pink for LSD, and blue for cocaine. (James King-Holmes/Science Photo Library/Photo Researchers, Inc.)

Contrary to what you may see in Hollywood movies, detectives do not taste a substance to identify it. This practice would endanger their lives and would not provide reliable or useable evidence. It is standard practice in laboratories to forbid not only the tasting of chemicals but the consumption of any food or beverages within the laboratory because these items could become contaminated with laboratory chemicals.

Chemical reactions are represented symbolically by writing a chemical equation. The compounds that react with each other are collectively known as the **reactants** and are always listed on the left side of the chemical equation. The compounds that are formed as a result of the reaction are called the **products** and appear on the right side of the equation. An arrow separates the reactants from the products.

The chemical equation for the detonation of gunpowder is shown in Reaction 4.7 (next page). After the formula for each compound, abbreviations for the physical state may be, but are not always, included: (s) = solid, (l) = liquid, (g) = gas.

$$\underbrace{4KNO_3(s) + 7C(s) + S(s)}_{\text{Reactants}} \rightarrow \underbrace{3CO_2(g) + 3CO(g) + 2N_2(g) + K_2CO_3(s) + K_2S(s)}_{\text{Products}} \qquad \textbf{(Reaction 4.7)}$$

Written in words rather than chemical symbols, Reaction 4.7 would be stated: "Four potassium nitrate formula units react with seven carbon atoms and one sulfur atom to produce three molecules of carbon dioxide, three molecules of carbon monoxide, two molecules of nitrogen gas, one formula unit of potassium carbonate, and one formula unit of potassium sulfide."

Recall that the law of conservation of mass states that mass is neither created nor destroyed during a chemical reaction. Also recall that Dalton described a chemical reaction as the rearrangement of atoms to form new compounds. If you carefully examine Reaction 4.7, you will see that each atom that makes up the reactants is accounted for in the products. The key to writing proper chemical equations is to make sure the reaction follows the law of conservation of mass. This means that chemical equations must be *balanced*.

4.6 Balancing Chemical Equations

Learning Objective
Balance chemical equations systematically.

Balancing a chemical equation involves manipulating the **coefficients** in front of each compound. The equation cannot be balanced by altering the subscripts in the formulas of compounds because this would change the identity of the compounds and create an incorrect representation of the reaction. The balancing of a chemical equation can be done in a systematic process outlined below.

1. List each element found in the reactants in a column beneath the reactant side of the equation. If oxygen and hydrogen are present, always place them at the bottom of the list.
2. Copy the identical list of elements in a column beneath the product side.
3. Record the number of atoms of each element found on the reactant side. Repeat this process on the product side.

4. Starting at the top of the list, attempt to balance the equation by placing coefficients in front of compounds to increase the number of atoms. Note: The coefficient modifies the number of *all* atoms in a compound; therefore, in your list, readjust the number of each atom in the compound.
5. Repeat until all elements are balanced in the equation.

Worked Example 13

Arsenic compounds have been used as poisons for most of recorded history. An early method for the detection of arsenic was the reaction of As_2O_3 with soot (carbon) to produce metallic arsenic and carbon monoxide. Balance the following equation:

$$\_\_\_\_As_2O_3 + \_\_\_\_C \rightarrow \_\_\_\_As + \_\_\_\_CO$$

SOLUTION

Step 1. List each element beneath each side of the reaction. Record the number of atoms of each element on both sides of the reaction.

$$\_\_\_\_As_2O_3 + \_\_\_\_C \rightarrow \_\_\_\_As + \_\_\_\_CO$$

As: 2	As: 1
C: 1	C: 1
O: 3	O: 1

Step 2. Balance As by placing the coefficient 2 in front of the product As.

$$\_\_\_\_As_2O_3 + \_\_\_\_C \rightarrow \underline{2}As + \_\_\_\_CO$$

As: 2	As: ~~1~~ 2
C: 1	C: 1
O: 3	O: 1

Step 3. Balance O by placing the coefficient 3 in front of the product CO. Note: This changes the number of carbon atoms also.

$$\_\_\_\_As_2O_3 + \_\_\_\_C \rightarrow \underline{2}As + \underline{3}CO$$

As: 2	As: ~~1~~ 2
C: 1	C: ~~1~~ 3
O: 3	O: ~~1~~ 3

Step 4. Balance C by placing the coefficient 3 in front of the reactant C.

$$\_\_As_2O_3 + \underline{3}C \rightarrow \underline{2}As + \underline{3}CO$$

As: 2	As: ~~1~~ 2
C: ~~1~~ 3	C: ~~1~~ 3
O: 3	O: ~~1~~ 3

Practice 4.13

The Marsh test for arsenic was based on the production of arsine gas (AsH_3) through the following chemical reaction: Hydrogen gas and arsenic(III) oxide react to produce arsine gas and water. Write the equation for this reaction and balance it.

ANSWER

$$6H_2 + As_2O_3 \rightarrow 2AsH_3 + 3H_2O$$

Worked Example 14

Toluene is a major component in gasoline and is also found in many flammable products such as paint thinners, fingernail polish remover, and adhesives. Balance the equation for the combustion reaction of toluene (C_7H_8) with oxygen gas to produce carbon dioxide and water.

SOLUTION

Step 1. List each element beneath each side of the equation. Record the number of atoms of each element on both sides of the equation.

$$\_\_\_\_\_C_7H_8 + \_\_\_\_\_O_2 \rightarrow \_\_\_\_\_CO_2 + \_\_\_\_\_H_2O$$

C: 7	C: 1
H: 8	H: 2
O: 2	O: 3

Step 2. Balance C by placing the coefficient 7 in front of the product CO_2.

$$\_\_\_\_\_C_7H_8 + \_\_\_\_\_O_2 \rightarrow \underline{7}CO_2 + \_\_\_\_\_H_2O$$

C: 7	C: ~~1~~ 7
H: 8	H: 2
O: 2	O: ~~3~~ 15

Step 3. Balance H by placing the coefficient 4 in front of the product H_2O.

$$\_\_\_\_\_C_7H_8 + \_\_\_\_\_O_2 \rightarrow 7CO_2 + \underline{4}H_2O$$

C: 7	C: ~~1~~ 7
H: 8	H: ~~2~~ 8
O: 2	O: ~~5~~ ~~15~~ 18

Step 4. Balance O by placing the coefficient 9 in front of the reactant O_2.

$$C_7H_8 + \underline{9}O_2 \rightarrow 7CO_2 + 4H_2O$$

C: 7	C: ~~1~~ 7
H: 8	H: ~~2~~ 8
O: ~~2~~ 18	O: ~~5~~ ~~15~~ 18

Practice 4.14

Mercury fulminate, $Hg(OCN)_2$, is an explosive used in detonators since the 1860s. Mercury fulminate forms elemental mercury, carbon monoxide, and nitrogen gas when detonated. Write and balance the equation.

ANSWER

$$Hg(OCN)_2 \rightarrow Hg + 2CO + N_2$$

4.7 Mathematics of Chemical Reactions: Mole Calculations

Learning Objective

Determine the molar mass of compounds and do basic gram–mole conversion problems.

A balanced chemical reaction can be viewed as a recipe. It lists the ingredients to make a set of products. Just as a recipe requires that we measure out a certain quantity of each ingredient, in the laboratory it is necessary to know how to measure out appropriate quantities of reactants. However, the individual atoms and molecules that are represented in the chemical formulas of an equation are far too small to measure on a laboratory balance. We need a system for measuring manageable quantities of atoms and molecules that will also tell us how many atoms or molecules we have in our sample.

A simple system has been devised for relating a measured mass of an element or compound to the number of atoms or molecules contained in the mass. The system is based on the fact that the atomic mass of an element, expressed in grams, contains a known quantity of atoms: 6.022×10^{23}. This number is called **Avogadro's number** (N_A) (named for the scientist who discovered the principle) and represents one **mole** (mol) of a substance. Just as a dozen represents a quantity of 12, a mole represents a quantity of 6.022×10^{23}. For example, one carbon atom has an atomic mass of 12.011 amu. This means if we measure 12.011 grams of carbon on a balance, this amount is one mole of carbon and contains 6.022×10^{23} carbon atoms.

The mass of one mole of any element is equal to the mass listed for that element, in grams, on the periodic table. The mass of one mole of any compound is equal to the combined masses of each component in the formula. One mole of water (H_2O) has a mass of 18.015 g because it contains one mole of oxygen (15.999 g) and two moles of hydrogen (2×1.0079). Figure 4.6 shows one mole of several elements and several compounds.

Figure 4.6 The mass of one mole of each compound is determined by adding together the molar mass of each element found in the compound. The total mass of each compound pictured here varies, yet each beaker contains the same number of molecules or formula units—Avogadro's number (6.022×10^{23}), which represents one mole of the compound. (Charles D. Winters/Photo Researchers, Inc.)

Worked Example 15

What is the mass of the following elements?

(a) 1.00 mol of Fe

(b) 2.00 mol of O

(c) 0.500 mol of Ca

SOLUTION

(a)

$$1.00 \cancel{\text{mol Fe}} \times \overbrace{\frac{55.847 \text{ g Fe}}{1 \cancel{\text{mol Fe}}}}^{\text{From the periodic table}} = 55.8 \text{ g Fe}$$

(b)

$$2.00 \cancel{\text{mol O}} \times \overbrace{\frac{15.999 \text{ g O}}{1 \cancel{\text{mol O}}}}^{\text{From the periodic table}} = 32.0 \text{ g O}$$

(c)

$$0.500 \cancel{\text{mol Ca}} \times \overbrace{\frac{40.08 \text{ g Ca}}{1 \cancel{\text{mol Ca}}}}^{\text{From the periodic table}} = 20.0 \text{ g Ca}$$

Practice 4.15

What is the mass of the following elements?

(a) 0.750 mol of Cu

(b) 3.00 mol of K

(c) 2.00 mol of Al

ANSWER

(a) 47.7 g of Cu

(b) 117 g of K

(c) 54.0 g of Al

Worked Example 16

Calculate the mass of 1 mol of the following compounds:

(a) As_2O_3

(b) C_7H_8

(c) KNO_3

SOLUTION

(a)

$$\begin{array}{rll} & \text{mass of As} & = 2 \text{ mol} \times 74.922 \text{ g/mol} \\ + & \underline{\text{mass of O}} & = + \underline{3 \text{ mol} \times 15.999 \text{ g/mol}} \\ & \text{mass of } As_2O_3 & = 197.841 \text{ g/mol} \end{array}$$

(b)

$$\begin{array}{rll} & \text{mass of C} & = 7 \text{ mol} \times 12.011 \text{ g/mol} \\ + & \underline{\text{mass of H}} & = + \underline{8 \text{ mol} \times 1.008 \text{ g/mol}} \\ & \text{mass of } C_7H_8 & = 92.141 \text{ g/mol} \end{array}$$

(c)

$$\begin{array}{rl} \text{mass of K} & = 1 \text{ mol} \times 39.098 \text{ g/mol} \\ \text{mass of N} & = 1 \text{ mol} \times 14.007 \text{ g/mol} \\ + \text{ mass of O} & = + 3 \text{ mol} \times 15.999 \text{ g/mol} \\ \hline \text{mass of } KNO_3 & = 101.102 \text{ g/mol} \end{array}$$

Practice 4.16

Calculate the mass of 1 mol of the following compounds:

(a) N_2

(b) K_2CO_3

(c) K_2S

ANSWER

(a) 28.014 g/mol

(b) 138.204 g/mol

(c) 110.261 g/mol

In the laboratory, it is rare that we need exactly one mole of a substance. More often a smaller or larger amount of a reactant is measured out and then the number of moles is calculated. The critical step in this calculation is the use of the molecular mass of the compound as the conversion factor between grams and moles.

Worked Example 17

Calculate how many moles of each element or compound is present given the mass below.

(a) 30.0 g of C

(b) 0.575 g of As_2O_3

(c) 85.5 g of NaCN

SOLUTION

(a) $$30.0 \cancel{\text{g}}\ C \times \frac{1 \text{ mol C}}{12.011 \cancel{\text{g}}} = 2.50 \text{ mol C}$$

(b) $$0.575 \cancel{\text{g}}\ As_2O_3 \times \frac{1 \text{ mol } As_2O_3}{\underbrace{197.841 \cancel{\text{g}}}_{\substack{2As \times 74.922 \text{ g/mol} \\ + 3O \times 15.999 \text{ g/mol}}}} = 0.00291 \text{ mol } As_2O_3$$

(c) $$85.5 \cancel{\text{g}}\ NaCN \times \frac{1 \text{ mol NaCN}}{\underbrace{49.008 \cancel{\text{g}}}_{\substack{1Na \times 22.990 \text{ g/mol} \\ +1C \times 12.011 \text{ g/mol} \\ +1N \times 14.007 \text{ g/mol}}}} = 1.74 \text{ mol NaCN}$$

Practice 4.17

Calculate how many moles of each element or compound is present given the mass below.

(a) 14.5 g of CaO

(b) 115 g of KBr
(c) 0.822 g of $Zn_3(PO_4)_2$

ANSWER
(a) 0.259 mol of CaO
(b) 0.966 mol of KBr
(c) 0.00213 mol of $Zn_3(PO_4)_2$

4.8 Mathematics of Chemical Reactions: Stoichiometry Calculations

Learning Objective

Use balanced chemical equations to determine the relationship between quantities of reactants and products.

Stoichiometry is the study and use of balanced chemical equations for quantifying the amount of reactants and products in a given chemical reaction under a specific set of conditions. In combination with a balanced chemical equation, mole calculations make it possible to determine the maximum amount of a compound that can be produced from particular quantities of reactants. Recall the chemical reaction that takes place when gunpowder is detonated:

$$4KNO_3(s) + 7C(s) + S(s) \rightarrow 3CO_2(g) + 3CO(g) + 2N_2(g) + K_2CO_3(s) + K_2S(s) \qquad \textbf{(Reaction 4.7)}$$

The coefficients in this equation represent how many molecules, formula units, or atoms of reactants and products are involved in the reaction. But the coefficients can also represent the number of moles of each reactant and product. Reaction 4.7 can be interpreted as, "Four moles of potassium nitrate react with seven moles of carbon and one mole of sulfur to produce three moles of carbon dioxide, three moles of carbon monoxide, two moles of nitrogen gas, one mole of potassium carbonate, and one mole of potassium sulfide."

Two calculations often used in the laboratory are: (1) calculation of the number of grams of product that can be obtained from a certain number of grams of reactant; and (2) calculation of how many grams of reactant are needed to produce a given amount of product. The steps required to solve these problems are the same.

1. Convert grams to moles of the given compound using the molecular weight.
2. Use the balanced chemical reaction to find the conversion factor between the given compound and the desired compound.
3. Multiply by the conversion factor to get moles of desired compound.
4. Convert from moles to grams of desired compound using molecular weight.

This procedure is also summarized in Figure 4.7.

$$\text{grams A} \xrightarrow{\text{MW of A}} \text{moles A} \xrightarrow[\text{coefficients}]{\text{Reaction}} \text{moles B} \xrightarrow{\text{MW of B}} \text{grams B}$$

Figure 4.7 Mathematical map for calculations relating grams of reactant A to grams of product B.

Worked Example 18

Calculate how many grams of carbon monoxide are formed when 10.0 g of KNO_3 are consumed in the following reaction:

$$4KNO_3(s) + 7C(s) + S(s) \rightarrow 3CO_2(g) + 3CO(g) + 2N_2(g) + K_2CO_3(s) + K_2S(s)$$

SOLUTION

$$10.0 \text{ g } KNO_3 \xrightarrow{\text{MW } KNO_3} \text{mol } KNO_3 \xrightarrow{4:3} \text{mol CO} \xrightarrow{\text{MW CO}} \text{g CO}$$

$$10 \text{ g } \cancel{KNO_3} \times \frac{1 \text{ } \cancel{\text{mol } KNO_3}}{101.1 \text{ g } \cancel{KNO_3}} \times \frac{3 \text{ } \cancel{\text{mol CO}}}{4 \text{ } \cancel{\text{mol } KNO_3}} \times \frac{28.01 \text{ g CO}}{1 \text{ } \cancel{\text{mol CO}}} = 2.08 \text{ g CO}$$

Practice 4.18

Calculate how many grams of sulfur are needed to form 4.50 g of K_2S.

ANSWER

1.31 g of S

Worked Example 19

Calculate how many grams of carbon must react to produce 0.100 g of arsenic according to the reaction:

$$As_2O_3 + 3C \rightarrow 2As + 3CO$$

SOLUTION

$$0.100 \text{ g As} \xrightarrow{\text{Atomic mass As}} \text{mol As} \xrightarrow{2:3} \text{mol C} \xrightarrow{\text{Atomic mass C}} \text{g C}$$

$$0.100 \text{ g } \cancel{\text{As}} \times \frac{1 \text{ } \cancel{\text{mol As}}}{74.922 \text{ g } \cancel{\text{As}}} \times \frac{3 \text{ } \cancel{\text{mol C}}}{2 \text{ } \cancel{\text{mol As}}} \times \frac{12.011 \text{ g C}}{1 \text{ } \cancel{\text{mol C}}} = 0.0240 \text{ g C}$$

Practice 4.19

Calculate the number of grams of water produced in the following Marsh test assuming the reaction consumes 1.55 g of As_2O_3.

$$6H_2 + As_2O_3 \rightarrow AsH_3 + H_2O$$

ANSWER

0.141 g of H_20

4.9 Types of Reactions

Learning Objective

Identify and classify chemical reactions.

There are several broad categories used to classify different types of reactions. The first class is **precipitation reactions,** which can be distinguished by the formation of a solid from two aqueous reactants. The abbreviation *aq,* which stands for *aqueous,* is used in equations to designate reactants that are dissolved in water. An excellent example of a precipitation reaction is the formation of AgCl to develop fingerprints located on rough surfaces such as wood:

$$\underbrace{\mathrm{NaCl}(aq)}_{\substack{\text{Sweat from}\\ \text{fingerprints}}} + \underbrace{\mathrm{AgNO_3}(aq)}_{\substack{\text{Sprayed on by}\\ \text{investigators}}} \rightarrow \mathrm{NaNO_3}(aq) + \underbrace{\mathrm{AgCl}(s)}_{\substack{\text{White solid, turns black}\\ \text{when exposed to sunlight}}} \qquad \textbf{(Reaction 4.8)}$$

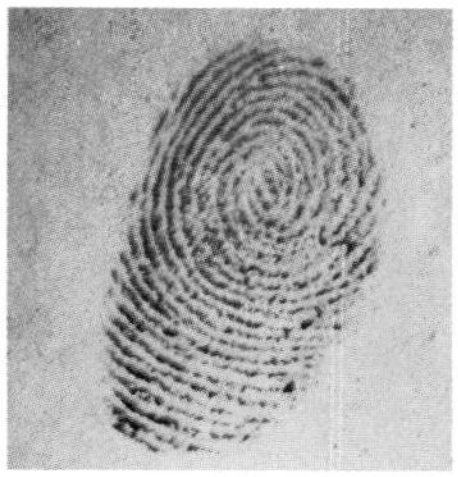

Fingerprint made visible with AgCl. (David Aubrey/Corbis)

A second class of reactions is **combustion reactions**. The most familiar combustion reactions are distinguished by the presence of oxygen gas reacting with an organic compound (a compound containing carbon and hydrogen) to produce carbon dioxide and water. Arsonists use the combustion of flammable organic compounds such as kerosene, gasoline, or lighter fluid to fuel the fires they set. One of the components of gasoline is octane (C_8H_{18}), which reacts with oxygen gas in a combustion reaction according to the following equation:

$$\underbrace{2\mathrm{C_8H_{18}}(l)}_{\text{Organic compound}} + \underbrace{25\mathrm{O_2}\ (g)}_{\substack{\text{Oxygen gas}\\ \text{is required}}} \rightarrow \underbrace{18\mathrm{H_2O}(g)}_{\text{Water vapor}} + \underbrace{16\mathrm{CO_2}(g)}_{\substack{\text{Carbon dioxide}\\ \text{(CO and C can}\\ \text{also be produced)}}} \qquad \textbf{(Reaction 4.9)}$$

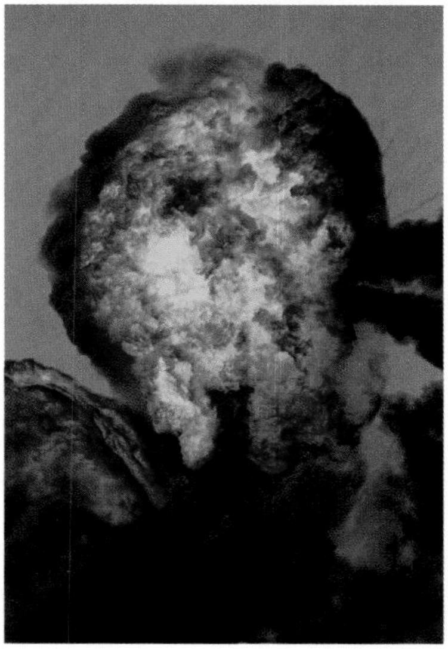

Combustion of gasoline. (Guy Motil/Corbis)

Neutralization reactions are characterized by the reaction of an acid with a base to produce a salt and water. A simple definition of an **acid** is any compound that can release the hydrogen ion, H^+, when dissolved in water. In common acids, hydrogen is usually listed first in the formula. A simple definition of a **base** is any compound that produces the hydroxide ion, OH^-, in solution. The common bases are identified by the presence of the hydroxide ion at the end of the chemical formula. The following reaction is representative of acid-base neutralizations:

$$\underbrace{\mathrm{HCl}(aq)}_{\text{Acid}} + \underbrace{\mathrm{NaOH}(aq)}_{\text{Base}} \rightarrow \underbrace{\mathrm{NaCl}(aq)}_{\text{Salt}} + \underbrace{\mathrm{H_2O}(l)}_{\text{Water}} \qquad \textbf{(Reaction 4.10)}$$

Another class of reactions is **redox reactions**. Redox is an abbreviation for reduction-oxidation, which involve the gain and loss of electrons between atoms in the reacting compounds. **Oxidation** is the loss of electrons; **reduction** is the gain of electrons. Oxidation and reduction processes occur simultaneously.

Many forms of redox reactions can be found, a common one being the reaction of a metal with an acid. This particular type of redox reaction is used to restore serial numbers that have been altered on guns or vehicle parts. The acid reacts with iron in the steel to produce Fe^{2+} ions and hydrogen gas, as in Reaction 4.11. Serial numbers can be visualized because the hydrogen bubbles tend to form faster on the steel that was compressed beneath the serial numbers as compared with the rest of the surface. The hydrogen gas bubbles form in the shape of the numbers that have been removed from the steel. In the reaction below, Fe(s) becomes Fe^{2+} (in the ionic compound $FeCl_2$), which involves the loss of two electrons that go to the hydrogen ion (H^+) to form hydrogen gas.

$$\underbrace{2\mathrm{HCl}(aq)}_{\text{Acid}} + \underbrace{\mathrm{Fe}(s)}_{\substack{\text{Major steel}\\ \text{component}}} \rightarrow \mathrm{FeCl_2}(aq) + \underbrace{\mathrm{H_2}(g)}_{\text{Bubbles}} \qquad \textbf{(Reaction 4.11)}$$

(a) (b)

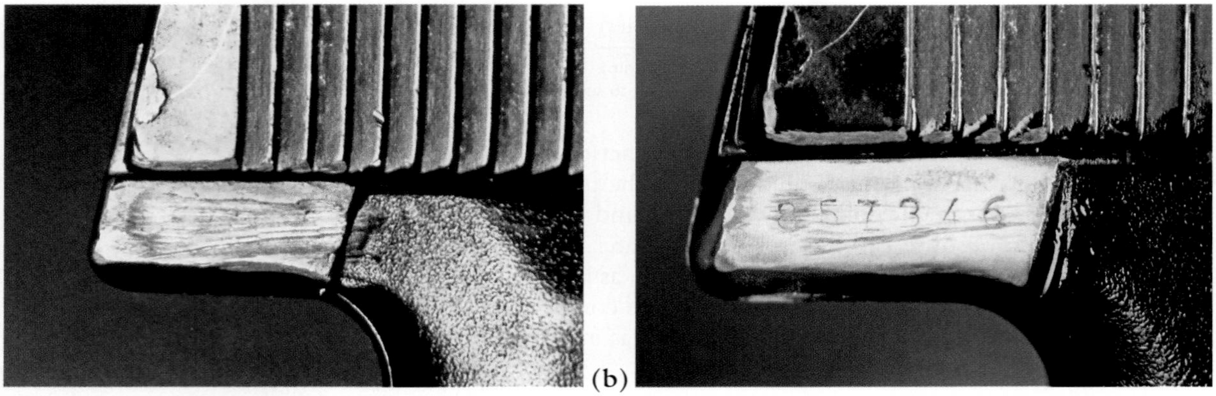

Photo (a) shows a gun with its serial number filed off in an attempt to conceal the gun's origin. Acid preferentially reacts with the iron that has been compressed in the stamping of the serial numbers, thus restoring the pattern of the serial number etchings, as shown in photo (b). (Courtesy of Sirchie Fingerprint Laboratories)

Redox reactions also occur when a substance combines with oxygen atoms during a reaction. An alternate definition of *oxidation* is the gain of oxygen by a substance. It should be noted that combustion reactions are a type of redox reaction because the fuel reacts with oxygen.

Worked Example 20

List the key features that distinguish each of the reaction types: precipitation, combustion, neutralization, and redox.

SOLUTION

Precipitation: solid forming from two aqueous reactants

Combustion: organic (carbon and hydrogen) compound reacting with oxygen gas to produce carbon dioxide and water

Neutralization: acid reacting with base

Redox: reactants gain and lose electrons

Practice 4.20

Classify each of the following reactions as either neutralization, combustion, reduction-oxidation, or precipitation. More than one reaction type may apply.

(a) Test for chloride presence: $BaCl_2(aq) + Na_2SO_4(aq) \rightarrow 2NaCl(aq) + BaSO_4(s)$

(b) Gas line explosion: $C_3H_8(g) + 5O_2(g) \rightarrow 3CO_2(g) + 4H_2O(g)$

(c) Battery acid and zinc metal: $H_2SO_4(aq) + Zn(s) \rightarrow H_2(g) + ZnSO_4(aq)$

ANSWER

(a) Precipitation

(b) Combustion and reduction-oxidation

(c) Reduction-oxidation

4.10 Mathematics of Chemical Reactions: Limiting Reactants and Theoretical Yields

Learning Objective

Apply stoichiometry calculations in solving limiting reactant problems.

In many laboratory experiments, it is usual for at least one of the reactants to be present in an excess amount—beyond the amount that could be completely consumed by the other reactants. An alternate perspective is that one of the reactants will be completely consumed before the others are used up, at which point the reaction will come to a stop. The reactant that is completely consumed is called the **limiting reactant** (or limiting reagent). It is critical to know which reactant is the limiting reactant because the maximum amount of product that can be theoretically produced, the **theoretical yield**, is dictated by the limiting reagent. Firefighters use this same idea when approaching a fire fight. The goal of the firefighter is to prevent either further fuel or further oxygen from reaching the flames. When firefighters are successful, the fire will stop once the existing fuel or oxygen is consumed.

Limiting reagent calculations are modifications of the stoichiometry problems that were introduced earlier in this chapter. The essence of the limiting reagent calculation is to start with a mass of a reactant (it doesn't matter which one you choose) and calculate whether or not the other reactant is present in a sufficient amount to react completely with the first substance. Once the limiting reagent is identified, it is common to calculate the theoretical yield of the reaction for a given product. There are many approaches that can be used to solve limiting reactant problems. The method presented below combines determining the limiting reagent with determining the theoretical yield of the reaction.

1. Determine the product of interest in the problem.
2. Convert grams of each reactant to grams of the product of interest. (This involves two separate stoichiometry calculations.)
3. The smaller amount calculated previously is the theoretical yield of the product produced by the limiting reactant.

Worked Example 21

Postmortem analysis of bodily fluids can provide investigative leads. For example, low levels of chloride in the blood serum may indicate that the victim vomited repeatedly, had diarrhea, or possibly suffered from congestive heart failure. The presence of Cl^- can be determined by the formation of AgCl(*s*) according to the reaction below. Determine the theoretical amount of AgCl that can be formed if a solution containing 0.10 g of NaCl is mixed with a solution containing 0.10 g of $AgNO_3$.

$$AgNO_3(aq) + NaCl(aq) \rightarrow AgCl(s) + NaNO_3(aq)$$

SOLUTION

AgCl is the product of interest.

Reactant 1: 0.10 g NaCl → moles NaCl → moles AgCl → g AgCl

$$0.10 \text{ g NaCl} \times \frac{1 \text{ mol NaCl}}{58.443 \text{ g NaCl}} \times \frac{1 \text{ mol AgCl}}{1 \text{ mol NaCl}} \times \frac{143.321 \text{ g AgCl}}{1 \text{ mol AgCl}} = 0.25 \text{ g AgCl}$$

Reactant 2: 0.10 g $AgNO_3$ → moles $AgNO_3$ → moles AgCl → g AgCl

$$0.10 \text{ g AgNO}_3 \times \frac{1 \text{ mol AgNO}_3}{169.872 \text{ g AgNO}_3} \times \frac{1 \text{ mol AgCl}}{1 \text{ mol AgNO}_3} \times \frac{143.321 \text{ g AgCl}}{1 \text{ mol AgCl}} = 0.084 \text{ g AgCl}$$

The silver nitrate is the limiting reagent because it would be completely used up by producing 0.084 g of silver chloride. The sodium chloride is the excess reagent because there is enough present to make up to 0.25 g of silver chloride; however, there is not sufficient silver nitrate.

Practice 4.21

The presence of sulfate ion can be detected in urine by precipitation with the barium ion to produce barium sulfate. Determine the limiting reagent and the theoretical yield of barium sulfate given that a solution containing 1.60 g of sodium sulfate is mixed with a solution containing 2.40 g of barium nitrate.

$$Ba(NO_3)_2(aq) + Na_2SO_4(aq) \rightarrow BaSO_4(s) + 2NaNO_3(aq)$$

ANSWER

The limiting reagent is barium nitrate and the theoretical yield is 2.14 g of $BaSO_4$.

Evidence Analysis | Spectrophotometry

One method for detecting the presence and quantity of compounds in a sample is a versatile laboratory technique called **spectrophotometry.** This method is based on the absorption of light that occurs when light is passed through a solution containing molecules of the substance under investigation. White light contains all of the colors of the spectrum (the rainbow). Some molecules have the ability to absorb certain colors of light. For example, FD&C Blue Dye No. 1, shown in Figure 4.8, is the dye responsible for the blue color in blue raspberry Kool-Aid. When white light strikes a molecule of the dye, the molecule absorbs orange and red light. The remaining light passes through the solution and creates the blue color you observe. The absorption of light is due to the electrons in the

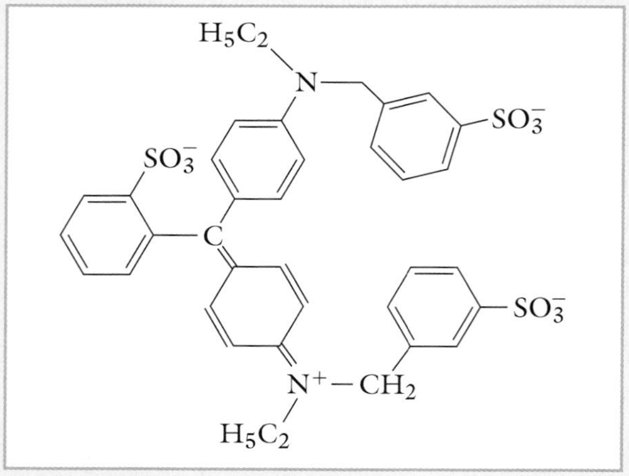

Figure 4.8 FD&C Blue Dye No. 1.

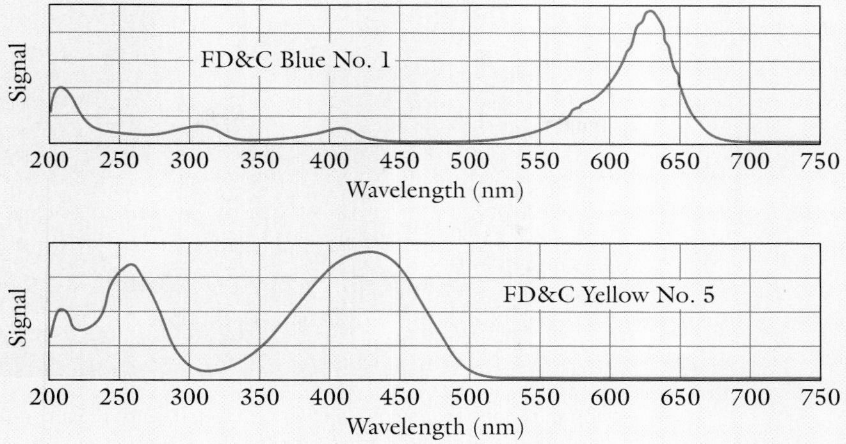

Figure 4.9 Absorption spectra for FD&C Blue Dye No. 1 and FD&C Yellow Dye No. 5.

dye molecule absorbing the energy from certain frequencies of light and going from a lower energy level to a higher energy level.

An unknown substance can tentatively be identified by measuring the wavelengths of light that are absorbed, called an *absorption spectrum*, when a solution containing the unknown compound is placed in a spectrophotometer. Shown in Figure 4.9 is the absorption spectrum for FD&C Blue Dye No. 1 and FD&C Yellow Dye No. 5. The two dyes clearly absorb different wavelengths of light and are clearly different substances.

The major role of spectrophotometry in the crime laboratory is to determine the amount of a compound dissolved in solution. You know that if you dissolve $\frac{1}{4}$ of a package of blue raspberry Kool-Aid in 2 quarts water, the solution would have a much lighter color than if you dissolved four packages of the Kool-Aid in 2 quarts. The more concentrated solution would be a much darker blue because there are more dye molecules in the same volume of solution. The higher the concentration of a colored compound, the less light passes through the solution; the solution thus appears darker (see Figure 4.10). This phenomenon allows us to measure how much dye is in the water by measuring how much light is transmitted through a solution and comparing it with solutions of known concentrations.

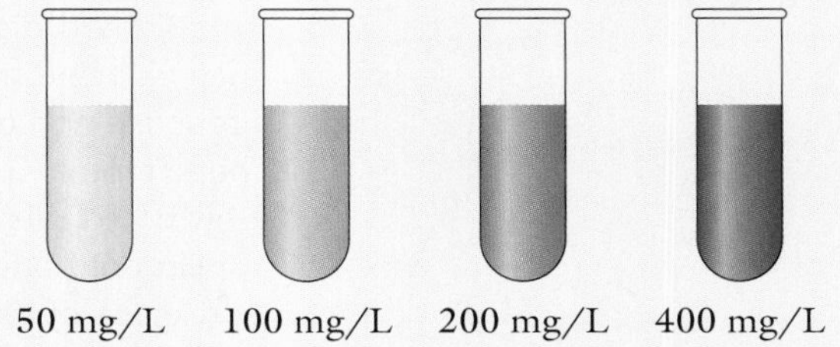

Figure 4.10 The amount of a colored compound dissolved in solution can be directly measured by comparing the amount of light it absorbs with the amount of light absorbed by a series of solutions of known concentration.

4.11 Case Study Finale: A Killer Headache

Richard Keyworth, a firefighter for Elk Grove Village, Illinois, stopped by the firehouse to check his mail one morning in 1982. While he was there, he heard about an odd case in which a 12-year-old girl named Mary Kellerman died of an apparent stroke. This seemed very unusual;

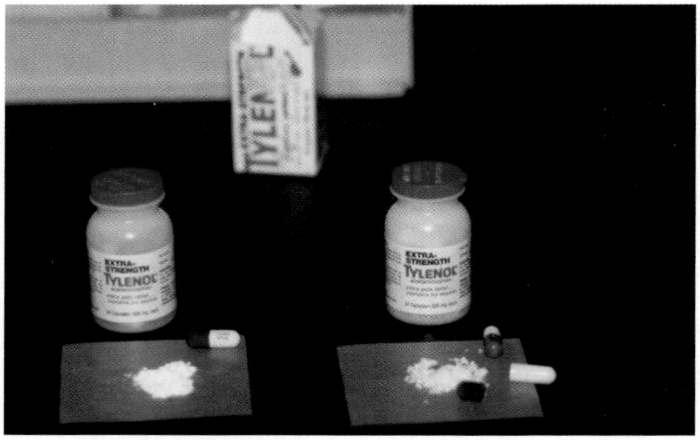

A normal Tylenol capsule is shown on the left, and a cyanide-filled capsule is shown on the right. (Bettmann/Corbis)

12-year-old girls don't typically suffer strokes. That night Richard received a call from Lt. Philip Cappitelli, a close friend of his from the Arlington Heights fire department. Philip was curious about the Mary Kellerman case because a member of his family had worked with Mary's mother and wanted to know if Richard had heard any of the details. During their conversation, Philip mentioned a strange case that happened in Arlington Heights, where three members of one family had mysteriously died in a single night.

Richard didn't know all the details of the Mary Kellerman case, other than what he had overheard, so he stopped back at the Elk Grove Village firehouse and got the report for Philip. As the two men went over the case, the symptoms seemed to match those that occurred at the Janus home. How could this be? The report in the Kellerman case stated that Mary had a mild cold and took an Extra Strength Tylenol before she was found unconscious on the bathroom floor. Richard wondered out loud if the Tylenol could have something to do with it. Neither of them knew about a bottle of Tylenol at the Janus home. Their suspicions were relayed to Dr. Thomas Kim at Northwest Community Hospital. Dr. Kim knew that the Janus family members had all taken Tylenol. When he heard a 12-year-old in a nearby suburb also mysteriously died after taking the same medicine, he immediately notified police. Soon the world would know that, somehow, someone was poisoning the Tylenol.

The symptoms were quickly diagnosed by a poison control center physician and the news was made public. Innocent people were dying from the ingestion of a simple ionic compound, sodium cyanide. All Tylenol in the Chicago metropolitan area was ordered off store shelves and people were encouraged to turn in unused Tylenol to the police. Forensic scientists analyzed every capsule of Tylenol in the region. In the end, seven people from the Chicago area died from cyanide poisoning.

A lethal dose of sodium cyanide or potassium cyanide, the most common forms of cyanide, is estimated to be between 100 and 300 mg, and death can occur within minutes of exposure. There is an antidote to cyanide

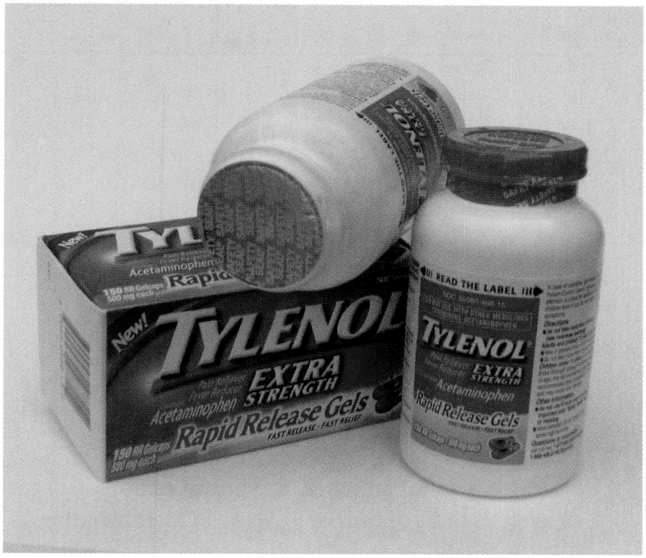

Figure 4.11 Tylenol safety seal. (Elyse Rieder)

poisoning, but unless it is administered very quickly after exposure, the damage cyanide inflicts on the body is too great to overcome. Cyanide interferes with a person's respiratory system and prevents the body from utilizing oxygen.

When sodium cyanide or potassium cyanide is dissolved in water, it forms a colorless solution that does not absorb visible light. However, cyanide will react with pyridine and pyralozone (two large organic molecules) to produce a blue-colored molecule with the formula $C_{25}H_{24}N_4O_2$. The solution concentration can then be measured using spectrophotometry. By comparing with known solutions the intensity of the color produced from a solution containing an unknown amount of cyanide, it is possible to determine the amount of cyanide in an unknown sample.

The early 1980s became filled with copycat tampering of consumer products, which led to stricter regulations on the safety sealing of products. As a direct result of the tampering, McNeil Consumer Products, a subsidiary of Johnson & Johnson, stopped manufacturing Tylenol capsules of granulated powder and instead limited sales to pressed tablets. In response to the tampering of their product, Johnson & Johnson also introduced the triple safety seal system shown in Figure 4.11, which consisted of a glued box, a plastic bottle cap seal, and a foil bottle seal.

Cyanide compounds are not available to the general public. However, individuals who work in manufacturing processes such as photography, metal plating, polymer manufacturing, research laboratories, and mining may have access to it. The police determined that the person or persons responsible had placed the cyanide-filled tablets across a large section of the Chicago metropolitan area, but law enforcement agents were never able to build a solid enough lead to make an arrest. This case remains unsolved.

CHAPTER SUMMARY

- The columns of the periodic table are called groups or families. Elements within a group undergo similar chemical reactions. The groups are numbered 1–18 from left to right, but some groups also have common names such as alkali metals, alkaline earth metals, halogens, and noble gases.
- The rows of the periodic table are called periods and are numbered 1–7. The only periods with common names are the lanthanide and actinide periods, which make up the inner transition metals located beneath the main body of the periodic table.
- Ionic compounds consist of two oppositely charged ions. The cation has a positive charge that results when an atom loses at least one electron. The anion has a negative charge that results when an atom gains at least one electron. Metals tend to form cations; nonmetals tend to form anions.
- Ionic compounds tend to have high melting points and form rigid structures. The strength of the attraction between ions is directly related to the magnitude of their charges.
- Ionic compounds are named by listing the cation first, followed by the anion. If the cation does not have a predictable ionic charge, roman numerals are used to indicate the charge.
- Covalent compounds are created when two nonmetal atoms form a bond by sharing electrons. The formulas of covalent compounds cannot be predicted. Therefore, the number of atoms of each element in the compound is indicated by prefixes in the name.
- The ability of an atom to pull electrons toward its nucleus when bonded to another atom is called electronegativity. The electronegativity of the elements increases from the bottom to the top of a group and from the left to the right of a period.
- Chemical equations are a shorthand method for quickly, efficiently, and clearly communicating information about a reaction. The coefficient in front of each compound in a chemical equation is used to balance the equation.
- One mole of any compound is equal to the molecular mass of the substance expressed in grams and contains 6.022×10^{23} molecules.
- Reactions can be classified into groups based on the types of substances in the reaction and how they react. Common reaction types are precipitation, combustion, neutralization, and redox reactions. A given reaction may fit into multiple classifications.
- A limiting reagent is one that is completely consumed in a reaction, leaving an excess amount of one or more of the other reactants. Determination of the limiting reagent is important because it controls the maximum amount of a product that can form.
- The concentration of a compound in a solution can be measured by determining how much light passes through a sample and how much of the light is absorbed by the solution.

KEY TERMS

CONTINUING THE INVESTIGATION Additional Readings, Resources, and References

Gerber, S. M., and R. Saferstein, *More Chemistry and Crime*, Washington D.C.: American Chemical Society, 1997.

For information about the Tylenol murders, see numerous articles in the *Chicago Tribune* from 10/01/82 through 10/06/82. Also, go to: www.crimelibrary.com/terrorists_spies/terrorists/Tylenol_murders/

For information about the forensics in another case of murder by poison, there is information on the Internet concerning several lawsuits titled *George James Trepal* vs. *State of Florida*.

REVIEW QUESTIONS AND PROBLEMS

Questions

1. Make a sketch of the periodic table and label each of the following areas: alkali metals, alkaline earth metals, transition metals, halogens, and noble gases.
2. Which elements tend to form cations? Which elements tend to form anions? Where is each of these groups of elements located on the periodic table?
3. What do elements within the same group have in common?
4. What happens to an atom when it becomes a cation? An anion?
5. What is a monatomic ion? Provide several examples.
6. What is a polyatomic ion? Provide several examples.
7. How does the nature of the crystal lattice affect the melting point of an ionic solid?
8. Why does MgS have a much higher melting point than NaF?
9. How do two atoms share electrons?
10. What type of elements form covalent compounds?
11. What is electronegativity and how does it affect covalent bonds?
12. Many compounds have common names that are regularly used. Choose two examples and provide the common name, scientific name, and formula.
13. What is the importance of a balanced chemical equation?
14. When balancing a chemical equation, why is it important to change the coefficients in front of compounds but not the subscripts in the formulas of compounds?
15. Why is the concept of the mole necessary?
16. What are the main features of the precipitation, combustion, neutralization, and redox reactions?
17. Why is it necessary to determine the limiting reagent in chemical reactions?
18. The calculated theoretical yield of a reaction is seldom obtained in a laboratory experiment. Can you think of anything that might lower the actual yield to a value less than the theoretical yield?
19. Why does a solution containing a greater amount of FD&C Red Dye No. 40 appear darker red than a solution that contains a smaller amount of the dye?
20. Sodium cyanide, when dissolved in water, forms a colorless solution that does not absorb visible light. What can be done experimentally to the solution so that visible light can be used to measure the amount of cyanide with a spectrophotometer?

Problems

21. Identify all halogens from the following four lists of elements. Write *none* if there are no halogens listed.
    (a) Cl, P, Al, I, Mg, Rb, Cr
    (b) Ge, Os, Y, Ca, K, O, Br
    (c) Ga, Mn, Cr, Sr, Li, Na, Ne
    (d) Cu, O, F, Ar, N, W, S
22. Identify all alkaline earth metals from the following four lists of elements. Write *none* if there are no alkaline earth metals listed.
    (a) Cl, P, Al, I, Mg, Rb, Cr
    (b) Ge, Os, Y, Ca, K, O, Br
    (c) Ga, Mn, Cr, Sr, Li, Na, Ne
    (d) Cu, O, F, Ar, N, W, S

23. Identify the group, group name (if any), and period for the following elements.
(a) Mn
(b) Cr
(c) As
(d) Pb

24. Identify the group, group name (if any), and period for the following elements.
(a) Li
(b) Cl
(c) S
(d) Au

25. Identify the following elements.
(a) Period 6, Group 12
(b) Halogen, Period 3
(c) Noble gas, Period 2
(d) Period 4, Group 16

26. Identify the following elements.
(a) Period 5, Group 7
(b) Alkali metal, Period 3
(c) Period 5, Group 16
(d) Period 3, Group 13

27. Using the periodic table as a guide, determine the charge that each of the following atoms will have in an ionic compound. Indicate *multiple* for any atom that can commonly have more than one charge.
(a) Ca
(b) P
(c) Br
(d) Ag

28. Using the periodic table as a guide, determine the charge that each of the following atoms will have in an ionic compound. Indicate *multiple* for any atom that can commonly have more than one charge.
(a) S
(b) N
(c) Na
(d) Fe

29. Write the formulas for the ionic compounds that will form between the following ions.
(a) Magnesium ion and chloride ion
(b) Lithium ion and fluoride ion
(c) Aluminum ion and oxide ion
(d) Zinc ion and iodide ion

30. Write the formulas for the ionic compounds that will form between the following ions.
(a) Calcium ion and phosphide ion
(b) Barium ion and oxide ion
(c) Aluminum ion and bromide ion
(d) Cadmium ion and fluoride ion

31. What is the formula and charge for each of the following polyatomic ions?
(a) Nitrate ion
(b) Hydroxide ion
(c) Cyanide ion
(d) Phosphate ion

32. What is the formula and charge for each of the following polyatomic ions?
(a) Ammonium ion
(b) Acetate ion
(c) Carbonate ion
(d) Sulfate ion

33. What is the formula for the compounds formed between the following ions?
(a) Ammonium ion and acetate ion
(b) Sodium ion and carbonate ion
(c) Magnesium ion and phosphate ion
(d) Lithium ion and permanganate ion

34. What is the formula for the compounds formed between the following ions?
(a) Calcium ion and carbonate ion
(b) Barium ion and hydroxide ion
(c) Strontium ion and cyanide ion
(d) Beryllium ion and nitrate ion

35. Determine the charge of the cation in the following ionic compounds.
(a) $FeCl_3$
(b) $CoSO_4$
(c) CuF_2
(d) CrS

36. Determine the charge of the cation in the following ionic compounds.
(a) FeO
(b) $CoPO_4$
(c) $CuNO_3$
(d) $Cr(CN)_3$

37. Name the following ionic compounds.
(a) CaS
(b) Mg_3N_2
(c) LiF
(d) $Al(OH)_3$

38. Name the following ionic compounds.
(a) Na_2S
(b) $Ba(NO_3)_2$
(c) $BaCl_2$
(d) KBr

39. Name the following compounds.
(a) $FeCl_3$
(b) $CoSO_4$
(c) CuF_2
(d) CrS

40. Name the following compounds.
(a) FeO
(b) $CoPO_4$
(c) $CuNO_3$
(d) $Cr(CN)_3$

41. Name the following covalent compounds.
(a) SF_6
(b) CCl_4
(c) NF_3
(d) PCl_5

42. Name the following covalent compounds.
(a) PCl_3
(b) CO_2
(c) OF_2
(d) SO_2

43. Write the proper formula for the following covalent compounds.
(a) Disulfur dichloride
(b) Dinitrogen pentasulfide
(c) Sulfur tetrafluoride
(d) Sulfur trioxide

44. Write the proper formula for the following covalent compounds.
(a) Phosphorus pentachloride
(b) Diphosphorus triiodide
(c) Bromine monofluoride
(d) Tribromine octoxide

45. Write the chemical equation that corresponds to the following description: Three moles of sodium cyanide react with one mole of iron(III) nitrate to form one mole of iron(III) cyanide and three moles of sodium nitrate.

46. Write the chemical equation that corresponds to the following description: One mole of calcium metal reacts with two moles of water to form one mole of calcium hydroxide and one mole of hydrogen gas.

47. Balance the following equations.
(a) $BaCl_2 + Na_3PO_4 \rightarrow Ba_3(PO_4)_2 + NaCl$
(b) sodium sulfide + iron (II) nitrate → sodium nitrate + iron (II) sulfide
(c) $C_3H_8 + O_2 \rightarrow CO_2 + H_2O$
(d) calcium acetate + potassium hydroxide → potassium acetate + calcium hydroxide

48. Balance the following equations.
(a) $AgNO_3 + LiBr \rightarrow AgBr + LiNO_3$
(b) ammonium chloride + potassium permanganate → ammonium permanganate + potassium chloride
(c) $H_2O_2 \rightarrow H_2O + O_2$
(d) H_2SO_4 +NaOH → water + Na_2SO_4

49. How many moles of each substance are present in the following quantities?
(a) 15.0 g of $AgNO_3$
(b) 0.141 g of C_3H_8
(c) 100.0 g of iron(II) nitrate
(d) 1.04 g of CO_2

50. How many moles of each substance are present in the following quantities?
(a) 32.0 g of $CuNO_3$
(b) 150.0 g of Ca
(c) 3.58 g of O_2
(d) 14.45 g of fluorine gas

51. What is the mass in grams of the following quantities of substances?
(a) 2.00 mol of N_2O_3
(b) 0.750 mol of chlorine gas
(c) 1.24 mol of iron(II) oxide
(d) 3.25 mol of $NH_4C_2H_3O_2$

52. What is the mass in grams of the following quantities of substances?
(a) 0.0525 mol of KBr
(b) 0.126 mol of NH_3
(c) 1.00 mol of sulfur trioxide
(d) 0.667 mol of sulfur hexafluoride

53. Calculate the number of grams of barium sulfate produced if 10.0 g of barium chloride reacts completely according to the reaction:

$$BaCl_2(aq) + Na_2SO_4(aq) \rightarrow 2NaCl(aq) + BaSO_4(s)$$

54. Calculate the number of grams of sodium sulfate consumed if 57.5 g of barium sulfate is produced according to the reaction:

$$BaCl_2(aq) + Na_2SO_4(aq) \rightarrow 2NaCl(aq) + BaSO_4(s)$$

55. Calculate the number of grams of sodium chloride produced if 75.3 g of barium chloride reacts completely according to the reaction:

$$BaCl_2(aq) + Na_2SO_4(aq) \rightarrow 2NaCl(aq) + BaSO_4(s)$$

56. Calculate the number of grams of propane (C_3H_8) consumed if 17.7 g of water is produced according to the reaction:

$$C_3H_8(g) + 5O_2(g) \rightarrow 3CO_2(g) + 4H_2O(g)$$

57. Calculate the number of grams of carbon dioxide produced if 25.0 g of propane (C_3H_8) is consumed according to the reaction:

$$C_3H_8(g) + 5O_2(g) \rightarrow 3CO_2(g) + 4H_2O(g)$$

58. Identify the type of chemical reaction based on the description given below:
    (a) The explosion last night at the railroad terminal was linked to a small leak in the tanker carrying diesel fuel.
    (b) If the swimming pool pH is greater than 8.0, add muriatic acid to lower the pH.
    (c) Carbon dioxide from the atmosphere will react with concentrated sodium hydroxide to form an insoluble carbonate.
    (d) Warning: The whitewall tire cleaner should not be used on metal because it contains hydrofluoric acid.

59. Identify the type of chemical reaction based on the description given below:
    (a) Hydrogen gas is generated by the action of hydrochloric acid on zinc metal.
    (b) Guncotton is a nitrated cellulose fiber that leaves almost no ash after ignition.
    (c) Latent fingerprints can be visualized by spraying a dilute silver nitrate solution onto a fingerprint containing the chloride ion. The resulting reaction forms solid silver chloride.
    (d) Take two teaspoons of milk of magnesia for occasional acid reflux episodes.

60. Determine the limiting reagent and the theoretical yield of $Fe(OH)_3$ for the following reaction, given a solution containing 15.0 g of $Fe(NO_3)_3$ and 15.0 g of KOH.

$$Fe(NO_3)_3(aq) + 3KOH(aq) \rightarrow Fe(OH)_3(s) + 3KNO_3(aq)$$

61. Determine the limiting reagent and the theoretical yield of $Fe(OH)_3$ for the following reaction, given a solution containing 75.0 g of $Fe(NO_3)_3$ and 25.0 g of KOH.

$$Fe(NO_3)_3(aq) + 3KOH(aq) \rightarrow Fe(OH)_3(s) + 3KNO_3(aq)$$

Forensic Chemistry Problems

62. The following ions can all be found in fingerprints: Na^+, K^+, Ca^{2+}, Mg^{2+}, Cl^-, Br^-, NO_3^-.
    (a) Name each ion.
    (b) How many different ionic compounds could be formed from the ions above? Write the name and formula for each possible compound.

63. Iodine, phosphorus, and lithium metal can all be found in their elemental states in clandestine drug labs. What group number, group name (if any), and period correspond to each element?

64. Calculate the number of grams of carbon monoxide produced from the detonation of 200.0 g of TNT ($C_7H_5N_3O_6$) given the following reaction:

$$2C_7H_5N_3O_6(s) \rightarrow 12CO(g) + 5H_2(g) + 3N_2(g) + 2C(s)$$

65. Calculate the number of grams of nitrogen gas produced from the detonation of 125.0 g of TNT ($C_7H_5N_3O_6$) given the following reaction:

$$2C_7H_5N_3O_6(s) \rightarrow 12CO(g) + 5H_2(g) + 3N_2(g) + 2C(s)$$

66. Calculate the number of grams of C4 military explosive ($C_3H_6N_6O_6$) detonated if the resulting explosion produces 247.4 g of carbon monoxide gas.

$$C_3H_6N_6O_6(s) \rightarrow 3CO(g) + 3H_2O(g) + 3N_2(g)$$

67. Calculate the number of grams of C4 military explosive ($C_3H_6N_6O_6$) detonated if the resulting explosion produces 117.4 g of nitrogen gas.

$$C_3H_6N_6O_6(s) \rightarrow 3CO(g) + 3H_2O(g) + 3N_2(g)$$

68. In the Marsh test for arsenic, why would it be imperative for hydrogen gas to be the excess reagent and As_2O_3 to be the limiting reagent? The initial reaction is:

$$6H_2 + As_2O_3 \rightarrow AsH_3 + H_2O$$

69. Cyanide analysis is done using a spectrophotometer. However, the human eye is very sensitive to changes in color. Estimate the concentration of cyanide in the following solution based on the solutions of known concentration in Figure 4.10.

Unknown

Case Study Problems

70. One form of evidence investigators looked for in the Tylenol poisoning case was fingerprints. How might the detectives have been able to link fingerprints to the killer who laced the Tylenol with cyanide? *Hint:* A large number of people from the manufacturing plant, the store where the Tylenol was sold, potential customers, etc., could have handled the Tylenol. Consider where their fingerprints would be found and how the killer could be identified. (The killer's fingerprints were never actually found.)

71. George Trepal was convicted and sentenced to death for the poisoning of a neighbor with the poisonous compound thallium(I) nitrate laced into bottles of soda. He was found to have a bottle of thallium(I) nitrate in his garage. If another thallium(I) salt (chloride or sulfate) had been used, Trepal could claim innocence. Samples of the tampered soda were sent to the laboratory for analysis. A preliminary screening method for the presence of the chloride ion is the addition of silver nitrate, because a precipitate forms with the chloride ion. Write a balanced chemical equation for the reaction of thallium(I) chloride with silver nitrate to produce thallium(I) nitrate and silver chloride. If a sample contained 0.54 g of thallium(I) chloride and 1.18 g of silver nitrate, determine the limiting reactant and the theoretical yield of silver chloride.

72. The Trepal case continued with the screening for the sulfate ion by the addition of barium chloride, as a precipitate forms with the sulfate ion. Write a balanced chemical equation for the reaction of thallium(I) sulfate with barium chloride to produce thallium(I) chloride and barium sulfate. If a sample contained 4.06 g of thallium(I) sulfate and 4.41 g of barium chloride, determine the limiting reactant and the theoretical yield of barium sulfate.

CHAPTER 5

Properties of Solutions I: Aqueous Solutions

An estimated 5 million fish weighing over 187 tons were killed between December 1999 and January 2000 along a 50-mile stretch of the White River that runs through central Indiana between the cities of Anderson and Indianapolis. *(Photodisc Green/Getty Images)*

Case Study: An Aquatic Apocalypse

The White River runs through central Indiana providing drinking water and fishing and boating recreation for many communities, including the capital city of Indianapolis. In Anderson, Indiana, treated wastewater from both residential and industrial sources is discharged into the river. The Anderson Publicly Owned Treatment Works (Anderson POTW), a wastewater treatment plant, is responsible for ensuring that the water discharged into the White River meets all state and national regulations for quality. To monitor the quality of the discharge, workers at the facility take samples at regular intervals and analyze them for common pollutants. If a pollutant level is too high, the treatment system is adjusted and the source of the pollutant investigated.

One compound of concern is the ammonium ion (NH_4^+), which is particularly toxic to aquatic life and comes from the ammonia (NH_3) present in human and animal waste. Ammonia is eliminated from the water by a biological treatment process that relies on *Nitrosomonas* bacteria to convert the ammonia to the nitrite ion (NO_2^-). The nitrite ion is also toxic to aquatic life, but a second genus of bacteria called *Nitrobacter* converts it to nitrate ion NO_3^-, which is nontoxic.

On December 13, 1999, the analysis of a water sample showed the concentration of ammonia to be 8 parts per million (ppm). The normal value is 0.1 ppm. Clearly, something was not working correctly at the treatment facility. Little did anyone know that multiple compounds deadly to fish were being released.

The first reports of a fish kill started on December 16 and continued well into the new year. Over the next several weeks, 117 tons of dead fish were hauled out of the river and disposed of in a landfill. A 50-mile stretch of the White River from Anderson to Indianapolis was devoid of any fish life. The final estimate was that 5 million fish weighing over 187 tons were killed. Fish kills of this magnitude do not occur naturally. How could such unprecedented devastation occur?

To find the agents responsible for devastating a 50-mile stretch of river would require a thorough understanding of the chemistry of aqueous solutions . . .

5.1 Aqueous Solutions

Learning Objective

Distinguish between a solvent and a solute.

From everyday observations, you are undoubtedly familiar with the process of dissolving one substance into another. Take a teaspoon of ordinary sugar (sucrose), stir it into a cup of coffee and what happens? The white crystals of sugar are no longer visible in the mixture of sugar and coffee and the resulting solution now has a sweet taste that the coffee alone did not have. But try to dissolve a teaspoon of sand (SiO_2) or oil in water, and you will notice something different. The sand settles to the bottom of the container after stirring and the oil floats to the top. Neither sand nor oil is incorporated into the water in the same way as sugar. Why is there a difference between the behavior of the sugar, sand, and oil? To understand why some substances dissolve in water and others do not, it is necessary to go beyond what can be seen with the naked eye.

In this chapter, we will go into some detail about the process of dissolution—that is, the interactions that take place on a molecular level when one substance dissolves into another. We will also discuss some properties of solutions, focusing on **aqueous** (water-based) solutions because of their importance in chemistry and forensic science. In an aqueous solution, water is the **solvent** (the liquid that dissolves a chemical compound) and is the major component of the solution. The substance that dissolves into the water is called the **solute** and is the minor component of the solution. In the example given above, sugar was the solute. We do not refer to the sand or oil as solutes because they do not form solutions with water.

The many samples of river water and wastewater from the Anderson POTW that were collected and analyzed to determine the cause of the White River fish kill are examples of aqueous solutions. Typical solutes in river water are minerals (calcium, iron, and sodium cations as well as nitrate, phosphate, and sulfate anions), dissolved gases (oxygen and carbon dioxide), and nutrients (agricultural fertilizers washed by rains into the river). The investigators needed to identify the toxic solutes that were killing off the White River fish and find the sources of those toxins.

There are many different types of aqueous solutions that are gathered at crime scenes and during autopsies. Consider the following:

- Alcoholic beverages such as wine, beer, and whiskey contain an amount of alcohol that depends on the beverage, but in most cases water is present in greater quantity than alcohol.
- Blood is a complex mixture of red and white blood cells and platelets suspended in blood plasma. Blood plasma is composed of water and dissolved solutes such as proteins, ions (Na^+, K^+, Cl^-), nutrients (sugars), and hormones. Any pharmaceuticals, illegal drugs, or alcohol consumed before death will be found as a solute in the blood plasma.
- Vitreous humor is the fluid of the eyeball and normally contains mostly dissolved proteins. It is sampled postmortem to determine whether alcohol is present. While alcohol is transferred throughout a person's entire body during consumption and will be present in all bodily fluids, it is not sufficient to find alcohol in other fluid or tissue samples alone because alcohol can be formed by the microbial decomposition of human tissues, especially near the stomach. However, vitreous humor does not contain the ingredients needed to form alcohol.

5.2 The Process of Dissolution

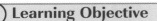

Learning Objective

Explain how water dissolves certain ionic and molecular compounds.

The key factor in the formation of an aqueous solution is the interaction of water molecules with the solute in such a way that the solute molecules or ions are pulled apart and dispersed throughout the water. The nature of this interaction relates to electronegativity, a concept initially covered in Section 4.3. Recall that electronegativity is the ability of an atom to pull shared electrons toward it. In a water molecule (H_2O), the oxygen atom is more electronegative than the hydrogen atoms. Therefore, the shared electrons between O and H are pulled toward the oxygen atom.

Figure 5.1 is a three-dimensional map of water that shows the effect of the large difference in electronegativity between the hydrogen and oxygen atoms within a bond. The region where electrons are most often found is colored red and the region where electrons are rarely found is colored blue. It is important to remember that water is neutral—it has no net negative or positive charge. However, the red area of the molecule represents a negative region near the oxygen atom and the blue area represents a positive region near the hydrogen atoms.

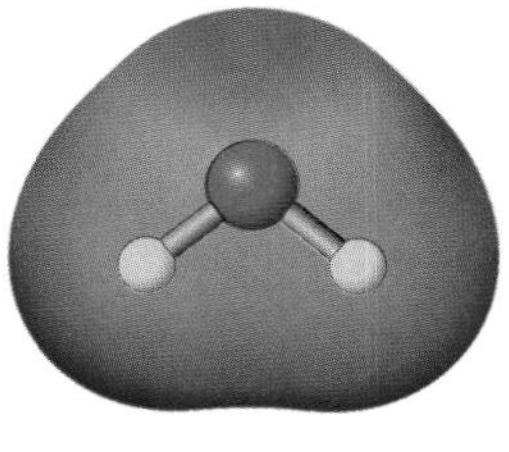

Figure 5.1 Electron density of water. The electrons within the hydrogen-oxygen bond are pulled toward the oxygen atom due to the difference in the electronegativity of the two atoms. The result is a polar molecule with the negative region (red) on the oxygen atom and the positive region (blue) located at the hydrogen atoms.

Molecules that contain a positive region and a negative region, as water does, are termed **polar** molecules. One factor that is important in the determination of which compounds are polar is the three-dimensional geometric shape of the molecule. (This topic is explored in more detail in Chapter 8.) Because opposite charges attract, the negative region of water will be attracted to any substance that has a positive charge or a positive polar region. The positive region of water is attracted to negative charges or negative polar regions.

To illustrate how water interacts with ions, let's examine how potassium cyanide dissolves when placed into water. Potassium cyanide is an ionic compound with the formula KCN. The potassium ion has a +1 charge and the cyanide polyatomic ion (CN^-) has a −1 charge. As the KCN crystalline powder is placed into water, water molecules surround the surface of the crystal. Figure 5.2 shows that the K^+ and CN^- ions separate as each ion becomes surrounded by water molecules, or *solvated.* The water molecules orient in such a way that negative regions (at the oxygen end) cluster around the K^+ ions while positive regions (at the hydrogen end) cluster around the CN^- ions. The attractive forces between water molecules and the ions are sufficient to cause a breaking of the ionic bond between K^+ and CN^- ions that held the solid crystal together.

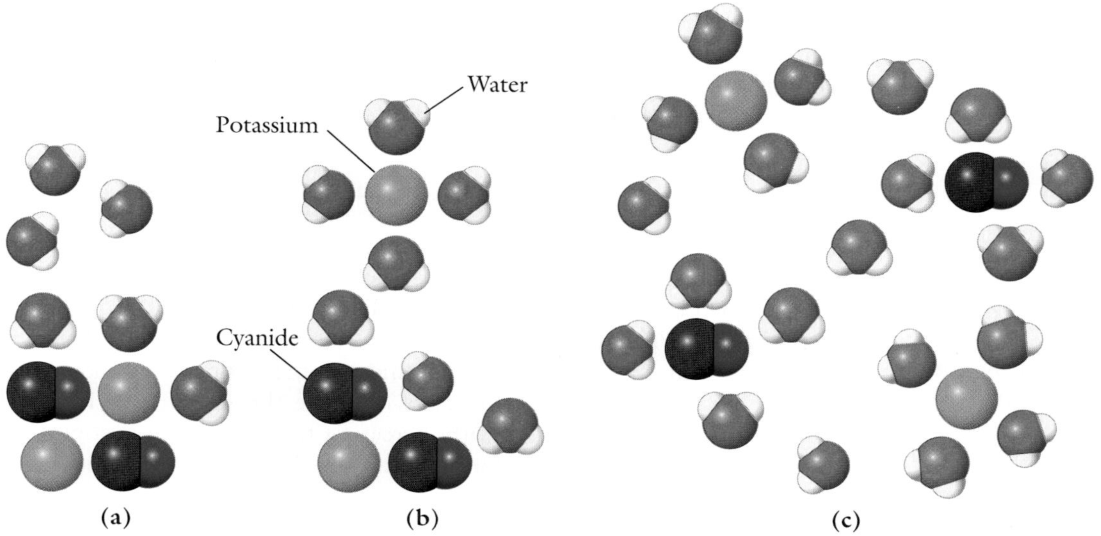

Figure 5.2 Dissolution of potassium cyanide in water. (a) When solid potassium cyanide begins to dissolve, the water molecules arrange themselves on the surface of the ionic solid in a way that the polar regions of water are lined up with the oppositely charged ion. (b) When enough water molecules surround an ion, it will be completely removed from the crystal lattice. (c) The process continues as new water molecules approach the crystal until the crystal has been fully dissolved.

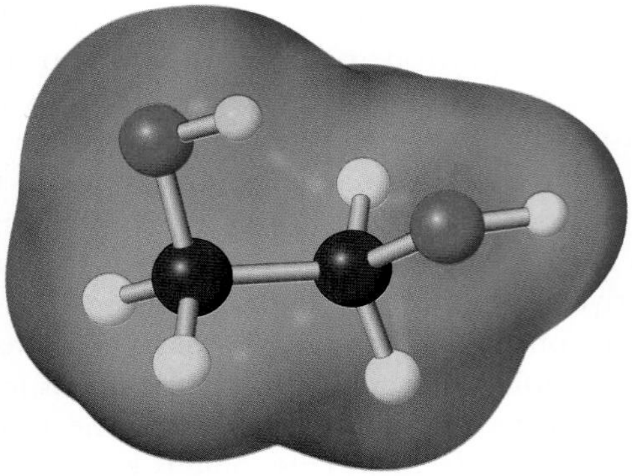

Figure 5.3 Electron density of ethylene glycol. Ethylene glycol ($C_2H_6O_2$) is a polar compound that dissolves in water and is the major ingredient in antifreeze.

Not all ionic compounds dissolve in water. For example, copper(II) cyanide is not soluble in water. The attractive force that water has for the ions is not strong enough to overcome the electrostatic attraction between the ions in the ionic bond. The solubility of ionic compounds is examined further in Section 5.5.

Molecular compounds do not form ions when they dissolve in water. However, water molecules still have to interact with the solute molecules if dissolution is to occur. For example, ethylene glycol, the main ingredient in antifreeze, is a molecular compound that is capable of dissolving in water. When you examine the model of ethylene glycol in Figure 5.3, it is apparent that, like water, it is a polar molecule. The positive regions of ethylene glycol molecules are attracted to the negative regions of water molecules.

There is a common saying in the laboratory that "like dissolves like," which means that, in general, polar compounds will dissolve in a polar solvent and nonpolar compounds will dissolve in nonpolar solvents. Ethylene glycol is soluble in water because both are polar compounds.

5.3 Rate of Dissolving Soluble Compounds

Learning Objective

Describe how temperature, surface area, and concentration affect the rate at which a compound dissolves.

The fact that a compound will form an aqueous solution does not mean that the dissolving process will occur rapidly. Several factors influence the rate at which a compound will dissolve. Figure 5.4 is a graph of temperature versus the maximum amount of each of several compounds that can dissolve in solution. For most compounds, as the temperature increases, the solubility of a substance increases. Because the rate at which compounds dissolve also increases with increasing temperature, it is common to speed the dissolution of a compound in the laboratory by heating the sample.

If a sugar cube is dropped into a cup of water, it will slowly dissolve. But if the same amount of powdered sugar is added to a cup of water, it will dissolve almost immediately. This occurs because the surface area of the solid compound affects the rate of dissolution. By increasing the surface area of the solid, more solvent molecules interact

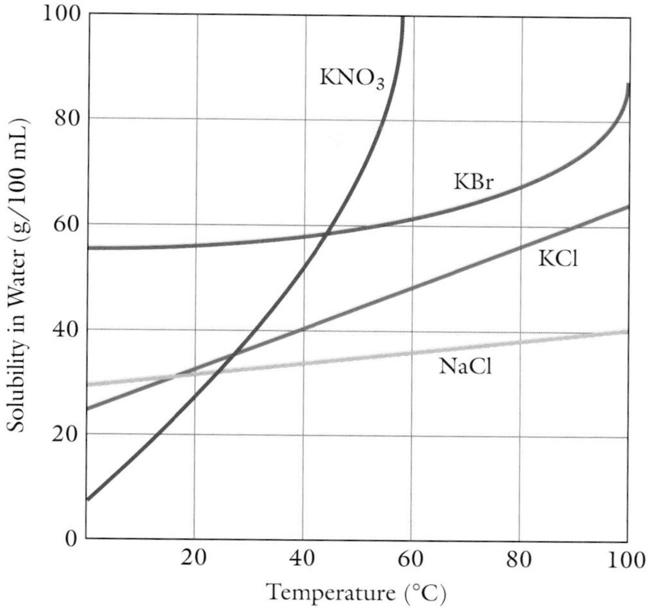

Figure 5.4 Solubility as a function of temperature. The amount of a solute that can dissolve in a solvent increases as a function of temperature. It is a common procedure to heat a solution that is being prepared in the laboratory.

In the lab, solutions are often simultaneously stirred and heated to increase the rate of dissolution. (Tom Pantages Stock Photos)

with the solid and the speed of the dissolution process increases. A solid cube that measures 1 cm × 1 cm × 1 cm has a surface area of 6 cm^2 (1 cm^2 for each of the six sides of the cube). By simply halving the cube, the surface area increases to 8 cm^2 and, as shown in Figure 5.5, the surface area increases to 12 cm^2 by splitting the original cube into eight smaller cubes.

The rate at which a solute is dissolved by a solvent is also influenced by the amount of solute that has already been dissolved. When

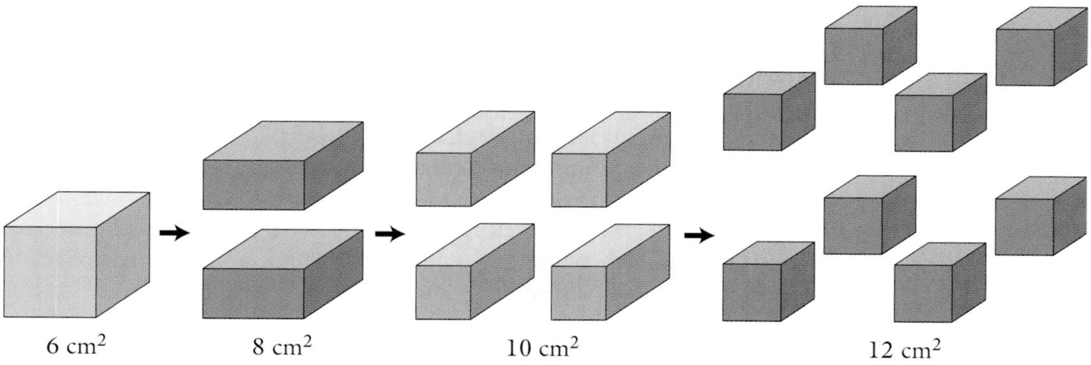

Figure 5.5 Total surface area as a function of particle size. The surface area of a solute increases dramatically with decreasing particle size. Each division of the original cube results in an increase in the total surface area of the original solid. A large surface area allows a larger amount of solvent to reach the solute, thus increasing the rate of dissolution.

the solute is first added, there is only pure solvent surrounding the solid particles at the interface of the solute and solvent, and the solid interacts with the solvent quickly. However, as the solute continues to dissolve, the solvent molecules surround those particles of the solute that are already dissolved. In order for the solid solute to dissolve, fresh solvent is needed to interact with the solid. Stirring the solution will bring fresh solvent to the surface. When solutions are prepared, it is common to both heat and stir solutions in the laboratory to speed up the dissolving process.

5.4 Solution Properties

Learning Objective

Distinguish between electrolytic and nonelectrolytic solutions.

Aqueous solutions exhibit different kinds of properties, depending on the type of solute present. One of these properties is the ability or inability of the solution to conduct an electric current. A compound that can conduct electricity when dissolved in water is called an **electrolyte**. All compounds that dissociate into ions in water are electrolytes—examples are sodium chloride and hydrochloric acid. A compound that cannot conduct electricity when dissolved in water is called a **nonelectrolyte**—examples are nonionic compounds such as sugar and ethanol.

Solutions in which ionic substances are dissolved in water conduct electricity by completing an electrical circuit. This occurs because the dissolved ions are no longer held together—they are free to move throughout the solution. When two electrodes are placed into the solution, the cations are attracted to the negative electrode and move toward it; the anions are attracted to the positive electrode and move toward it. The movement of ions between the two electrodes constitutes a flow of charge through the solution.

Even though all electrolytes conduct electricity in solution, they do not all do so to the same extent. Figure 5.6 shows an experimental apparatus in which a light bulb is connected to a power supply and to two electrodes in an open circuit. When the electrodes are immersed in solution, the circuit will be closed and electricity will flow, provided that the solution contains an electrolyte.

Figure 5.6a shows the electrodes immersed in pure water: Electricity cannot flow between the electrodes, so the light bulb remains dark. Figure 5.6b shows an electrode immersed in a **strong electrolyte**: The light bulb is bright because the ionic compound is fully dissociated into free ions and a large amount of electricity is flowing. Figure 5.6c shows an electrode immersed in a **weak electrolyte**: The light bulb is dim because the ionic compound is only partially dissociated into free ions and therefore only a small amount of electricity can pass through the solution. Examples of weak electrolytes are acetic acid and ammonium hydroxide. If a nonelectrolyte is dissolved in water, ions do not form and electricity will not pass through the light bulb.

The previous chapter provided many examples of ions that are important in forensic science investigations. The ability of ions to conduct electricity is one method that is used to detect the presence of ions after they have been separated from a mixture by chromatography. DNA analysis is also based on the movement of ions between electrodes.

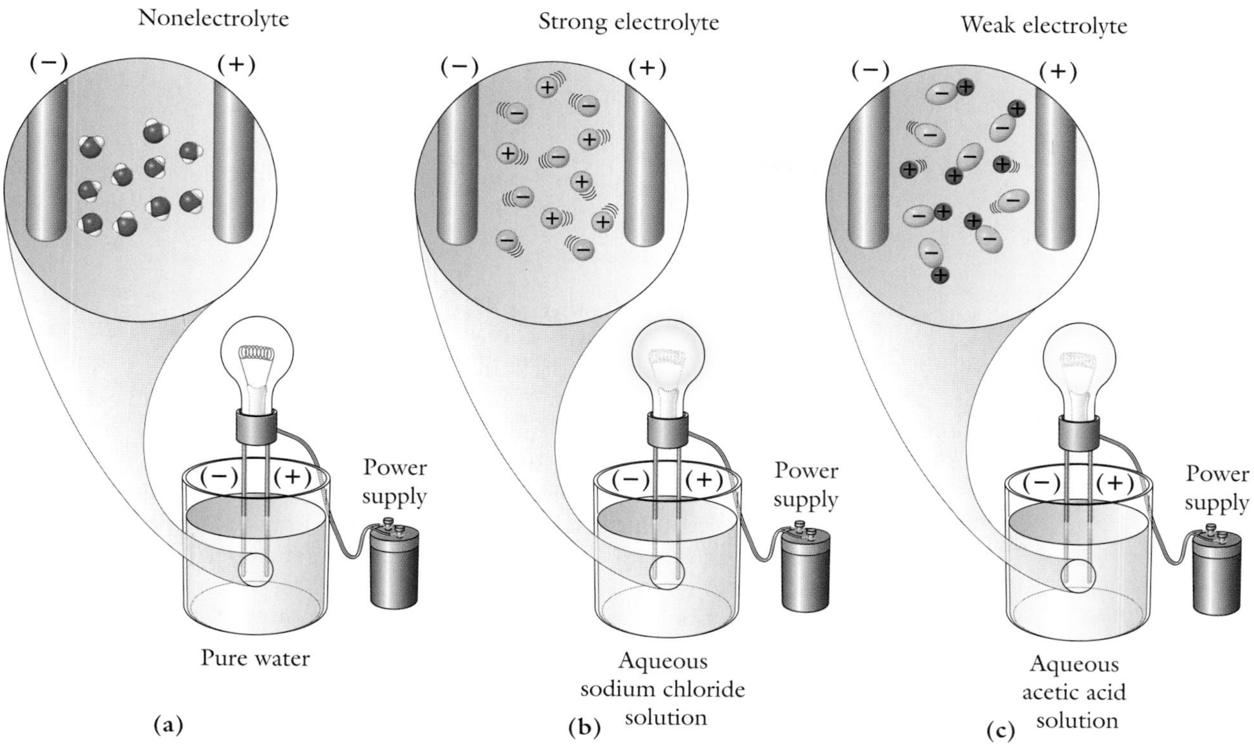

Figure 5.6 The ability of a solution to conduct electricity is directly related to the ability of the solute to dissociate into component ions during the dissolution process. Nonelectrolytes will not conduct electricity, but strong electrolytes are excellent conductors of electricity. Weak electrolytes are poor conductors, as they exist in both a dissociated ionic form and in an undissociated form.

Worked Example 1

Which of the following compounds are electrolytes?

(a) N_2O

(b) $Fe(NO_3)_2$

(c) C_3H_8

SOLUTION

(a) N_2O is a covalent compound that doesn't form ions: nonelectrolyte.

(b) $Fe(NO_3)_2$ is an ionic compound: electrolyte.

(c) C_3H_8 is a covalent compound that doesn't form ions: nonelectrolyte.

Practice 5.1

Given below is a partial list of ingredients found in a sports drink, listed by both name and chemical formula. Determine which compounds are nonelectrolytes and which are electrolytes.

Water, H_2O
Sucrose syrup, $C_{12}H_{22}O_{11}$

Sodium citrate, $Na_3C_6H_5O_7$
Potassium citrate, $K_3C_6H_5O_7$
Sucralose, $C_{12}H_{19}C_{13}O_8$

ANSWER
Electrolytes: Sodium and potassium citrate are ionic compounds. Nonelectrolytes: Water (solvent), sucrose, and sucralose are covalent compounds.

5.5 Solubility and Solubility Rules

Learning Objective

Determine when a solution is saturated and whether or not a compound is soluble.

The maximum amount of solute that can be dissolved into a certain quantity of solvent varies from one compound to another. Table 5.1 shows the solubility of several cyanide compounds in water. Notice that 50 g of potassium cyanide can dissolve in 100 mL of water but only 1.7 g of cadmium cyanide will dissolve in the same volume. Solutions of the cyanide salts that are highly soluble (KCN and NaCN) can be dangerous because a very small quantity of solution can supply a lethal dose of cyanide—less than 0.1 g.

A solution that contains the maximum amount of solute that can be dissolved is considered **saturated** (the values listed in Table 5.1 correspond to saturated solutions). A solution that contains less solute than the solubility value is referred to as an **unsaturated** solution. A more mathematical approach to expressing concentration is discussed later in this section.

Figure 5.4 shows that the temperature of the solvent changes the solubility of compounds. If a saturated solution is cooled, the solubility of the compound decreases, and under normal circumstances the excess amount of solute will come out of solution as a solid. But an interesting situation can arise when a saturated solution is made at an elevated temperature and is then carefully and slowly cooled down. The excess solute may remain in solution, resulting in a **supersaturated** solution. There is more solute dissolved in a supersaturated solution than the solvent should be able to hold at the now-lowered temperature. Supersaturated solutions are not stable and will often return to the saturated level by crystallizing out the excess dissolved solute, as shown in Figure 5.7.

Table 5.1 Solubility of Cyanide Compounds in Water at 25°C

Compound	Solubility (g/100 mL)
Potassium cyanide	50
Sodium cyanide	48
Cadmium cyanide	1.7
Copper(II) cyanide	0

Worked Example 2

A 500.0-mL solution contains 225 g of potassium cyanide. Is the solution unsaturated, saturated, or supersaturated?

Figure 5.7 A supersaturated solution. A solution of supersaturated sodium acetate (left) is unstable and will crystallize out of solution the excess dissolved solid (right). (Richard Megna/Fundamental Photographs)

SOLUTION Using Table 5.1, a saturated solution contains 50 g of KCN per 100 mL. Using this as a conversion factor,

$$\frac{50\text{ g KCN}}{100\ \cancel{\text{mL}}} \times 500.0\ \cancel{\text{mL}} = 250.0\text{ g KCN}$$

The solution is unsaturated because it contains only 225 g of KCN.

Practice 5.2

How many grams of NaCN are needed to make 25 mL of the saturated solution?

ANSWER

12 g of NaCN

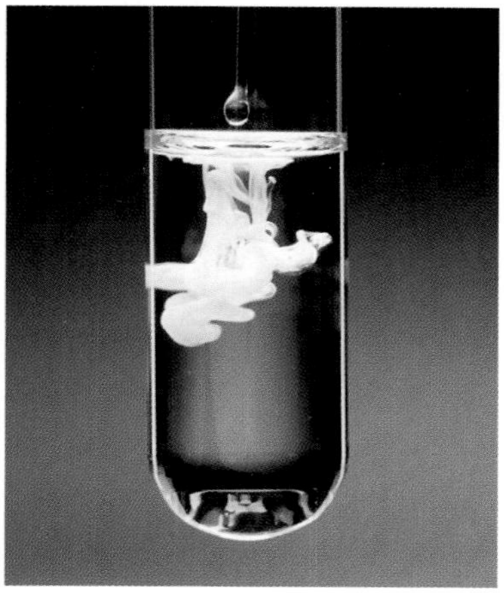

The reaction of silver and chloride ions to form a precipitate of solid silver chloride. (Charles D. Winters/Photo Researchers)

As we saw in Chapter 4, forensic scientists use a variety of precipitation reactions. The presence of the chloride ion in urine can be determined by the addition of the silver ion:

$$Ag^+(aq) + Cl^-(aq) \rightarrow AgCl(s) \qquad \textbf{(Reaction 5.1)}$$

All precipitation reactions share one common trait: the formation of an insoluble product called the *precipitate*. Which ions will form insoluble precipitates and which ions will remain dissolved in water? That depends on the solubility of each compound. A list of solubility rules is given below, but it should be noted that many of the rules have exceptions.

Rules for Soluble Compounds

1. All compounds containing ammonium ions (NH_4^+)
2. All compounds containing alkali metal ions (Li^+, Na^+, and K^+)
3. All compounds containing acetate ions ($C_2H_3O_2^-$)

4. All compounds containing nitrate ions (NO_3^-)
5. Most compounds containing chloride (Cl^-), bromide (Br^-), or iodide (I^-) ions, with the exceptions of compounds with these cations: Ag^+, Hg_2^{2+}, or Pb^{2+}
6. Most compounds containing sulfate ions (SO_4^{2-}), with the exception of compounds with these cations: Ca^{2+}, Ba^{2+}, Sr^{2+}, or Pb^{2+}

Rules for Insoluble Compounds

7. All carbonate (CO_3^{2-}), phosphate (PO_4^{3-}), and chromate (CrO_4^{2-}) ions, except when the cation comes from those listed in rules 1 and 2
8. All sulfide ions (S^{2-}), except when the cation is from those listed in rules 1 and 2 or is Ca^{2+}, Sr^{2+}, or Ba^{2+}

Worked Example 3

Consulting the solubility rules, identify each of the following compounds as soluble or insoluble in water.

(a) $Pb(C_2H_3O_2)_2$

(b) $CaSO_4$

(c) $CuCO_3$

(d) LiF

SOLUTION

(a) Soluble (rule 3)

(b) Insoluble (rule 6, exception)

(c) Insoluble (rule 7)

(d) Soluble (rule 2)

Practice 5.3

What ion would you use to form an insoluble compound with the following ions?

(a) Cl^-

(b) SO_4^{2-}

(c) CO_3^{2-}

(d) NH_4^+

ANSWER

(a) Ag^+, Pb^{2+}, or Hg_2^{2+}

(b) Ca^{2+}, Ba^{2+}, Sr^{2+}, or Pb^{2+}

(c) Any cation other than those listed in rules 1 and 2

(d) Can't make an insoluble ammonium compound.

Worked Example 4

A beverage is suspected to have been tampered with, by the addition of a winter ice remover. Ice removers are typically made up of sodium chloride, calcium chloride, magnesium chloride, or a combination of these three salts. Using the solubility rules, create a strategy for adding reagents that would cause precipitation to determine which salt(s), if any, were in the beverage. Indicate whether any ion cannot be determined by precipitation.

SOLUTION There is more than one correct solution to this problem.

1. All three salts share the chloride anion. The presence of the chloride ion can be verified by adding Ag^+.
2. The sodium ion does not form a precipitate with any of the reagents listed in the solubility rules, so the presence of sodium cannot be determined with this method.
3. The calcium ion would form a precipitate with the addition of SO_4^{2-}.
4. The magnesium ion would form a precipitate with the addition of S^{2-}.
5. As the presence of the calcium, magnesium, and chloride ion is not proof of tampering, a control sample of the beverage that is known not to have been tampered with should also be tested.

Practice 5.4

Why would you not want to use Na_2SO_4 or Na_2S to precipitate calcium and magnesium in Worked Example 4?

ANSWER

Sodium is one of the ions that are suspected to have been added, so future tests on the sample would be compromised by the addition of sodium during the preliminary testing of the sample.

5.6 Net Ionic Reactions

Learning Objective

Explain how the driving force behind a precipitation reaction is the formation of an insoluble compound.

"I've found it! I've found it," he shouted to my companion, running towards us with a test-tube in his hand. "I have found a re-agent which is precipitated by hemoglobin, and by nothing else." Had he discovered a gold mine, greater delight could not have shone upon his features. . . . As he spoke, he threw into the vessel a few white crystals, and then added some drops of a transparent fluid. In an instant the contents assumed a dull mahogany colour, and a brownish dust was precipitated to the bottom of the glass jar.

—Sir Arthur Conan Doyle, *A Study in Scarlet*

Thus far, we have been considering examples of solutions that have only one solute present. What happens if we mix solutions together, as Sherlock Holmes did in *A Study in Scarlet*? The answer depends on whether any of the solutes are capable of reacting with one another, or if they merely remain in solution without reacting. For example, if we mix aqueous solutions of lead(II) nitrate and potassium iodide, a bright yellow substance, lead(II) iodide, forms and eventually settles out of solution as a precipitate. The equation for this reaction is

The reaction of lead(II) nitrate with potassium iodide to form a precipitate of lead(II) iodide. (David Taylor/Photo Researchers)

$$Pb(NO_3)_2(aq) + 2KI(aq) \rightarrow PbI_2(s) + 2KNO_3(aq) \qquad \textbf{(Reaction 5.2)}$$

Recall that when ionic compounds dissolve in water, the ions are solvated (surrounded by solvent molecules) and become independent of one another. The Pb^{2+} ions move through solution independently of the two

NO_3^- ions. The same is true for the K^+ and I^- ions found in the reactant side of the reaction.

The equation for Reaction 5.2 can be rewritten so that all of the aqueous ionic compounds are written as free ions, as shown below in a **total ionic equation**, which is a more accurate depiction of what is occurring in solution. Notice that the precipitate, PbI_2, is not written as free ions because it is an insoluble compound according to solubility rule 5.

$$Pb^{2+}(aq) + 2NO_3^-(aq) + 2K^+(aq) + 2I^-(aq) \rightarrow PbI_2(s) + 2K^+(aq) + 2NO_3^-(aq)$$

A close examination of this equation shows that the potassium ion exists on both the reactant and product sides of the equation without undergoing a change. The nitrate ion is also found on both sides of the equation and therefore does not undergo any reaction or change. Ions that do not actively form a new compound in a chemical reaction are called **spectator ions**. The spectator ions in this equation can be cancelled out to produce the **net ionic equation**, as shown below.

$$Pb^{2+}(aq) + \overbrace{2NO_3^-(aq)}^{\text{Spectator ion}} + \overbrace{2K^+(aq)}^{\text{Spectator ion}} + 2I^-(aq) \rightarrow PbI_2(s) + \overbrace{2K^+(aq)}^{\text{Spectator ion}} + \overbrace{2NO_3^-(aq)}^{\text{Spectator ion}}$$

$$Pb^{2+}(aq) + 2I^-(aq) \rightarrow PbI_2(s) \qquad \textbf{(Reaction 5.3)}$$

Reaction 5.3 shows the net effect of Reaction 5.2, which is the combination of lead(II) ions with iodide ions to form the solid precipitate lead(II) iodide. The spectator ions, while they do not react, play a vital role in the reaction: They provide charge balance for the reacting ions because it is impossible to add just Pb^{2+} or I^- ions to a reaction.

Worked Example 5

A scientist wishes to detect the presence of lead ions by precipitation with potassium iodide. However, the laboratory does not have any potassium iodide in stock. Suggest an alternative compound that could be used to produce lead(II) iodide.

SOLUTION For this reaction, the important part of potassium iodide is the iodide ion. The potassium serves as a spectator ion. If the stock room has another soluble iodide salt such as NaI or NH_4I, the identical net ionic reaction will occur.

Practice 5.5

The precipitation of AgCl(s) calls for the addition of $AgNO_3(aq)$ to $CaCl_2(aq)$. Write the overall balanced equation, the total ionic equation, and the net ionic equation. If $CaCl_2(aq)$ is not available, suggest another alternative chloride salt that could be used.

ANSWER

The overall balanced equation is

$$2AgNO_3(aq) + CaCl_2(aq) \rightarrow Ca(NO_3)_2(aq) + 2AgCl(s)$$

The total ionic equation is

$$2Ag^+(aq) + 2NO_3^-(aq) + Ca^{2+}(aq) + 2Cl^-(aq) \rightarrow Ca^{2+}(aq) + 2NO_3^-(aq) + 2AgCl(s)$$

The net ionic equation is

$$2Ag^+(aq) + 2Cl^-(aq) \rightarrow 2AgCl(s)$$

which must be simplified to

$$Ag^+(aq) + Cl^-(aq) \rightarrow AgCl(s)$$

NaCl could be used as an alternative chloride salt that is soluble in water.

Worked Example 6

What is the net ionic equation for the precipitation of magnesium phosphate from the addition of a magnesium nitrate solution to a sodium phosphate solution?

SOLUTION The overall balanced equation is

$$3Mg(NO_3)_2(aq) + 2Na_3PO_4(aq) \rightarrow Mg_3(PO_4)_2(s) + 6NaNO_3(aq)$$

The total ionic equation is

$$3Mg^{2+}(aq) + 6NO_3^-(aq) + 6Na^+(aq) + 2PO_4^{3-}(aq) \rightarrow Mg_3(PO_4)_2(s) + 6Na^+(aq) + 6NO_3^-(aq)$$

The net ionic equation is

$$3Mg^{2+}(aq) + 2PO_4^{3-}(aq) \rightarrow Mg_3(PO_4)_2(s)$$

Practice 5.6

What are the spectator ions in the reaction of an ammonium chromate solution with a copper(II) bromide solution?

ANSWER
NH_4^+ and Br^-

5.7 Mathematics of Solutions: Concentration Calculations

Learning Objective

Determine the concentration of a solution.

For most purposes in the laboratory, it is important to know how much solute is dissolved in the solvent. A solution that contains 50.0 g of potassium cyanide per liter is quite different from a solution that contains 0.05 g of potassium cyanide per liter.

The solubility values in Table 5.1 indicate concentration by giving the number of grams of solute present in 100 mL of solution, and represent the mass of solute required to produce a saturated solution. However, most solutions used in the laboratory are unsaturated solutions. The concentration unit most commonly used in chemistry is **molarity** (M), which is defined as the number of moles of solute per liter of solution.

$$\text{Molarity} = \frac{\text{moles of solute}}{\text{liters of solution}}$$

You have already learned to calculate the number of moles of a compound from the atomic mass of the compound. For example, the molecular mass of NaCN is 49.01 amu. Therefore, one mole of NaCN has a mass of 49.01 g. If 49.01 g are dissolved into 1.00 liter, the concentration of the resulting solution would be 1.00 M. Expressing the concentration in units of molarity accurately communicates the concentration of a solution. Even if the mass of solute given is not equal to the molar mass or the volume is not expressed in liters, it is possible to calculate the molarity of a solution using conversion factors, as shown below.

Worked Example 7

What is the molarity of a solution prepared by dissolving 24.5 g of NaCN in water to a final volume of 250.0 mL?

SOLUTION It may be easier to do this calculation in two steps. First, calculate the moles of NaCN:

$$24.5\ \cancel{g}\ \text{NaCN} \times \frac{1\ \text{mol NaCN}}{49.01\ \cancel{g}\ \text{NaCN}} = 0.500\ \text{mol NaCN}$$

Then, calculate the volume of solution in liters:

$$250.0\ \cancel{mL} \times \frac{1\ \text{L}}{1000\ \cancel{mL}} = 0.2500\ \text{L}$$

The final step is to combine these two values in the formula for molarity:

$$\text{M} = \frac{0.500\ \text{mol NaCN}}{0.2500\ \text{L}} = 2.00\ \text{M NaCN}$$

Practice 5.7

What is the molarity of a saturated solution of sodium cyanide?

ANSWER
9.8 M NaCN

Worked Example 8

A silver nitrate solution is used to develop fingerprints on porous objects such as wood by precipitating any chloride ions in the fingerprint. The solution is prepared by dissolving 30.0 g of silver nitrate in water to a final volume of 1.00 L. What is the molarity of this solution?

SOLUTION

$$M = \frac{\text{moles of AgNO}_3}{\text{liters of solution}}$$

$$30.0\ \cancel{g}\ AgNO_3 \times \frac{1\ mol\ AgNO_3}{169.87\ \cancel{g}\ AgNO_3} = 0.177\ mol\ AgNO_3$$

$$M = \frac{0.177\ mol\ AgNO_3}{1.00\ L} = 0.177\ M\ AgNO_3$$

Practice 5.8

Fingerprints that have been deposited in blood can be developed using an aqueous mixture with the dye amido black. One of the main ingredients in this mixture is a citric acid solution prepared by dissolving 38.0 g of citric acid ($C_6H_8O_7$) in water, to a final volume of 2.00 L. What is the concentration of the citric acid solution?

ANSWER

0.198 M $C_6H_8O_7$

Dilution Calculations

The forensic scientist has to work with solution dilutions regularly. Recall that the spectrophotometry experiment discussed in Chapter 4 required solutions of known concentrations. These standards are made from diluting a **stock solution**, an accurately prepared solution of known concentration that can be diluted to make solutions of lower concentration. Dilution calculations are all determined using the formula

$$\underbrace{M_1V_1}_{\text{Initial}} = \underbrace{M_2V_2}_{\text{Final}}$$

The initial solution concentration (M_1) is multiplied by the initial volume of solution (V_1) and set equal to the final solution concentration (M_2) multiplied by the final volume of solution (V_2). It should be noted that as long as the units for the concentrations and volumes are the same on both sides of the equation, any units may be used—be it liters, milliliters, molarity, or percentages.

Worked Example 9

A presumptive test for LSD is the Erlich test that requires a 3.25 M HCl solution. Concentrated hydrochloric acid—the form in which most laboratories purchase HCl—is 12.1 M. If a 100.0-mL sample of 3.25 M HCl is needed, how many mL of the concentrated HCl should be diluted?

SOLUTION

$$M_1V_1 = M_2V_2$$

$$\left.\begin{aligned} M_1 &= 12.1\ M \\ V_1 &= ? \end{aligned}\right\}\text{Initial solution}$$

$$\left.\begin{aligned} M_2 &= 3.25\ M \\ V_2 &= 100.0\ mL \end{aligned}\right\}\text{Final solution}$$

$$(12.1\text{ M})(V_1) = (3.25\text{ M})(100\text{ mL}) \Rightarrow$$
$$V_1 = \frac{(3.25\text{ M})(100\text{ mL})}{12.1\text{ M}} = 26.9\text{ mL}$$

Practice 5.9

Concentrated hydrochloric acid is 37.2% HCl. Calculate the final percentage of HCl of the diluted solution used in the Erlich test from Worked Example 9.

ANSWER

10.0% HCl

Worked Example 10

A presumptive test is usually a quick, cheap test that, with a positive result, indicates the drug of interest is most likely present; a confirmatory test is required to definitively prove the presence of the drug. For heroin, the presumptive test involves adding a mercury(II) chloride solution, which forms a rosette of needle-shaped crystals if heroin is present. The directions for preparing the solution call for 13.9 mL of a 1.00 M stock solution to be diluted to 75.0 mL. What is the final molarity of the mercury(II) chloride test solution?

SOLUTION

$$\left.\begin{aligned} M_1V_1 &= M_2V_2 \\ M_1 &= 1.00\text{ M} \\ V_1 &= 13.9\text{ mL} \end{aligned}\right\}\text{ Initial solution}$$
$$\left.\begin{aligned} M_2 &= ? \\ V_2 &= 75.0\text{ mL} \end{aligned}\right\}\text{ Final solution}$$

$$(1.00\text{ M})(13.9\text{ mL}) = (M_2)(75.0\text{ mL}) \Rightarrow$$
$$M_2 = \frac{(1.00\text{ M})(13.9\text{ mL})}{(75.0\text{ mL})} = 0.185\text{ M }HgCl_2$$

Practice 5.10

The drug PCP will form crystals that look like branching needles when it reacts with a potassium iodide solution. What concentration (M) should the stock solution be if the procedure calls for 25.0 mL of stock solution to be diluted to 100.0 mL for a 0.600 M KI solution?

ANSWER

2.40 M KI

5.8 Acid Chemistry

Learning Objective

Distinguish between strong and weak acids.

In previous sections of this chapter, we have considered some examples of aqueous solutions, ways of expressing their concentrations, and factors that affect the solubility of compounds. In this section and the next, we turn our attention to two common types of aqueous solutions—acids and bases—that are important in the chemistry laboratory, in forensic science, and in consumer products.

Table 5.2 Names, Formulas, and Sources of Various Acids

Name	Formula	Strength	Common Use
Hydrochloric acid	HCl	Strong	Pool chemicals
Nitric acid	HNO_3	Strong	Acid rain
Sulfuric acid	H_2SO_4	Strong	Car batteries
Acetic acid	$HC_2H_3O_2$	Weak	Vinegar
Carbonic acid	H_2CO_3	Weak	Soft drink
Hydrofluoric acid	HF	Weak	Wheel cleaner
Phosphoric acid	H_3PO_4	Weak	Hair coloring

Historically, criminals would pour highly corrosive battery acid on the faces and hands of a murder victim in an attempt to disfigure the body beyond identification. With the advent of DNA analysis, disfigurement no longer prevents identification. Officers and investigators often come across various acids at the scene of clandestine drug labs that create a potentially hazardous situation. The possession of key chemical precursors, including acids, for the manufacture of illegal drugs is sufficient in most jurisdictions to secure an arrest for the intent to produce illegal drugs. However, although acids play a significant role in criminal cases such as these, acids are more commonly used for legitimate purposes in the laboratory and in consumer products. A simple definition of an **acid** is any compound that can release the hydrogen ion, H^+, into solution. Most acids are easily identified by the fact that the formula starts with the hydrogen atom. Table 5.2 lists some commonly used acids and products that contain them.

Acids can be classified as **weak** or **strong**, indicating whether the acid is a weak or strong electrolyte. For example, a weak acid such as acetic acid only partially dissociates into its ions (H^+ and $C_2H_3O_2^-$) in solution. Therefore, a solution of acetic acid is a weak conductor of electricity. A strong acid, such as hydrochloric acid (HCl), completely dissociates into ions in solution and is a strong conductor.

Hydrofluoric acid is an example of a weak acid, but lest you think weak acids are not dangerous, hydrofluoric acid can easily dissolve glass! It is important to understand that the *weak* and *strong* qualifiers do *not* refer either to concentration or danger level.

Acids undergo a reduction-oxidation reaction with certain metals. For example, the hydrogen ion (H^+) will be reduced to hydrogen gas (H_2) when iron metal is added to an acid. The iron metal is in turn oxidized to form iron(II) ions. This reaction, shown in Figure 5.8, can be used to visualize serial numbers that have been removed from a firearm. When a serial number is stamped into the metal, it causes deformations in the solid structure of metal that extend deeper than the numbers. When acids are placed on the area where a removed serial number was located, the reaction occurs faster in the area that was deformed. The iron is removed where the serial numbers had been, and the numbers are visually restored. A kit containing acid solutions for various types of metals is shown in Figure 5.8.

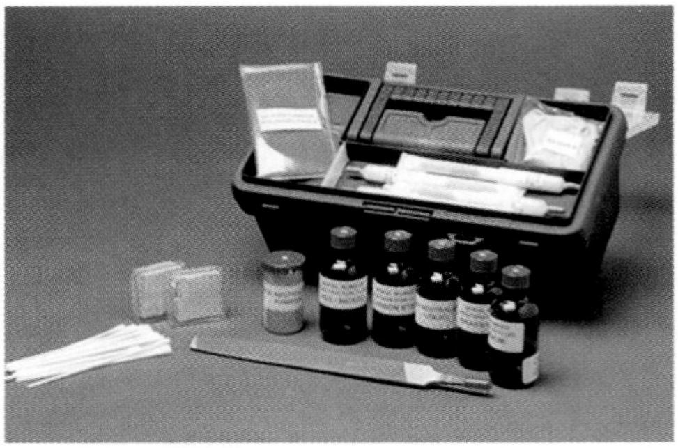

Figure 5.8 A serial number restoration kit contains a collection of acid solutions used to restore the serial number etchings on a variety of metal surfaces. (Courtesy Wisconsin Dept. of Justice, Crime Labs Division)

5.9 Base Chemistry

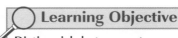

Learning Objective

Distinguish between strong and weak bases.

A simple definition of a **base** is any compound that produces the hydroxide ion, OH^-, in solution. In neutralization reactions, a base reacts with an acid to form a salt and water. Table 5.3 is a list of common bases that can be found in everyday consumer products. The terms *caustic, lye, alkaline,* and *alkali* are all associated with compounds that are bases.

Ammonium hydroxide is also called *aqueous ammonia* because it is formed by the reaction of ammonia and water, as illustrated in the following reaction:

$$NH_3(g) + H_2O(l) \rightarrow NH_4OH(aq) \quad \textbf{(Reaction 5.4)}$$

From Table 5.3 it is clear that basic compounds are mostly used as cleaning agents. The reason for this is the ability of bases to react with grease and fats to produce water-soluble soap. In fact, the traditional recipe for making soap is to combine animal fat with lye (NaOH). This type of reaction is called a **saponification** reaction. When a kitchen sink is clogged, it is due to an accumulation of grease on the inside of the pipes. Pouring concentrated sodium hydroxide down the drain converts the grease into soap, which then dissolves in the water, releasing the clog.

The action of soap depends on the "like dissolves like" principle. The soap molecule has a very long, nonpolar carbon chain with an ionized group at one end, as shown in Figure 5.9. The ionized region of the soap molecule associates with the positive polar region of water molecules. This interaction is strong enough to keep the nonpolar region of the molecule in solution. When a long nonpolar grease molecule comes in contact with

Table 5.3 Names, Formulas, and Sources of Various Bases

Name	Formula	Strength	Common Use
Ammonium hydroxide	NH_4OH	Weak	All purpose cleaners
Sodium hydroxide	NaOH	Strong	Drain cleaners
Potassium hydroxide	KOH	Strong	All purpose cleaners
Calcium hydroxide	$Ca(OH)_2$	Strong	Toilet bowl cleaners

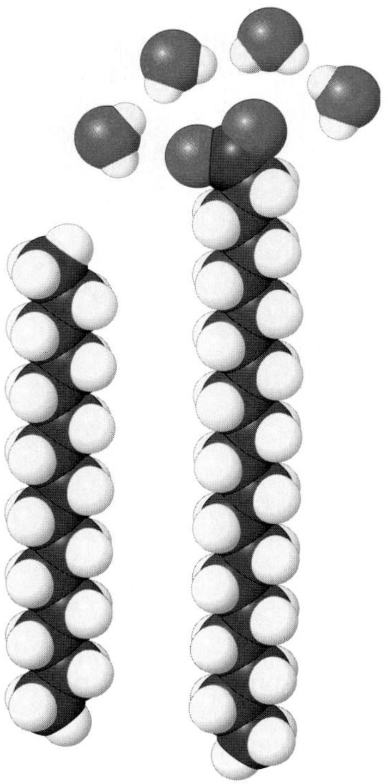

Figure 5.9 How soap dissolves oil and grease. (a) Oil and grease molecules consist of long chains of carbon and hydrogen atoms. (b) The soap molecule has a similar long-chain structure but also has an ionized polar end. The polar end interacts strongly with water, serving to bridge the solubility gap between water and oil.

the soap molecule, the nonpolar carbon chains associate with one another and the grease dissolves in the water. The nature of the interactions between molecules will be explored in greater detail in Chapter 6.

An unusual situation arises when a human body, which in the average person contains 10 to 30% body fat, is disposed of in alkaline soil. The normal decomposition process does not occur. Instead the saponification reaction occurs and the body is converted into a waxy, soaplike substance called *adipocere* or *grave wax*.

5.10 The pH Scale and Buffers

Learning Objective

Describe what the pH scale measures and what a buffer does.

In June 1964, Dr. Martin Luther King, Jr., came to St. Augustine, Florida, and was promptly arrested and given a 10-day jail sentence for attempting to eat at the Monson Motor Lodge Restaurant, a segregated restaurant. Dr. King's arrest initiated a string of protests at the motor lodge, resulting in more arrests and larger protests. By June 18, a large number of civil rights activists had joined the sit-in at the Monson Motor Lodge.

At the same time, protestors were having wade-ins at the segregated beaches of St. Augustine. Perhaps inspired by the idea of a wade-in, a group of protestors at the motor lodge decided to go swimming in the segregated motor lodge pool. The hotel manager, James Brock, reacted

Dr. Martin Luther King, Jr., serving a 10-day jail sentence for attempting to eat at a segregated restaurant in St. Augustine, Florida, in June 1964. (Bettmann/Corbis)

to this new development by going into the swimming pool chemical supply area and returning with two jugs of muriatic acid, more commonly known in the chemistry laboratory as hydrochloric acid. Mr. Brock proceeded to pour two jugs of the concentrated acid into the pool in order to drive the protestors out, an incident captured on film as shown below.

Concentrated hydrochloric acid is an extremely dangerous compound that can easily cause severe chemical burns, which was exactly why Mr. Brock chose to add it to the swimming pool. What effect did the hydrochloric acid

James Brock, manager of the Monson Motor Lodge, pours concentrated hydrochloric acid into the hotel pool in an attempt to drive civil rights protestors out. (Bettmann/Corbis)

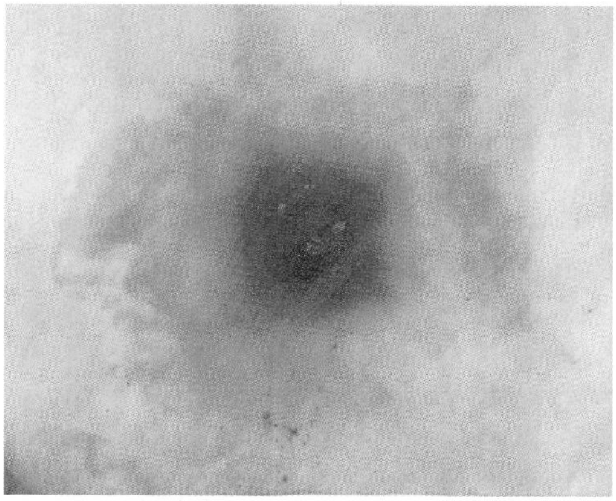

The presence of gunshot residue can be determined by the application of sodium rhodizonate at a pH of 2.8. A positive test will produce a scarlet color. (Jeffrey Scott Doyle/www.FirearmsID.com)

have when added to such a large volume of water? Were the swimmers in danger? How could the effect be determined? The answer lies in understanding a measure of acidity known as the *pH scale*. What happened to the protesters in the swimming pool is described at the end of Section 5.11.

You may have encountered the term *pH* if you have heard or seen commercials for "pH-balanced" consumer products. The term also is used in environmental regulations that specify the legal pH range for discharge of industrial wastewater into the environment. When forensic scientists look for gunpowder patterns, they will pretreat the area by spraying it with a solution that has a pH of 2.8, and will then spray the area with a compound called *sodium rhodizonate* ($Na_2C_6O_6$). A chemical reaction occurs that turns gunpowder residue a pink-red color, but only after the pH of the area has been properly adjusted. Many chemical reactions require a specified pH range as one of the reaction conditions.

But what does **pH** mean and why is it used? The concept of pH was developed as a way of expressing the acidity of a solution. An acid is any compound that produces hydrogen ions in solution. However, a strong acid, such as hydrochloric acid, will produce more free hydrogen ions in solution than will a weak acid, such as acetic acid—even if the strong and weak acids are at equal molar concentrations.

In 1909 Søren Sørenson, a Danish scientist, published work in which he determined the acidity of a sample by measuring how much free hydrogen ion was in solution. He developed the concept of pH to express the acidity: The *p* refers to the negative logarithm to the power of 10, the *H* refers to the hydrogen ion. For every unit change in pH, the concentration of hydrogen ions changes by a factor of 10, as pH is a logarithmic scale. A solution with a pH of 6 has 10 times the amount of hydrogen ion as a solution with a pH of 7.

The pH scale goes from 0 to 14; a pH of 7 is neutral. This is the pH of pure water, which has a very low concentration of hydrogen ion (0.0000001 M) naturally present. Pure water also has the same concentration of hydroxide ions (0.0000001M). If the pH is less than 7, the solution is acidic and the hydrogen ion concentration is greater than the hydroxide concentration. If the pH is greater than 7, the solution is basic and the concentration of hydroxide ions is greater than the concen-

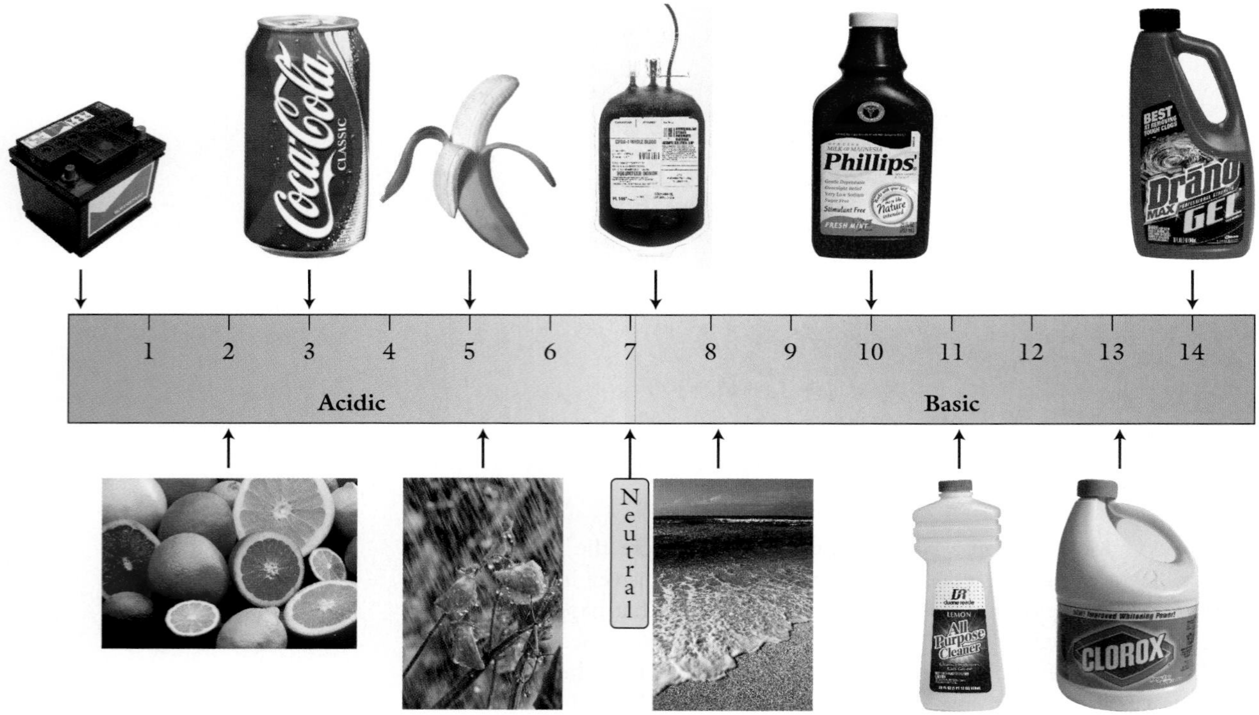

Figure 5.10 The pH scale. The pH of common household solutions can range over the entire pH scale. (a: Ingram Publishing/Alamy; b: The Coca-Cola Company; c: Brand X Pictures/Getty; d: Davies & Starr/The Image Bank/Getty; e, f, j, k: The Photo Works; g: Photodisc Green/Getty Images; h: Tony Cortazzi/Alamy; i: Corbis)

tration of hydrogen ions. Figure 5.10 illustrates the pH scale and the pH of common solutions.

5.11 Mathematics of Solutions: Calculating pH

Learning Objective

Calculate the pH of solutions.

The concentration of free hydrogen ions must be known in order to calculate the pH of a solution. For strong acids, the free hydrogen ion concentration will correspond to the molar concentration of the acid because of the full dissociation of the strong acids. The pH is calculated according to the following equation:

$$\text{pH} = \underbrace{-\log}_{\text{p}}\ \underbrace{[\text{H}^+]}_{\text{H}}$$

The brackets mean the molarity of H^+ in the solution.

Worked Example 11

What is the pH of a 0.10 M HCl solution?

SOLUTION HCl is a strong electrolyte that fully dissociates into H^+ and Cl^-. Therefore, $[H^+] = 0.10$ M. The logarithm of 0.01 to the base 10 is

$$\log [\text{H}^+] = \log (0.1) = -1$$

The pH of the HCl is

$$pH = -\log [H^+] = -[-1] = 1$$

Practice 5.11

What is the pH of a 0.010 M HNO_3 solution?

ANSWER
pH = 2

Worked Example 12

A can of soda is suspected of being tampered with, by the addition of sodium hydroxide. The concentration of H^+ is determined to be 0.00094 M. Does this support the suspicion of tampering by the addition of sodium hydroxide?

SOLUTION Determine the pH of the soda: pH = $-\log$ (0.00094) = 3.03. This is an acidic pH, reasonable for the pH of soda according to Figure 5.10. If sodium hydroxide had been added, the pH should have been higher. Therefore, the pH does not support the suspicion of tampering with sodium hydroxide.

Practice 5.12

If $[H^+]$ of the soda from Worked Example 12 were determined to be 0.000000014 M, would this support the suspicion of tampering by sodium hydroxide?

ANSWER
The pH of the soda is 7.85, so yes, tampering has occurred. Further tests would be needed to confirm whether NaOH was the substance added.

Returning to the swimming pool incident at the Monson Motor Lodge, we can now estimate the effect of the acid on the pH of the water. Recall that hydrochloric acid is a strong acid; therefore, it fully dissociates in water. The final concentration of the hydrogen ion in the swimming pool would correspond to the final concentration of the hydrochloric acid. To calculate the pH of the water in the swimming pool, a few assumptions about the size of the swimming pool will have to be made, because the motor lodge no longer exists. Capacities of typical in-ground swimming pools range from 20,000 to 40,000 gallons. A capacity of 20,000 gallons would represent a worst-case scenario, as this provides the least dilution of the acid. The concentration of hydrochloric acid used for swimming pools is typically 30 to 32%. However, the concentration in the worst-case scenario is 37.2% (12.1 M), the most concentrated form of hydrochloric acid. Assuming 2 gallons of HCl were poured into the pool, the resulting pH is calculated below.

$$\left.\begin{aligned} M_1 &= 12.1 \text{ M} \\ V_1 &= 2 \text{ gal} \end{aligned}\right\} \text{Initial solution}$$

$$\left.\begin{aligned} M_2 &= ? \\ V_2 &= 20{,}000 \text{ gal} \end{aligned}\right\} \text{Final solution}$$

$$M_1V_1 = M_2V_2$$
$$(12.1\ \text{M})(2\ \text{gal}) = M_2\ (20{,}000\ \text{gal}) \Rightarrow$$
$$M_2 = \frac{(12.1\ \text{M})(2\ \text{gal})}{20{,}000\ \text{gal}} = 0.00121\ \text{M} \Rightarrow$$
$$\text{pH} = -\log 0.00121 = 2.92$$

The pH of the swimming pool at the motor lodge would probably not have been this low because most swimming pools are maintained at a pH of 7.2 to 7.8 by a buffer. A **buffer** is a combination of a weak acid and a soluble salt that contains the same anion as the weak acid. A buffer solution has the ability to consume both hydrogen ions and hydroxide ions, which prevents, to an extent, a dramatic change in pH.

If a strong acid is added to a solution buffered with a weak acid and its salt, the hydrogen ions will combine with the anion of the salt to form a molecule of the weak acid. This has the effect of removing the hydrogen ion from the solution. If a strong base is added to a buffered solution, it is neutralized by the weak acid; the products of this reaction are water and the salt of the weak acid. If excessive amounts of an acid or base are added, the capacity of the buffer is overwhelmed and the pH will change. The ability of a buffer to resist a change in pH is illustrated below.

$$\text{Swimming pool buffer:}\quad \underbrace{HCO_3^-}_{\substack{\text{Reacts with}\\ OH^-}} + \underbrace{CO_3^{2-}}_{\substack{\text{Reacts with}\\ H^+}}$$

$$CO_3^{2-} + H^+ \rightarrow HCO_3^- \qquad \textbf{(Reaction 5.5)}$$
$$HCO_3^- + OH^- \rightarrow H_2O + CO_3^{2-} \qquad \textbf{(Reaction 5.6)}$$

Reaction 5.5 illustrates the neutralization of a strong acid by reacting with the carbonate ion, forming the bicarbonate ion (HCO_3^-), a weak acid. A strong base is neutralized in Reaction 5.6 by reacting with the bicarbonate ion to produce water and the carbonate ion. In both situations, the pH of the solution is stabilized because neither the hydroxide ion nor the hydrogen ions remain in solution.

There are many situations in which it is desirable to hold the pH close to a particular value. Blood, for example, has to have a pH value of between 7.35 and 7.45. The medical terms *acidosis* and *alkalosis* refer to conditions in which the buffered pH of blood is overwhelmed, producing serious medical consequences.

The final pH of the swimming pool after the muriatic acid was added would not have been overly dangerous. But the civil rights activists who were near James Brock as he poured the acid into the pool faced a very real danger because the concentrated acid was not diluted instantly.

5.12 Case Study Finale: An Aquatic Apocalypse

Fish kills are notoriously hard to investigate because they can be caused by either criminal dumping of pollutants or by natural means such as a reduction in dissolved levels of oxygen caused by the heat of summer.

However, the magnitude of the White River fish kill simultaneously occurring with the high levels of ammonia at the Anderson POTW led investigators to focus on the Anderson POTW and the industrial companies that discharge wastewater to the Anderson POTW.

Why were elevated ammonia levels reaching the river? The bacteria that should have been converting waste ammonia to nitrite ions and then to harmless nitrate ions were found to be dead within the entire biological wastewater treatment system. But elevated ammonia levels alone could not explain so extensive a fish kill. Also, what killed the bacteria? Something was entering the Anderson POTW that was wreaking havoc on the White River ecosystem.

The Anderson Public Owned Treatment Works, Anderson, Indiana. (The Indianapolis Star)

Samples of wastewater flowing from industrial sites to the Anderson POTW were periodically analyzed, and on December 16, 1999, a sample was evaluated from Guide Corporation, a company that plated copper, chromium, and nickel onto the plastic structures for headlights. The analysis identified a number of highly toxic substances that were poisoning the river and the wastewater treatment system. Upon investigation, the following scenario came into focus.

Guide Corporation had been leasing the manufacturing plant from General Motors with the agreement that the plant would be shut down and decommissioned no later than December 22, 1999. This process required disposing of hundreds of thousands of gallons of plating solutions and metal-containing sludge that had been accumulating at the bottom of the tanks for decades.

To dispose of the plating solution and sludge, the metal ions had to be removed from the solutions. This process was complicated by the fact that plating requires the metal ions in the plating solution to be tightly bound to another compound that keeps them in solution—a process called **chelation**. The first step in removing metal ions from solution is to reverse the chelation process to free the metal ions by providing an alternative ion that binds more strongly to the chelating compound than to the metal ion. Once the metal ions have been released, sodium hydroxide is added to raise the pH and precipitate out the insoluble metal hydroxides.

The final step is to remove any trace amount of metal ions remaining in solution by adding sodium dimethyldithiocarbamate, also known as HMP2000, a compound that is extremely toxic to aquatic life. It must be used with great care in wastewater treatment, because it decomposes to form a compound called *thiram* that is even more toxic. After all of these steps have been properly completed, the precipitated solids have to be allowed to settle out by gravity in a holding tank for 8 to 10 days before the water can be returned to the river.

Guide Corporation was under internal pressure to decommission the facility quickly. In the company's rush, mistakes and blatant violations of protocol ran rampant. The final stage of allowing solutions to settle for 8 to 10 days was reduced to 2 to 4 hours. Under normal conditions, 20 to 30 gallons of HMP2000 would be used to treat a 100,000-gallon batch of solution. During the decommissioning, the amount of HMP2000 needed was miscalculated, and each batch was being treated with 1000 gallons. This meant that over 10,000 gallons were used in 10 days! Vast amounts of unreacted HMP2000 flowed into the Anderson POTW, killing all bacteria in their biological treatment facility and causing the elevated ammonia readings. The plant was now discharging toxic levels of ammonia, HMP2000, and thiram. Illegal levels of the metal ions were also being discharged because some of the batches of plating solution were still chelated when the sodium hydroxide was added, and therefore, the ions did not precipitate.

After extensive legal battles resulting from state and federal lawsuits against Guide Corporation, the company agreed to pay $14 million in fines and penalties to restore the river and restock the fish. The Anderson POTW agreed to make improvements in the treatment facility to prevent another such release. The long-term prospects for the ecological state of the White River are positive because of the decomposition and natural removal of the pollutants from the river.

CHAPTER SUMMARY

- A solution is made up of a solvent (the major constituent) and a solute (the minor component). An aqueous solution specifically refers to the use of water as the solvent.
- Ionic compounds and polar molecular compounds dissolve in water because of the polar nature of water. The difference in the electronegativity of hydrogen and oxygen is such that shared electrons reside closer to the oxygen atom, providing a negative polar region. The relative absence of electrons near the hydrogen atoms results in a positive polar region.
- The water molecules orient their polar regions to attract the oppositely charged region of the solute. By surrounding the solute particle, the water is able to overcome the attractive forces holding the solute particle in the solid.
- The temperature of the solvent, the surface area of the solute, and the concentration of the solute within a solution all affect the rate at which a substance dissolves. A fine powder that is simultaneously heated and stirred will dissolve at the fastest rate.
- Electrolytes are compounds that, when dissolved in water, can conduct electricity since they dissociate into component ions that are free to move between oppositely charged electrodes. Nonelectrolytes are compounds that will not conduct electricity because they do not dissociate into ions upon dissolving.
- Compounds that fully dissociate into ions in solution are called strong electrolytes; those that partially dissociate are called weak electrolytes.
- Saturated solutions contain the maximum concentration of dissolved solute that the solution can hold at a given temperature. Solutions with less than this concentration are termed unsaturated, and those unstable solutions containing more dissolved solute are termed supersaturated.
- Net ionic equations show the ionic species in solution that come together to form an insoluble precipitate. The other ions in solution shown in the total ionic equation provide charge balance but are interchangeable with other ions.
- Concentration of solute in solutions is commonly expressed as molarity (M), which is mathematically equal to the moles of solute divided by the liters of solution. Stock solutions are carefully prepared solutions of known concentration from which dilute solutions can be made.
- An acid is a compound that will produce hydrogen ions when dissolved in water. Acids can be either strong (fully dissociated) or weak (partially dissociated). A base is a compound that will produce hydroxide ions when dissolved in water. Bases can be either strong (fully dissociated) or weak (partially dissociated).
- The pH of a solution is a logarithmic measurement of the free hydrogen ion concentration. The pH scale goes from 0 (acidic) to 14 (basic), with a pH of 7 representing a neutral pH value.
- A buffer is a mixture of a weak acid and a soluble salt of the weak acid. This mixture is capable of reacting with both strong acids and strong bases to prevent a substantial change in the pH of the original solution.

KEY TERMS

aqueous, p. 136
solvent, p. 136
solute, p. 136
polar, p. 137
electrolyte, p. 140
nonelectrolyte, p. 140
strong electrolyte, p. 140
weak electrolyte, p. 140
saturated, p. 142
unsaturated, p. 142
supersaturated, p. 142
total ionic equation, p. 146
spectator ion, p. 146
net ionic equation, p. 146
molarity (M), p. 148
stock solution, p. 149
acid, p. 151
weak acid, p. 151
strong acid, p. 151
base, p. 152
saponification, p. 152
pH, p. 155
buffer, p. 158
chelation, p. 159

CONTINUING THE INVESTIGATION Additional Readings, Resources and References

Doyle, A. C. *Sherlock Holmes: The Complete Novels and Stories*, vol. 1, New York: Random House, 2003.

Lane, Brian. *The Encyclopedia of Forensic Science*, London: Magpie Books, 2004.

Siegel, J. A., ed. *Encyclopedia of Forensic Sciences*, San Diego: Academic Press, 2000.

U.S. Department of Justice, *Processing Guide for Developing Latent Prints*, Washington, D.C.: U.S. Department of Justice, 2000.

For more information about the swimming pool incident in St. Augustine, Florida: www.drbronsontours.com/bronsonhistorypageamericancivilrights.html

For an article on the Food and Drug Administration's efforts to enforce the Federal Anti-Tampering Act: www.fda.gov/fdac/features/695_forensic.html

For a copy of the Environmental Protection Agency's civil complaint against Guide Corporation: www.in.gov/idem/mycommunity/wrcac/whiteriver/complaint/epacomplaint.pdf

For additional information about the fish kill in White River: www2.indystar.com/library/factfiles/environment/white_river/fish_kill.html

REVIEW QUESTIONS AND PROBLEMS

Questions

1. What is a solvent? What is the solvent used in most solutions?
2. What is a solute? How can you tell the difference between the solvent and solute?
3. Sketch a beaker containing an aqueous K_2S solution. Show the water molecules interacting properly with the solute particles. Label the solvent and the solute.
4. Modify your sketch from Question 3 to include a cathode and anode. Indicate with arrows the direction in which the cations and anions will move through solution.
5. How is the dissolution process different for ionic compounds such as table salt (NaCl) compared with polar molecular compounds such as sugar?
6. Historically, mechanics would use gasoline to clean grease from their hands and arms (a very dangerous practice that should not be done!). Explain why they did this.
7. Explain the difference between electrolytes and nonelectrolytes. Give examples of both.
8. Explain the difference between strong and weak electrolytes. Give examples of both, and list what compounds/ions would be found in a solution of each example.
9. Explain the difference between a saturated solution and an unsaturated solution.
10. How could you determine whether a solution is saturated or not?
11. Explain how a supersaturated solution is prepared.
12. Which compound from Figure 5.4 has the greatest solubility at 20°C? At 50°C?

13. What is the role of spectator ions in a precipitation reaction? Can a precipitation reaction occur without spectator ions?

14. You may have heard of the old saying, "If you're not part of the solution, you're part of the problem." Explain the old chemistry joke, "If you're not part of the solution, you're part of the precipitate."

15. Water, a polar molecule, is labeled 1 in the following figure. Hexane, a nonpolar organic solvent, is labeled 2. Which compounds will mix together? Molecule (a) is ethanol, (b) is cyclohexane, and (c) is decane.

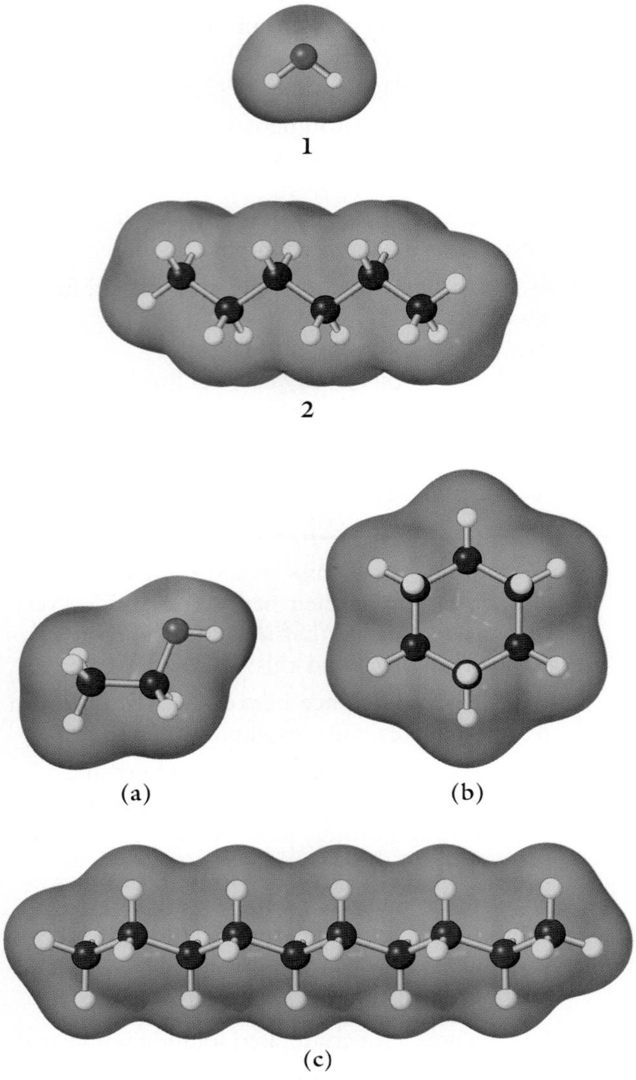

Figure for Question 15.

16. What is the total ionic equation attempting to show? Why is it useful?

17. What is the net ionic equation attempting to show? Why is it useful?

18. What factors affect the rate of dissolving a soluble compound?

19. What three things can be done to increase the rate at which a substance is dissolved?

20. What is the definition of an acid? Of a base?

21. What is the difference between a weak acid and a strong acid? A weak base and a strong base?

22. Are weak acids inherently safer than strong acids?

23. Name two types of reactions that acids can undergo.

24. Name two types of reactions that bases can undergo.

25. What is the purpose of the pH scale? Why was it developed?

26. If the pH of a solution increases from pH = 5.0 to pH = 8.0, what happens to the concentration of hydrogen ions in the solution?

27. What is the purpose of a buffer? How does it work?

28. What could be done experimentally to determine whether a solution contains a buffer?

Problems

29. Identify each of the following compounds as either an electrolyte or a nonelectrolyte.
(a) $NH_4C_2H_3O_2$
(b) CH_3OH
(c) HCl
(d) CO_2

30. Identify each of the following compounds as either an electrolyte or a nonelectrolyte.
(a) $MgCO_3$
(b) S_2O_3
(c) $NaOH$
(d) $C_2H_4O_2$

31. Identify each of the following as either a strong electrolyte or a weak electrolyte.
(a) $CaCl_2$
(b) HCl

(c) NH_4OH
(d) H_3PO_4

32. Identify each of the following as either a strong electrolyte or a weak electrolyte.
(a) HNO_3
(b) NaOH
(c) LiF
(d) H_2CO_3

33. The solubility of sodium cyanide is 48.0 g/100 mL at 25°C. Determine whether the following solutions are saturated, unsaturated, or supersaturated.
(a) 120.0 g of NaCN dissolved in 250.0 mL
(b) 15.0 g of NaCN dissolved in 25.0 mL
(c) 4.00 g of NaCN dissolved in 8.33 mL
(d) 192 mg of NaCN dissolved in 0.400 mL

34. The solubility of cadmium cyanide is 1.70 g/100 mL at 25°C. Determine whether the following solutions are saturated, unsaturated, or supersaturated.
(a) 5.661 g of $Cd(CN)_2$ in 330.0 mL
(b) 348.5 g of $Cd(CN)_2$ in 20.5 L
(c) 3.40 g of $Cd(CN)_2$ in 250.0 mL
(d) 85.0 mg of $Cd(CN)_2$ in 5.00 mL

35. How many grams of potassium cyanide would be needed to make a saturated solution for the volumes indicated below? The solubility of potassium cyanide is 50.0 g/100 mL.
(a) 1.50 L
(b) 2.75 mL
(c) 0.250 L
(d) 500.0 mL

36. How many grams of potassium cyanide would be needed to make a saturated solution for the volumes indicated below? The solubility of potassium cyanide is 50.0 g/100 mL.
(a) 3.4 L
(b) 175 mL
(c) 1.80 mL
(d) 500.0 μL

37. Based on the graph shown in Figure 5.4, determine whether a 40.0-g sample of each salt in solution at a temperature of 40°C is unsaturated, saturated, or supersaturated.

38. Based on the graph shown in Figure 5.4, determine whether a 50.0-g sample of each salt in solution at a temperature of 50°C is unsaturated, saturated, or supersaturated.

39. Based on the graph shown in Figure 5.4, determine the solubility of KCl at the following temperatures.
(a) 20°C
(b) 45°C
(c) 100°C
(d) 70°C

40. Based on the graph shown in Figure 5.4, determine the temperature at which the following amounts of KNO_3 constitute a saturated solution.
(a) 90.0 g
(b) 10.0 g
(c) 50.0 g
(d) 75.0 g

41. Calculate the molarity for each of the solutions listed below.
(a) 31.45 g of NaCl in 1.50 L
(b) 14.41 g of MgS in 0.750 L
(c) 0.4567 g of $CuSO_4$ in 825 mL
(d) 25.5 mg of NaCN in 5.00 mL

42. Calculate the molarity for each of the solutions listed below.
(a) 121.45 g of KOH in 100.0 mL
(b) 23.49 g of NH_4OH in 150.0 mL
(c) 217.5 g of $LiNO_3$ in 1.50 L
(d) 15.25 g of $Pb(C_2H_3O_2)_2$ in 50.0 mL

43. How many moles of calcium carbonate are in the following solutions?
(a) 25.0 mL of 0.997 M $CaCO_3$
(b) 10.0 mL of 2.50 M $CaCO_3$
(c) 525.0 mL of 0.501 M $CaCO_3$
(d) 1.25 L of 3.42 M $CaCO_3$

44. How many moles of barium nitrate are in the following solutions?
(a) 45.05 mL of 2.21 M $Ba(NO_3)_2$
(b) 75.51 mL of 8.5×10^{-2} M $Ba(NO_3)_2$
(c) 1.89 L of 0.0250 M $Ba(NO_3)_2$
(d) 2.25 L of 1.55 M $Ba(NO_3)_2$

45. Determine the number of grams of calcium carbonate in each solution described in Problem 43.

46. Determine the number of grams of barium nitrate in each solution described in Problem 44.

47. What would the final concentration be of a hydrochloric acid solution prepared by diluting 100.0 mL of concentrated HCl (12.1 M) to each of the following volumes?

(a) 250.0 mL
(b) 1.00 L
(c) 5.00 L
(d) 750.0 mL

48. What would the final concentration be of a sodium hydroxide solution prepared by diluting 50.00 mL of concentrated NaOH (19.3 M) to each of the following volumes?
(a) 1.50 L
(b) 500.0 mL
(c) 2.55 L
(d) 4.50 L

49. How many milliliters of concentrated HCl (12.1 M) are needed to make the following amounts of acid?
(a) 5.00 L of 0.100 M
(b) 1.00 L of 6.00 M
(c) 2.00 L of 3.00 M
(d) 100 mL of 0.100 M

50. How many milliliters of concentrated NaOH (19.3 M) are needed to make the following amounts of base?
(a) 8.00 L of 0.250 M
(b) 300 mL of 3.00 M
(c) 0.250 L of 0.125 M
(d) 10.0 mL of 0.01000 M

51. Determine whether each of the following compounds is soluble or insoluble.
(a) Ammonium carbonate
(b) Calcium carbonate
(c) Calcium sulfate
(d) Sodium sulfate

52. Determine whether each of the following compounds is soluble or insoluble.
(a) Sodium sulfide
(b) Iron(II) sulfide
(c) Iron(II) chloride
(d) Silver chloride

53. Write the balanced equation, total ionic equation, and net ionic equation for the following unbalanced chemical reaction.

$$Na_2SO_4(aq) + MgCl_2(aq) \rightarrow NaCl(aq) + MgSO_4(s)$$

54. Write the balanced equation, total ionic equation, and net ionic equation for the following unbalanced chemical reaction.

$$Cu(C_2H_3O_2)_2(aq) + Na_3PO_4(aq) \rightarrow$$

55. Write the balanced equation, total ionic equation, and the net ionic equation for the following reaction.

Iron(III) acetate + sodium sulfide →
Iron(III) sulfide + sodium acetate

56. Write the balanced equation, total ionic equation, and net ionic equation for the following unbalanced chemical reaction.

Silver nitrate + aluminum chloride →

57. Name each of the following acids. Indicate whether it is a strong or weak acid.
(a) H_2CO_3
(b) HNO_3
(c) H_3PO_4
(d) H_2SO_4

58. Name each of the following bases. Indicate whether it is a strong or weak base.
(a) NH_4OH
(b) NaOH
(c) $Ca(OH)_2$
(d) KOH

59. For each of the consumer products pictured at the top of page 165, indicate whether the product contains an acid or base, provide the compound name and formula, and determine whether it is a strong or weak electrolyte.

60. For each of the consumer products pictured at the bottom of page 165, indicate whether the product contains an acid or base, provide the compound name and formula, and determine whether it is a strong or weak electrolyte.

61. Calculate the pH of the following solutions of HCl.
(a) 0.034 M
(b) 0.0054 M
(c) 0.15 M
(d) 2.04×10^{-4} M

62. Calculate the pH of the following solutions of HNO_3.
(a) 0.24 M
(b) 0.00087 M
(c) 0.25 M
(d) 0.000001 M

the pain. Each patient has a different psychological response to pain as well as a different biochemical response to pain medication—factors that can further complicate the issue of pain management. It is only through the skilled care and monitoring provided by health professionals that the appropriate dosage and type of pain medicine are determined for a particular patient.

One type of medicine used to control pain is meperidine, more commonly known as Demerol. Meperidine belongs to a class of drugs called the *opioids,* which also includes morphine, oxycodone, and methadone. These narcotics serve an important role in pain relief. However, they are also drugs that are commonly abused illegally.

Hospitals have had to incorporate security procedures for handling and dispensing these powerful narcotics to their patients to prevent the drugs from being stolen or abused by hospital staff. Periodically, newspapers report cases of hospital employees who have been caught stealing from the narcotics locker.

One case that was particularly challenging occurred at a hospital in upstate New York. Some patients began reporting that they were not receiving any pain relief after the administration of pain medication. Although hospital staff are aware that sometimes patients really do not experience pain relief from medication, the staff at this hospital believed something was wrong. Patients who had responded well to Demerol previously were now saying that the pain was not abating after administration of the medication. When the same patients were later administered another dose of Demerol, they experienced pain relief.

The hospital staff and administration suspected that someone was stealing Demerol and replacing it with another substance in the syringes used for injections. However, they had no suspects, nor did they know how this switch was being made. Hospital officials contacted the local police department, and police eventually turned the investigation over to the U.S. Drug Enforcement Administration (DEA).

The first challenge confronting the DEA agent assigned to this case was to develop a list of possible suspects from the several hundred employees on any given shift. The agent started to narrow in on a suspect by tracking all incidents of failed pain medication occurring on each shift. The next step was to rotate small groups of nurses from one shift to another systematically. By doing so, it soon became evident that a particular group of nurses had an unusually high number of complaints. This small group was further split and moved across shifts individually until it became clear which nurse was responsible for the missing Demerol.

The guilty person had been identified, but two questions remained: How was the nurse stealing the Demerol and what substance was being substituted for it in the syringes used to deliver medication to the patients? The DEA agent turned to forensic chemist James Wesley of the Monroe County Forensic Laboratory for answers.

Wesley had years of experience analyzing both legal and illegal narcotics. When presented with the hospital's problem, he applied his knowledge of the properties of solutions to solve this case.

By measuring the mass and the freezing point of a solution that had been tampered with, the modus operandi of the nurse who had been stealing the Demerol was revealed . . .

6.1 Intermolecular Forces and Surface Tension

Learning Objective

Describe intermolecular forces and how they create the property called surface tension.

It is a common misconception that a drop of liquid will assume the shape of a teardrop as it falls. A falling drop of liquid actually takes the shape of a sphere. Why does it do this? To answer this question we must explore the forces that exist between the molecules of a substance in the liquid state.

In the liquid, solid, or aqueous phases, particles (molecules, atoms, or ions) are located in close proximity to one another, and several types of attractive forces, called **intermolecular forces**, can develop between adjacent particles. Intermolecular forces are not as strong as the covalent or ionic bonds that exist within the particles. However, intermolecular forces significantly affect a substance's physical properties such as its boiling point and melting point.

Intermolecular forces also control the shape of a drop of liquid. Any molecule within the center of a droplet will be attracted to neighboring molecules through intermolecular forces in all possible directions, as shown by the blue arrows in Figure 6.1. However, if a molecule is at the surface of the droplet, it is attracted only to molecules located adjacent

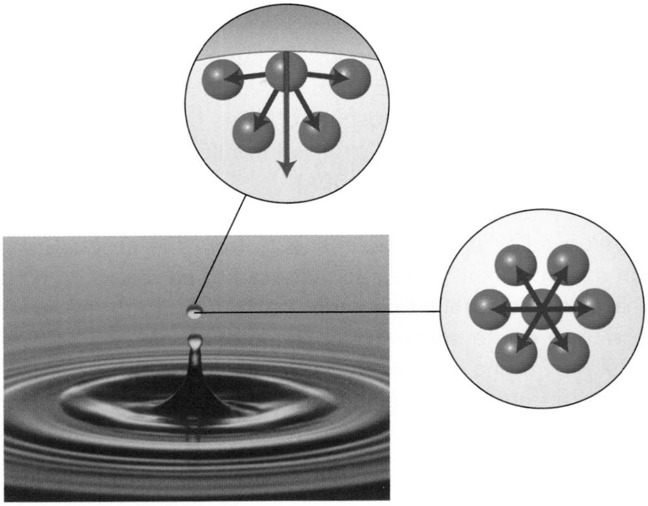

Figure 6.1 Intermolecular forces within a liquid drop. The attractive forces surrounding a molecule on the surface of a drop pull the molecule inward. Molecules within the liquid drop are pulled equally in all directions. (Photo by John Gillmoure/Corbis)

to it or beneath it, as shown by the blue arrows. The result is that the surface molecule is being pulled with a net force toward the center of the droplet, as shown by the brown arrow. Each molecule on the surface is being pulled inward in the same way. This effect minimizes the surface area of the liquid, and the shape that minimizes surface area for a given volume is the sphere.

Surface tension is a property of liquids that is a measure of how much force is needed to overcome the pull of the intermolecular forces on molecules at the surface of the liquid: The higher surface tension of a liquid, the stronger the intermolecular forces.

Worked Example 1

Calculate the surface area of a sphere with a volume of 1 cm^3, given the formulas for a sphere: volume $= \frac{4}{3}\pi r^3$ and surface area $= 4\pi r^2$.

SOLUTION The formulas for the volume and surface area of a sphere are

$$V = \frac{4}{3}\pi r^3 \qquad \text{and} \qquad A = 4\pi r^2$$

First, you need to solve for r given that $V = 1\ cm^3$. We have

$$V = \frac{4}{3}\pi r^3 \Rightarrow 1\ cm^3 = \frac{4}{3}\pi r^3$$

$$r^3 = 1\ cm^3 \times \frac{3}{4\pi} \Rightarrow r = \sqrt[3]{1\ cm^3 \times \frac{3}{4\pi}} = 0.620\ cm$$

Now, the formula for the surface area can be used:

$$A = 4\pi r^2 = 4\pi(0.620\ cm)^2 = 4.84\ cm^2$$

Practice 6.1

Calculate the surface area of a cube with a volume of 1 cm^3, given formulas for the cube: volume = length^3 and surface area = 6 × length^2. What percent increase in surface area does a cube have compared with a sphere of the same volume?

ANSWER
Surface area = 6 cm^2, 124% larger than a sphere of the same volume.

The fact that droplets of liquid will form spheres provides investigators with a major tool for solving violent crimes: the interpretation of blood spatter. The basic principle of blood spatter analysis is that a spherical droplet of blood will leave a stain on a surface in a shape that depends on the angle at which it strikes the surface. If a blood droplet falls straight down onto a smooth surface at a 90° angle, it forms a circular pattern. However, if a droplet strikes a smooth surface at a different angle, it forms an oval shape. Figure 6.2 shows that the length and width of the oval are determined by the angle at which the droplet strikes the surface.

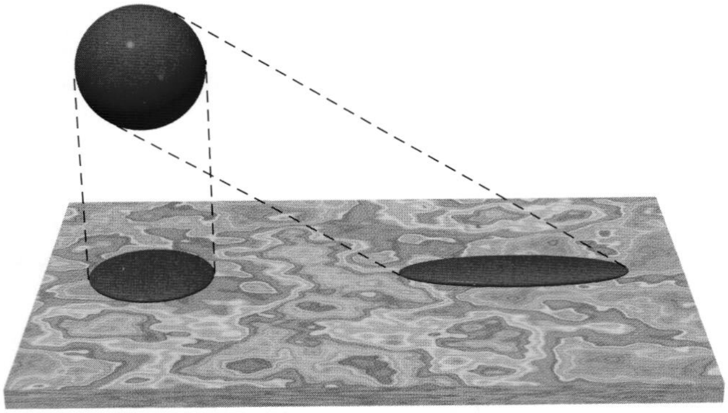

Figure 6.2 Blood drop striking a surface. If a blood droplet falls straight down at a 90° angle to a surface, it forms a circular mark. If it falls at a different angle, it forms an oval mark. The angle at which a drop of blood strikes a smooth surface can be determined by measuring the length and width of the resulting oval shape.

Blood spatter evidence literally provides a blow-by-blow account of what occurred at the scene of a violent crime. The essence of blood spatter analysis is translating the two-dimensional pattern backwards into three-dimensional space to determine where and how the blood originated. An investigator will measure the dimensions of a number of these spatter marks to determine where the drops of blood originated.

6.2 Types of Intermolecular Forces

Learning Objective

Identify the different intermolecular forces and how they affect the physical properties of a substance.

Electrons located in the orbitals of an atom form what is called the **electron cloud**. Electrons are in continual motion within the electron cloud. Because this motion is random, a majority of the electrons will occasionally be on one side of the molecule. When electrons are bunched on one side, a temporary dipole is set up in which the molecule has one region that is more negative and another region that is more positive.

The temporary dipole of a molecule will cause a change in the molecule next to it. The temporarily negative end of the molecule will repel the electrons in a neighboring molecule, distorting its electron cloud. The neighboring molecule then has an **induced dipole**, which is also tempo-

Fighting a fire in the winter is complicated by the formation of ice on equipment. Intermolecular forces determine both the freezing point and boiling point of water. (Michael J. Coppola/onscenephotos.com)

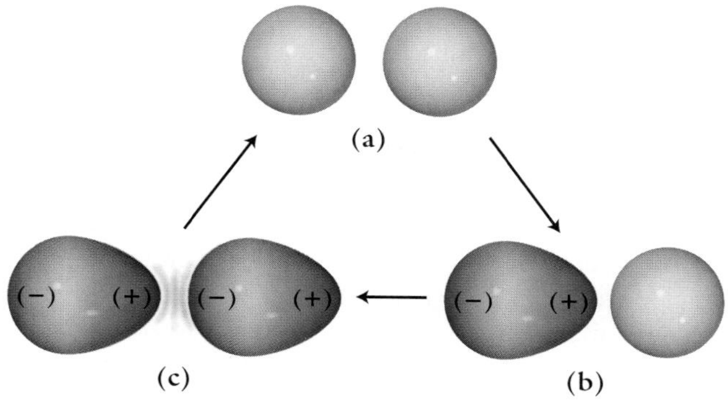

Figure 6.3 Induced dipole–induced dipole intermolecular force. (a) In a nonpolar molecule, the electron density is evenly dispersed. (b) However, due to the random motion of electrons, a temporary dipole may form. (c) The temporary dipole induces a dipole in a neighboring molecule, creating a weak attractive force.

rary. The two temporary dipoles will attract each other with their oppositely charged ends. The existence of these dipoles is very short-lived because the electrons keep moving, as shown in Figure 6.3.

The attractive forces created in this manner are called **dispersion forces,** also known as **Van der Waals forces** or **London forces**, and these are the weakest of the forces that can exist between molecules. All substances will have dispersion forces present.

Although dispersion forces are the weakest of the intermolecular forces, for nonpolar substances (such as the halogens shown in Table 6.1) they are the *only* attractive forces present and play a vital role in determining the physical properties of the compounds.

For example, the melting point of a nonpolar compound is affected by the intermolecular forces present. The molecules of a nonpolar solid are held together in a rigid structure with attractive dispersion forces between neighboring molecules. To melt a solid compound, it is necessary to provide enough energy for each molecule to overcome the attractive intermolecular forces holding it in place. The weaker the intermolecular force, the less energy is needed by the molecules to break free of the solid structure and move about in the liquid state.

The strength of the temporary dipole depends on how easily the electron cloud is distorted; this is called the **polarizability** of a molecule. Table 6.1 lists the physical properties of the nonpolar halogens. Notice that fluorine gas, the lightest of the halogens, has the lowest melting

Table 6.1 Physical Properties of Nonpolar Halogens

Halogen	Molecular Mass (g/mol)	Melting Point (°C)	Physical State at Room Temperature
Fluorine, F_2	38.0	−219.6	Gas
Chlorine, Cl_2	70.9	−101.5	Gas
Bromine, Br_2	179.8	−7.3	Liquid
Iodine, I_2	253.8	113.7	Solid

point. As the molecular mass of the diatomic elements increases, so does the melting point of the compound. Iodine, the diatomic element with the highest mass, has the highest melting point. The trend illustrates that the more massive the compound, the higher the melting point, which implies that polarizability increases as size increases.

Why are electrons more easily polarized in larger compounds? Recall that electrons are attracted to the nucleus of each atom. In larger atoms, electrons in the outer orbitals are much farther away from the nucleus than in smaller atoms and are held less tightly by the attractive force of the nucleus. This makes it easier for the electrons to be pushed aside, creating a stronger dipole.

Worked Example 2

Place the following compounds in order of increasing polarizability: CBr_4, CF_4, CCl_4, and CI_4.

SOLUTION The polarizability increases as the molecular mass increases: $CF_4 < CCl_4 < CBr_4 < CI_4$.

Practice 6.2

Place the noble gas elements in order from lowest melting point to highest melting point.

ANSWER
The order is He < Ne < Ar < Kr < Xe < Rn.

A second type of intermolecular force that can exist between molecules is the **dipole-dipole force**. This is present in polar compounds that have a permanent dipole due to differences in the electronegativity of the atoms bonded together, as explained in Section 4.3. In molecules that have two or more bonds, the three-dimensional arrangement of the bonds also determines whether a molecule is polar. (The influence of the geometric factor in determining the polarity of a molecule will be discussed further in Section 8.6.)

The positive region of a polar molecule will be attracted to the negative region of a neighboring polar molecule, as shown in Figure 6.4. The strength of the dipole-dipole force is approximately the same as or slightly greater than the strength of the dispersion force.

Worked Example 3

Which of the bonds will have the largest dipole?

(a) H—F (b) H—H (c) H—O (d) H—S

SOLUTION The magnitude of the dipole in a bond is a function of the difference in electronegativity of the two atoms. The trend for electronegativity on the periodic table increases from bottom to top in groups (columns) and increases from left to right in periods (rows). The largest difference in electronegativity is between the atoms in bond (a).

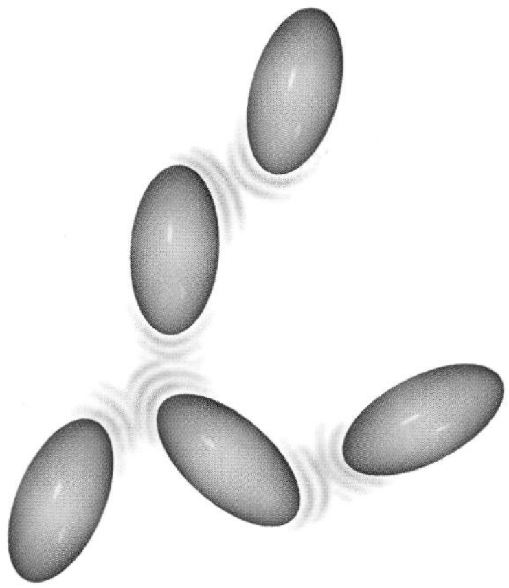

Figure 6.4 Dipole-dipole intermolecular force. The oppositely charged regions of polar molecules are attracted to one another through electrostatic attraction.

Practice 6.3

Two compounds often identified in cases of accidental poisoning are ethanol (CH_3CH_2OH), from ingestion of alcoholic beverages, or ethylene glycol ($HOCH_2CH_2OH$), the main ingredient in antifreeze. Which compound has the highest melting point?

ANSWER
Ethylene glycol

A third type of intermolecular force is called **hydrogen bonding**, which is a particularly strong form of a dipole-dipole interaction. Hydrogen bonding is not to be confused with the covalent bond that hydrogen atoms form with other atoms *within* a molecule. Intermolecular forces always refer to forces *between* molecules of a substance.

Hydrogen bonding between molecules occurs for compounds in which a hydrogen atom is bonded directly to a nitrogen, oxygen, or fluorine atom, as in NH_3, H_2O, or HF. The hydrogen bond results from the highly electronegative atoms (N, O, or F) pulling the shared electrons toward themselves. Recall that hydrogen has only one proton and one electron; if the electron is pulled away from the hydrogen atom in a bond, the result is a bare proton exposed on the end of the bond. This highly positive region is then attracted very strongly to the electron-rich electronegative element (N, O, or F) on adjacent molecules.

Worked Example 4

Which of the following molecules will form hydrogen bonds?

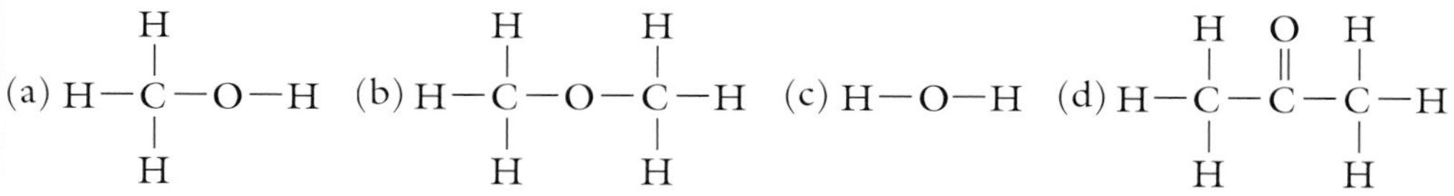

SOLUTION Compounds (a) and (c) form hydrogen bonds between molecules. They are the only structures that have hydrogen atoms directly bonded to the highly electronegative oxygen atoms as required to form hydrogen bonds.

Practice 6.4

Place the molecules in Worked Example 4 in order of increasing boiling point.

ANSWER

The order is compound (b) < compound (d) < compound (c) < compound (a).

6.3 Mixed Intermolecular Forces

Learning Objective

List and describe other intermolecular forces that form in mixtures.

Thus far, intermolecular forces have been explained by comparing how one molecule will interact with an identical molecule in a pure substance. It is important to realize that intermolecular forces occur between all molecules that are in proximity to one another—whether they constitute a pure compound or a complex mixture.

Forensic scientists are often presented with complex substances such as blood that has solid components (red blood cells, white blood cells, and platelets) suspended in a liquid (water) that has polar, nonpolar, and ionic substances dissolved in it. The polar components include water, glucose, and urea; nonpolar compounds include oxygen and carbon dioxide gas. The proteins albumin, hemoglobin, and immunoglobulin are large polar molecules that are suspended in the solution. Electrolyte ions such as sodium, potassium, and chloride ions are also present in blood.

Oxygen gas exists in two forms in blood: oxygen bound to hemoglobin within the red blood cells, which accounts for 99% of the oxygen, and 1% in the form of O_2 dissolved in the plasma. Interactions occur between all of the molecules and ions present in blood. Do water molecules in blood interact differently with dissolved oxygen gas than with sugar? The intuitive answer is that water interacts with a neutral, nonpolar molecule in different ways than it interacts with a polar molecule. To clarify these differences, we must examine some additional types of attractive forces that occur in mixtures.

The interaction of water with glucose, a polar molecule, is the type of dipole-dipole interaction covered in the previous section—where the example discussed sucrose dissolving in water, as covered in Section 5.2. Another prominent intermolecular force initially discussed in Section 5.2 is the **ion-dipole force**. There we saw that an ionic compound dissolves in water because the water molecules surround the ions of an ionic solid so that opposite charges of the ion and the polar water molecules align. The ion-dipole intermolecular force is critical for the dissolution of an ionic compound.

Worked Example 5

Which components in blood exhibit ion-dipole interactions?

SOLUTION Water and urea, both polar compounds, exhibit ion-dipole interactions with the electrolyte ions (Na^+, K^+, Cl^-).

Practice 6.5

What intermolecular force plays a dominant role in an aqueous solution of alcohol?

ANSWER
Dipole-dipole force

Another intermolecular force present in a solution is the **dipole-induced dipole force**, which originates when a polar molecule causes the electrons of an adjacent nonpolar molecule to be distorted. This occurs because the negative region of a polar molecule will repel the evenly distributed electrons of a nonpolar molecule. Likewise, the positive region of the polar molecule will attract the evenly distributed electrons of a nonpolar molecule. Either source of distortion induces a temporary dipole in the nonpolar molecule. The induced dipole of the nonpolar molecule is weakly attracted to the permanent dipole of the polar molecule. However, the distorted electron cloud will return to normal and the attraction ends.

The dipole-induced dipole intermolecular force is weaker than the dipole-dipole intermolecular force. Despite being a weak force, the dipole-induced dipole force allows a nonpolar gas such as oxygen to dissolve in a polar solvent such as water—a phenomenon that is necessary for life to exist. The solubilities of several gases in water are listed in Table 6.2.

Worked Example 6

Which components in blood exhibit dipole-induced dipole intermolecular forces?

SOLUTION The polar molecules such as water and urea interact with nonpolar solutes such as the dissolved gases CO_2 and O_2.

Practice 6.6

Is nitrogen gas more or less soluble in blood than oxygen gas? Explain why.

ANSWER
The dipole-induced dipole intermolecular force is the main interaction between the gases and the water in blood. The oxygen gas is more soluble because it is a larger molecule and therefore is more polarizable.

Table 6.2 Solubility of Nonpolar Gases in Water

Gas	Solubility (g/100 mL water)
H_2	0.000160
N_2	0.000190
O_2	0.000434

John C. Kotz and Paul Treichel, *Chemistry & Chemical Reactivity*, 5th ed., Brooks/Cole Publishing, 2002.

The boiling point of a pure solvent changes with the addition of a solute to form a solution. (A. Pasieka/Photo Researchers, Inc.)

6.4 Colligative Properties: Boiling Point of Solutions

Learning Objective

Explain on a molecular level why solutions have higher boiling points than the pure solvent.

In determining how the syringe solutions from the case study should be analyzed, we can start with the basic question, "What do we know about pure water and what do we know about solutions?" We know the physical properties of pure water such as its boiling point (100°C under standard conditions), freezing point (0°C), and density (1.0 g/mL). Do these values change when a solute is added to the water to make a solution? The density of water will obviously change if a substance is dissolved in it, and a solution of Demerol has a density of 1.037 g/mL. Analysis of a sample often starts with investigating its physical properties.

A common physical property that is easily tested is the boiling point of a liquid. When a substance boils, it changes from the liquid to the gas state. The temperature at which this occurs differs from one substance to another. For example, the standard boiling point for rubbing alcohol is 83°C, while water boils at 100°C. Why? To answer this, we need both an understanding of the intermolecular forces discussed in the previous sections and a more complete understanding of what is occurring at the molecular level when a liquid boils.

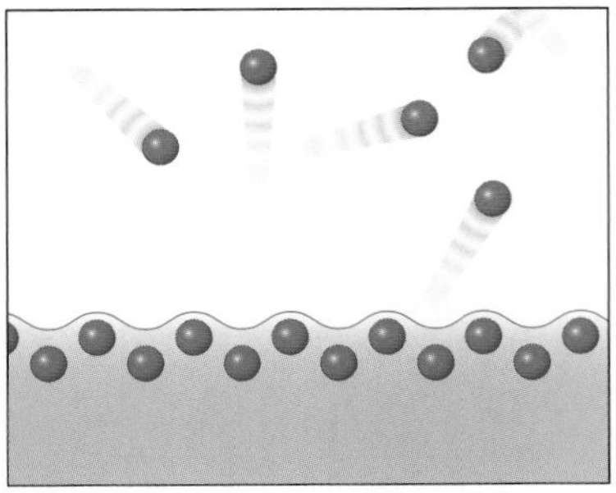

Figure 6.5 Vapor pressure of a liquid. A small percentage of molecules in a liquid will have sufficient energy to go into the gas phase.

The individual molecules that make up a liquid move through it at various speeds. Some of the molecules have sufficient kinetic energy (the energy of motion) to overcome the intermolecular forces of the surrounding molecules at the surface of the liquid and escape into the gas phase. The term **vapor** is used to describe the gas phase of a compound that is normally a liquid at room temperature. The **pressure** of a gas is due to the particles colliding with walls of the container that holds the gas and the liquid. The **vapor pressure** of a liquid is the pressure exerted by the vapor molecules, as illustrated in Figure 6.5.

The vapor pressure of a liquid results from molecules that have sufficient energy to escape the surface of the liquid and collide with the con-

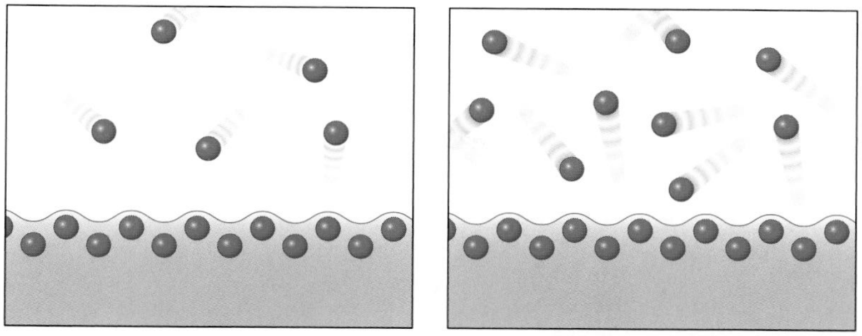

Figure 6.6 Vapor pressure as a function of temperature. Because the kinetic energy of molecules increases at higher temperatures, a larger number of molecules will have sufficient energy to exist in the gas phase.

tainer walls. As a liquid is heated, the molecules gain kinetic energy. The increased kinetic energy results in more molecules that have the ability to escape the surface of the liquid and change into the vapor state, as illustrated in Figure 6.6.

When a liquid boils, the liquid molecules have enough kinetic energy to go directly into the vapor state and form a bubble anywhere in the liquid. The liquid molecules no longer have to be at the surface to go into the gaseous state. Therefore, the **boiling point** is the temperature at which the vapor pressure of a liquid is equal to the external pressure exerted on the liquid. The term *external pressure* usually refers to the weight of the atmosphere pushing down on the liquid in an open container, and boiling points are typically determined with 1 atmosphere (atm) of pressure on the liquid. The formation of a vapor bubble within the liquid can occur only when the vapor pressure is equal to the external pressure. If the vapor pressure is less than the external pressure, bubbles of vapor do not form, as the atmospheric pressure would crush them.

How can this information be used to analyze the syringe solutions of the case study? Does the boiling point change when a solute is added to water? When a substance is dissolved in water, it forms a homogeneous solution, so there is an even distribution of solute particles throughout the solution, including the surface. Figure 6.7 illustrates that if the solute is occupying some sites at the surface, those sites are unavailable for the liquid to use in forming vapor molecules.

By adding a solute, the vapor pressure is lowered, but in order for the liquid to boil, the vapor pressure must be equal to the external pressure. The boiling point of the solution is then elevated beyond that of the pure solvent because additional heat must be added to compensate for the vapor pressure decrease.

Worked Example 7

A sports drink has been sent to a crime lab as part of an investigation into suspected tampering. If the boiling point of the sample matches the boiling point of an unaltered sample of the sports drink, does this result mean that the sample was not tampered with?

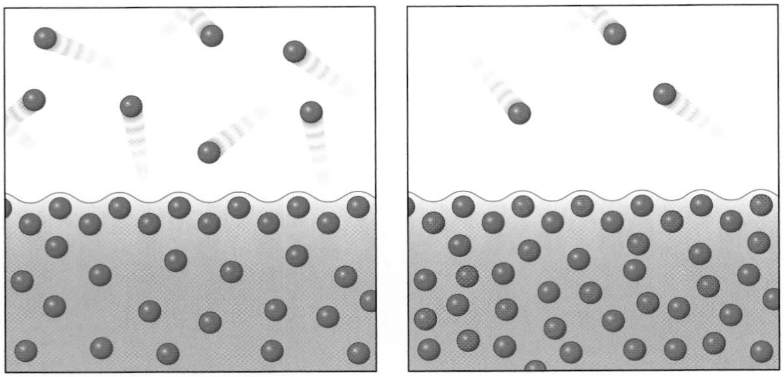

Figure 6.7 Vapor pressure of pure liquids versus solutions. Solute molecules replace solvent molecules at the surface of a solution, as in part (b). When fewer solvent molecules can escape into the vapor phase, vapor pressure is lowered.

SOLUTION If the boiling point of the suspected solution matches an unaltered solution, this would indicate that it is unlikely the sample was tampered with. However, several possibilities could still exist. A toxic substance might be present at such extremely low levels that the boiling point is not significantly altered, or the toxic substance may have evaporated from solution or decomposed.

Practice 6.7

If the suspect sports drink sample does not match the boiling point of an unaltered sample, does it mean that it has been tampered with?

ANSWER

If the physical properties of the two solutions do not match, there is definitely a difference between the two samples. It does not mean the sample was maliciously tampered with. Further testing of the sample would be needed to determine the nature of the difference, and additional investigative work would be needed to account for the difference in the solutions.

The boiling point of an aqueous solution containing 1 mole of sugar (a molecular compound) dissolved in 1 kilogram of water is 100.512°C, an elevation of 0.512°C above the boiling point of pure water. The boiling point of a solution containing 1 mole of NaCl (an ionic compound) in the same quantity of water is 101.024°C, an elevation twice as great as the increase observed for the sugar solution. Why is this the case?

The extent to which **boiling point elevation** of a solution occurs (in comparison to the boiling point of the pure solvent) depends strictly on the number of particles of solute in solution, *not* their identity. Properties that depend only on the number of particles, not their identity, are called **colligative properties**. Colligative properties include vapor pressure lowering, boiling point elevation, freezing point depression, and osmotic pressure, all of which are covered in this chapter. The examples used within this text will focus on water as the solvent; however, the colligative properties apply to any solvent–solute combination.

When sodium chloride dissolves in water, the salt dissociates into its component ions—two particles for each unit of sodium chloride. One mole of sodium chloride has twice the effect on the boiling point elevation as dissolving the same number of moles of sugar because sugar is a molecular compound that does not dissociate into multiple ions.

A key factor in boiling point elevation measurements is that the solute must be nonvolatile, which in this case means that the solute does not boil at a temperature lower than that of the solvent. If the solute has a low boiling point, our model of boiling point elevation is no longer completely accurate. Ethanol, for example, is volatile because it will boil at 78°C. In a mixture of ethanol and water, alcohol will be lost during the heating of the solution before the boiling point of water is reached. Because the compound in the solution might be volatile, boiling point measurements are not typically used to determine possible adulterants in liquids. An alternative to measuring boiling point is to measure the freezing point of a solution, as will be discussed in Section 6.6.

6.5 Mathematics of Boiling Point Elevation

Learning Objective

Illustrate how much the boiling point of a solution will change because of the concentration of the solute.

The change in the boiling point of a solution compared with the boiling point of the pure solvent can be calculated using the following equation:

$$\Delta T_{bp} = K_{bp} m_{particles} \quad (1)$$

ΔT_{bp} is the change in the boiling point temperature, K_{bp} is the boiling point constant that is unique to each solvent, and $m_{particles}$ is the molality of the solute particles. **Molality** (m) is a method of expressing a solute concentration that is similar to, but not the same as, molarity. The molality of a solution is the number of moles of solute per kilogram (1000 grams) of solvent:

$$\text{Molality} = \frac{\text{moles of solute}}{\text{kilograms of solution}}$$

The reason molarity (moles/liter) is not suitable is that the volume of liquids changes as a function of temperature. Therefore, the molarity of a solution at room temperature is different from the molarity of the same solution near the boiling point. Molality, however, is based strictly on the mass of the solvent, and the mass does not change. The following equation shows how the molality of the solute particles is calculated:

$$m_{particles} = \frac{\text{moles of solute} \times \text{number of particles per solute}}{\text{kilograms of solvent}}$$

Worked Example 8

What is the boiling point of a solution made from 0.200 g of KCN dissolved in 10.0 mL of water? The K_{bp} for water is 0.512°C/m.

SOLUTION The first step is to calculate the moles of KCN:

$$0.200 \text{ g KCN} \times \frac{1 \text{ mol KCN}}{65.12 \text{ g}} = 0.00307 \text{ mol}$$

The second step is to determine the number of particles: KCN is an ionic compound made up of K^+ and CN^- particles, so there are two particles. The third step is to calculate the mass of water in kilograms:

$$10.0 \text{ mL } H_2O \times \underbrace{\frac{1 \text{ g } H_2O}{1 \text{ mL } H_2O}}_{\text{Density of water}} \times \frac{1 \text{ kg}}{1000 \text{ g}} = 0.0100 \text{ kg}$$

The fourth step is to calculate the molarity of the solute particles:

$$m_{\text{particles}} = \frac{0.00307 \text{ mol KCN} \times 2 \text{ particles/mol KCN}}{0.0100 \text{ kg}} = 0.614\ m$$

The fifth step is to calculate the increase in the boiling point:

$$\Delta T_{\text{bp}} = K_{\text{bp}} m_{\text{particles}} = 0.512°\text{C}/m \times 0.614\ m = 0.314°\text{C}$$

Finally, we calculate the elevated boiling point by adding the increase to the normal boiling point:

$$T_{\text{bp}} = 100°\text{C} + 0.314°\text{C} = 100.314°\text{C}$$

Practice 6.8

Determine the molality of a 100.0-mL water sample to which antifreeze (ethylene glycol, $C_2H_6O_2$) has been added, if the boiling point of the solution rises to 104.5°C. From the molality, determine the number of grams of ethylene glycol added to the sample.

ANSWER
8.8 m ethylene glycol and 55 g of ethylene glycol

Worked Example 9

Which of the following aqueous solutions has the highest boiling point?

Solution A: 0.10 mol of NaCN dissolved in 100.0 mL of water

Solution B: 0.25 mol of $C_6H_{12}O_6$ (sugar) dissolved in 100.0 mL of water

Solution C: 0.20 mol of $NaNO_3$ dissolved in 200.0 mL of water

SOLUTION The boiling point change for each solution could be calculated by using equation 1 (on page 181). However, this question can be answered in a simpler way by noting that the solution that has the most particles per unit volume will have the highest boiling point.

Solution A: NaCN has 2 particles, therefore 0.20 mol of particles per 100 mL.

Solution B: $C_6H_{12}O_6$ has 1 particle, therefore 0.25 mol of particles per 100 mL.

Solution C: $NaNO_3$ has 2 particles, therefore 0.40 mol of particles per 200 mL, which means 0.20 mol of particles for 100 mL.

Solution B has the highest boiling point.

Practice 6.9

Which of the following aqueous solutions has the highest boiling point?

Solution A: 40.0 g of NH_4NO_3 dissolved in 100.0 mL of water

Solution B: 90.0 g of $C_6H_{12}O_6$ (sugar) dissolved in 100.0 mL of water

Solution C: 39.0 g of Na_2S dissolved in 100.0 mL of water

ANSWER
Solution C

6.6 Colligative Properties: Freezing Point of Solutions

Learning Objective

Explain why solutions have lower freezing points than the pure solvent.

In 1867, Alfred Nobel developed a method for stabilizing nitroglycerin, a highly explosive liquid compound, into a more stable and commercially viable product that he named *dynamite*. Dynamite helped spur an industrial and construction boom by providing engineers with the ability to excavate large areas of land with explosives.

The problem with using pure nitroglycerin as an explosive is that it is sensitive to shock. Just jarring or dropping a container of nitroglycerin can detonate the material. However, nitroglycerin becomes even more dangerous as it freezes and starts to form a solid—the friction of the crystals rubbing against each other can trigger an explosion. Because the freezing point of nitroglycerin is only 13.5°C (56.3°F), nitroglycerin becomes particularly dangerous to handle in cold weather.

Alfred Nobel (1833–1896), inventor of dynamite and founder of the Nobel Prize. (Bettmann/Corbis)

In **freezing point depression**, the temperature at which a liquid freezes is lowered by adding a nonvolatile solute to form a solution. The stabilization of nitroglycerin to make commercial dynamite takes advantage of freezing point depression by creating a solution of nitroglycerin with ethylene glycol dinitrate, which is another explosive compound, as shown in Figure 6.8. The freezing point of ethylene glycol dinitrate is −22.8°C (−9.04°F). The final product is an equal mixture of the two explosive compounds and results in a freezing point of approximately −20°C (−4°F), which greatly reduces the risk of shock detonation.

The making of dynamite is a good example of how the principle of freezing point depression, a colligative property, was used for the production of a more stable explosive. This same principle is used in winter to melt ice from roads and to prevent radiator fluid from freezing. But what

Nitroglycerin	Ethylene glycol dinitrate
CH_2-ONO_2 $CH-ONO_2$ CH_2-ONO_2	CH_2-ONO_2 CH_2-ONO_2

Figure 6.8 Nitroglycerin and ethylene glycol dinitrate.

is occurring on a molecular scale to lower the freezing point in these circumstances?

There are two processes occurring when the temperature of a liquid is at its freezing point. Some of the molecules in the liquid slow down and are captured by the solid. However, some of the solid molecules have enough energy to enter the liquid. As shown in Figure 6.9a, the two processes are in **equilibrium.** There is no net change in the number of molecules in either phase: For every one molecule that joins the solid, another joins the liquid.

When a solute such as NaCl is dissolved in a solvent, it blocks part of the liquid solvent from interacting with the molecules in the solid phase. However, it does not prevent the solid particles from entering the liquid phase, as shown in Figure 6.9b. The result is that the rate at which molecules bind to the solid decreases, but the rate at which molecules enter the liquid remains unchanged. As a result, the solid begins to melt.

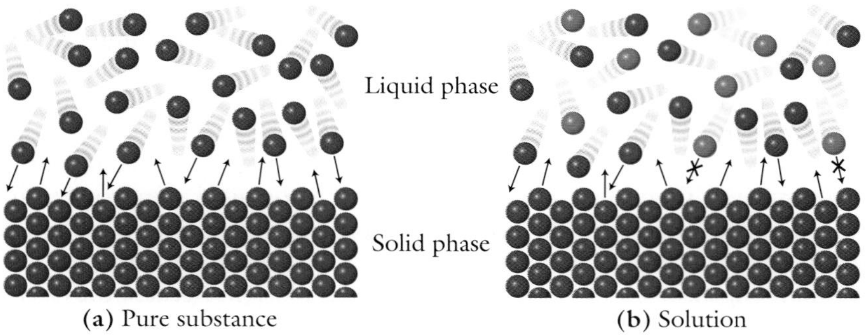

Figure 6.9 Molecular view of freezing point depression. (a) In the pure substance at the freezing point, molecules in the solid phase escape the surface at the same rate that molecules in the liquid phase return to the surface. (b) In the solution, solute molecules replace solvent molecules interacting with the solid surface, thereby decreasing the rate at which solvent molecules enter the solid phase. The presence of solute molecules does not affect the rate of melting.

Worked Example 10

What is the advantage of measuring the freezing point instead of the boiling point in determining whether a liquid sample contains a solute?

SOLUTION If volatile compounds are present in the solution, they will not be lost during the freezing process, whereas they are preferentially lost in measuring the boiling point of water.

Practice 6.10

Place the following compounds in order from the greatest change to the least change in the freezing point of a liquid, assuming that an equal number of molecules are added to each solution: NaCN, $MgCl_2$, and $Al(NO_3)_3$.

ANSWER

The order is $Al(NO_3)_3 > MgCl_2 > NaCN$.

6.7 Mathematics of Freezing Point Depression

Learning Objective

Illustrate how the freezing point of a solution will change because of the concentration of the solute.

The change in the freezing point temperature of a solution can be calculated using equation 2. You will notice that this equation is strikingly similar to equation 1 for boiling point elevation. However, it is important to note that the freezing point constant K_{fp}, which is unique to each solvent, is *not* the same as the K_{bp} value for the same solvent. The K_{fp} for water is $-1.86°C/m$, whereas the K_{bp} for water is $0.512°C/m$.

$$\Delta T_{fp} = K_{fp} m_{particles} \quad (2)$$

Worked Example 11

What is the freezing point of a solution made from 0.200 g of KCN dissolved in 10.0 mL of water?

SOLUTION The concentration of the KCN solution in this example is the same as the one in Worked Example 8. We can use the molality calculated there to get the decrease in freezing temperature.

$$\Delta T_{fp} = K_{fp} m_{particles} = -1.86°C/m \times 0.614\ m = -1.14°C$$

Hence, the freezing point is $0.00°C - 1.14°C = -1.14°C$.

Practice 6.11

In Practice 6.8, it was determined that 55 g of ethylene glycol ($C_2H_6O_2$) added to 100 mL of water produced a solution with a molality of 8.8. Determine the freezing point of that same solution.

ANSWER

$-16°C$

6.8 Colligative Properties: Osmosis

Learning Objective

Describe the process of osmosis through a semipermeable membrane and its effect on pressure and concentration.

To understand osmosis on a molecular level, it is helpful to first consider the concept of diffusion. **Diffusion** is the process in which solute particles move by purely random motion from a region of high concentration to a region of lower concentration within a solution. As a solid solute dissolves, the solute particles (molecules or ions) will move randomly, colliding with one another, solvent molecules, and the walls of the container. This process continues until the solute is evenly distributed throughout the solution.

Osmosis differs from diffusion in that the movement in solution is of the water molecules across a semipermeable membrane, such as a cell wall. A **semipermeable membrane** is a thin membrane with extremely

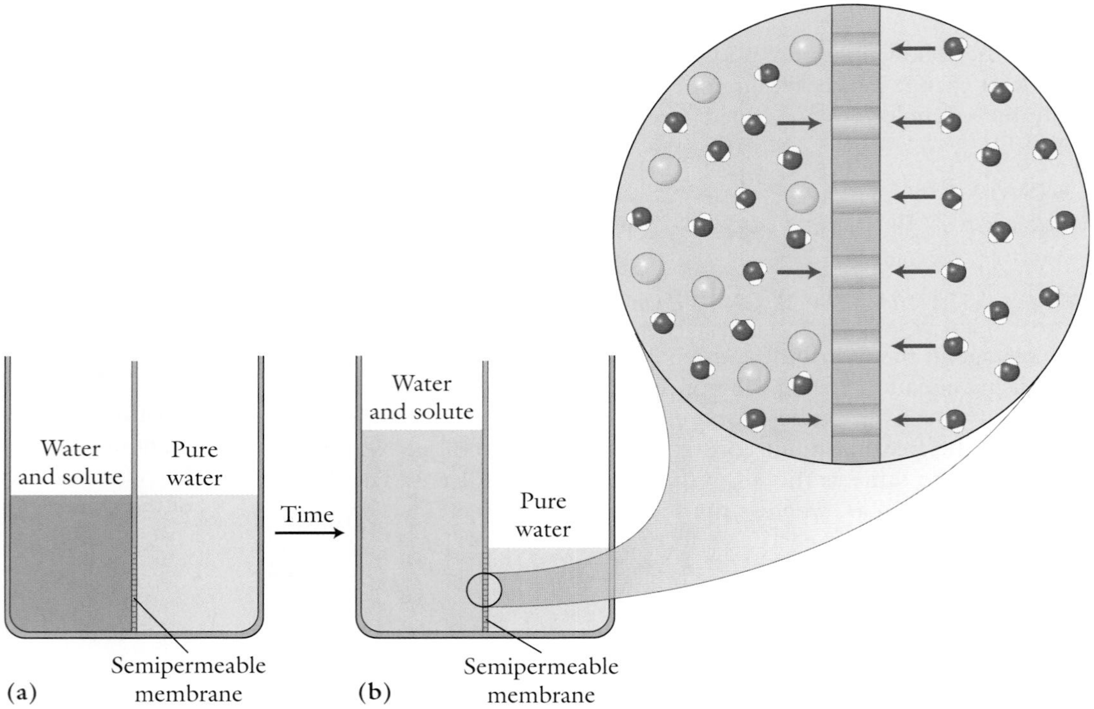

Figure 6.10 Molecular view of osmosis. Water molecules move across a semipermeable membrane in either direction. However, in (a), solute molecules replace water molecules on the solution side of the membrane, thereby decreasing the rate at which water molecules cross the membrane toward the pure solvent side. This results in a net gain in the volume of solution on the solute side, as shown in (b).

small holes that allow small molecules such as water to move back and forth through the membrane but prevent larger molecules or ions from passing through. Water will move from a region of low solute concentration to a region of high solute concentration, as illustrated in Figure 6.10.

The molecular view of osmosis is that water molecules from both sides of the membrane can pass through it. However, the solute particles on the concentrated side cannot pass through. If pure water were on both sides, the same number of water molecules would strike and pass through the membrane in each direction. When a solution is on one side, there are fewer water molecules per volume of solution, so fewer water molecules will strike the membrane and a net gain of water occurs on the concentrated side of the membrane.

The resulting difference in volumes on either side of the membrane introduces a new force to consider—the additional pressure pushing down the higher column of solution. This pressure starts to counteract the process of osmosis. Osmosis continues until the pressure pushing down on the higher column of water is enough to create an equal transfer of water molecules across the membrane. The **osmotic pressure** of a solution is the pressure needed to prevent a net change in water volume across the membrane.

Water purification can be accomplished by a process called **reverse osmosis**. In this process, a pressure greater than the osmotic pressure is applied to impure water, forcing water molecules to move across the membrane preferentially toward the pure water side.

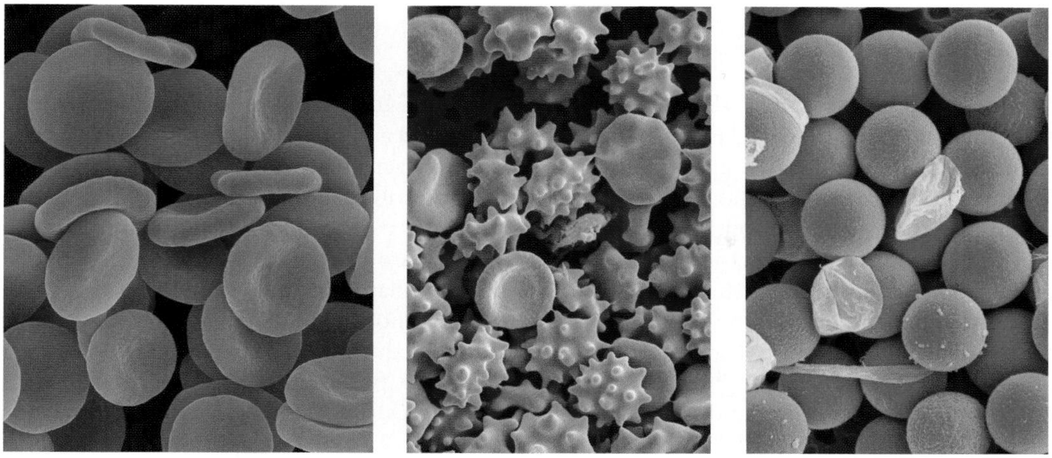

(a) (b) (c)

Figure 6.11 Red blood cells and osmosis. Red blood cells can undergo osmosis if exposed to a solution with a different concentration of solute molecules. The images show blood cells exposed to (a) an isotonic solution, (b) a hypertonic solution, and (c) a hypotonic solution. (a: Dennis Kunkel/Phototake; b, c: Kalab/Custom Medical Stock)

When an intravenous (IV) fluid is administered to a patient, it is critical that the concentration of solutes in the IV fluid match the concentration of solutes in the blood plasma. An IV fluid that meets this condition is called an **isotonic solution**. If pure water were used in an IV fluid, the concentration of solutes within the body cells would become greater than the concentration of solutes in the blood plasma. Water would then move by osmosis into the body cells, causing them to swell and possibly rupture. The same problem can result if a **hypotonic solution**—one that has a lower solute concentration than blood—is used as an IV fluid. A different problem results if the IV fluid is too concentrated. The water within the body cells would move by osmosis to the outside of the cell, causing the cell to shrivel to the point of being destroyed. A solution that is more concentrated than blood is called a **hypertonic solution.** The effect of three solution conditions on red blood cells is illustrated in Figure 6.11.

Many times suspicious situations are the result of unusual, but not criminal, circumstances. Medical doctors will often run a battery of tests while attempting to ascertain the cause of an individual's condition. When these situations are investigated, it is critical to determine whether anything unusual is present in a person's blood and, if so, what it is. The **osmolality** of the blood is a measure of how many particles are dissolved in blood.

The *osmol gap* is a measure of the difference between the expected and actual osmotic pressures of a person's blood. A higher than normal value for the blood's osmotic pressure indicates the presence of an unusual compound or unusually high concentration of a blood component. Further blood work must be done to identify the unknown substance. Most modern medical laboratories will use devices designed to measure the freezing point depression of samples, from which the osmol gap can be calculated.

Evidence Analysis | HPLC

The Drug Enforcement Administration (DEA) is charged with enforcing the federal laws related to controlled substances. This includes preparing criminal and civil cases against anyone who is involved with the transportation, distribution, manufacturing, or growing of controlled substances or their chemical precursors. At first glance, the mission of the DEA might appear to be nothing more than a crime unit for drug offenses, but that is an oversimplification of its role. The DEA is literally the command center for the war on drugs, and plays a role similar to that of the Pentagon. A crucial step in the war on drugs, as in any war, is the gathering and analysis of intelligence on the enemy.

The DEA gathers information and intelligence on all major drug cartels in the world. Each cartel specializes in certain drugs and uses specific methods for the manufacturing of its drugs. Each drug cartel has developed extensive distribution systems for its illegal products and is constantly looking for new methods of smuggling the products.

The DEA develops chemical informants that can provide information on how the illegal drugs are synthesized, the location of the clandestine drug lab, how drugs are smuggled into the United States, and which drug cartels control particular cities. The chemical informant comes from analyzing a sample of the illegal substance with an instrument called a **high performance liquid chromatograph (HPLC)**.

HPLC is a more sophisticated version of chromatography than the thin-layer chromatography (TLC) introduced in Chapter 1 but is based on the same principles. In TLC the stationary phase is a glass plate; in HPLC the stationary phase consists of small polymer or SiO_2 particles contained in a stainless steel tube that is about 1 cm in diameter and 20 cm long. In TLC, the mobile phase was added to the bottom of a beaker and allowed to move by capillary action across the stationary phase; in HPLC, the solvent is mechanically pumped through the tube containing the stationary phase. The sample mixture of solute molecules is injected into the mobile phase.

The various molecules that make up the mixture are attracted through intermolecular forces to both the stationary phase and the mobile phase, but to different degrees. Those molecules that are most attracted to the mobile phase pass through the system first; those molecules that are attracted to the stationary phase pass through the system last. The stationary material used in HPLC can be coated with different compounds. The coating is usually chosen to optimize the differences in intermolecular forces of the compounds in the mixture.

If the stationary phase is coated with an amino-type compound (shown in the figure), will polar or nonpolar compounds come out of the column first?

Amino-type compound

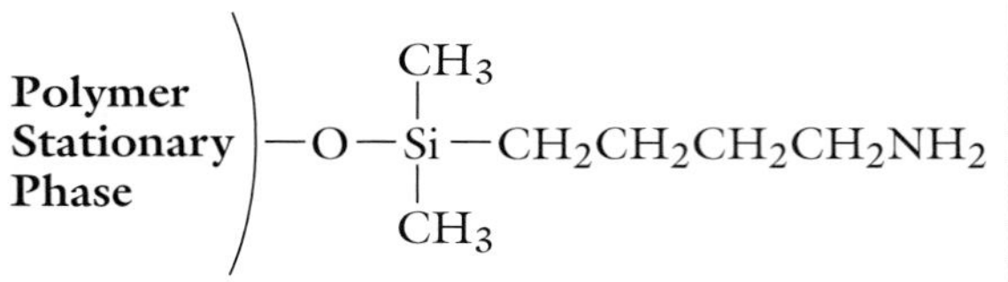

The nitrogen group at the end of the amino-type stationary phase molecule is very electronegative, creating a polar stationary phase; therefore, polar molecules will be most attracted to the stationary phase. The nonpolar compounds will not be attracted to the stationary phase and will be

6.9 Case Study Finale: Something for the Pain

James Wesley knew that if the contents of a syringe containing Demerol were being replaced with another solution, the colligative properties of the solutions would be different. He determined the molality and density of a known sample of Demerol, the liquid in a syringe that had been tampered with, and samples of solutions that were possible replacements for the missing Demerol. The results are provided in Table 6.3 (on page 190). The

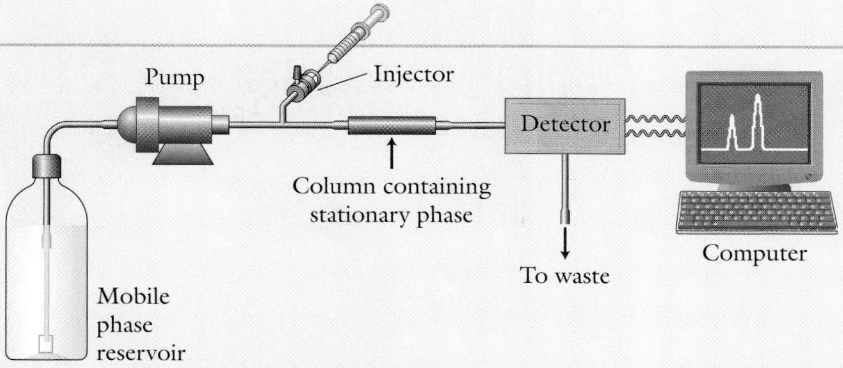

Figure 6.12 High performance liquid chromatography.

pushed out of the column very quickly with the mobile phase.

If the stationary phase is coated with an octyl-type compound, will polar or nonpolar compounds come out first?

Octyl-type compound

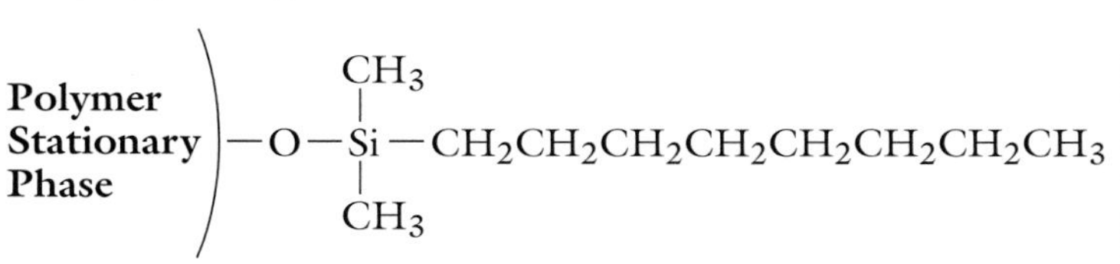

The octyl-type stationary phase molecule contains mostly C—H bonds, which produce a nonpolar stationary phase. Therefore, polar compounds will come out first because the nonpolar compounds will be more attracted to the nonpolar stationary phase.

The detectors work in HPLC by measuring how much light each compound absorbs as it leaves the column and enters the detector. A schematic of an HPLC system is shown in Figure 6.12. The advantage of HPLC is that it can be used to analyze and quantify extremely small samples and detect trace amounts of impurities, whereas TLC is useful only for identifying the main components of a mixture.

The DEA uses HPLC to determine which trace impurities are present in cocaine or heroin. DEA scientists can then match the profiles of impurities to a specific method used by one of the drug cartels to make the drug. In this manner, agents can trace illegal drugs seized in a raid directly to the cartel that manufactured it. When a cartel tries a new method of smuggling, the DEA can determine from seized samples which cartel is responsible. If a new distributor of illegal drugs is coming into a town, the DEA can determine who it is. The intelligence gained from trace impurity analysis provides vast amounts of information that would be impossible to gather without informants at the highest levels of each cartel.

molality is measured in units called *osmols,* which represent the amount of a substance that depresses the freezing point of water by 1.86°C.

Wesley was able to combine the information from these two sets of data to determine that the syringe most likely contained the bacteriostatic saline solution. The density in this solution was very similar to the one in the suspect syringe. The molality of the suspect syringe's contents differed from all of the samples simply because there was some leftover Demerol in the syringe, which was then diluted with the saline solution, causing a molality higher than the saline but lower than pure Demerol. When the nurse was confronted with the evidence of her tampering and with the knowledge of exactly how she was stealing the Demerol, she confessed to the crime.

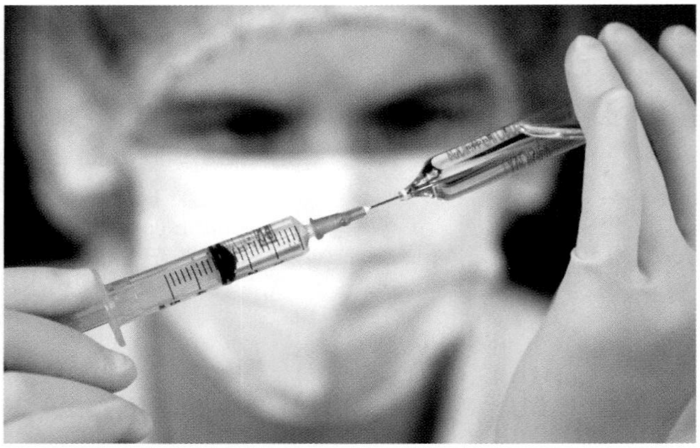

A nurse administering pain medication. (Dominique Platriez/Photo Researchers, Inc.)

Table 6.3 Physical Properties of the Case Study Evidence

Sample	Osmolality [mOsm/kg]	Density (g/cm^3)
75 mg Demerol, control syringe	429	1.037
75 mg Demerol, suspect syringe	381	1.011
Abbott bacteriostatic saline	374	1.010
Lyphomed saline	291	1.004
Quad bacteriostatic water	93	1.005
Abbott sterile water	1	1.000

James F. Wesley, "Osmolality—A Novel and Sensitive Tool for Detection of Tampering of Beverages Adulterated with Ethanol, -Butyrolactone, and 1,4-Butanediol, and for Detection of Dilution-Tampered Demerol Syringes," *Microgram Journal* 1 (2003): 8.

Worked Example 12

Based on just the density data of the samples, can investigators determine whether the suspect syringe had been tampered with? Can the identity of the substance in the syringe be determined?

SOLUTION Yes, it is possible to tell that the syringe had been tampered with because the density of Demerol is significantly higher than that of the contents in the suspect syringe. No, the identity cannot be determined from the data because the salines and the bacteriostatic water have density values that are too similar to allow for easy differentiation between the substances.

Practice 6.12

Based on just the molality data of the samples, can investigators determine whether the suspect syringe had been tampered with? Can the identity of the suspect syringe's contents be determined?

ANSWER

Yes, the molality of Demerol is sufficiently different from the other known samples to determine that it had been replaced with another solution. Yes, the identity of the syringe's contents can be determined because the molality of all the samples are different.

CHAPTER SUMMARY

- Intermolecular forces are electrostatic attractions between positive and negative regions that develop between the molecules of a compound. Intermolecular forces affect the physical properties of solutions.
- The spherical shape of a liquid droplet is due to the net inward pull of the molecules found on the surface of the droplet by the molecules in proximity to the surface molecules. A sphere has the smallest surface area for a given volume of liquid.
- Dispersion forces, also called Van der Waals forces, result from a distortion of the electron cloud surrounding the atoms. This distortion sets up a temporary dipole within the molecule, which induces a dipole on molecules surrounding it.
- The ease of distorting an electron cloud is called the polarizability of the molecule. The larger the atoms or molecules, the easier it is to distort the electron cloud and form an induced dipole.
- The dipole-dipole intermolecular force is established between polar molecules and increases with the strength of the dipole within a molecule. A particularly strong form of dipole-dipole interaction is the hydrogen bond, which occurs only when a sufficiently electronegative element (N, O, or F) is bonded directly to a hydrogen atom.
- In solutions, other intermolecular attractions can develop such as ion-dipole and dipole-induced dipole interactions.
- Diffusion is the passive mixing of solute and solvent particles by random molecular motion from a region of high solute concentration to a region of low solute concentration.
- The melting point, boiling point, vapor pressure, and osmotic pressure of solutions are called colligative properties—those for which the important factor is not the identity of the solute but simply how many solute particles are present in solution.
- HPLC relies on the intermolecular forces of attraction between the compounds of a mixture to the stationary phase and mobile phase to separate the components.

KEY TERMS

intermolecular forces, p. 170
surface tension, p. 171
electron cloud, p. 172
induced dipole, p. 172
dispersion forces, p. 173
Van der Waals forces, p. 173
London forces, p. 173
polarizability, p. 173
dipole-dipole force, p. 174
hydrogen bonding, p. 175
ion-dipole force, p. 176
dipole-induced dipole force, p. 177
vapor, p. 178
pressure, p. 178
vapor pressure, p. 178
boiling point, p. 179
boiling point elevation, p. 180
colligative properties, p. 180
molality, p. 181
freezing point depression, p. 183
equilibrium, p. 184
diffusion, p. 185
osmosis, p. 185
semipermeable membrane, p. 185
osmotic pressure, p. 186
reverse osmosis, p. 186
isotonic solution, p. 187
hypotonic solution, p. 187
hypertonic solution, p. 187
osmolality, p. 187
high performance liquid chromatography (HPLC), p. 188

CONTINUING THE INVESTIGATION Additional Readings, Resources, and References

Wesley, James F. "Osmolality—A Novel and Sensitive Tool for Detection of Tampering of Beverages Adulterated with Ethanol, -Butyrolactone, and 1,4-Butanediol, and for Detection of Dilution-Tampered Demerol Syringes," *Microgram Journal* 1 (2003): 8.

For more information about DEA's Office of Forensic Sciences: www.dea.gov/programs/forensicsci/microgram/index.html

For more information about osmotic pressure: www.nlm.nih.gov/medlineplus/ency/article/003463.htm

For more information about Alfred Nobel: nobelprize.org/nobel/alfred-nobel/index.html

REVIEW QUESTIONS AND PROBLEMS

Questions

1. What types of intermolecular forces exist in a polar solvent?
2. What type of intermolecular force exists in a nonpolar solvent?
3. What type of intermolecular forces will develop when ionic compounds are dissolved in a polar solvent?
4. What type of intermolecular forces will develop when polar molecules are dissolved in a polar solvent?
5. Athough oxygen gas is nonpolar, it will dissolve to a very small extent in the polar solvent water. Which intermolecular force(s) is/are responsible for the interaction between the two compounds?
6. Will nitrogen gas be more or less soluble in water as compared with oxygen gas?
7. Why is the induced dipole–induced dipole (dispersion) intermolecular force the weakest?
8. Why is the hydrogen bond the strongest intermolecular force?
9. What are the conditions required for the formation of the hydrogen bond?
10. What is the difference between the movement of alcohol through the human body and the movement of nutrients throughout the body?
11. What is meant when a system is said to be in equilibrium?
12. Explain what the vapor pressure of a liquid is and what happens to the vapor pressure as the temperature of the liquid increases. Sketch a molecular view of the same liquid at two different temperatures that illustrates the differences in vapor pressure.
13. Explain what is meant by the boiling point of a solution and how vapor pressure plays a role.
14. Explain why two different liquids will have different boiling points in terms of the types of intermolecular forces present.
15. Explain how the vapor pressure of a solution is lowered by the presence of a solute, and sketch a molecular view of this process.
16. Explain how the freezing point of a solution is lowered by the presence of a solute, and sketch a molecular view of this process.
17. If a saltwater sample has a freezing point of $-4°C$, draw a molecular view of what happens when the temperature is lowered to $-5°C$.
18. In winter it is common to spread sodium chloride on highways to melt ice. However, salt is corrosive and not suitable for use on airplanes. Using the Internet as a resource, determine what is contained in the solutions used to deice airplanes.
19. Calcium chloride is a more expensive alternative to sodium chloride that can be used to melt ice during winter and is usually sold as "Super Deicer!" Is calcium chloride more efficient than sodium chloride at melting ice?
20. Explain why freezing point depression is the preferred method for the detection of tampered solutions.
21. Until the twentieth century, a common method for preserving food was to use salt in high concentration brines (pickling) or to simply store the meat in barrels lined with salt. Explain how these methods used the principle of osmosis to prevent bacteria or mold growth.
22. Explain how drinking water can be obtained from seawater based on the process of osmosis.
23. Explain the role of intermolecular forces in HPLC analysis used to trace the point of origin of a heroin sample seized by the DEA.

Problems

24. Calculate the surface area of a regular tetrahedron with a volume of 1 cm^3. For the regular tetrahedron, the volume and surface area are given by:

$$V = \frac{\sqrt{2}}{12}(L)^2 \quad \text{and} \quad A = (L)^2\sqrt{3}$$

where L is the length of an edge. What percent increase in surface area does a regular tetrahedron have as compared with a sphere of the same volume?

25. Calculate the surface area of a regular octahedron with a volume of 1 cm^3. For the

regular octahedron, the volume and surface area are given by:

$$V = \frac{\sqrt{2}}{3} a^3 \quad \text{and} \quad A = 2a^2\sqrt{3}$$

where a is the length of an edge. What percent increase in surface area does a regular tetrahedron have as compared with a sphere of the same volume?

26. Place the following list of atoms in order of increasing polarizability: Br, K, Fe, As.

27. Place the following list of atoms in order of increasing polarizability: B, Al, Ga, In.

28. List all of the intermolecular forces that will be present in the following liquids, and underline the most dominant force for each liquid. The type of compound is indicated in parentheses.
(a) NH_3 (polar)
(b) C_3H_8 (nonpolar)
(c) N_2 (nonpolar)

29. List all of the intermolecular forces that will be present in the following liquids, and underline the most dominant force for each liquid. The type of compound is indicated in parentheses.
(a) H_2O (polar)
(b) CH_3OH (polar)
(c) Br_2 (nonpolar)

30. List all of the intermolecular forces that will be present in the following aqueous solutions, and underline the most dominant force for each liquid. The type of compound is indicated in parentheses.
(a) NaCl (ionic)
(b) CO_2 (nonpolar)
(c) $C_6H_{12}O_6$ (polar)

31. List all of the intermolecular forces that will be present in the following aqueous solutions, and underline the most dominant force for each liquid. The type of compound is indicated in parentheses.
(a) O_2 (nonpolar)
(b) CH_3OH (polar)
(c) $NH_4C_2H_3O_2$ (ionic)

32. Which compound will have a higher boiling point, HI or HCl? What is the most prominent intermolecular force present?

33. Which compound will have a higher boiling point, HF or HBr? What is the most prominent intermolecular force present?

34. Place the following solutions in order of their decreasing vapor pressure of water.

Solution A: 0.1 M $(NH_4)_2SO_4$
Solution B: 0.2 M $CuSO_4$
Solution C: 0.15 M KI

35. Place the following solutions in order of their decreasing vapor pressure of water.

Solution A: 0.5 M $CaCl_2$
Solution B: 0.6 M Na_3PO_4
Solution C: 0.11 M LiI

36. Why does the vapor pressure of a liquid increase as the temperature increases?

37. Why would it not be a good idea to calibrate thermometers by using boiling water?

38. Will carbon monoxide or carbon dioxide diffuse faster through blood, assuming they are present at equal concentrations?

39. Place the following solutions in order of highest boiling point to lowest boiling point: 0.1 m NaCl, 0.05 m Na_2S, and 0.08 m glucose ($C_6H_{12}O_6$).

40. Place the following solutions in order of highest boiling point to lowest boiling point: 0.25 m KI, 0.05 m Na_3PO_4, and 0.10 m $NH_4C_2H_3O_2$.

41. Calculate the boiling point for each solution in Problem 39, given that the K_{bp} = 0.512°C/m for water.

42. Calculate the boiling point for each solution in Problem 40, given that the K_{bp} = 0.512°C/m for water.

43. Calculate the change in the boiling point of water (K_{bp} = 0.512°C/m) under the following conditions.
(a) 21.00 g of KI dissolved in 250.0 g of water
(b) 13.3 g of NH_4NO_3 dissolved in 50.0 g of water
(c) 125.0 g of $CO(NH_2)_2$ (urea) dissolved in 225 g of water

44. Calculate the change in the freezing point of water (K_{fp} = −1.86°C/m) under the following conditions.

(a) 32.3 g of $C_2H_6O_2$ (ethylene glycol) dissolved in 100.0 g of water
(b) 78.8 g of $NH_4C_2H_3O_2$ dissolved in 125.0 g of water
(c) 25.8 g of NaOH dissolved in 75.8 g of water

45. Calculate the freezing point for each solution listed in Problem 39, given that the $K_{fp} = -1.86°C/m$ for water.

46. Calculate the freezing point for each solution listed in Problem 40, given that the $K_{fp} = -1.86°C/m$ for water.

47. Calculate the change in the freezing point of water ($K_{fp} = -1.86°C/m$) under the following conditions.
(a) 8.24 g of $FeCl_3$ dissolved in 50.00 g of water
(b) 273.2 g of $NaNO_3$ dissolved in 350.0 g of water
(c) 377.2 g of $(NH_4)_2SO_4$ dissolved in 225.0 g of water

48. Calculate the change in the freezing point of water ($K_{fp} = 1.86°C/m$) under the following conditions.
(a) 8.24 g of $FeCl_3$, dissolved in 50.00 g of water
(b) 3.22 g of of caffeine ($C_8H_{10}N_4O_2$) dissolved in 75.0 g of water
(c) 74.83 g of KCN dissolved in 1000.0 g of water

49. Determine the molality of an aqueous glucose ($C_6H_{12}O_6$) solution under the following conditions.
(a) Freezing point = −3.4°C
(b) Boiling point = 103.3°C
(c) Boiling point = 102.5°C

50. Determine the molality of an aqueous ethylene glycol solution ($C_2H_6O_2$) under the following conditions.
(a) Boiling point = 120.0°C
(b) Freezing point = −10.0°C
(c) Freezing point = −15.5°C

51. Assuming each solution from Problem 49 contained 1.00 kg of solvent, determine the moles of glucose and the mass of glucose dissolved in each solution.

52. Assuming each solution from Problem 50 contained 1.00 kg of solvent, determine the moles of ethylene glycol and the mass of ethylene glycol dissolved in each solution.

53. Draw a system showing the natural osmotic process of seawater being placed on one side of the membrane and pure water on the opposite side of the membrane. Be sure to indicate the movement of all solute and solvent particles.

54. Draw a membrane system showing the reverse osmosis of seawater to pure water, with the seawater placed on one side of the membrane and pure water on the opposite side of the membrane. Be sure to indicate the movement of all solute and solvent particles.

Forensic Chemistry Problems

55. In the case of a fatal drunk driving accident, the alcohol consumed throughout the evening can be determined by sampling the vitreous humor. However, moments before the fatal crash, the deceased consumed an extremely large portion of alcohol. Will this be reflected in the vitreous humor or not?

56. A liquid sample is sent to a laboratory for analysis. The contents are suspected to be either propanol (boiling point = 97.2°C) or heptane (boiling point = 98.4°C). Explain whether simply measuring the boiling point would be sufficient for identifying the compound.

57. For Problem 56, will the additional information that the density of propanol is 0.80 g/cm^3 and the density of heptane is 0.68 g/cm^3 make a difference in your answer? Explain.

Case Study Problems

58. A jogger is found collapsed and unresponsive along a beach in southern Wales. The police are called and an ambulance takes the woman to the local hospital. The jogger does not appear to be dehydrated, and her medical history provides no clues as to her present condition. Toxicology screening is ordered, and the contents of a half-full water bottle found with the jogger are sent to the laboratory for analysis.

The toxicology screen comes back negative for alcohol and the commonly abused classes of drugs. However, the preliminary freezing point measurement of the water bottle contents reveals it is a solution, not pure water. Does this imply the jogger was poisoned? List several compounds that could be present in water that the jogger herself may have added. How would you determine whether more than one compound had been added to the water?

59. A young man in his early twenties was found dead with several cans of petroleum-based cleaners and a plastic bag by his side. Investigators made the initial hypothesis that the young man had died of asphyxiation from inhaling the volatile fumes from the cleaners. The estimated time of death was 16 hours before discovery. When the autopsy was conducted several days later, only trace amounts of volatile petroleum-based compounds could be detected in his lungs, well beneath the lethal level. Is the original hypothesis of death by asphyxiation from inhalants still valid? Explain why only trace levels would be found.

CHAPTER 7

Drug Chemistry

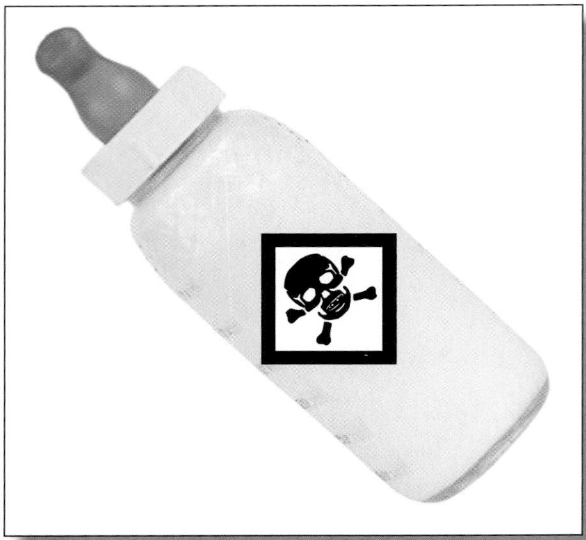

What could drive a mother to poison her own child's baby formula with antifreeze? (Bottle: Hemera Technologies/Alamy; warning: Fotosearch)

Case Study: The Experts Agreed

The summer of 1989 found the Stallings family living the American dream. Their first child, Ryan, was born earlier that spring, and they moved into a new lakefront home at the beginning of the summer. Patricia Stallings was a typical new mother, always checking on her baby and tending to his needs.

One Friday in early July, she noticed that Ryan had vomited after drinking his bedtime bottle of milk. She made a mental note to ensure that he was getting enough to eat the next day and hoped he would be feeling better in the morning. As Saturday came, Ryan seemed to be feeling better and was holding down his food. However, on Sunday, Ryan could hardly keep any food down. He was acting sluggish and having difficulty breathing. His mother immediately called an area hospital and took her son in to see the doctor.

The emergency room doctor ordered a battery of blood tests and was shocked when he received the results. Ryan Stallings had 1,2-ethanediol—

the main ingredient in antifreeze and better known as ethylene glycol—present in his blood! Ethylene glycol poisoning is extremely dangerous and often deadly. The Missouri Division of Family Services was called, and Ryan was immediately put into their custody. The Stallings family was stunned. How had their son been exposed to 1,2-ethanediol? The police believed that a member of the Stallings family had intentionally poisoned Ryan.

Family members were allowed supervised visits with Ryan, who was staying in foster care. It was during a supervised visit in early September that Patricia Stallings would spend her last few hours with her son feeding him a bottle. Within three days of Patricia's last visit, her son would be dead. Once again the laboratory results on Ryan's blood came back positive for 1,2-ethanediol. The police wasted no time arresting a suspect: Patricia Stallings.

How could a mother poison her own child? The prosecutor theorized that she suffered from Munchausen's syndrome by proxy. This is a mental illness in which a parent subjects his or her child to unnecessary medical procedures in a desperate attempt to gather attention and sympathy. Laboratory tests showed that the baby's bottle that Patricia used had traces of 1,2-ethanediol, and a gallon of antifreeze was found in the basement. Furthermore, the autopsy revealed that calcium oxalate crystals had formed in Ryan's brain tissue, a finding indicative of 1,2-ethanediol poisoning. It didn't take a jury long to convict Patricia Stallings of first degree murder and to sentence her to life in prison.

Although it would seem Patricia's life could not be worse, it was complicated by the fact that she was three months pregnant with her second son, David Jr., when Ryan died. The Missouri Division of Family Services took custody of David Jr. as soon as he was born and placed him into foster care. No one realized at the time that David Jr. would be the key to proving his mother innocent of murder—and the experts wrong. He would provide evidence that the substance in Ryan's blood had not been 1,2-ethanediol but rather propanoic acid that had formed from 2-methylpropanedioic acid.

What exactly are these compounds and how could a newborn prove his mother's innocence? To answer these questions, we must first discuss organic chemistry . . .

7.1 Introduction to Organic Chemistry

Learning Objective
Differentiate between organic and inorganic compounds.

In the simplest terms, an **organic compound** is composed primarily of carbon and hydrogen atoms. Conversely, an **inorganic compound** is composed of elements from the rest of the periodic table. You are probably

Table 7.1 Carbon Prefixes

Prefix	Carbon Atoms
meth-	1
eth-	2
prop-	3
but-	4
pent-	5
hex-	6
hept-	7
oct-	8
non-	9
dec-	10

familiar with several of the simplest of the organic compounds because of their use as fuels: methane (CH_4), propane (C_3H_8), and butane (C_4H_{10}). The name of the compounds provides information about the number of carbon atoms in each compound, as shown in Table 7.1. More information about how to determine the formulas and structures of organic compounds will be provided in later sections.

The chemical compounds from the case study fit the definition of organic compounds, because the chemical formula of 1,2-ethanediol is $C_2H_6O_2$. Propanoic acid has the formula $C_3H_6O_2$, and 2-methylpropanedioic acid has the formula $C_4H_6O_4$. Other elements such as oxygen, nitrogen, phosphorus, and sulfur can be found as minor components of organic compounds.

Until the early nineteenth century, organic compounds had to be isolated from living or once-living sources (organisms) such as plants and animals, because scientists were unable to synthesize organic compounds in the laboratory. However, in 1828 Friedrich Wöhler was able to successfully synthesize urea, an organic compound, in the laboratory. Modern organic chemistry is no longer limited to compounds found in living systems. Organic compounds now include a seemingly endless array of substances resulting from the discovery of new natural molecules and from the synthesis of new molecules in the laboratory.

It might seem that not too many compounds would result from combinations with just carbon, hydrogen, and a few other elements as building blocks. But as you will learn in this chapter, the variety of ways to combine these elements into organic compounds is almost limitless. Nearly 4000 new substances are added each day to the registry of the Chemical Abstracts Service, an internationally used database of chemical compounds. Most of the new substances are organic compounds. Of the more than 25 million known compounds, over 90% contain carbon.

One of the most important uses of organic chemistry today is the development of new compounds that can potentially be used as pharmaceuticals. For some pharmaceuticals, such as narcotic painkillers, the desired medical effect lends itself to abuse. The chemical structure of illegal narcotics and prescription drugs can be very similar. Notice, for example, the similarities between heroin, an illegal narcotic, and Vicodin, a drug prescribed for pain, shown in Figure 7.1.

As you progress through the chapter, you will learn how to interpret these structures, but the similarities between heroin and Vicodin are strik-

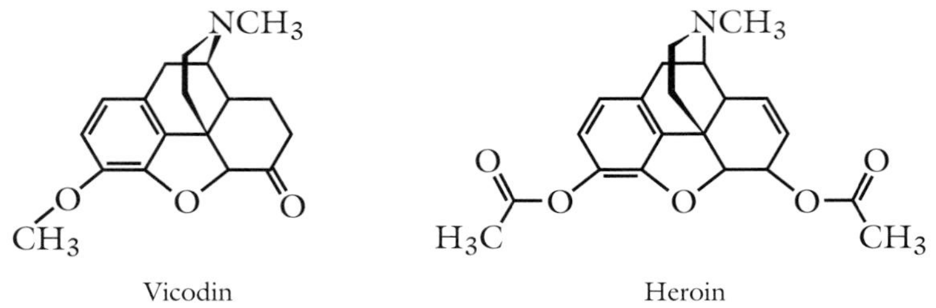

Figure 7.1 Chemical structures. The central portions of Vicodin ($C_{18}H_{21}NO_3$) and heroin ($C_{21}H_{23}NO_5$) drug molecules are very similar. The molecules differ only by three carbon, two hydrogen, and two oxygen atoms.

ingly apparent just by casual observation. In fact, it is common for pharmaceutical companies to study slight modifications of a basic drug structure to determine whether the new molecule will be more effective and have fewer side effects than the original drug. The example of Vicodin and heroin shows that simply adding or removing a few atoms of carbon or oxygen changes how that molecule interacts with the human body.

7.2 Alkanes

Learning Objective

Name alkanes and draw their structures.

The structures of cocaine and crack cocaine are shown in Figure 7.2. Cocaine is an ionic compound (notice the positive charge of the nitrogen ion and the negative charge of the chloride ion) whereas crack cocaine is a neutral compound. Because cocaine is soluble in water, drug users can snort (breathe in through the nose) finely powdered cocaine, which dissolves in the moisture of the nasal cavity and passes into the bloodstream. Crack cocaine is not soluble in water and must be smoked rather than snorted, because it does not dissolve in the nasal passage.

When a drug sample suspected of being some form of cocaine is brought into the forensic laboratory, it is necessary to determine which form is present. Because cocaine and crack cocaine have different physical properties, the two can easily be distinguished by adding the sample to hexane, a nonpolar organic solvent. In Chapters 5 and 6, the process of dissolution and the role of intermolecular forces were discussed. Determining whether a sample is cocaine or crack cocaine is simply an application of *like dissolves like*. The neutral, nonpolar, crack cocaine readily dissolves in the nonpolar hexane whereas cocaine, due to its ionic nature, does not. Cocaine is easily dissolved by water through dipole-ion attraction. The solubility difference quickly indicates to a forensic chemist which form of cocaine is present. The next step is to obtain a positive identification of the substance and its purity by using chromatography.

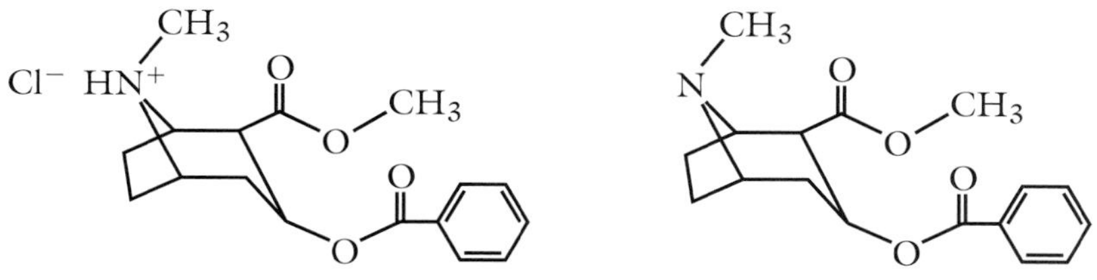

Cocaine　　Crack cocaine

Figure 7.2 Forms of cocaine. Cocaine is an ionic salt that is soluble in water. Crack cocaine is a neutral molecule that is insoluble in water but is soluble in nonpolar solvents such as hexane.

The three-dimensional shape of molecules is extremely important in determining the physical properties of a compound; many compounds discussed in this chapter are thus shown as three-dimensional models to help you visualize the molecule. Chapter 8 will focus on determining the three-dimensional shapes of small molecules and how the local geometry at each carbon atom affects the overall shape of larger molecules.

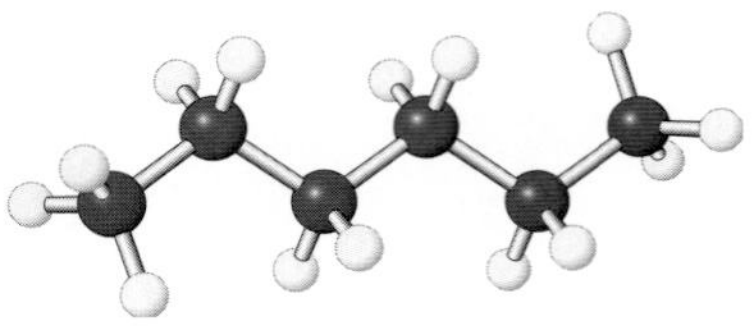

Hexane (C_6H_{14})

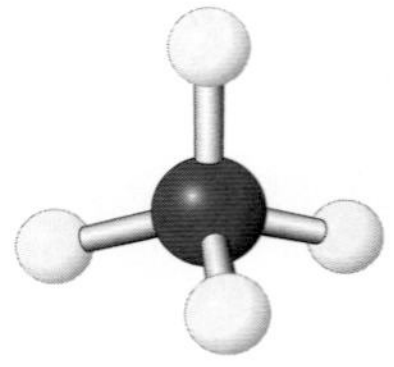

Methane (CH_4)

The simplest class of organic compounds is the **alkanes**, an example of which is hexane (see top left). Alkanes are commonly used in the laboratory as solvents because they are fairly nonreactive; that is, they dissolve nonpolar compounds without undergoing chemical reactions with the compounds. Alkanes are also commonly used as fuel sources and, for this reason, can become a tool for arsonists. The analysis of alkanes relates to forensic investigations of arson and will be discussed in later chapters.

The simplest of all alkanes is the compound methane (CH_4), which consists of a carbon atom bonded to four hydrogen atoms (see lower left). The next simplest compound is ethane (C_2H_6). The alkane class of organic compounds has chemical formulas that follow the pattern C_nH_{2n+2} where n is the number of carbon atoms. Ethane consists of two carbon atoms bonded together, with each carbon having three bonds to hydrogen atoms. This information is not readily apparent by just looking at the formula C_2H_6. For this reason, structural formulas are often written for organic compounds. A **structural formula** shows not only the type and number of atoms present in a molecule but also the way the atoms are arranged with respect to one another. Writing out the structural formula of alkanes is fairly simple if you follow these basic rules:

Rules for Organic Structures

1. Carbon atoms bond together to form a chain.

2. Carbon *always* has a total of four bonds.

3. Hydrogen has *only* one bond.

The structural formula for ethane (C_2H_6) can be determined by applying these rules:

1. The two carbon atoms must be bonded together: C—C.

2. Each carbon atom needs three more bonds.

3. There are six hydrogen atoms, three bonded to each carbon atom:

$$\begin{array}{ccc} & H & H \\ & | & | \\ H- & C- & C-H \\ & | & | \\ & H & H \end{array}$$

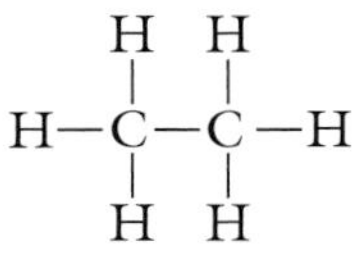

Worked Example 1

Given the rules for organic compounds, what is the formula and structure for the compound propane that contains three carbon atoms?

SOLUTION Using the formula, $n = 3$ and $C_3H_{2(3)+2} = C_3H_8$. If the carbon atoms are bonded together, the basic structure must be: C—C—C. Since a carbon atom forms a total of four bonds, each end carbon has three hydrogen atoms and the center carbon has two hydrogen atoms, for a total of eight hydrogen atoms.

$$\begin{array}{cccc} & H & H & H \\ & | & | & | \\ H- & C- & C- & C-H \\ & | & | & | \\ & H & H & H \end{array}$$

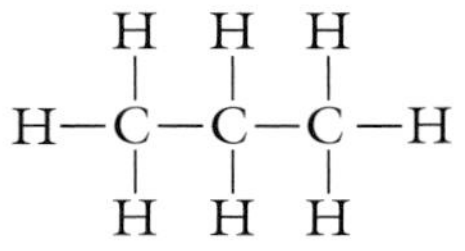

Practice 7.1

Write the formula and draw the structure for the compound butane, which contains four carbon atoms.

ANSWER

C_4H_{10} 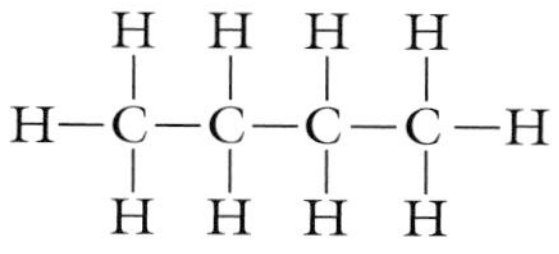

As the number of carbon atoms increases, writing the structural formula showing all the bonds can become quite cumbersome. The **condensed structural formula** of a compound shows the atoms in the same order as the structural formula but leaves out the lines representing bonds. For example, the condensed structural formula for butane is $CH_3CH_2CH_2CH_3$. The formulas and condensed structural formulas for the first ten alkane compounds are given in Figure 7.3. As stated earlier, the names also indicate the number of carbon atoms present: *meth*- represents 1, *eth*- represents 2, *prop*- represents 3, and so on. You should commit these prefixes to memory.

Methane	CH_4	CH_4
Ethane	C_2H_6	CH_3CH_3
Propane	C_3H_8	$CH_3CH_2CH_3$
Butane	C_4H_{10}	$CH_3CH_2CH_2CH_3$
Pentane	C_5H_{12}	$CH_3CH_2CH_2CH_2CH_3$
Hexane	C_6H_{14}	$CH_3CH_2CH_2CH_2CH_2CH_3$
Heptane	C_7H_{16}	$CH_3CH_2CH_2CH_2CH_2CH_3$
Octane	C_8H_{18}	$CH_3CH_2CH_2CH_2CH_2CH_2CH_2CH_3$
Nonane	C_9H_{20}	$CH_3CH_2CH_2CH_2CH_2CH_2CH_2CH_2CH_3$
Decane	$C_{10}H_{22}$	$CH_3CH_2CH_2CH_2CH_2CH_2CH_2CH_2CH_2CH_3$

Figure 7.3 Alkane formulas.

The structures of cocaine shown in Figure 7.2 illustrate another method of drawing the structures of organic compounds. In this method, lines represent the bonds between carbon atoms that are understood to be at the beginning and end of the lines and at the places where the lines change direction. To simplify the **line structure**, hydrogen atoms are not typically shown when bonded to carbon. For example, butane has four carbon atoms bonded together in a chain, as represented by the line structure below.

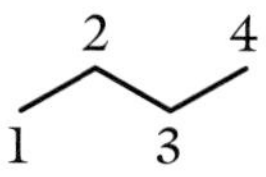

Worked Example 2

Write the line structure for hexane, the solvent commonly used to extract neutral drugs in the laboratory.

SOLUTION

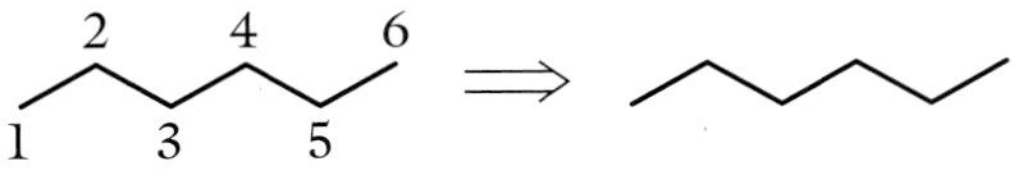

Practice 7.2

Write the line structure for heptane.

ANSWER

Worked Example 3

What is the name and formula for the following compound?

SOLUTION

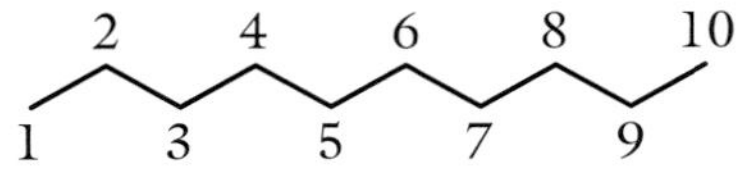

The number of carbon atoms is 10, so the compound is decane and the formula is $C_nH_{2n+2} = C_{10}H_{22}$.

Practice 7.3

What is the name and formula for the following compound?

ANSWER

Octane (C_8H_{18})

A dangerous form of substance abuse is "huffing," the inhalation of vapors of chemical substances, such as the solvent mixtures used in many commercial products. Huffing can cause serious health problems—such as irreversible damage to the heart, kidneys, lung, liver, and brain—and even death. Chronic huffing can also cause personality changes, resulting from permanent brain damage that will not be resolved, even if the user stops abusing inhalants.

The damage in huffing is caused by toxic organic compounds, such as alkanes and compounds called *alkenes,* as well as *alkynes* that are closely related to alkanes. Alkenes and alkynes, discussed in the next section, are created when carbon atoms form multiple bonds to another carbon atom.

7.3 Alkenes and Alkynes

Learning Objective

Name alkenes and alkynes and draw their structures.

Alkenes

The carbon-carbon bonds in alkanes are single bonds in which two carbon atoms share one pair of electrons. Compounds that contain a carbon-

carbon double bond (a sharing of two pairs of electrons) belong to the **alkene** class of organic compounds. Alkenes have the general formula C_nH_{2n} and the *-ene* ending to their names. The simplest alkene is $CH_2{=}CH_2$, which is called *ethene*. The rules for drawing the structural formula, condensed structural formula, and line structure still apply.

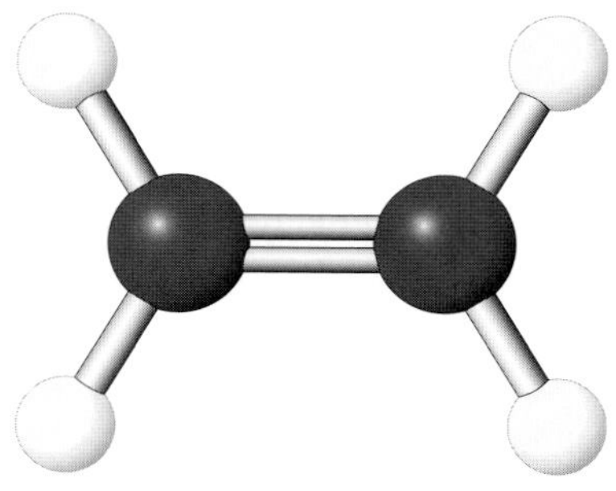

Ethene (C_2H_4)

As the length of the carbon chain increases, there are different possible locations for the double bond, as shown in Figure 7.4 of butene. Does the placement of the double bond within the molecule change the nature of the compound? If the answer is yes, then the physical properties of the three compounds shown in Figures 7.4a, 7.4b, and 7.4c will differ. In fact they do: The boiling point of the compounds in Figures 7.4a and 7.4c is −6.3°C, but that of the compound in Figure 7.4b is 4°C.

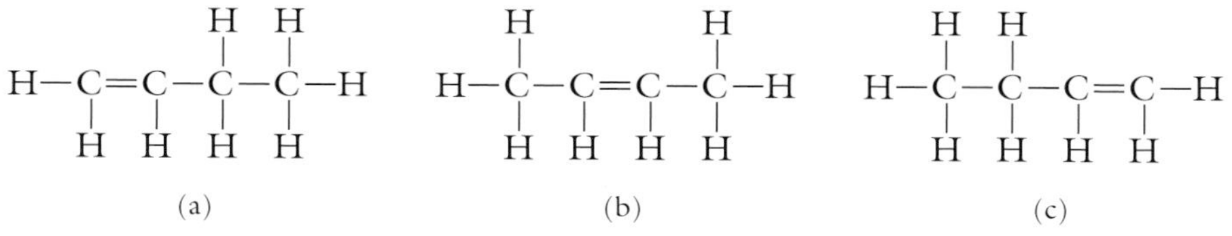

Figure 7.4 Double bond placement in butane.

The compounds shown in Figures 7.4a and 7.4c are, in fact, the same compound, just flipped end for end. The compound in Figure 7.4b is different. Two compounds that have the identical chemical formula but different structural formulas are called **isomers.** Distinguishing one isomer from another requires a systematic method for naming the compounds.

Isomers play a critical role in investigating the origin of narcotics and explosives. During the synthesis of organic compounds, it is common for very small amounts (sometimes referred to as *trace amounts*) of isomers to be formed. Furthermore, the relative amounts and types of impurities usually vary from one production facility to another due to slight variations in conditions such as temperature, pressure, and the purity of the chemicals used in the reaction. The DEA uses this information to trace heroin and cocaine to their countries of origin and even to the specific drug cartel. The FBI has assisted in researching isomer analysis to trace the source of TNT to the individual country and manufacturing facility.

One benefit of this type of intelligence gathering is that the drug cartels and explosive manufacturers can do nothing to hide the trace amounts in their products. Although it is theoretically possible to purify the product to a degree that this information would be lost, the cost of doing so is prohibitive.

Historically, there was no agreed-upon method for naming organic compounds. Today the naming of compounds follows suggested rules from the International Union of Pure and Applied Chemistry (IUPAC), the same organization that regulates changes or additions to the periodic table. There are many compounds that are commonly referred to by their old names, such as ethylene glycol in the opening case study. The IUPAC name for this poison is 1,2-ethanediol.

The IUPAC rules for naming alkenes are listed as follows, with an example to illustrate their application.

1. Number the carbons in order forward and backward:

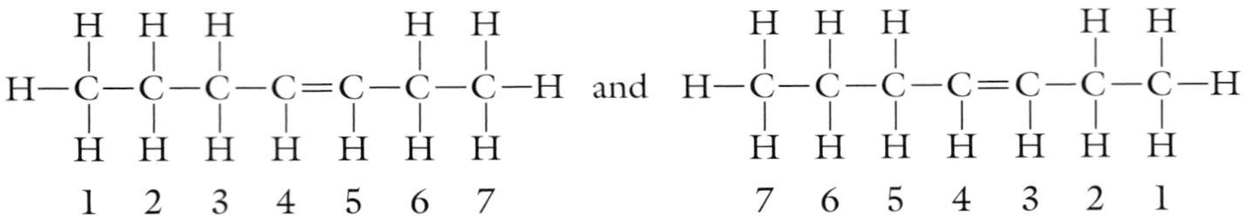

2. Indicate the length of the carbon chain by using the prefixes in Table 7.1: *hept-*.
3. Add the *-ene* ending to the name: heptene.
4. Place in front of the name the lowest value for the carbon atom with a double bond: 3-heptene.

The correct name for this compound is 3-heptene, *not* 4-heptene, which illustrates why it is important to number the alkene chain in both directions. Applying these naming rules to the structures drawn in Figure 7.4, the compound in Figures 7.4a and 7.4c is 1-butene, and the compound in Figure 7.4b is 2-butene.

Worked Example 4

Write the condensed structural formula and the line structure for 2-pentene.

SOLUTION

1. Recall that the prefix *pent-* indicates a five-carbon chain. Start by drawing the carbon chain and place the double bond after the second carbon: C—C=C—C—C.
2. Apply the rule that all carbon atoms need four bonds, with hydrogen atoms making up the remaining atoms: $CH_3CH{=}CHCH_2CH_3$.
3. Confirm that the chemical formula follows the C_nH_{2n} pattern for alkenes from the condensed structural formula in step 2. Because there are five carbon atoms and ten hydrogen atoms, the chemical formula is C_5H_{10}.
4. Draw the line structure based on step 1 with a double line to show the double bond:

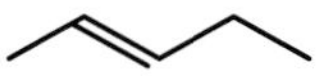

Practice 7.4

Write the condensed structural formula and the line structure for 3-octene.

ANSWER

$CH_3CH_2CH{=}CHCH_2CH_2CH_2CH_3$ 

Alkynes

Two carbon atoms can form not only single and double bonds but also triple bonds in which three pairs of electrons are shared. Compounds containing a carbon-carbon triple bond are known as **alkynes** and have the general formula C_nH_{2n-2}. The simplest of the alkynes is $H{-}C{\equiv}C{-}H$, which is called *ethyne*. With the exception of using the *-yne* ending to name alkynes, the same considerations and rules apply to naming alkynes and drawing their structures as applied to alkenes. Just as alkenes form isomers in which the double bond is located in different places along the carbon chain, alkyne isomers result from different placements of the triple bond.

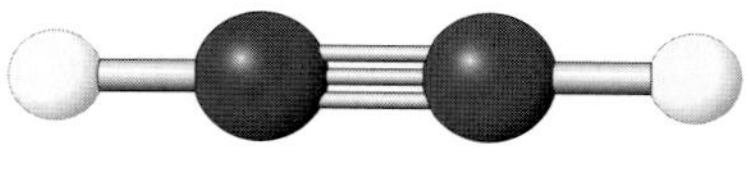

Ethyne (C_2H_2)

Worked Example 5

Write the line structure for the compound $CH_3CH_2CHCHCH_2CCCH_3$.

SOLUTION

1. Draw the expanded structure to clarify where the hydrogen atoms belong and how many bonds are on each atom.

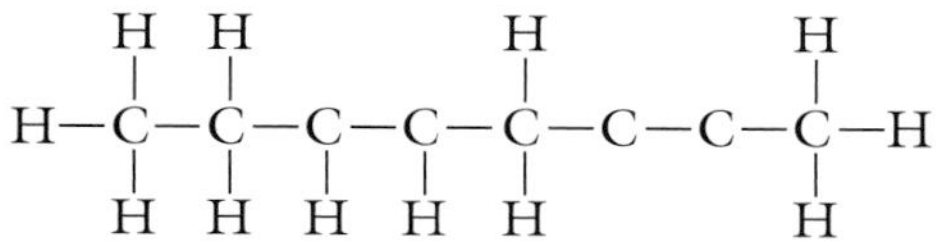

2. Identify carbon atoms with fewer than four bonds. Add double and triple bonds as needed:

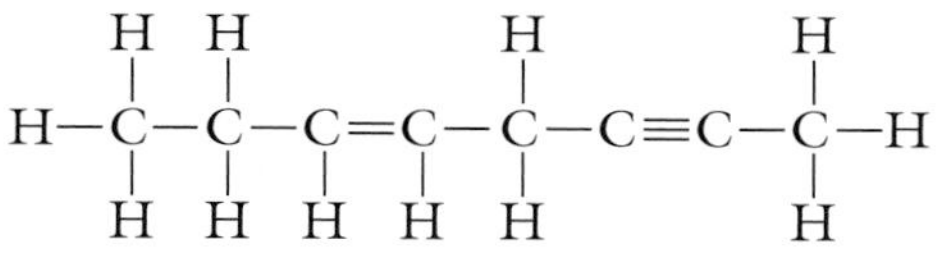

Practice 7.5

Write the condensed structural formula for 3-hexyne.

ANSWER

$CH_3CH_2CCCH_2CH_3$

7.4 Branched Isomers

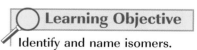

Learning Objective

Identify and name isomers.

The isomers we have discussed up to now are compounds that have the same chemical formula but different structural formulas due to different placements of the double or triple bond within the carbon chain. Another kind of isomer formed by alkanes, alkenes, and alkynes results from the branching of the carbon atoms, as shown in Figure 7.5. All three structures share the same chemical formula, C_5H_{12}. However, each compound has properties different from the others. For example, the boiling point of the compound in Figure 7.5a is 36°C, that of Figure 7.5b is 28°C, and that of Figure 7.5c is 9.5°C.

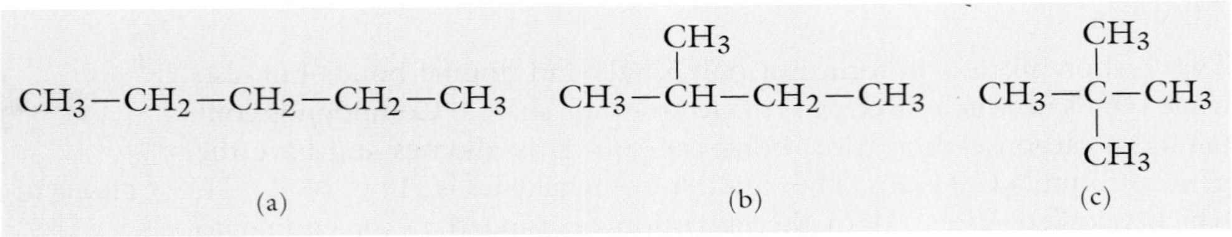

Figure 7.5 Isomers of pentane.

The name of each isomer must accurately communicate its structure. The IUPAC rules used to indicate where double or triple bonds are located along the carbon chain are also used with minor modifications for branched-chain isomers:

Naming Branched-Chain Isomers

1. Find the longest unbranched carbon chain.
2. Number the carbon atoms so that the carbon atom with the branch attached has the lowest value.
3. Indicate the branch length by using the prefixes in the table below (based on the Table 7.1 prefixes, modified with a *-yl* ending).
4. Indicate the name of the compound with the prefixes in Table 7.1 and the *-ane* ending applied to the number of carbon atoms in the longest branch.
5. If a branch appears more than once, use the prefixes *di-*, *tri-*, *tetra-*, etc.
6. In front of the name, indicate the location of the branches by writing the numbers of the carbon atoms at which the branches are located, separated by commas.

Prefix	Branch Length
methyl-	1
ethyl-	2
propyl-	3
butyl-	4
pentyl-	5
hexyl-	6
heptyl-	7
octyl-	8
nonyl-	9
decyl-	10

The compound in Figure 7.5a has no branches and is named *pentane*, or sometimes *n*-pentane for "*normal* (unbranched) pentane." The compound in Figure 7.5b has a four-carbon (4-C) chain with a 1-C branch on carbon atom number 2. Therefore, the name is 2-methylbutane. The compound in Figure 7.5c has a 3-C chain with two 1-C branches on carbon atom number 2. Therefore, the name is 2,2-dimethylpropane. The complexity and nuances of naming organic compounds can increase dramatically for compounds that have multiple branches and multiple double or triple bonds. We will focus on naming simple branched compounds in this textbook.

Worked Example 6

Name the following compounds:

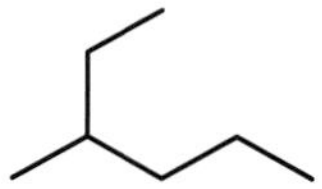

(a)

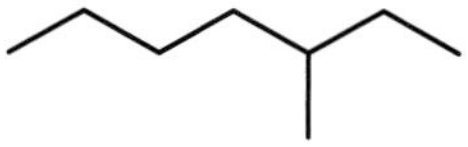

(b)

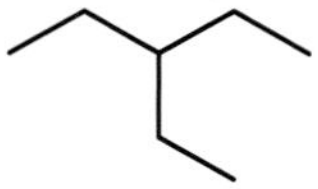

(c)

SOLUTION

(a) The longest chain is a 6-C chain and the branch is a 1-C chain. The branch is located at carbon atom number 3, so the name is 3-methylhexane.

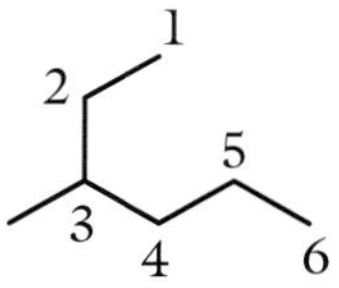

(b) The longest chain is a 7-C chain and the branch is a 1-C chain. The branch is located at carbon atom number 3, so the name is 3-methylheptane.

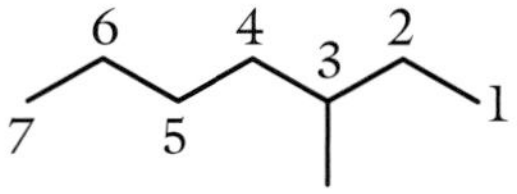

(c) The longest chain is a 5-C chain and the branch is a 2-C chain. The branch is located at carbon atom number 3, so the name is 3-ethylpentane.

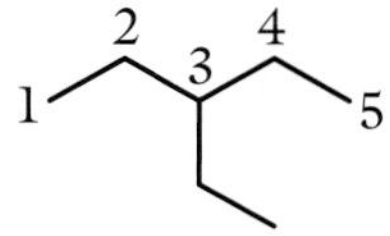

Practice 7.6

Draw the following molecules:

(a) 4-propylheptane

(b) 2-methylhexane

ANSWER

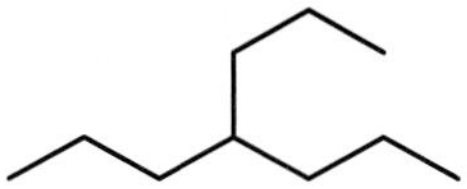

(a)

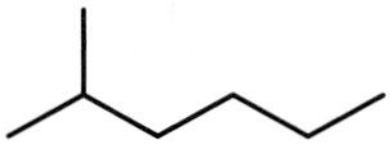

(b)

7.5 Cyclic Compounds

Learning Objective

Learn about cyclic compounds and how to name them.

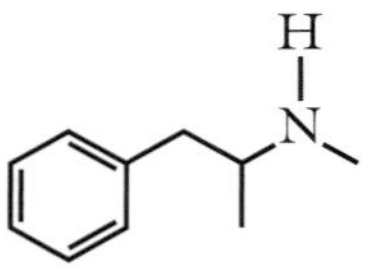

Figure 7.6 Methamphetamine.

Methamphetamine, commonly called *meth,* is a powerfully addictive stimulant that affects the central nervous system. The structure of methamphetamine is shown in Figure 7.6. Use of this drug has become a serious problem across the nation. Although a widespread methamphetamine problem has developed only in the last 15 years, this illegal substance has roots in World War II, when field laboratories were set up to make methamphetamine for German soldiers. The drug is capable of keeping a person awake for days, even weeks, at a time! It also creates a sense of paranoia and edginess that the German command decided to exploit to stretch their troops further. Hence, one of the street names used for this drug today is "Nazi dope."

Today, do-it-yourself methamphetamine labs are found in every state. Recipes for making methamphetamine are passed on from one group of users to another, and the ingredients are fairly easy to obtain. This situation should be a concern to every citizen because the clandestine laboratories pose serious safety risks not only to the manufacturers but to the public. The synthesis of methamphetamine involves toxic, flammable, and potentially explosive materials that are handled by untrained individuals who often have not slept for days while on the drug. Manufacturing the average batch of methamphetamine creates five pounds of toxic waste, which is being dumped into city sewer systems, backyards, roadsides, parking lots, and playgrounds. Labs are found in homes, hotels, dorm rooms, and even cars driving down the highway.

Firefighters and police officers are also put in an extremely dangerous position when dealing with the hazards of these laboratories. Many have suffered injury and even death when entering clandestine drug labs. At the writing of this book, nineteen Illinois State Police officers are on permanent disability from injuries suffered in methamphetamine labs. Every clandestine drug lab that is raided also requires a hazardous-materials cleanup at the taxpayers' expense—on average, nearly $10,000 for each cleanup.

The structure of methamphetamine includes a prominent ring, a feature common to many organic compounds that deserves further explana-

Law enforcement officials disposing of the hazardous materials from a meth lab. (Wes Pope/Aurora Photos)

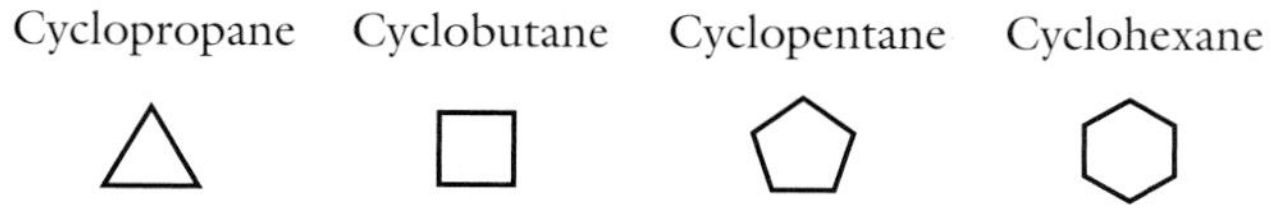

Figure 7.7 Cycloalkanes.

tion. But before discussing the particular type of ring in methamphetamine, we start with the simplest of the ring structures, the **cycloalkanes.** As the name implies, the cyclic rings consist of carbon atoms connected with single bonds. Figure 7.7 shows the structures for several cycloalkane compounds. Cyclohexane is a common solvent used in laboratory settings and is an ingredient in several latent-fingerprint developer solutions.

Worked Example 7

What is the chemical formula for each of the cycloalkanes in Figure 7.7?

SOLUTION In alkane ring structures, each carbon forms single bonds to two other carbons. Since carbon always forms four bonds in total, the other two bonds are to hydrogen atoms. Therefore, the formula for cyclopropane is C_3H_6, for cyclobutane is C_4H_8, for cyclopentane is C_5H_{10}, and for cyclohexane is C_6H_{12}.

Practice 7.7

The rule for alkane formulas is C_nH_{2n+2}. What is the rule for the formulas of cylcoalkanes?

ANSWER
C_nH_{2n}

In Figure 7.6, the methamphetamine ring structure shows alternating double bonds and single bonds around a six-carbon ring, called a *benzene ring*. **Benzene** is a very commonly used industrial and laboratory solvent belonging to a class called **aromatic compounds**. In the early days of discovery of organic compounds, those that were fragrant were grouped together and classified as aromatic. Most of these aromatic compounds contain a benzene ring. You will notice that most of the prescription drugs and illegal drugs that we will study in this textbook contain one or more benzene rings.

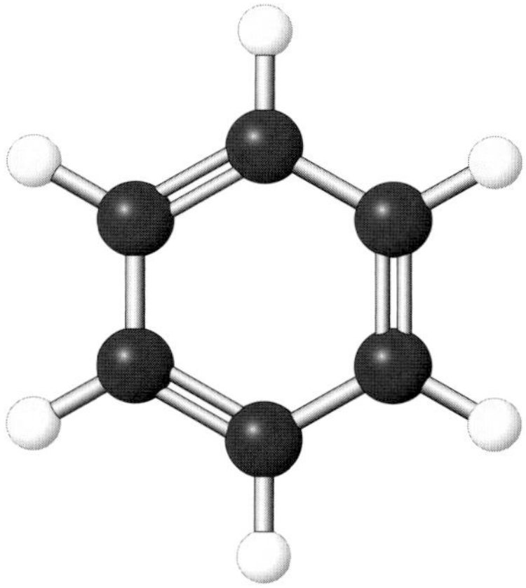

Benzene (C_6H_6)

It is appropriate to take a closer look at the bonding of benzene, as illustrated in Figure 7.8. The difference between the structure in Figure 7.8a and the structure in Figure 7.8b is that the double bonds have been shifted by one carbon. Does this mean that the two structures represent different compounds? Shifting the bonds in these two structures does not really change the molecule, but the bonding in benzene turns out to be more complex than alternating double and single bonds. The length of a double bond is, on average, shorter than a single bond. However, when the bonds of benzene are measured, the length of carbon-carbon bonds are the same! The length of the bond in benzene is longer than a double bond but shorter than a single bond.

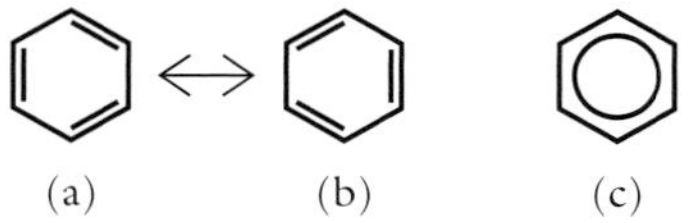

Figure 7.8 Benzene ring representations.

The actual structure of benzene is not accurately represented by either Figure 7.8a or Figure 7.8b—it is somewhere between the two. The structures in Figure 7.1a and Figure 7.8b are called **resonance structures**, a term used when more than one equivalent structure can be drawn for a compound. (Resonance structures will be discussed more fully in the next chapter.) The structure in Figure 7.8c is actually a shorthand version of benzene that chemists often use to represent the resonance structures of benzene.

Worked Example 8

LSD (lysergic acid diethylamide) is a potent hallucinogenic drug. How many benzene rings are in LSD? How many other ring structures are present?

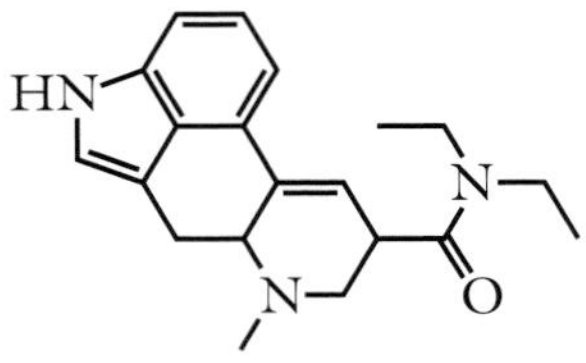

SOLUTION In the structure below, the benzene ring is colored blue. The yellow ring has four carbon atoms and one nitrogen atom. The green ring is cyclohexane, and the orange ring has five carbon atoms and one nitrogen atom.

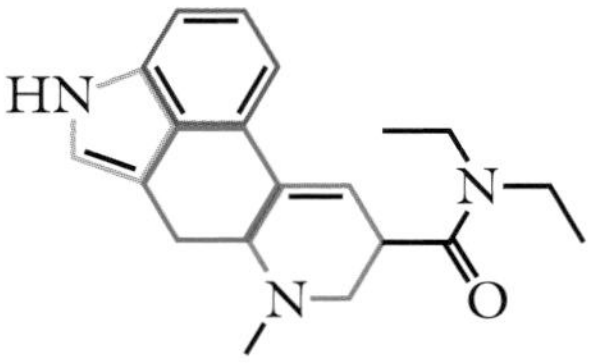

Practice 7.8

What is the formula for benzene?

ANSWER

C_6H_6

Several other common compounds that are simple derivatives of benzene are listed in Figure 7.9. Each of them plays an interesting role in forensic science. For example, phenol is used to extract DNA and is an ingredient in some fingerprint-developer formulations. While phenol has been used as a poison, it is present at low concentrations in several common mouthwashes. Toluene is a common solvent found in paint, printing and automotive cleaners, and many industrial solvents. Toluene is abused by individuals who inhale the vapors of such products. Since it is flammable, toluene can also be used as an accelerant for arson. The last structure, aniline, can be used to detect the presence of the chlorate ion, an ingredient in many explosives. Aniline forms an intensely colored compound when it reacts with chlorate. Many fabric dyes are also based on the aniline structure.

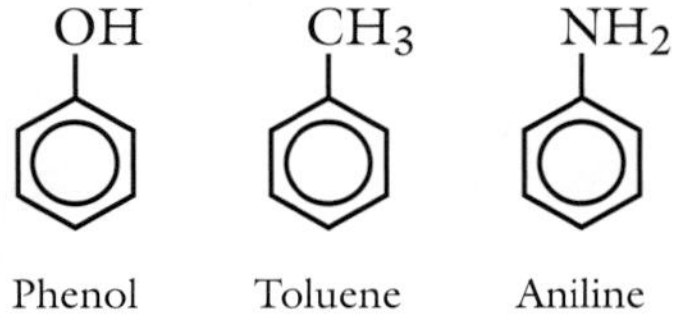

Figure 7.9 Compounds based on benzene.

7.6 Ethers, Ketones, and Esters

Ethers

It is apparent from the chemical structures shown in Figure 7.10 that oxygen atoms play an important role in drug molecules such as oxycodone and heroin. Oxycodone is a very addictive pain-relieving drug sold under the name OxyContin, and heroin is an illegal narcotic. It is also apparent that oxygen is found in several different bonding arrangements. In some cases, an oxygen atom is simply bonded between two carbon atoms. This arrangement is called the **ether** functional group and is often indicated as R—O—R, where R represents any carbon functional group. A **functional group** is an atom or group of atoms in an organic compound that imparts a distinct set of physical and chemical properties to the molecule.

Learning Objective

Recognize, draw, and name compounds containing the ether, ketone, and ester functional groups.

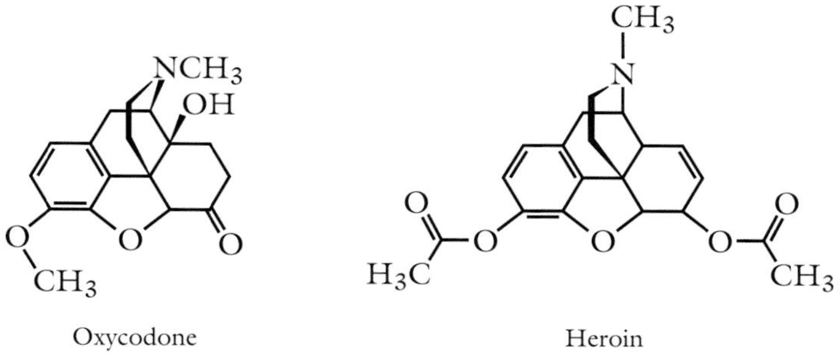

Figure 7.10 Chemical structures of oxycodone and heroin.

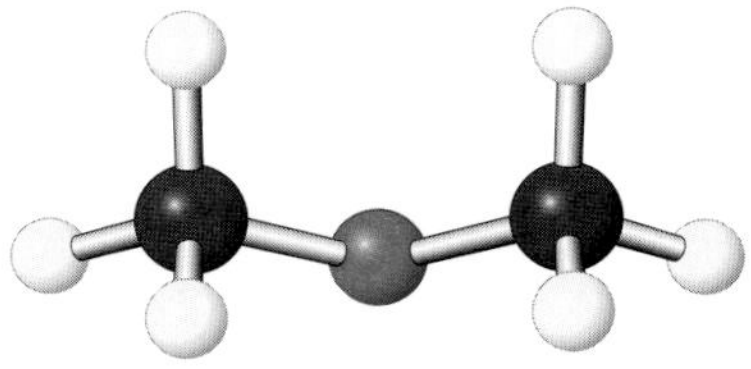

Methoxymethane (C_2H_6O) is a member of the ether functional group. All ethers share the R—O—R structure.

The rules for naming an ether compound are summarized below using the compound $CH_3OCH_2CH_3$ as an example:

Naming Ethers

1. The name of the ether starts with the name of the shortest R group, modified with the ending *-oxy*: methoxy.
2. The name of the longest R group makes up the remainder of the name: methoxyethane.

Worked Example 9

Ethoxyethane is one of the extremely flammable compounds used in the production of methamphetamine. It is historically known as the first surgical anesthetic and revolutionized surgery in 1846. It was used for the next 80 years until safer, more effective substances were identified. What is the structural formula for ethoxyethane?

SOLUTION

1. The first part of *ethoxy*ethane represents a 2-C chain bonded to the oxygen atom: CH_3CH_2O.
2. The second part of ethoxy*ethane* represents a 2-C chain: CH_2CH_3.
3. Hence, the structural formula for ethoxyethane is $CH_3CH_2OCH_2CH_3$.

Practice 7.9

Give the condensed structural formula for the following compounds:

(a) Methoxypropane

(b) Butoxypentane

(c) Ethoxybutane

ANSWER

(a) $CH_3OCH_2CH_2CH_3$

(b) $CH_3CH_2CH_2CH_2OCH_2CH_2CH_2CH_2CH_3$

(c) $CH_3CH_2OCH_2CH_2CH_2CH_3$

The boiling point of the ethers, shown in Table 7.2, increases as the size of the molecule increases because of increased Van der Waals forces between adjacent molecules. Methoxypropane and ethoxyethane are isomers of each other; notice that the placement of the oxygen atom changes the boiling point of the two compounds by 5°C.

Table 7.2 Properties of Ethers

Compound	Formula	Molecular Weight	Boiling Point
Methoxymethane	C_2H_6O	46.1	−22°C
Methoxyethane	C_3H_8O	60.1	7°C
Methoxypropane	$C_4H_{10}O$	74.1	39°C
Ethoxyethane	$C_4H_{10}O$	74.1	34°C

Ketones

A second functional group observed in the drug molecules listed in Figure 7.10 is the **ketone** group, in which an oxygen atom is double-bonded to a central carbon atom. The general formula for a ketone group is $R{-}\overset{\overset{\displaystyle O}{\|}}{C}{-}R$, where R again represents any carbon group. The simplest ketone is 2-propanone, a three-carbon ketone often called by its more common name, *acetone.*

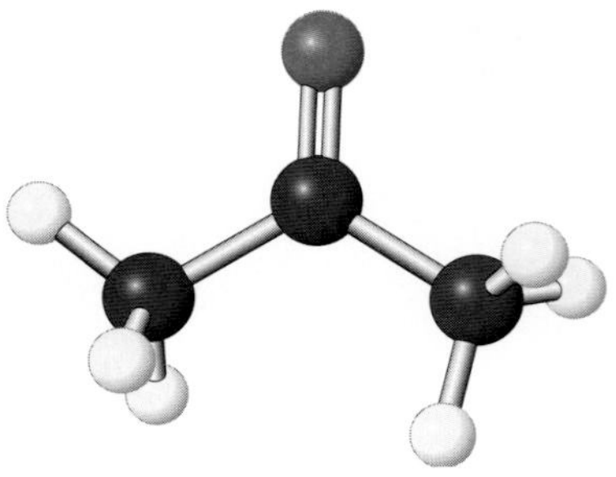

2-propanone (C_3H_6O) is a member of the ketone functional group. All ketones share the $R{-}\overset{\overset{\displaystyle O}{\|}}{C}{-}R$ structure.

Glucose undergoes a series of oxidation-reduction reactions within living cells to release the stored chemical energy. Insulin is a hormone that helps transport glucose from the blood into the cell to undergo this metabolic process. Diabetic patients require supplemental insulin because their bodies no longer produce sufficient amounts. One indicator that a diabetic patient hasn't been taking his or her insulin shots is the characteristic smell of acetone on the breath, a by-product of the incomplete oxidation of glucose. Clandestine drug labs often use 2-propanone in the preparation of methamphetamine, cocaine, and heroin, and it is readily available because it is a common solvent found in paint thinner and fingernail polish remover. The rules for naming ketones are listed below, using the following compound as an example:

$$CH_3CH_2\overset{\overset{\displaystyle O}{\|}}{C}CH_2CH_2CH_3$$

Naming Ketones

1. Count the number of carbon atoms in the longest continuous chain: 6.
2. Using the prefixes in Table 7.1, apply the ending *-none*: hexanone.
3. Identify the number of the carbon atom with the C=O group, always choosing the lowest possible value, and place it at the beginning of the name: 3-hexanone.

The ketone functional group is more electronegative than the ether functional group. Therefore, a ketone has stronger intermolecular forces than an ether of comparable size. This results in higher boiling points for ketones, as shown in Table 7.3. The position of C=O on isomers 2-pentanone and 3-pentanone causes a difference, albeit a slight one, in the boiling point of each compound.

Table 7.3 Properties of Ketones

Compound	Formula	Molecular Weight	Boiling Point
2-propanone	C_3H_6O	58.1	56°C
2-butanone	C_4H_8O	72.1	80°C
2-pentanone	$C_5H_{10}O$	86.1	100°C
3-pentanone	$C_5H_{10}O$	86.1	102°C

Worked Example 10

The compound 2-butanone is used in clandestine drug laboratories. Draw the condensed formula structure of this compound.

SOLUTION

1. The prefix (2-*buta*none) indicates a 4-C chain: C—C—C—C.
2. The ending (2-buta*none*) indicates that one of the carbon atoms is double bonded to an oxygen atom: C=O.
3. The number in front (*2*-butanone) indicates that the second carbon has the C=O.
4. Hydrogen atoms make up the remainder of the bonds:

$$CH_3\overset{\overset{\displaystyle O}{\|}}{C}CH_2CH_3$$

Practice 7.10

Draw and name all straight-chain isomers of 3-heptanone.

ANSWER

$$CH_3CH_2\overset{\overset{\displaystyle O}{\|}}{C}CH_2CH_2CH_2CH_3$$

3-heptanone

$$CH_3\overset{\overset{\displaystyle O}{\|}}{C}CH_2CH_2CH_2CH_2CH_3$$

2-heptanone

$$CH_3CH_2CH_2\overset{\overset{\displaystyle O}{\|}}{C}CH_2CH_2CH_3$$
4-heptanone

Esters

Another oxygen-containing functional group found in the drug molecules shown in Figure 7.10 is the **ester** functional group, represented by $R{-}O{-}\overset{\overset{\displaystyle O}{\|}}{C}{-}R$. Esters are commonly found in nature—they are the molecules that give us fragrance and flavor. Esters are also used in developing fingerprints using superglue fumes, which polymerize in the presence of moisture found in an invisible fingerprint, leaving a visible grey fingerprint.

The name of an ester compound has two parts, which are derived from the two separate R groups. The rules for naming esters are listed below, using the following compound as an example:

$$CH_3CH_2CH_2O\overset{\overset{\displaystyle O}{\|}}{C}CH_3$$

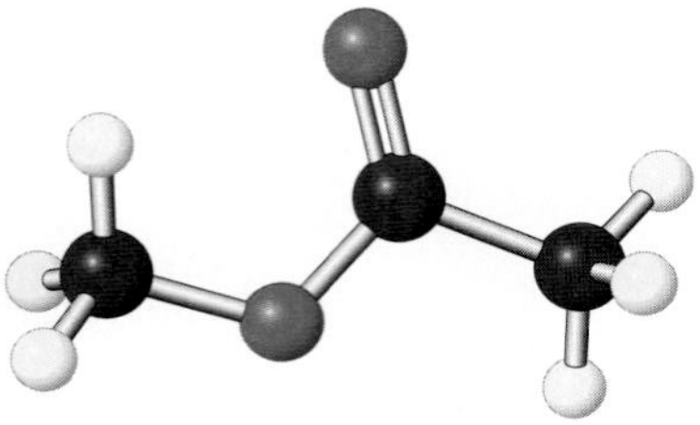

Methylethanoate ($C_3H_6O_2$) is a member of the ester functional group. All esters share the $R{-}O{-}\overset{\overset{\displaystyle O}{\|}}{C}{-}R$ structure.

Naming Esters

1. The first R group is the one not directly attached to the C=O. Count the number of carbon atoms in it (using the prefixes in Table 7.1) and apply the *-yl* ending: propyl.
2. Count the number of carbon atoms in the second R group, including the carbon atom in the C=O group (using the prefixes in Table 7.1), and apply the *-oate* ending: propyl ethanoate.

Worked Example 11

Draw the condensed structure formula for ethyl butanoate.

SOLUTION

1. The ethyl R group is a 2-C chain: C—C.
2. The butanoate R group is a 4-C chain: C—C—C—C.
3. The ending *-oate* indicates an ester functional group connecting the two R groups:

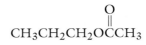

$$C{-}C{-}O{-}\overset{\overset{\displaystyle O}{\|}}{C}{-}C{-}C{-}C$$

4. Include hydrogen atoms as needed to give each carbon atom a total of four bonds:

$$CH_3CH_2O\overset{\overset{\displaystyle O}{\|}}{C}CH_2CH_2CH_3$$

Practice 7.11

Name all possible straight carbon chain isomers of ethyl butanoate.

ANSWER

The isomers are methyl pentanoate, propyl propanoate, butyl ethanoate, and pentyl methanoate.

As would be expected, the boiling point of the ester compounds increases with molecular weight, as shown in Table 7.4. When comparing esters with other organic compounds of similar molecular mass, the boiling point of esters is greater than that of ethers, but less than that of ketones. The electronegativity of the functional groups follows this same pattern: ketones > esters > ethers.

Table 7.4 Properties of Esters

Compound	Structure	Molecular Weight	Boiling Point
Methyl methanoate	$C_2H_4O_2$	60.1	32°C
Methyl ethanoate	$C_3H_6O_2$	74.1	56°C
Ethyl methanoate	$C_3H_6O_2$	74.1	54°C
Ethyl ethanoate	$C_4H_8O_2$	88.1	77°C

Worked Example 12

Draw the condensed structural formulas for the following compounds and place them in order from lowest boiling point to highest.

(a) 2-butanone

(b) Methoxypropane

(c) Methyl ethanoate

(d) Pentane

SOLUTION The boiling points of the compounds are based on the total intermolecular forces present in each molecule. The electronegativity of the functional groups, which directly affects the strength of the intermolecular forces present, follows the pattern: ketones > esters > ethers. Because pentane does not have an oxygen atom as the other compounds do, it would have the weakest intermolecular forces present.

$$\text{(d) } CH_3CH_2CH_2CH_2CH_3 < \text{(b) } CH_3OCH_2CH_2CH_3 < \text{(c) } CH_3O\overset{\overset{\displaystyle O}{\|}}{C}CH_3 < \text{(a) } CH_3\overset{\overset{\displaystyle O}{\|}}{C}CH_2CH_3$$

Practice 7.12

Sketch oxycodone and heroin from Figure 7.10, circle each ketone functional group, draw a box around each ether functional group, and draw a dashed circle around the ester functional groups.

ANSWER

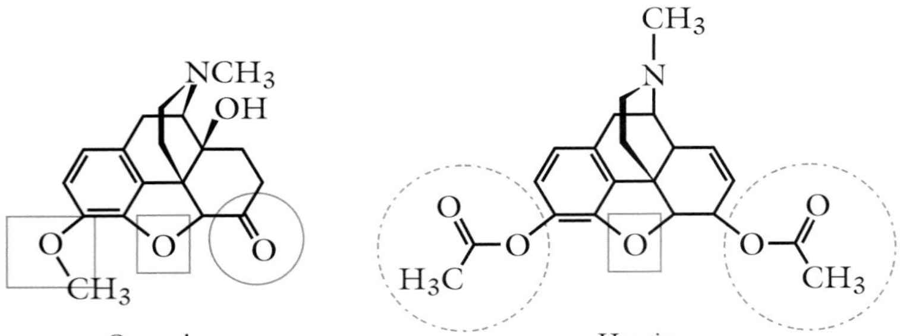

Oxycodone Heroin

7.7 Amines

Learning Objective

Recognize, draw, and name amines.

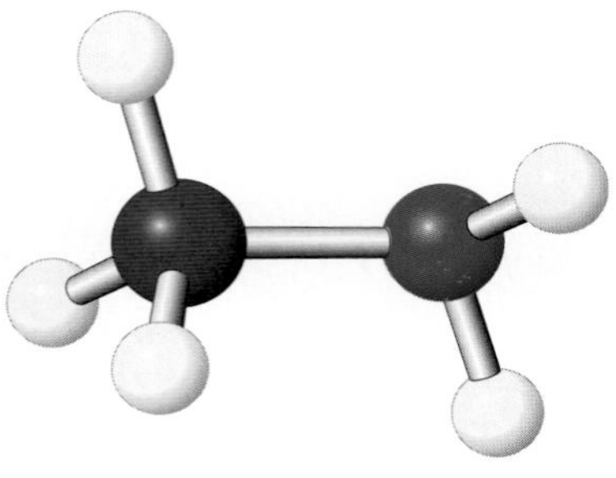

Methylamine (CH_5N), a primary amine.

Organic molecules containing nitrogen belong to a category of compounds called **amines.** Nitrogen differs from carbon in that it forms three bonds to other atoms. In **primary amines**, the nitrogen atom is bonded to one carbon atom and two hydrogen atoms. A primary amine has the general formula R—NH_2. Figure 7.11a shows the structure of the primary amine MDA, which is one of the main ingredients in ecstasy tablets. The compound in Figure 7.11b is MDMA and is an example of a **secondary amine.** MDMA is also an ingredient in ecstasy tablets. In secondary amines, the nitrogen atom is bonded to two carbon groups and one hydrogen atom. The general formula for secondary amines is R_2NH. The compound in Figure 7.11c is cocaine, an example of a **tertiary amine.** Tertiary amines have three carbon atoms bonded directly to the nitrogen atom, and the general formula is R_3N.

Amines are named by listing in alphabetical order the carbon groups attached to the nitrogen atom. The simplest primary amine is methylamine, CH_3NH_2. The simplest secondary amine is dimethylamine $(CH_3)_2NH$, and the simplest tertiary amine is trimethylamine, $(CH_3)_3N$.

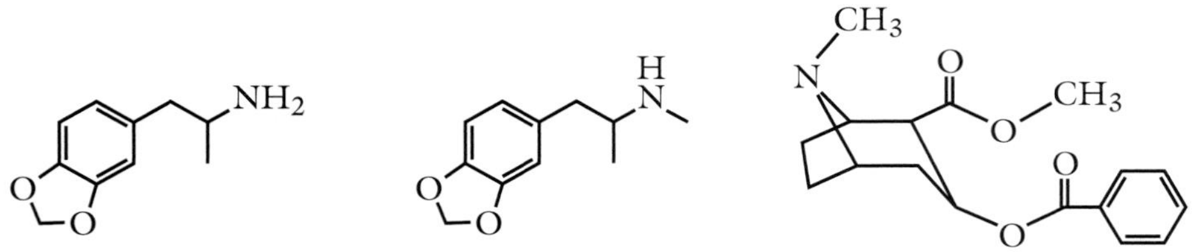

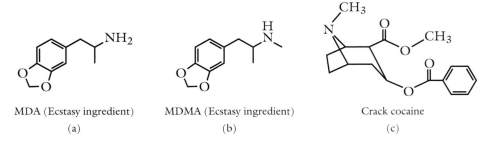

MDA (Ecstasy ingredient) MDMA (Ecstasy ingredient) Crack cocaine

(a) (b) (c)

Figure 7.11 Examples of (a) primary, (b) secondary, and (c) tertiary amines.

Worked Example 13

Draw the structures for the following compounds:

(a) Propylamine

(b) Butylmethylamine

(c) Ethylmethylpropylamine

SOLUTION

(a) Propylamine has a 3-C chain on a nitrogen atom with two hydrogen atoms: $CH_3-CH_2-CH_2-\overset{\overset{\displaystyle H}{|}}{N}-H$

(b) Butylethylamine has a 4-C chain and a 2-C chain on a nitrogen atom with one hydrogen atom:

$$CH_3-CH_2-CH_2-CH_2-\overset{\overset{\displaystyle H}{|}}{N}-CH_2-CH_3$$

(c) Ethylmethylpropylamine has a 2-C chain, a 1-C chain, and a 3-C chain on a nitrogen atom: $CH_3-CH_2-\overset{\overset{\displaystyle CH_3}{|}}{N}-CH_2-CH_2-CH_3$

Practice 7.13

Name the following molecules:

(a) $CH_3-CH_2-CH_2-\overset{\overset{\displaystyle H}{|}}{N}-CH_2-CH_2-CH_2-CH_2-CH_3$

(b) $CH_3-CH_2-\overset{\overset{\displaystyle CH_3}{|}}{N}-CH_2-CH_2-CH_2-CH_2-CH_2-CH_3$

(c) $CH_3-CH_2-\overset{\overset{\displaystyle H}{|}}{N}-CH_2-CH_2-CH_2-CH_3$

ANSWER

(a) Pentylpropylamine

(b) Ethylhexylmethylamine

(c) Butylethylamine

7.8 Alcohols, Aldehydes, and Carboxylic Acids

Learning Objective

Recognize, draw, and name alcohols, aldehydes, and carboxylic acids.

In Germany, a first offender for drunk driving is heavily fined and may lose his or her driver's license permanently. Initially, the severe penalties encouraged drivers to flee the accident scene if alcohol was involved. Hours later, when police apprehended a suspect with alcohol on the breath, the suspect might claim the alcohol was consumed after the accident to calm nerves. This is no longer a commonly used defense for a simple reason: A blood sample taken hours after an accident can be used to determine whether the suspect was drunk at the time of the accident. How this can be done requires an understanding of alcohol compounds and the products of their metabolism: aldehydes and carboxylic acids.

The **alcohol** functional group is $R-OH$. The simplest alcohol is methanol (CH_3OH), in which a methyl group ($-CH_3$) is bonded to the oxygen of the $-OH$ group. Ethanol (CH_3CH_2OH) is the main compound consumed in alcoholic beverages. It should be pointed out that

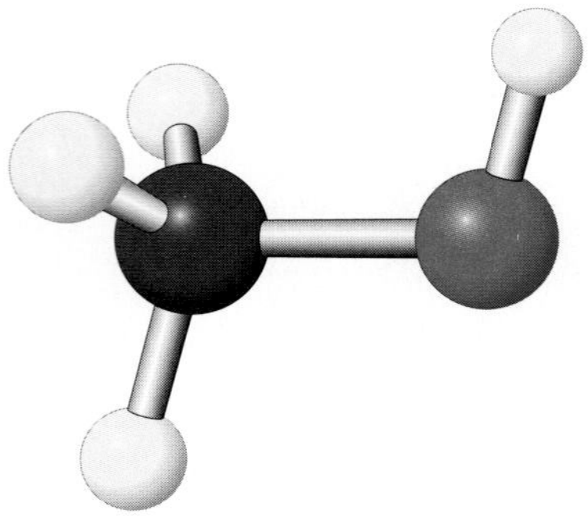

Methanol (CH_4O) is a member of the alcohol functional group. All alcohols share the R—OH structure.

almost all alcohol compounds are toxic to varying degrees and can easily lead to death if consumed in quantities that exceed the toxic threshold.

The term **congener** refers to any compound that shares the same functional group with another compound. More than 800 different compounds have been identified in various alcoholic beverages, most of which contribute to the color and taste of the beverage. Many of the alcohol congener compounds are poisonous if ingested at high enough concentrations. Furthermore, the metabolism of ethanol and alcohol congeners produces compounds with greater toxicity, called *aldehydes.*

Alcohols are named by modifying the ending of the carbon chain name with *-ol.* For alcohol compounds with at least three carbon atoms, there are isomers based on the location of the —OH group. The name of the compound then includes the carbon number for the —OH group. For example:

$$\text{1-propanol} = CH_3CH_2CH_2OH$$

$$\text{2-propanol} = CH_3\underset{\substack{|\\OH}}{C}HCH_3$$

If there are two —OH groups on the chain of carbon atoms, the ending is *-diol,* as in 1,2-ethanediol, the substance in the Patricia Stallings case study.

Worked Example 14

Write the structural formulas for the following compounds that are all alcohol congeners found in alcoholic beverages:

(a) 1-propanol

(b) 2-butanol

(c) Methanol

SOLUTION

(a) A 3-C chain with an —OH group on carbon atom number 1: $HO—CH_2—CH_2—CH_3$

(b) A 4-C chain with an —OH group on carbon atom number 2:

$$CH_3—\underset{\substack{|\\OH}}{CH_2}—CH_2—CH_3$$

(c) One carbon atom with an —OH group: $CH_3—OH$

Practice 7.14

Name the following compounds:

(a) 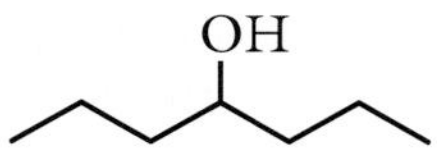 (b)  (c) 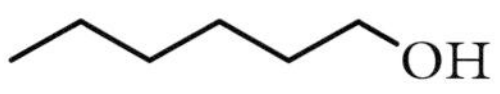

ANSWER

(a) 4-heptanol

(b) 1-propanol

(c) 1-hexanol

Aldehydes

The initial step in the metabolism of alcohol is to produce a compound with an **aldehyde** functional group. The aldehyde functional group consists of a terminal carbon atom having a double bond to an oxygen atom and a single bond to hydrogen. The aldehyde general formula is $R-\overset{\overset{\displaystyle O}{\|}}{C}H$. The simplest aldehyde is methanal (HCHO), which consists of one carbon atom with a double bond to oxygen and a single bond to each hydrogen. The common name for methanal is *formaldehyde,* the compound used in embalming fluid. Aldehydes are named by modifying the ending of the name of the carbon chain with *-al.*

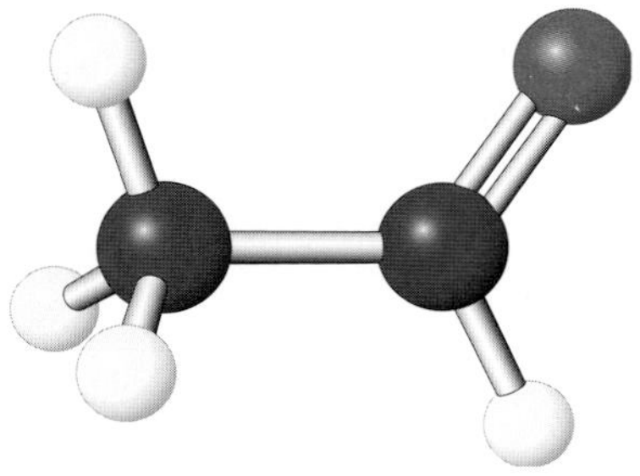

Ethanal (C_2H_4O) is a member of the aldehyde functional group. All aldehydes share the $R-\overset{\overset{\displaystyle O}{\|}}{C}H$ structure.

Worked Example 15

Write the structures of the following aldehydes, all of which are produced during the metabolism of alcoholic beverages.

(a) Propanal

(b) Butanal

(c) Methanal

SOLUTION

(a) A 3-C chain ending with C=O: $CH_3CH_2\overset{\overset{\displaystyle O}{\|}}{C}H$

(b) A 4-C chain ending with C=O: $CH_3CH_2CH_2\overset{\overset{\displaystyle O}{\|}}{C}H$

(c) A 1-C chain with C=O: $H\overset{\overset{\displaystyle O}{\|}}{C}H$

Practice 7.15

Name the following compounds, which are produced during decomposition of the human body and are studied as possible markers for electronic grave site detection.

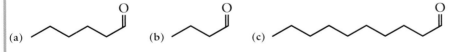

(a) 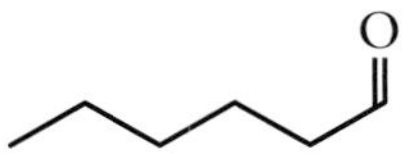 (b) 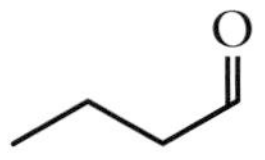 (c) 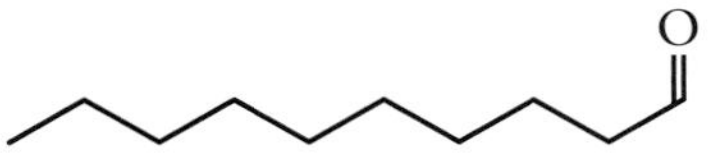

ANSWER

(a) Hexanal

(b) Butanal

(c) Decanal

Ethanal is the first compound produced by the metabolism of ethanol and is extremely poisonous. Ethanal is thought to be responsible for the flushed complexion, headaches, dizziness, and nausea that accompany alcohol consumption and hangovers. Hangovers are more severe when alcoholic beverages with a high level of alcohol congeners are consumed because the congeners remain in the body longer. Many of the alcohol congeners are soluble in fat due to the nonpolar carbon chains that interact with the nonpolar fat molecules. Therefore, the congeners accumulate in fat during periods of high alcohol consumption and are slowly released after the alcohol consumption ceases. As the congeners are released back into the blood stream, they are metabolized to their corresponding toxic aldehyde compounds.

Fortunately, the aldehyde compounds are further metabolized in the body to less troublesome compounds called *carboxylic acids.* The prescription compound Antabuse is used as a deterrent to drinking because it prevents the metabolism of aldehydes to carboxylic acids, permitting a buildup of aldehydes that very quickly cause an individual to become nauseated and have severe headaches.

Carboxylic Acids

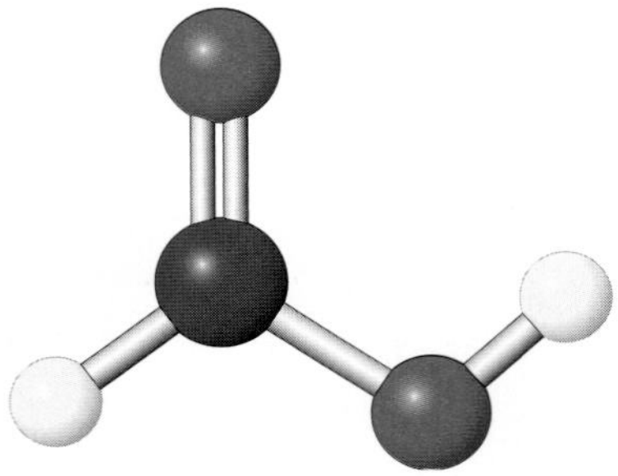

Methanoic acid (CH_2O_2) is a member of the carboxylic acid functional group. All carboxylic acids share the R—COOH or $R-\overset{\overset{\displaystyle O}{\|}}{C}-OH$ structure.

The **carboxylic acid** functional group consists of a terminal carbon atom that contains both a double-bonded oxygen atom and an —OH group. The general formula is represented as $R-\overset{\overset{\displaystyle O}{\|}}{C}-OH$ or R—COOH. The terminal —OH group is *not* an alcohol group, nor is it a hydroxide ion. Carboxylic acids are weak acids that partially dissociate to form H^+ and $R-COO^-$ ions in solution. The simplest carboxylic acid is methanoic acid and contains one carbon atom. The metabolism of ethanol (CH_3CH_2OH) first involves its conversion to ethanal (CH_3CHO), which is an aldehyde. The ethanal in turn is converted to ethanoic acid (CH_3COOH), which is a carboxylic acid. You previously learned the common name of ethanoic acid—*acetic acid,* the main ingredient in vinegar.

The name of the carboxylic acid compounds is created by modifying the ending of the name of the carbon chain with *-oic* and adding the word *acid*, as in *methanoic acid*.

Worked Example 16

Write the line structures for the compounds involved in the Patricia Stallings case study.

(a) 1,2-ethanediol

(b) Propanoic acid

(c) 2-methylpropanedioic acid

SOLUTION

(a) A 2-C chain with an —OH on each carbon atom: 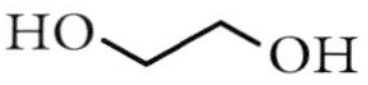

(b) A 3-C chain with a carboxylic acid group: 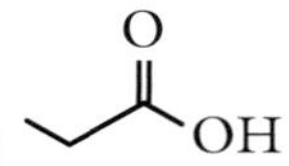

(c) A 3-C chain with $-CH_3$ on carbon number 2 and —COOH on carbon numbers 1 and 3: 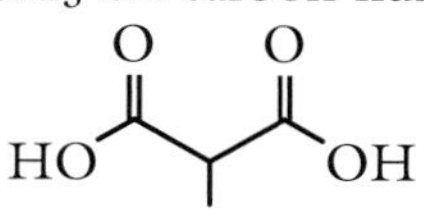

Practice 7.16

Why is it impossible to have the compound 3-hexanoic acid?

Table 7.5 Summary of Organic Compounds

Class	Formula	Example	Naming	Example Name
Alkanes	C_nH_{2n+2}	$CH_3CH_2CH_3$	C chain length + *-ane*	Propane
Alkenes	C_nH_{2n}	$CH_2{=}CH_2CH_3$	C=C location + C chain length + *-ene*	1-propene
Alkynes	C_nH_{2n-2}	$CH{\equiv}CCH_3$	C≡C location + C chain length + *-yne*	1-propyne
Ethers	R–O–R	$CH_3OCH_2CH_3$	Short C chain + *-oxy* + long C chain + *-ane*	Methoxyethane
Ketones	R–C(=O)–R	$CH_3C(=O)CH_3$	C=O location + C chain length + *-one*	2-propanone
Esters	R–O–C(=O)–R	$CH_3OC(=O)CH_3$	C chain not touching C=O, C chain length + *-oate*	Methyl ethanoate
Amines	R_3N	$CH_3NHCH_2CH_3$	Name each C chain (alphabetical order) + *-amine*	Ethylmethylamine
Alcohols	R–OH	CH_3OH	C chain length + *-ol*	Methanol
Aldehydes	R–C(=O)H	$CH_3C(=O)H$	C chain length + *-al*	Ethanal
Carboxylic acids	R–C(=O)–OH	$CH_3C(=O)OH$	C chain length + *-oic* + *acid*	Ethanoic acid

ANSWER

The carboxylic acid functional group must occur on a terminal carbon atom.

The final step in the metabolism of alcohol is to form carbon dioxide and water. The human body will metabolize ethanol, which is present at much higher concentrations than the alcohol congeners, before it will metabolize the congeners. Therefore, the amount of alcohol consumed hours earlier can be determined by measuring the relative amounts of the alcohol congeners remaining in the blood.

7.9 How to Extract Organic Compounds: Solubility and Acid-Base Properties

In the 1990s, crack cocaine came to dominate a large portion of the illegal drug market. In one interesting case, police officers were chasing a suspected crack dealer when the suspect entered a fast-food restaurant with the police closely following. The drug dealer threw his entire inventory of drugs, bags and all, directly into the deep fryer. The crack cocaine dissolved and was not recovered by police. The suspect was never prosecuted,

Learning Objective

Identify and describe how to separate acidic, basic, and neutral organic compounds dissolved in organic solvents.

and much to the dismay of forensic drug chemists across the nation, he inspired crack houses to keep a crock-pot full of cooking oil turned on at all times. When police would raid a crack house, the dealers would attempt to dump their inventory into the hot oil, thinking this would destroy the crack cocaine.

Because crack cocaine is a nonpolar compound, it dissolves in nonpolar solvents such as vegetable oil. Nevertheless, forensic chemists can recover the crack cocaine from oil using a series of steps that exploit the solubility of polar and nonpolar compounds and the acid-base properties of organic compounds. We will now explore the principles that make the recovery of the crack cocaine possible.

We have discussed carboxylic acids as examples of organic acids, but there are also organic compounds, such as organic amines, that function as bases. According to the definitions previously given, acids are compounds that produce free H^+ ions in an aqueous solution, and bases are compounds that produce free OH^- ions in an aqueous solution. This is often referred to as the **Arrhenius definition** of acids and bases.

However, we can describe acids and bases in a broader way by using the **Brønsted-Lowry definition:** An acid is any compound that can donate a hydrogen ion; a base is any compound that can accept a hydrogen ion. The Brønsted-Lowry acid-base definition expands the number of compounds that we consider to be acids or bases. All compounds that were identified as acids and bases from the Arrhenius definition still fit the new Brønsted-Lowry definition.

Why do amines fit the definition of a base? If we examine the valence electron arrangement of a nitrogen atom, we see that nitrogen has five valence electrons available to form bonds. However, in all of the amine structures, only three covalent bonds are found on nitrogen. Because each atom in a bond shares one electron, nitrogen is sharing only three of its five valence electrons. The other two are called **lone pair electrons**—electrons that are paired up in an orbital and are not part of a covalent bond. The lone pair electrons impart a highly negative region around the nitrogen atom that can attract a positive ion such as the H^+ ion. Figure 7.12 shows the ability of an amine to act as a Brønsted-Lowry base, and a carboxylic acid to act as a Brønsted-Lowry acid.

Returning to the crock-pot situation, the task of a forensic scientist is to determine whether crack cocaine is dissolved in an oil sample and, if so, how to separate the drug from the solution so that it can be used as evidence. The structure of crack cocaine (see Figure 7.2) shows that it is an amine because it contains a nitrogen atom bonded to three carbon groups. And, as we saw above, amines are organic bases. This property

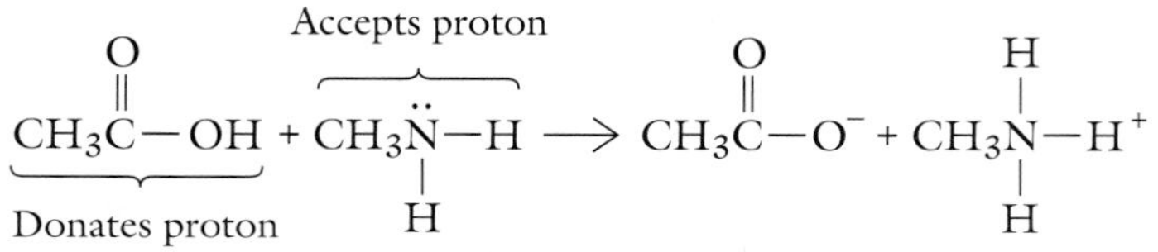

Figure 7.12 Brønsted-Lowry acid-base reaction.

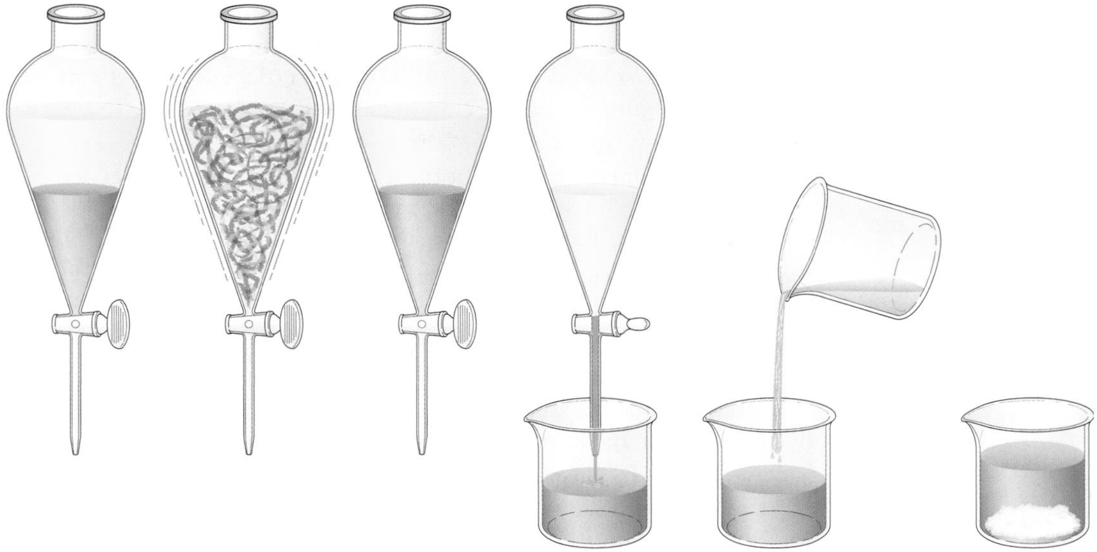

Figure 7.13 Extraction of acidic or basic organic compounds.

of amines is important in the separation of crack cocaine from oil. But organic acids also play a part.

Carboxylic acids, especially those with long carbon chains, tend to dissolve in nonpolar organic solvents such as vegetable oil. However, if you deprotonate (remove a proton from) the carboxylic acid and convert it into an anion, it becomes soluble in water and not in the organic solvent. The amine compounds also dissolve in organic solvents when in the neutral state, but when protonated (given a proton) to form a cation, amines become insoluble in organic solvents and soluble in water. Knowledge of these principles of solubility and acid-base chemistry allows a forensic scientist to use the pH of a solvent like a switch to control the solubility of compounds in either organic solvents or water.

A procedure for the extraction of acidic, basic, and neutral organic molecules from an organic solvent is shown in Figure 7.13. The pear-shaped container is called a *separatory funnel* and is specifically designed for this type of experimental procedure. All the molecules in the organic solvent are initially neutral.

- The first step is to add to the organic solvent either an acid such as HCl or a base such as NaOH. An acid will protonate an amine and a base will deprotonate a carboxylic acid. Because the HCl and NaOH are aqueous reagents, they will not mix with the nonpolar organic solvent and two layers will form, as shown in Figure 7.13a.
- For a weak acid or a weak base dissolved in the organic solvent to react with either the HCl or the NaOH in the separate aqueous layer, the solution must be vigorously shaken, as shown in Figure 7.13b.

- As the organic molecules are converted to their ionic form, they become insoluble in the organic solvent and soluble in the aqueous solvent. The two layers are then allowed to settle and separate, as depicted in Figure 7.13c.
- In the next step, the more dense solution (typically the aqueous solution) is drained from the bottom of the separatory funnel, as shown in Figure 7.13d. Recall that the aqueous solution now contains the ionic form of the organic compound.
- Figure 7.13e shows the neutralization of the solution (with acid or base as needed). This causes the ionic organic compound to become neutral.
- When the ionic form of the organic compound is neutralized, it is insoluble in aqueous solution. The neutral molecule will crystallize out of solution and can easily be filtered, as shown in Figure 7.13f.
- The neutral organic molecules in the organic solvent always stay there and can typically be recovered by evaporating the organic solvent and leaving the neutral organic molecules behind.

Worked Example 17

Describe how to determine whether a sample of vegetable oil contained crack cocaine.

SOLUTION Cocaine has a tertiary amine functional group. Therefore, if a sample of the oil were treated with aqueous hydrochloric acid, the cocaine salt would dissolve in the water. After the aqueous solution is separated from the organic solution, add NaOH to the aqueous solution until the cocaine salt is neutralized and the crack cocaine precipitates out of solution.

Practice 7.17

Many over-the-counter pain medicines contain aspirin and caffeine. How do you isolate the two compounds from a tablet? The structures of aspirin and caffeine are shown in the figure below.

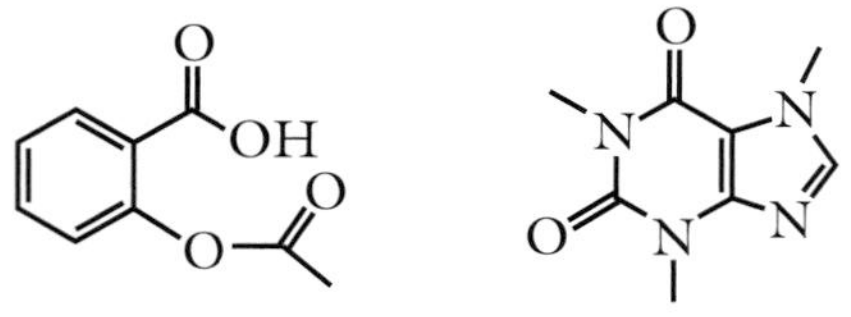

Aspirin   Caffeine

ANSWER
Dissolve the tablet in an organic solvent. The aspirin molecule can be extracted into an aqueous layer by the addition of NaOH, which deprotonates the carboxylic acid functional group. Caffeine can be extracted into an aqueous layer by the addition of HCl, which protonates the tertiary nitrogen atoms. The amine cations and the carboxylic acid anions can be removed by neutralizing the solutions.

Evidence Analysis | Infrared Spectroscopy

Infrared radiation can be used to positively identify a chemical compound and is commonly used in a forensic laboratory for analyzing organic compounds such as illegal and prescription drugs. Infrared radiation helps to secure convictions against thousands of drug dealers each year, and it is also applied to the analysis of documents, inks, plastic, and explosives. To understand how infrared radiation is used to identify compounds, it is necessary to look into what happens when infrared light strikes an object.

The infrared region of the spectrum begins where the red region of visible light ends. Infrared radiation is invisible to the human eye, but we sense it as heat when it strikes our skin. Obviously, the molecules that make up our skin are affected by infrared radiation, but how? When infrared radiation interacts with our skin, the bonds between the molecules of our skin absorb the energy, and the temperature of the skin increases.

Likewise, the bonds that form between atoms in a molecule are capable of absorbing infrared radiation. The bonds are not rigid, inflexible connections that force the atoms into a single fixed position. Rather, the bonds act more like springs that keep the two atoms together. The bonds can therefore be compressed, stretched, or bent just like a spring. Figure 7.14 shows the three principle vibrations that can occur in bonds between atoms when infrared radiation strikes a molecule. A molecule will absorb the infrared radiation if the energy of the radiation matches the energy of one of the vibrations.

There are two main regions of interest when measuring the absorption of infrared radiation: the **functional group region** (4000 cm^{-1} to

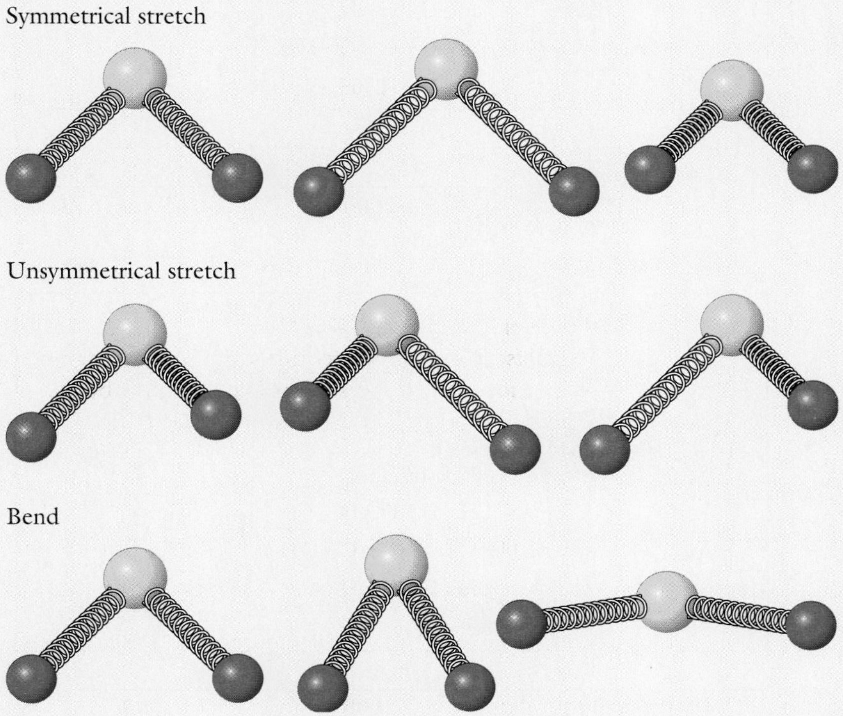

Figure 7.14 The infrared active vibrations of a molecule. The symmetrical stretch involves the simultaneous stretching and then compression of bonds. The unsymmetrical stretching involves the simultaneous compression of one bond and stretching of another. The bending motion involves two atoms flexing in toward each other and then outward away from each other.

1300 cm^{-1}) and the **fingerprint region** (1300 cm^{-1} to 900 cm^{-1}). Note that for infrared radiation, the traditional unit used to specify the frequency or wavelength of the radiation is the *wave number*, which is the number of waves per centimeter.

The functional group region shows characteristic absorption patterns or bands for each of the functional groups covered in this chapter. For example, a carbon atom with a single bond to a hydrogen atom (C—H) absorbs infrared radiation in the range of 2850 to 2980 cm^{-1}, whereas an oxygen atom single bonded to a hydrogen atom (O—H) absorbs radiation from 3300 to 3550 cm^{-1}.

The fingerprint region provides a pattern of absorption bands for each molecule and is created from all of the atoms within the molecule undergoing vibrations. This creates a unique pattern not reproduced by any other molecule.

The energy associated with each type of vibration from Figure 7.14 depends on what type of bond exists between the atoms (single, double, or triple) and on what types of atoms are involved in the bonding. Figure 7.15 shows the infrared spec-

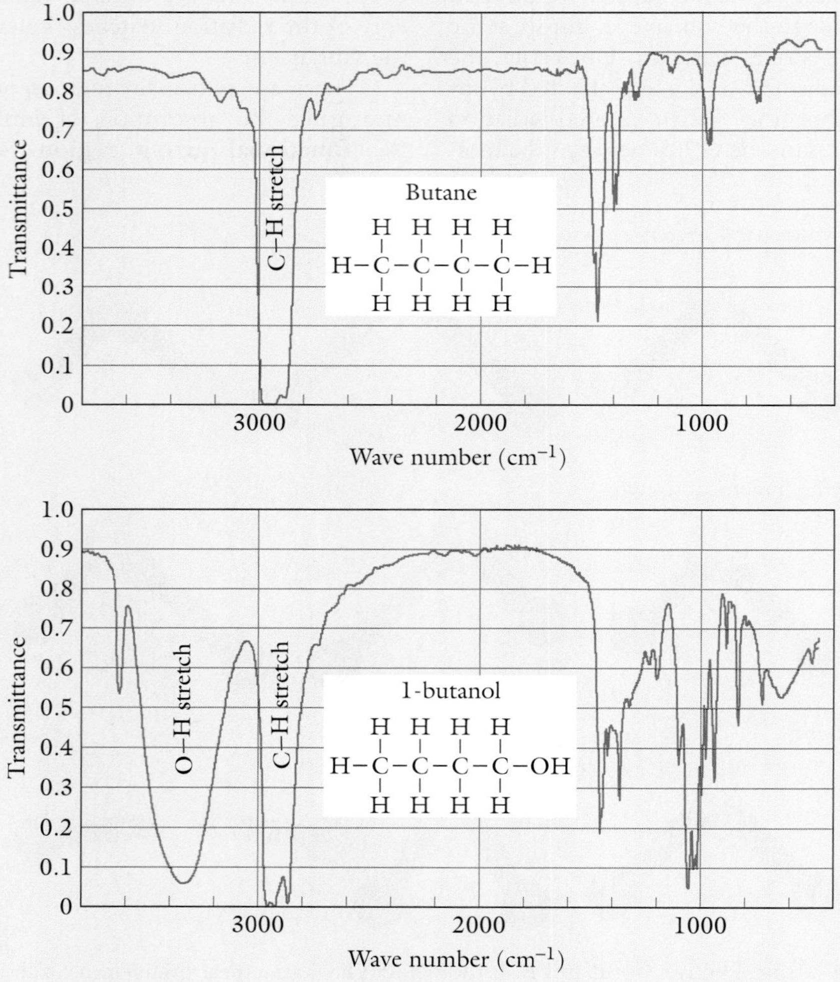

Figure 7.15 Infrared absorption spectrum of butane and 1-butanol. As the infrared radiation striking the sample is scanned through the functional group region (4000 cm^{-1} to 1300 cm^{-1}), any functional groups within the compound will absorb energy. This creates a distinctive pattern, as can be seen for both the C—H stretch and the O—H stretch. Each compound creates a unique pattern in the fingerprint region (1300 cm^{-1} to 900 cm^{-1}).

trum of butane and 1-butanol. Notice the change in both the functional group region and the fingerprint region, caused by the addition of a simple alcohol functional group.

In the laboratory, the infrared spectrum is obtained using an instrument called a **Fourier Transform Infrared Spectrometer (FTIR)**. The Fourier transform is a complex mathematical process (beyond the scope of this book) used to obtain the signal from the raw data. For analysis of a compound, the sample is placed inside the instrument and the spectrum is obtained within minutes. The experimentally obtained spectrum is compared with a library of standard spectra prepared by the manufacturer of the instrument. A tentative match can be made provided the spectra of the compound is found within the library. The FTIR system is versatile in that it can be used to measure the spectrum of solids, gases, and liquids. One limitation of FTIR analysis is that it does not work well for complicated mixtures, because the resulting spectrum reflects the absorption of all compounds within the mixture. FTIR analysis was not used in the Stallings case study most likely for this very reason.

A typical FTIR system. (Perkin Elmer)

7.10 CASE STUDY FINALE: The Experts Agreed

Patricia Stallings was convicted of poisoning her son Ryan with antifreeze after laboratory results showed the presence of 1,2-ethanediol in his blood. Could a newborn baby help to prove his mother innocent of the murder of his older brother? The first sign of her innocence came when Ryan's brother David Jr., who was being kept in foster care, started showing the same symptoms that Ryan had exhibited. His foster parents rushed him to the hospital, where a doctor diagnosed his sickness as methylmalonic acidemia (MMA), a rare genetic disease in which the body is unable to properly digest proteins.

When the body digests proteins, propanoic acid is produced. The propanoic acid is then converted to 2-methylpropanedioic acid, often called *methylmalonic acid.* When a person suffers from MMA and eats food high in protein, such as baby formula, the body can build up an excessive amount of propanoic acid. This build-up creates symptoms like those displayed by Ryan Stallings and then by his little brother.

(St. Louis Post-Dispatch)

Worked Example 18

Considering the new piece of information revealed in the case study, what question(s) should the defense attorneys for Patricia Stallings have investigated and why?

SOLUTION Since the symptoms Ryan experienced prior to his death were similar to those being experienced by David Jr., could Ryan also have had MMA? Could MMA have been responsible for his death?

Practice 7.18

Even if Ryan did have MMA as his younger brother did, this would not immediately vindicate Patricia Stallings. What evidence from the case study could and could not be explained by MMA?

ANSWER
Although Ryan's symptoms matched those of a person suffering from MMA, the laboratory samples of Ryan's blood showed 1,2-ethanediol, not propanoic acid.

The question in Practice 7.18 raises an interesting counterquestion: Could the laboratory have been mistaken when it said the samples showed the presence of 1,2-ethanediol and not propanoic acid? In Worked Example 16, you determined the structures of the two molecules, and it is evident that the similarity between the molecules is minimal. The molecules in question have a different number of carbon atoms and different functional groups. Thus, the two molecules should have substantially different properties. Furthermore, a second laboratory confirmed the results of the blood work.

Further testing on Ryan's blood clearly established that he was, in fact, suffering from MMA. Only 1 in 50,000 newborns is diagnosed with MMA, but Ryan and David Jr. had parents who were both carriers of the gene. This gave the children a 1 out of 4 chance of having MMA.

Dr. William Sly and his colleague Dr. James Shoemaker of St. Louis University had been following the case with some interest. The two biochemists couldn't accept the idea that a child with MMA was poisoned with a compound that would mimic MMA symptoms. Sly and Shoemaker were able to obtain a sample of Ryan's blood, on which they conducted a simple test. They added a reagent that would turn a purple color if ethylene glycol (1,2-ethanediol) were present. The blood sample did not turn purple. Bolstered by this finding, the two scientists came to the conclusion that the only rea-

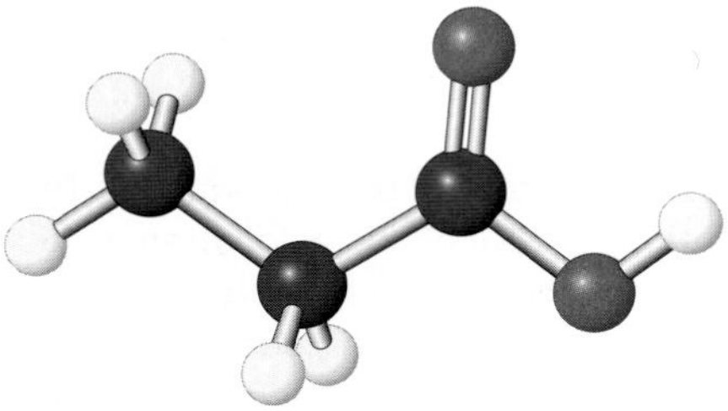

Propanoic acid

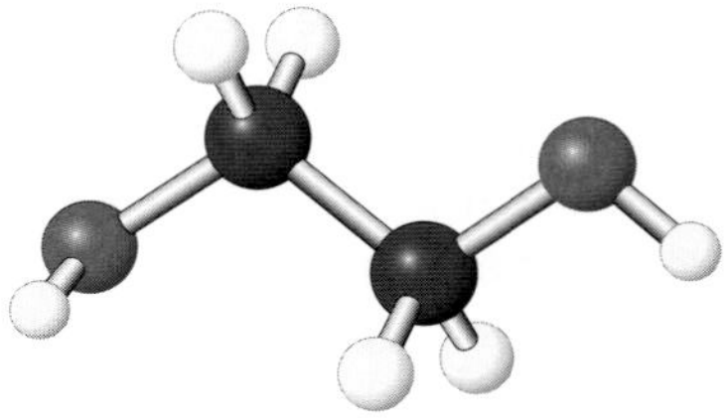

1,2-ethanediol

sonable explanation was that the laboratory analysis was incorrect and that ethylene glycol was never in Ryan's blood!

Sly and Shoemaker theorized that the laboratories had misidentified the two compounds. To test their hypothesis, Sly and Shoemaker sent blood samples that had been spiked with propanoic acid to seven certified laboratories. Three of the laboratories reported finding ethylene glycol instead. The method used to identify the two compounds was not foolproof and relied on the interpretation of data by a skilled analyst. Several of the analysts made critical mistakes.

At this point, even the prosecutors started to doubt their case against Patricia Stallings and consulted one of the leading experts on genetic disorders, Dr. Piero Rinaldo, who was at Yale University at the time. Dr. Rinaldo confirmed the findings of Dr. Sly and Dr. Shoemaker and even went so far as to indicate that the treatment the hospital gave to Ryan for antifreeze poisoning was actually responsible for his death. The hospital had administered ethanol to Ryan, which is the correct treatment for antifreeze poisoning. However, in this case the ethanol was converted into oxalate ions, which combined with calcium in the blood and precipitated out of solution. The resulting calcium oxalate crystals caused severe kidney and brain damage. Finally, an independent analysis of the bottle used to feed Ryan showed that there was no ethylene glycol present.

A new trial was ordered for Patricia Stallings. The prosecutors dropped all charges and she was freed. Little David Jr. had proven his mother's innocence and helped to solve the tragic events surrounding his brother's death.

CHAPTER SUMMARY

- Alkanes consist of chains of single-bonded carbon atoms with a sufficient number of hydrogen atoms to satisfy the generic formula C_nH_{2n+2}. The number of carbon atoms in the chain is expressed by beginning the name with *meth-* for one carbon, *eth-* for two, *prop-* for three, and so on, as shown in Table 7.1. The ending *-ane* is added to alkane names. Compounds in which a carbon atom forms a double bond to another carbon atom belong to the alkene class of compounds. A closely related class of compounds is the alkynes, in which two carbon atoms form a triple bond.
- Isomers are molecules that have the same chemical formulas but do not have the same chemical structures. One type of isomer results from different locations of a double or triple bond along a carbon chain. Another form occurs when the carbon chain branches. The names of isomers specify the carbon number where the double/triple bond or the carbon branch is located.
- Alkane compounds can also form ring structures called cycloalkanes, made entirely from single-bonded carbon atoms. Another type of common cyclic compound is benzene, which is represented by a six-carbon ring with alternating double and single bonds. The benzene ring can also be modified by the attachment of an $-OH$ group to form phenol, a $-CH_3$ group to form toluene, and an $-NH_2$ to form aniline.
- Organic compounds fall into various classes based on the presence of functional groups that include the ketone, ether, ester, amine, alcohol, aldehyde, and carboxylic acid groups.
- Amines are an example of an organic Brønsted-Lowry base (accepts a hydrogen ion) and carboxylic acids are an example of an organic Brønsted-Lowry acid (donates a hydrogen ion). These properties can be exploited to extract amines or carboxylic acids from mixtures. The ionic forms of the compounds are soluble in water whereas the neutral forms are soluble in nonpolar organic solvents.

KEY TERMS

organic compound, p. 197
inorganic compound, p. 197
alkane, p. 200
structural formula, p. 200
condensed structural formula, p. 201
line structure, p. 201
alkene, p. 203
isomer, p. 203
alkyne, p. 205
cycloalkane, p. 209
benzene, p. 209
aromatic compound, p. 209
resonance structures, p. 210
ether, p. 211
functional group, p. 211
ketone, p. 212
ester, p. 214
amine, p. 216
primary amine, p. 216
secondary amine, p. 216
tertiary amine, p. 216
alcohol, p. 217
congener, p. 218
aldehyde, p. 219
carboxylic acid, p. 220
Arrhenius acid, p. 222
Arrhenius base, p. 222
Brønsted-Lowry acid, p. 222
Brønsted-Lowry base, p. 222
lone pair electrons, p. 222
functional group region, p. 225
fingerprint region, p. 226
Fourier Transform Infrared Spectrometer (FTIR), p. 227

CONTINUING THE INVESTIGATION Additional Readings, Resources, and References

Drug Identification Bible, Grand Junction, CO: Amera-Chem, Inc., 2004.

Graham, Tim. "When Good Science Goes Bad," *Chemistry Matters,* October 2004, pp. 16–18.

For information about the Patricia Stallings case, see numerous articles in the *St. Louis Post-Dispatch* between September 1989 and October 1991.

The following Web sites show animations of molecules undergoing excitation by infrared radiation: http://chipo.chem.uic.edu/web1/ocol/spec/IR1.htm and http://www.chem.umass.edu/~nermmw/Spectra/irspectra/index.htm

For multiple articles regarding the analysis and use of alcohol congeners, see J. A. Siegel, ed., *Encyclopedia of Forensic Sciences*, San Diego: Academic Press, 2000.

REVIEW QUESTIONS AND PROBLEMS

Questions

1. Discuss the historical origins of the terms *organic* and *inorganic* for classifying chemical compounds.
2. Discuss the modern definition of the term *organic* for classifying chemical compounds.
3. What is the generic formula for all alkanes? What are some common uses of alkanes?
4. What is the difference between the alkanes, alkenes, and alkynes?
5. What are the generic formulas of the alkenes and alkynes?
6. What is the difference between the chemical formula and the condensed formula of organic compounds? What information does each provide?
7. Explain how to interpret the line structures for alkanes, alkenes, and alkynes.
8. What do isomers represent?
9. Will two isomers have the same physical properties? Explain why or why not.
10. What do the alkanes and cycloalkanes have in common? How do they differ?
11. What does it mean when a compound has several resonance structures?
12. Can you determine the structure of an organic molecule simply by looking at the chemical formula? Why or why not?
13. What is the amine functional group? What is the difference between a primary, secondary, and tertiary amine?
14. What is the alcohol functional group? Are all alcohols safe to consume?
15. What are congeners? Give several examples of alcohol congeners.

16. How are the aldehyde functional group and the carboxylic acid functional group similar? How are they different?
17. How are ketone, ether, and ester functional groups similar? How are they different?
18. Explain the difference between a Brønsted-Lowry acid and an Arrhenius acid. Give an example of each.
19. Explain the difference between a Brønsted-Lowry base and an Arrhenius base. Give an example of each.
20. Explain how you can use the acid-base properties of certain functional groups for extraction of an organic substance from a solution.

Problems

21. Name the following alkanes.
(a) CH_4
(b) C_5H_{12}
(c) C_3H_8
(d) C_6H_{14}
22. Name the following alkanes.
(a) C_7H_{16}
(b) C_2H_6
(c) C_8H_{18}
(d) C_4H_{10}
23. What is the chemical formula for each of the following compounds?
(a) Pentane
(b) Propane
(c) Octane
(d) Methane
24. What is the chemical formula for each of the following compounds?
(a) Decane
(b) Heptane
(c) Hexane
(d) Ethane
25. Draw the condensed structural formula for each of the following compounds.
(a) CH_4
(b) C_5H_{12}
(c) C_3H_8
(d) C_6H_{14}
26. Draw the condensed structural formula for each of the following compounds.
(a) C_7H_{16}
(b) C_2H_6
(c) C_8H_{18}
(d) C_4H_{10}
27. Draw the line structure for each of the following compounds.
(a) Pentane
(b) Propane
(c) Octane
(d) Hexane
28. Draw the line structure for each of the following compounds.
(a) Decane
(b) Heptane
(c) Hexane
(d) Ethane
29. Write the chemical formula, condensed structural formula, and line structure for each of the following compounds.
(a) 1-butyne
(b) 2-pentene
(c) 1-hexene
(d) 3-octyne
30. Write the chemical formula, condensed structural formula, and line structure for each of the following compounds.
(a) 1-propene
(b) 2-heptene
(c) 2-octyne
(d) 1-pentyne
31. How many unique isomers can be made by altering the placement of the multiple bond in each of the following compounds?
(a) Pentyne
(b) Hexene
(c) Butene
(d) Nonene
32. How many unique isomers can be made by altering the placement of the multiple bond in each of the following?
(a) Octyne
(b) Propene
(c) Hexyne
(d) Butyne
33. Name the following compounds.

(a) 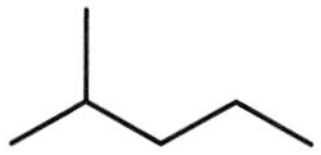

(b) 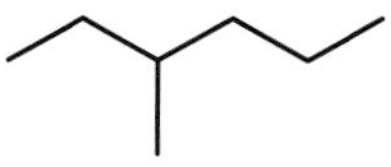

(c) 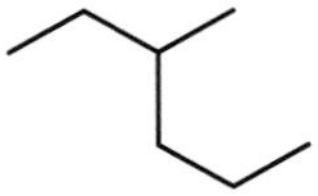

(d) 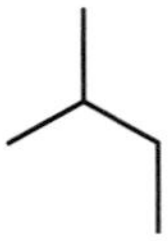

34. Name the following compounds.

(a) 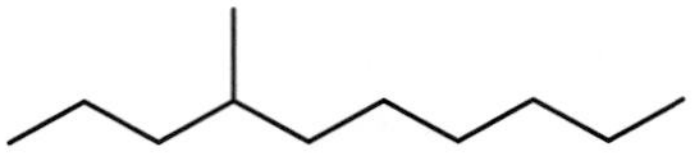

(b) 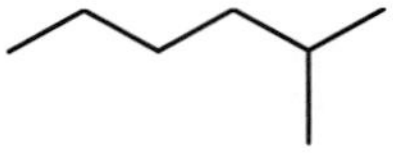

(c) 

(d) 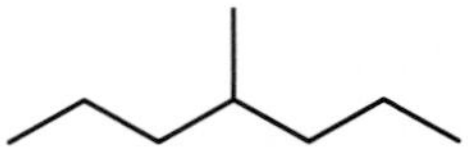

35. Draw the line structure for each of the following compounds.
(a) 3-ethylheptane
(b) 4-propyldecane
(c) 2-methylbutane
(d) 3-methyloctane

36. Draw the line structure for each of the following compounds.
(a) 2-methylhexane
(b) 3-ethyloctane
(c) 2-methylheptane
(d) 4-methylnonane

37. Write the chemical formula and line structure for each of the following compounds.
(a) Cyclopropane
(b) Benzene
(c) Cyclohexane
(d) Toluene

38. Write the chemical formula and line structure for each of the following compounds.
(a) Cyclobutane
(b) Phenol
(c) Cyclopentane
(d) Aniline

39. Draw the condensed structural formula for each of the following compounds.
(a) 2-propanone
(b) Ethoxypropane
(c) Methyl ethanoate
(d) 2-butanone

40. Draw the condensed structural formula for each of the following compounds.
(a) Methoxybutane
(b) Butyl pentanoate
(c) 3-pentanone
(d) Ethoxyethane

41. Name the following compounds.

(a) $CH_3-CH_2-C(=O)-CH_2-CH_2-CH_3$

(b) $CH_3-CH_2-O-CH_2-CH_2-CH_2-CH_3$

(c) $CH_3-CH_2-O-C(=O)-CH_2-CH_2-CH_2-CH_3$

(d) $CH_3-CH_2-CH_2-C(=O)-CH_2-CH_3$

42. Name the following compounds.

(a) $CH_3-C(=O)-CH_2-CH_2-CH_2-CH_3$

(b) $CH_3-CH_2-CH_2-CH_2-O-CH_2-CH_3$

(c) $CH_3-CH_2-CH_2-CH_2-CH_2-O-C(=O)-CH_2-CH_3$

(d) $CH_3-CH_2-CH_2-C(=O)-CH_2-CH_3$

43. Draw the condensed structural formula for each of the following compounds.
(a) Butylamine
(b) Ethylmethylamine
(c) Triethylamine
(d) Butylethylmethylamine

44. Draw the condensed structural formula for each of the following compounds.
(a) Ethylamine
(b) Diethylpropylamine
(c) Diethylamine
(d) Dimethylpropylamine

45. Draw the line structure for each of the compounds listed in Problem 43.

46. Draw the line structure for each of the compounds listed in Problem 44.

47. Draw the condensed structural formula for each of the following compounds.
(a) 2-butanol
(b) 1-propanol
(c) 3-pentanol
(d) 2-hexanol

48. Draw the condensed structural formula for each of the following compounds.
(a) 3-octanol
(b) 2-propanol
(c) Cyclohexanol
(d) 3-pentanol

49. Name the following compounds.

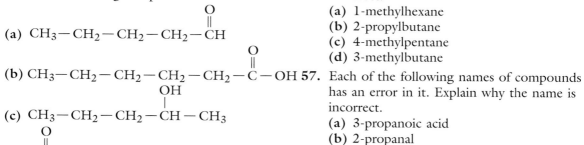

(a) $CH_3-CH_2-CH_2-CH_2-\overset{\overset{\displaystyle O}{\|}}{C}H$

(b) $CH_3-CH_2-CH_2-CH_2-CH_2-\overset{\overset{\displaystyle O}{\|}}{C}-OH$

(c) $CH_3-CH_2-CH_2-\overset{\overset{\displaystyle OH}{|}}{C}H-CH_3$

(d) $H\overset{\overset{\displaystyle O}{\|}}{C}-CH_2-CH_2-CH_2-CH_2-CH_3$

50. Name the following compounds.

(a) $CH_3-CH_2-\overset{\overset{\displaystyle O}{\|}}{C}-OH$

(b) $CH_3-CH_2-\overset{\overset{\displaystyle O}{\|}}{C}H$

(c) $H\overset{\overset{\displaystyle O}{\|}}{C}-CH_2-CH_2-CH_3$

(d) $CH_3-CH_2-CH_2-\overset{\overset{\displaystyle O}{\|}}{C}-OH$

51. Draw the condensed structural formula for each of the following compounds.
(a) Pentanoic acid
(b) Octanal
(c) Hexanoic acid
(d) Heptanal

52. Draw the condensed structural formula for each of the following compounds.
(a) Ethanoic acid
(b) Butanal
(c) Decanoic acid
(d) Propanal

53. Draw and name all of the isomers of hexanol made by altering only the placement of the alcohol functional group.

54. Draw and name all of the isomers of pentanol made by altering only the placement of the alcohol functional group.

55. Each of the following names of compounds has an error in it. Explain why the name is incorrect.
(a) 3-methylpropane
(b) 2-propylethane
(c) 4-ethylheptane
(d) 2-methylethane

56. Each of the following names of compounds has an error in it. Explain why the name is incorrect.
(a) 1-methylhexane
(b) 2-propylbutane
(c) 4-methylpentane
(d) 3-methylbutane

57. Each of the following names of compounds has an error in it. Explain why the name is incorrect.
(a) 3-propanoic acid
(b) 2-propanal
(c) 1-butanone
(d) 3-butanol

58. Each of the following compounds names has an error in it. Explain why the name is incorrect.
(a) 2-butanoic acid
(b) 2-pentanal
(c) 3-propanone
(d) 4-butanol

59. Identify each of the following compounds as an acid, base, or neutral compound according to the Brønsted-Lowry concept.
(a) 2-butanone
(b) Butanoic acid
(c) 1-propanol
(d) Methylamine

60. Identify each of the following compounds as an acid, base, or neutral compound according to the Brønsted-Lowry concept.
(a) Methoxyethane
(b) Methyl propanoate
(c) Butylamine
(d) Hexanal

Forensic Chemistry Problems

Problems 61–68 refer to Table 7.6, Drug Structures.

61. List all drugs that contain an alcohol functional group.

62. List all drugs that contain an amine functional group. Indicate whether the amine is a primary, secondary, or tertiary amine.

63. List all drugs that contain a ketone functional group.

Table 7.6 Drug Structures

Top 4 Illegal Drugs

Methamphetamine 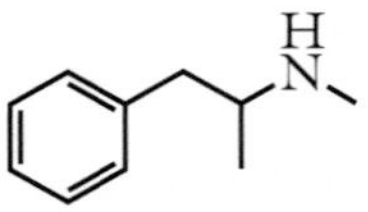	Cocaine 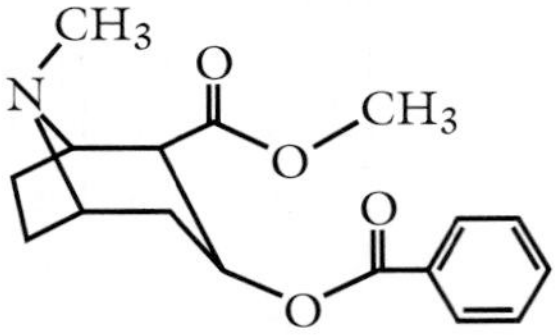	Heroin 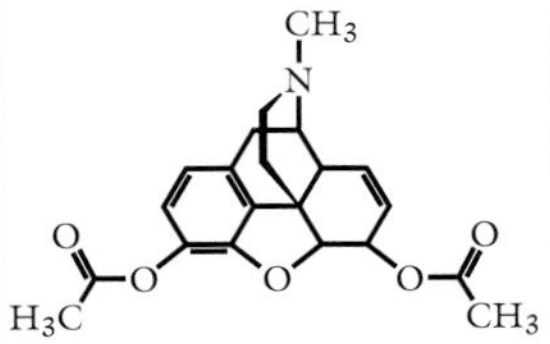	Marijuana 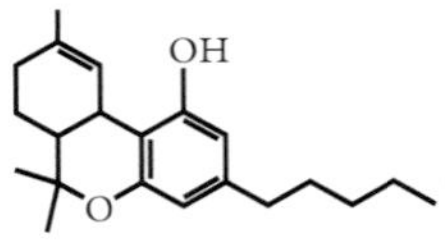

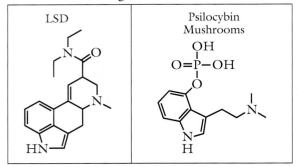

Top Club Drugs

MDMA 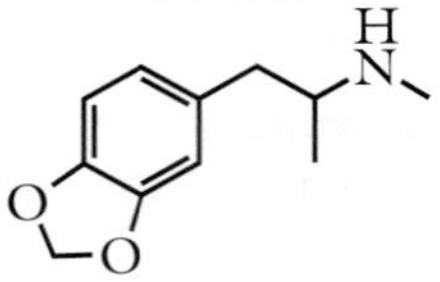	MDA 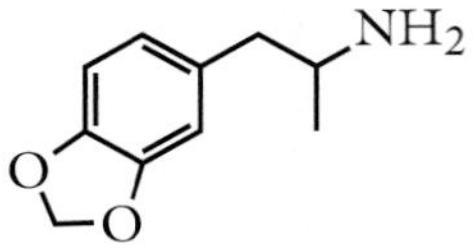	Ketamine 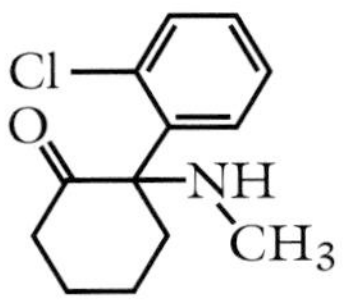	GHB 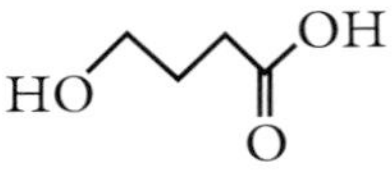
GBL 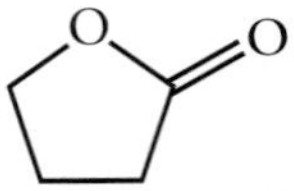	1,4-butanediol 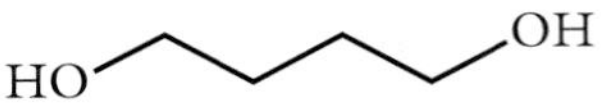		

Concert Scene Drugs

LSD 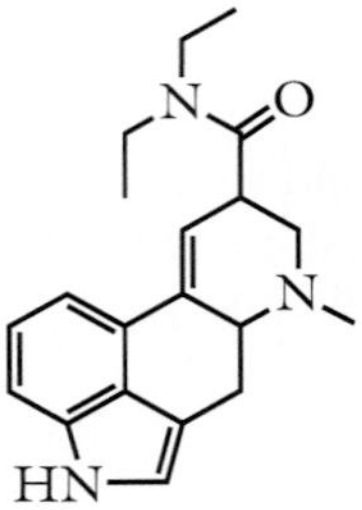	Psilocybin Mushrooms 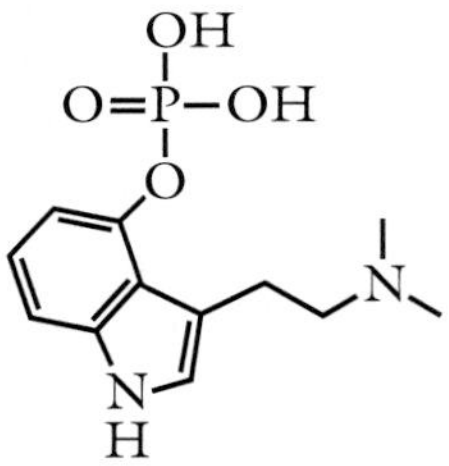

Top Abused Prescription Drugs

OxyContin 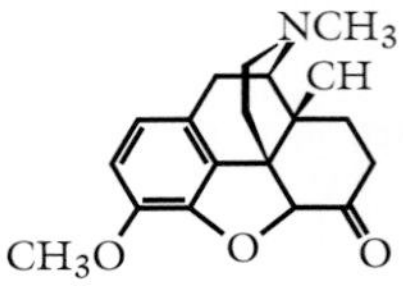	Vicodin 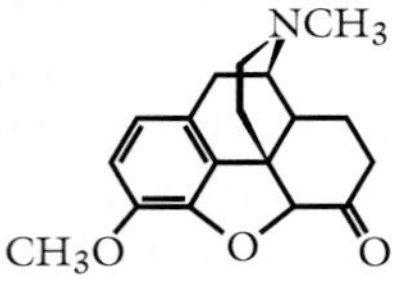	Ritalin 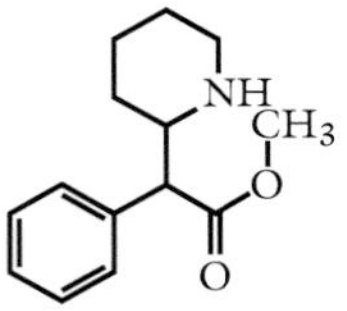	Xanax 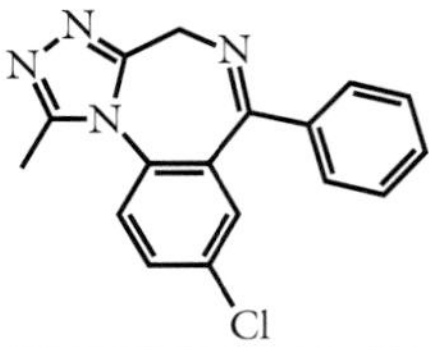
Lorazepam 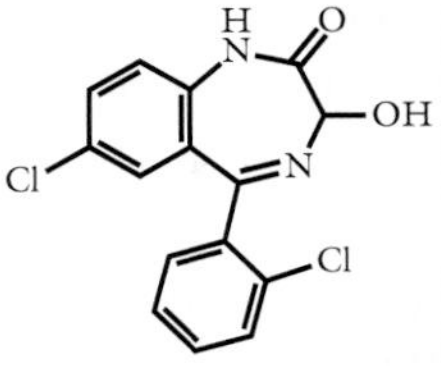	Clonazepam 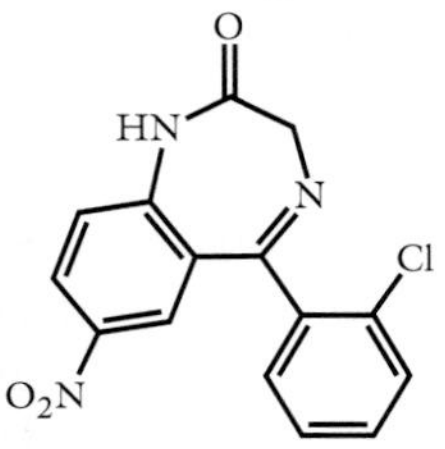	Codeine 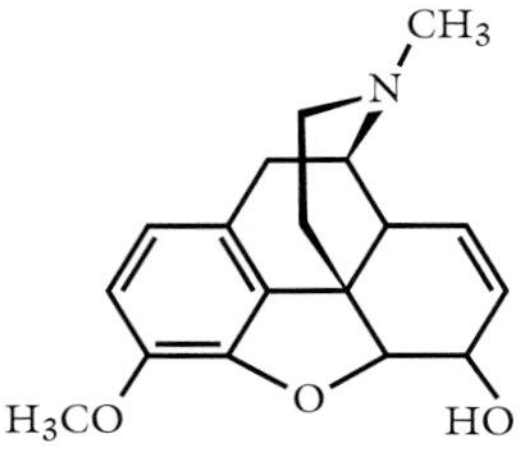	Morphine 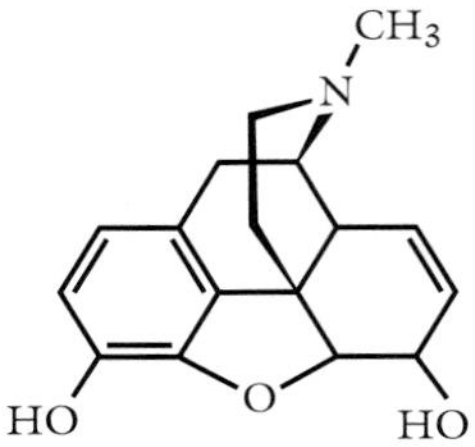

64. List all drugs that contain an aldehyde functional group.

65. List all drugs that contain an ether functional group.

66. List all drugs that contain an ester functional group.

67. List all drugs that contain a benzene ring.

68. List all drugs that contain a cycloalkane.

69. The taxes levied against ethanol intended for human consumption are significantly higher than the taxes on ethanol that is intended for use in industry or laboratories as solvents and reactants. To ensure that ethanol intended for the latter purposes is not being misused, it is *denatured,* a term meaning that methanol has been added to the ethanol. If the denatured alcohol is consumed, the methanol reacts to form the toxic product methanal (formaldehyde). Write the alcohol-aldehyde-carboxylic acid structures that are formed when ethanol and methanol are metabolized.

70. The congeners of alcohols present in alcoholic beverages are largely responsible for the effects of a hangover. Write the condensed structural formula for each of the following common congeners found in alcoholic beverages:
(a) Methanol
(b) 1-propanol
(c) 1-butanol
(d) 2-butanol
(e) 2-methyl-1-propanol
(f) 2-methyl-1-butanol
(g) 3-methyl-1-butanol

71. The smell of decaying flesh is due mainly to the production of putrescine and cadaverine from decaying proteins. Putrescine is 1,4-butanediamine and cadaverine is 1,5-pentanediamine. Write the condensed structural formulas for these two compounds.

72. The compounds methanal through decanal (1-carbon to 10-carbon aldehydes) are produced during decomposition processes and are currently being studied as possible marker compounds for locating clandestine grave sites. Write the condensed structural formulas for methanal through decanal.

Case Study Problem

73. The possession of 1,4-butanediol is illegal. The compound is a controlled substance that can be abused directly because it is metabolized to form GHB in the body. Shown in the figure below is the FTIR data obtained from analysis of an evidence sample. The defendant claimed the compound was actually the legal compound 1-butanol. Based on the FTIR data here and in Figure 7.15, could this defense work? Explain.

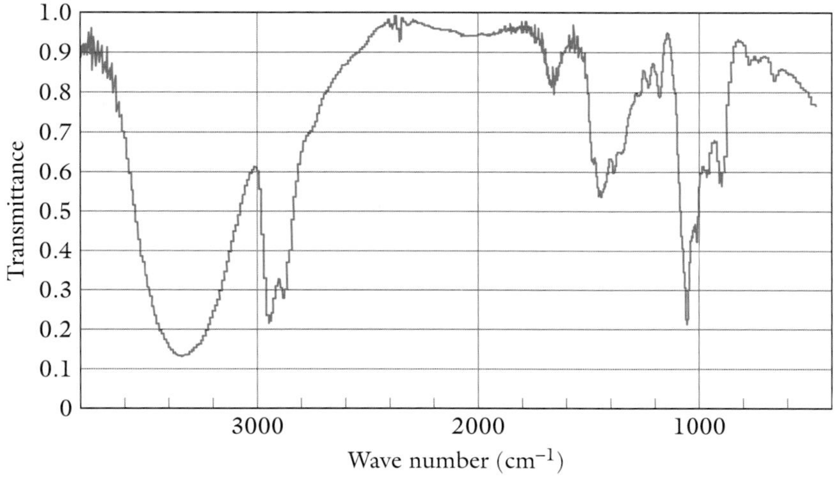

CHAPTER 8

Chemistry of Addiction

(Mike Couto/Courtesy **The Standard Times***)*

Case Study: Vigilante Jell-O

When somebody does serious harm to a child, a parent's instinct is to want the offender to pay for the crime. However, sometimes the desire for vengeance overwhelms the desire for justice, and people take the law into their own hands. This was the case in the early 1990s when Christina Martin's world was turned upside down by the revelation that her 14-year-old daughter was being sexually abused. This horrific crime had been perpetrated by someone Christina thought she could trust—her boyfriend Richard Alfredo. Christina had been living with Richard for many years when her daughter informed her that Richard had been abusing her for the last eight years. Whether Christina didn't believe the justice system would work, or her desire for personal revenge simply overtook any considerations of the legal process, she did not contact the police. Richard would never face a trial or even be confronted with these accusations. Instead, Christina started putting a deadly plan into place.

Richard Alfredo was 61 years old and had a history of serious heart disease. He had suffered at least one heart attack and had undergone coronary bypass surgery. Christina planned to murder Richard by slipping into his Jell-O a large dose of mescaline (3, 4, 5-trimethoxyphenethylamine), a hallucinogenic drug that is the active ingredient in peyote. She assumed that his heart would not be able to withstand the effects of the drug because of his poor health and, furthermore, that his death would be attributed to ill health. Christina made no secret of her plans, apparently telling several friends and family members that she intended to kill Richard.

Christina and her daughter were new to purchasing illegal drugs and approached several local teenagers with their request to buy mescaline. Christina also told them what she intended to do with it. The teenagers didn't want to be accessories to murder, but they did want to make some easy money. They sold her bogus drugs. However, even when this became evident to Christina, she was not deterred and sought out another source for the deadly ingredient of Richard's last meal.

On January 21, 1990, Richard Alfredo died shortly after eating. The death certificate stated that he died of a myocardial infarction, the medical term for a heart attack. The plan had worked, but the mother and daughter had told too many people of their intentions. Word soon reached the police from two anonymous callers. Thirty-one days after Richard's burial, his body was exhumed for an autopsy, from a coffin now filled with groundwater that had leaked into his grave. The pathologist collected samples from the bodily remains and sent them for analysis. The results came back negative for mescaline. However, one test came back positive for lysergic acid diethylamide (LSD), another hallucinogenic drug often sold on the streets as mescaline.

On November 14, 1992, a jury deliberated for five and a half hours before finding Christina guilty of first degree murder, despite her daughter's testimony that she, not her mother, laced Richard's Jell-O with LSD. Christina hired a new lawyer who filed an appeal, a common action in the case of a murder conviction. The new lawyer offered strong evidence that her original attorney had provided ineffective counsel. In July 1998, the Massachusetts Superior Court agreed and ordered a new trial. The newspapers jumped on the part of the appeal claiming ineffective counsel and presented the story as if the Superior Court had found a technicality on which to change the verdict. However, the actual basis for the Superior Court's decision was that the state's evidence did not prove the presence of LSD in Richard's body, and her lawyer had never challenged the evidence!

The test that showed positive results for LSD did not constitute adequate proof that LSD was present in the sample. How could the state convict a person for poisoning someone with LSD, but not have proof of the poison's presence in the body?

To understand what happened in the laboratory, we must explore chemical bonding in greater depth . . .

8.1 Nature of Covalent Bonds

Learning Objective

Explain the valence bond theory of covalent bonding.

The analysis of human blood or urine samples for drugs can be quite complex and time-consuming because of the many components in bodily fluids and the great variety of illegal drugs that could potentially be present in the sample. Drug molecules may pass through the body unchanged or break down into assorted new compounds, adding to the complexity of the analysis. Some drug molecules also tend to bind tightly with particular protein molecules present in blood. This binding can artificially lower the apparent concentration of the drug if only the unbound form of the drug is measured. Another complicating factor is that legal prescription drugs and their compounds can also be present in blood or urine. Many of these legal drugs are similar in structure to some illegal drugs, as illustrated in Chapter 7.

To narrow down the vast number of possible drugs that could be present in a sample, screening tests are used to determine, in general, what classes or types of drugs the sample may contain. Screening tests, which are fairly quick and inexpensive, give a positive result for a broad class of compounds.

One common method of drug screening is called an **immunoassay.** The basis of immunoassay is that all molecules of a particular compound have a unique three-dimensional shape. The immunoassay brings together the target molecule (the drug or substance being tested) with another molecule that is specifically chosen because its structure has an opening into which only the target molecule will be able to fit. This customized fit is very similar to the way illegal drugs or, for that matter, legal prescription drugs, function in the body. The three-dimensional shape of the drug molecule is such that it can bind to a receptor molecule in the body, triggering the drug's desired effect.

Figure 8.1 shows a large circular molecule of cyclodextrin with a benzene molecule within the center space. This example is a simplified illustration of how molecules can interact with one another based on their

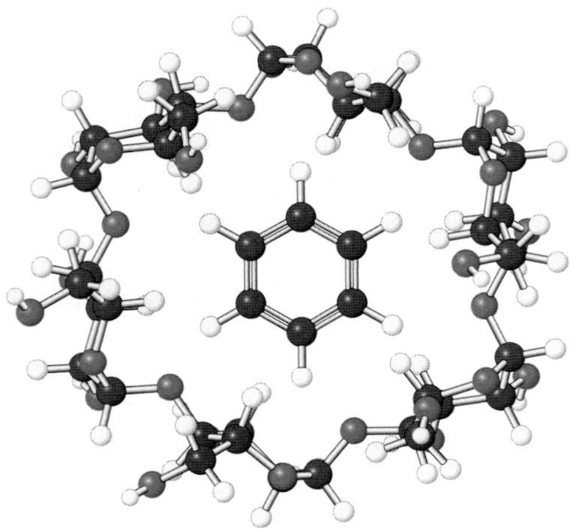

Figure 8.1 Molecules interact based on their size and three-dimensional shape. For example, cyclodextrin is a large circular molecule with an open center cavity into which the benzene molecule can easily enter.

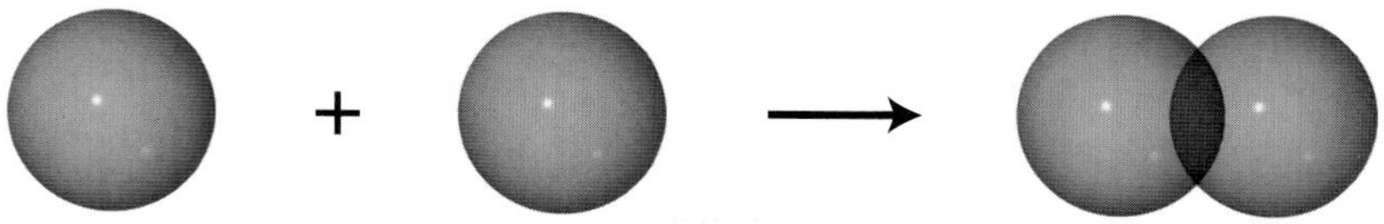

Figure 8.2 Overlap of atomic orbitals. The formation of a covalent bond is based on the sharing of electrons. Electrons are shared between two atoms by overlapping a region of three-dimensional space in which the electrons have a high probability of being found.

size and shape. The opening in the center of a molecule of cyclodextrin is of sufficient size and polarity that the benzene molecule fits inside the cyclodextrin molecule. The three-dimensional shape and bonding angles make it possible for this interaction to occur.

Cyclodextrin has been used as a stationary phase in liquid chromatography for the separation and detection of methamphetamine because the structure of methamphetamine is dominated by a benzene ring, as shown in Section 7.5. Understanding the three-dimensional shapes of molecules is vitally important in comprehending the chemistry of addiction and immunoassays. We will examine both of these topics in more depth as the chapter progresses.

The nature of the bonding between the individual atoms of a molecule determines the three-dimensional shape of the molecule. In our earlier discussions of chemical compounds, we saw that atoms form two types of bonds—ionic or covalent. Ionic bonds involve the gain or loss of electrons from neutral atoms to form positive and negative ions that are attracted to one another. Metal atoms form cations while nonmetal atoms form anions.

Covalent bonds form when electrons are shared between two atoms. The sharing of electrons occurs because the relative difference in electronegativities of the two atoms is not sufficiently large for either atom to remove an electron from the other atom. Covalent bonds form when two nonmetal atoms bond together and share electrons. But how exactly does a pair of atoms share electrons?

Electrons are located in atomic orbitals, which are the regions around the nucleus in which there is the greatest probability of finding an electron. For a covalent bond to form between two atoms, orbitals must overlap. The overlapped orbitals create a region of higher electron density in which two electrons are simultaneously located in the orbitals of both atoms.

This model of covalent bonding is called the **valence bond theory.** Figure 8.2 illustrates the simplest case of a covalent bond forming between two hydrogen atoms. Each hydrogen atom has an electron in a spherically shaped *s* orbital. When *p* orbitals are involved in the formation of a single bond between atoms, the overlap occurs at the end of one lobe of the orbital.

Worked Example 1

Draw the covalent bond between the *s* orbital of a hydrogen atom and the *p* orbital of a fluorine atom found in hydrofluoric acid (HF).

SOLUTION The H atom has one electron in an *s* orbital. The F atom has one unpaired electron in a *p* orbital (dumbbell shape). The overlap of the *s* and *p* orbitals to form a covalent bond is represented as follows:

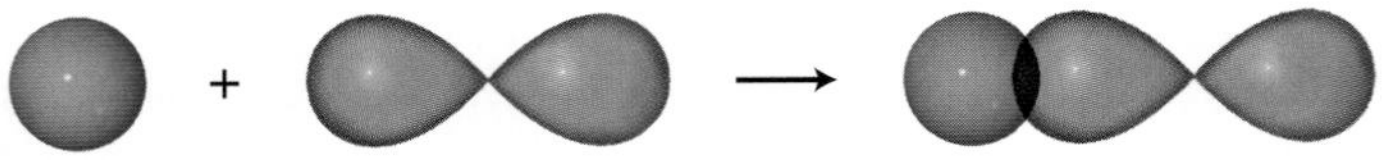

Practice 8.1

Draw the covalent bond between the F atoms in fluorine gas (F_2).

ANSWER

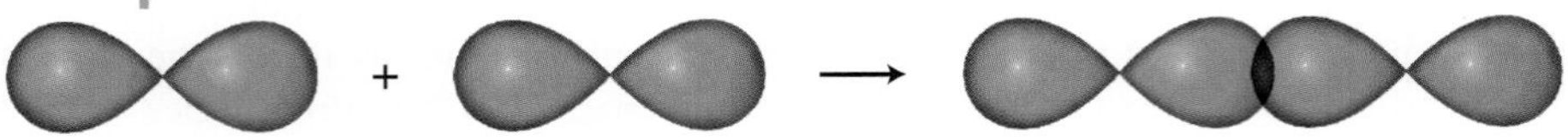

The valence bond theory works well for explaining how bonding occurs in a covalent compound. However, for determining the shapes of molecules, the theory works only in the simplest of cases, such as those illustrated above. For molecules that are even slightly more complex, another method for determining the geometry is necessary. For larger molecules such as LSD ($C_{20}H_{25}N_3O$), the overall three-dimensional shape of the molecule is a result of the local geometry at each atom influencing the possible location of the neighboring atoms.

There are two types of information needed to determine the three-dimensional shape of a molecule: the locations of all of the valence electrons and how those electrons interact with one another. The *Lewis theory of bonding*, covered in the next section, addresses the first issue. The *valence shell electron pair repulsion theory* addresses the second issue, as it determines the actual geometry of the molecules from the information obtained from the Lewis theory.

8.2 Lewis Structures of Ionic Compounds

Learning Objective

Construct the Lewis structures of ionic compounds.

In Chapter 4, we saw the predictive power of the periodic table in determining the formula for ionic compounds. For instance, all alkali metals (Group 1) form cations with a charge of +1, all alkaline earth metals (Group 2) form cations with a charge of +2, and the halogens (Group 7) form anions with a charge of −1.

The **Lewis theory** explains this trend, stating that the atoms are attempting to achieve the same electron configuration as the closest noble gas by gaining or losing electrons as needed. The noble gases are chemically unreactive and do not bond to other atoms under normal conditions. The reason for this unusual stability is that the valence shell of the noble gas atoms has been completely filled. When the other elements in the periodic table react, the individual atoms strive to fill their outer valence shells and attain the stable electronic structure of the noble gas elements. Put another way, the elements react in such a way as to become **isoelectronic** with the noble gas elements. For example, sodium has one more electron than neon, the closest noble gas. Sodium metal is extremely reactive, but the sodium ion, Na^+, is very stable. By losing one electron and forming the Na^+ ion, sodium becomes isoelectronic with neon and

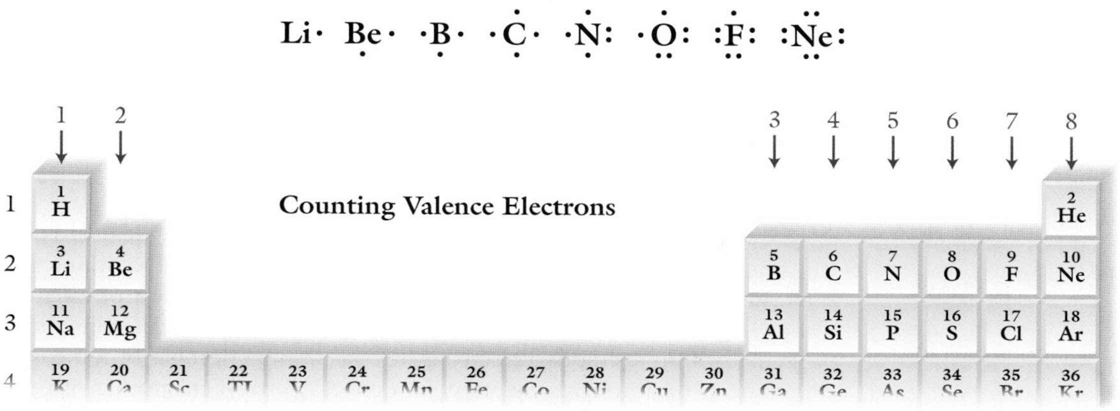

Figure 8.3 Lewis dot structures for lithium through neon.

has a filled valence shell. The sodium ion will not react further to gain or lose more electrons under standard conditions.

This pattern of elements forming bonds to become isoelectronic with the noble gases serves as the basis for the **octet rule.** This rule, sometimes called the *rule of eight*, states that elements react in order to attain a total of eight valence electrons, a configuration that corresponds to a filled *s*-orbital and *p*-orbital set. The exceptions to this rule are those elements that achieve an electron configuration identical to helium, which needs only two electrons.

The Lewis symbol for an element uses dots to represent the valence electrons for an atom. The electrons are placed around the elemental symbol, as shown at the top of Figure 8.3. The number of valence electrons is determined by counting the number of *s* and *p* electrons in the outermost level of the atom, as illustrated beneath the Lewis symbols in the figure.

Worked Example 2

Draw the Lewis dot structures for the following elements:

(a) H

(b) Ca

(c) As

SOLUTION

(a) Referring to the periodic table, H has only one *s* electron: H·

(b) Likewise, Ca has two *s* electrons: Ca·

(c) As has two *s* electrons and three *p* electrons (do not count *d* electrons!): ·As:

Practice 8.2

Draw the Lewis dot structures for the following elements:

(a) S

(b) Kr

(c) P

ANSWER

(a) $\cdot\ddot{P}:$ 

(b) $\cdot\ddot{\underset{..}{S}}:$ 

(c) $:\ddot{\underset{..}{K}}r:$ 

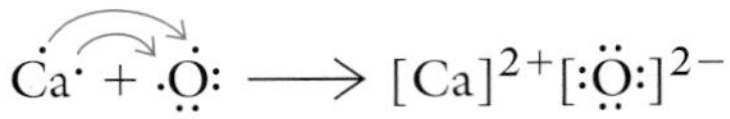

Lewis dot structures can be used to show the gain and loss of electrons in the formation of an ionic bond between a metal and a nonmetal, as illustrated in Figure 8.4. The reaction shows the formation of calcium oxide (CaO), in which the calcium atom loses two valence electrons to an oxygen atom. In losing two electrons, the calcium 2+ ion has become isoelectronic with argon, completing its octet. The oxygen atom, with six original valence electrons, gains the two electrons from calcium and becomes isoelectronic with neon. The reaction produces the stable ionic compound calcium oxide. The formulas and charges of simple ionic compounds can be predicted and understood by writing the Lewis dot structures.

$$Ca\cdot + \cdot\ddot{\underset{..}{O}}: \longrightarrow [Ca]^{2+}[:\ddot{\underset{..}{O}}:]^{2-}$$

Figure 8.4 Formation of an ionic compound.

Worked Example 3

Draw the Lewis dot structures for the ionic bonding in MgF_2.

SOLUTION The Mg atom will lose its two valence electrons, one to each F atom forming the ionic compound MgF_2.

$$:\ddot{\underset{..}{F}}\cdot + \cdot Mg\cdot + \cdot\ddot{\underset{..}{F}}: \longrightarrow [:\ddot{\underset{..}{F}}:]^{1-}[Mg]^{2+}[:\ddot{\underset{..}{F}}:]^{1-}$$

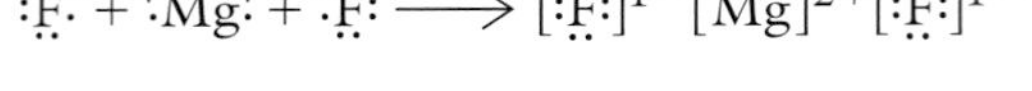

Practice 8.3

Draw the Lewis dot structures for the ionic bonding in Na_2S.

ANSWER

$$[Na]^{1+}\ [:\ddot{\underset{..}{S}}:]^{2-}\ [Na]^{1+}$$

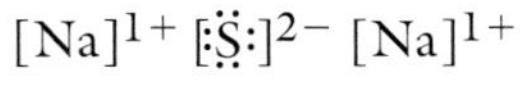

8.3 | Lewis Structures of Covalent Compounds

Learning Objective

Draw the Lewis structures of covalent compounds.

Lewis dot structures can also illustrate covalent bonds between atoms in simple molecules. Figure 8.5 shows the valence electrons and covalent bond in the molecules Cl_2 and O_2. In the chlorine reaction, each chlorine atom has seven valence electrons and is striving to have eight. Each chlorine atom gains one electron, becoming isoelectronic with argon. Unlike atoms that react to form ionic compounds, each chlorine atom in Cl_2 has an equal tendency to obtain an electron. Therefore, it is impossible for one chlorine atom to form an anion and the other to form a cation. The only way a chlorine atom can achieve an octet in forming Cl_2 is for each atom to share its one unpaired electron with the other atom. The **single bond** that forms between two chlorine atoms consists of the two shared

$$:\ddot{\underset{..}{Cl}}\cdot + \cdot\ddot{\underset{..}{Cl}}: \longrightarrow :\ddot{\underset{..}{Cl}}-\ddot{\underset{..}{Cl}}: \qquad :\ddot{\underset{..}{O}}\cdot + \cdot\ddot{\underset{..}{O}}: \longrightarrow :\ddot{\underset{..}{O}}-\ddot{\underset{..}{O}}: \longrightarrow :\ddot{O}=\ddot{O}:$$

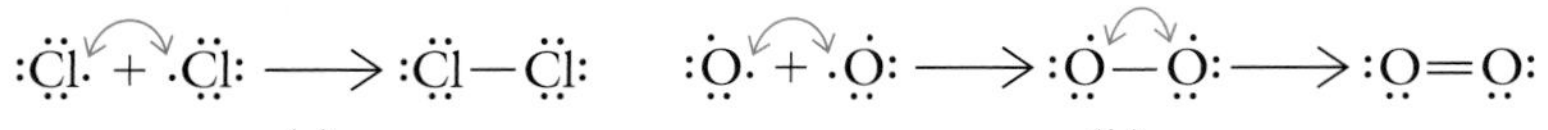

(a)   (b)

Figure 8.5 Lewis symbols for simple covalent compounds.

electrons. In the oxygen reaction, each oxygen atom has six valence electrons and is striving for eight. To achieve the octet, each oxygen atom must share two electrons with the other, thus forming a **double bond**. It is also possible to share three sets of electrons, forming a **triple bond** in order to achieve an octet.

Worked Example 4

Draw the Lewis structure for the covalent bonding in N_2.

SOLUTION Each N atom has five valence electrons and needs three to complete an octet. By sharing three electrons each and forming a triple bond, both N atoms obtain an octet of valence electrons.

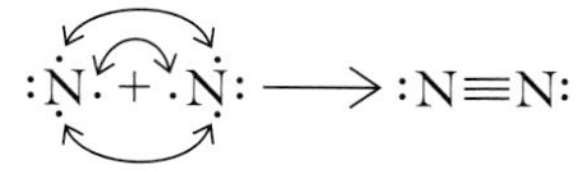

Practice 8.4

Draw the covalent bonding in ammonia, NH_3.

ANSWER

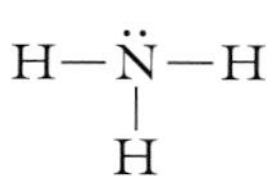

The Lewis structures of complex molecules cannot be determined by simply attempting to fill the vacant spots on the Lewis dot structure of individual atoms. A set of rules, listed below, must be followed to obtain the proper structure. Silicon dioxide (SO_2), the compound responsible for the characteristic odor of a burning match, is used to illustrate each step. It is not necessary to write the Lewis dot structure of each atom, as is done for simpler compounds.

Step 1. Determine the total number of valence electrons in the compound. For neutral molecules, add the valence electrons contributed by each atom. For polyatomic cations, *subtract* the charge magnitude of the ion from the total number of valence electrons, because electrons are *lost* in the formation of the cation. For polyatomic anions, *add* the charge magnitude of the ion to the total number of valence electrons, because electrons are *gained* in the formation of anions. In the example of SO_2, there are six valence electrons from sulfur and six valence electrons from each of the two oxygen atoms, for a total of $6 + 2 \times 6 = 18$.

Step 2. Draw a skeletal structure that consists of single bonds from the central atom to each of the outer atoms. Subtract two electrons for each single bond used from the total number of valence electrons. The first atom in a formula is generally the central atom. Hydrogen is *never* a central atom.

O—S—O

$$\begin{array}{r} 18 \\ -4 \\ \hline 14 \end{array}$$

Step 3. Place pairs of electrons around the outer atoms to give each atom an octet of electrons. Subtract the electrons used from the total available. Hydrogen needs only a duet (two electrons total). Because they are not part of a bond, these pairs of electrons are called a **lone pair** or a **nonbonding pair.**

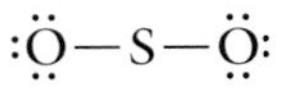

$$\begin{array}{r} 18 \\ \underline{-4} \\ 14 \\ \underline{-12} \\ 2 \end{array}$$

Step 4. If there are surplus electrons, place those electrons as lone pairs on the central atom.

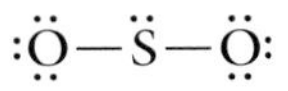

$$\begin{array}{r} 18 \\ \underline{-4} \\ 14 \\ \underline{-12} \\ 2 \\ \underline{-2} \\ 0 \end{array}$$

Step 5. If and only if the central atom does not have an octet of electrons resulting from steps 1 through 4, form a double bond by moving a lone pair of electrons from one of the outer atoms. (Use a triple bond only if a second double bond to another outer atom does not produce the desired structure with octets around all atoms.)

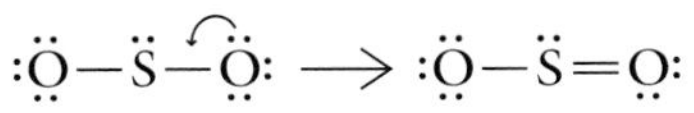

The structure obtained from this procedure could not have been determined by merely examining the simple Lewis structures of the individual atoms. In general, it is best to use these rules whenever the molecule has three or more atoms.

Worked Example 5

Draw the Lewis structure for SO_3.

SOLUTION

Step 1. The total number of valence electrons is $6 + 3 \times 6 = 24$.

Step 2. The skeletal structure has three single bonds, so subtract six electrons from the total number of valence electrons: $24 - 6 = 18$.

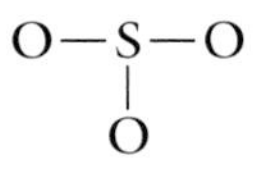

Step 3. Complete the octet of outer atoms and subtract the number of electrons used from the total number of valence electrons remaining: $18 - 18 = 0$.

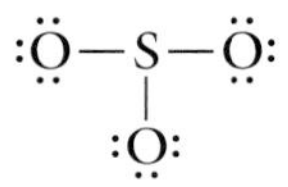

Step 4. Since there are no electrons left to place on the central atom, the diagram remains unchanged.

Step 5. Complete the octet of the central atom by forming multiple bonds:

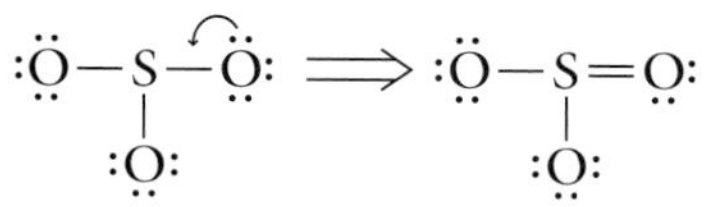

Practice 8.5

Draw the Lewis structure for the ammonium ion, NH_4^+.

ANSWER

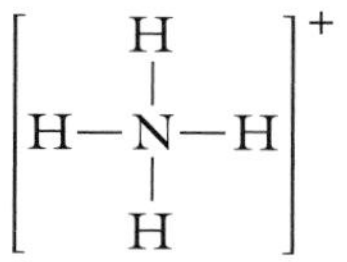

Worked Example 6

Examine these Lewis structures for any errors.

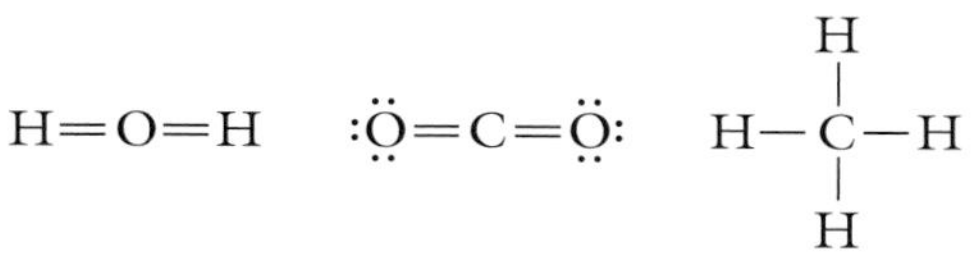

(a) (b) (c)

SOLUTION Structure (a) is incorrect because hydrogen atoms cannot have double bonds. Structure (b) is incorrect because oxygen atoms cannot have 10 valence electrons. Structure (c) is correct.

Practice 8.6

Draw the correct Lewis structures for Worked Example 6.

ANSWER

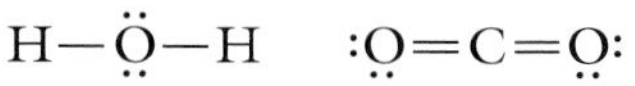

(a) (b)

8.4 Resonance Structures

In Section 7.5, the benzene ring was first introduced as a major structural component in organic chemistry. Benzene is one of the ingredients found in gasoline and can be present in a sample from an arson scene. However, the mere presence of benzene does not indicate arson. Benzene or compounds that contain a benzene ring are present in many household cleaners and certain plastics.

We saw that the benzene ring could be represented by two equivalent structures called *resonance structures.* To understand the nature of resonance structures in greater depth, we will now look more closely at those of benzene, shown in Figure 8.6. The carbon atoms are numbered for ease of discussion.

Learning Objective

Draw resonance structures for compounds that have multiple equivalent Lewis structures to accurately depict the bonding in a compound.

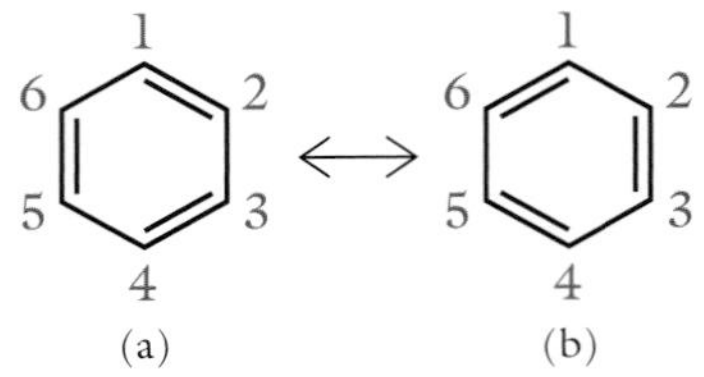

Figure 8.6 Resonance structures of benzene.

The only difference between structures (a) and (b) is that the location of the double bonds has shifted between adjacent carbon atoms. The structures are equivalent because we can simply rotate structure (a) clockwise and it would be indistinguishable from structure (b).

However, there is more to a resonance structure than the ability to rotate the molecule. Consider the bonds in structure (a) between C_1 and C_2 and between C_1 and C_6. If all carbon atoms have identical chemical reactivity, why would one carbon atom form a double bond and the other carbon atom form a single bond in the benzene molecule? The real structure of benzene is found somewhere between structures (a) and (b). There are no rotating double bonds within benzene. One might view the double bonds as secondary partial bonds throughout the entire ring. Another way to describe the bonding is to say that six electrons are delocalized throughout the ring and shared equally by all six carbons.

A computer-generated model of benzene depicting the bonding in the molecule is shown in Figure 8.7.

Worked Example 7

Draw the three Lewis structures representing the resonance structures of sulfur trioxide (SO_3).

SOLUTION The Lewis structure of sulfur trioxide was derived in Worked Example 5. The location of the double bond can occur at any of the oxygen atoms because they are all equivalent.

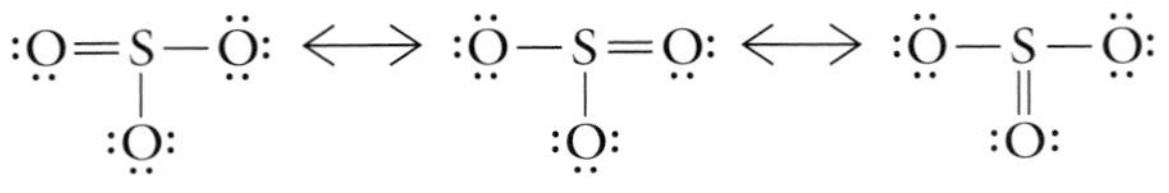

Practice 8.7

Draw the two Lewis structures representing the resonance structures for sulfur dioxide (SO_2).

ANSWER

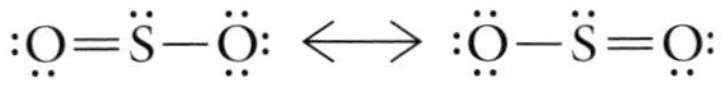

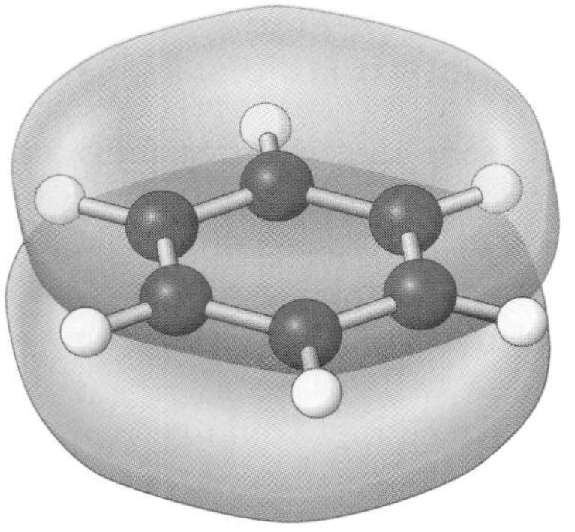

Figure 8.7 The benzene molecule is shown horizontally to reveal the delocalized shape of electron density above and below the plane of carbon atoms.

8.5 VSEPR Theory

Ammonium nitrate was one of the ingredients used in the bombing of the Murrah Federal Building in Oklahoma City and is a fairly simple ionic compound. The presence of ammonium nitrate can be detected by immunoassay tests, which rely on the shapes of molecules or ions. The geometry of more complex structures such as LSD (from the vigilante Jell-O case study) is dictated by the three-dimensional location of each component atom, which can be predicted by **valence shell electron pair repulsion theory (VSEPR theory).** This theory is based on the principle that electrons in bonds and lone pairs repel one another and, in doing so, move as far apart from one another as possible. To use the VSEPR theory, it is necessary to start with the proper Lewis structure. The molecule's shape can then be determined by applying the principles of the VSEPR theory. The shapes of complex molecules so derived are most often visualized through computer modeling software.

Learning Objective

Predict the three-dimensional shape of a molecule.

One common mistake students initially make is to look at the Lewis structure they have drawn and assume that the molecule's geometry is determined by the appearance of the Lewis structure. For example, because the Lewis structure for water is $H-\ddot{\underset{..}{O}}-H$, some students incorrectly state that it is a linear molecule. However, it is known from experimental evidence that the molecular shape of a water molecule is bent, not linear. This geometry can be predicted accurately by following the rules of the VSEPR theory.

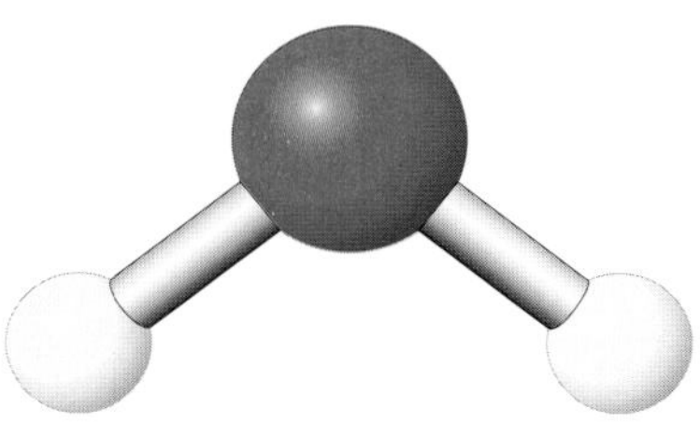

Water (H_2O) has a bent molecular geometry.

Another point of importance is that VSEPR theory first provides information on **electron geometry**—the location of valence *electrons*—and, based on this information, then determines **molecular geometry**—the location of *atoms* within the molecule. The electron geometry and the molecular geometry can be different. As we shall see, the difference depends on the presence or absence of lone pair electrons on the central atom of the molecule. The steps for determining electron geometries are provided below. The rules for determining molecular geometries will be given after the steps for determining electron geometries.

Step 1. Draw the Lewis structure of the molecule. (Starting with the correct Lewis structure is essential!)

Step 2. Determine how many regions of electron density surround the central atom. An **electron region** is one set of lone pair electrons, a single bond, a double bond, or a triple bond. The following examples show how to determine the number of electron regions:

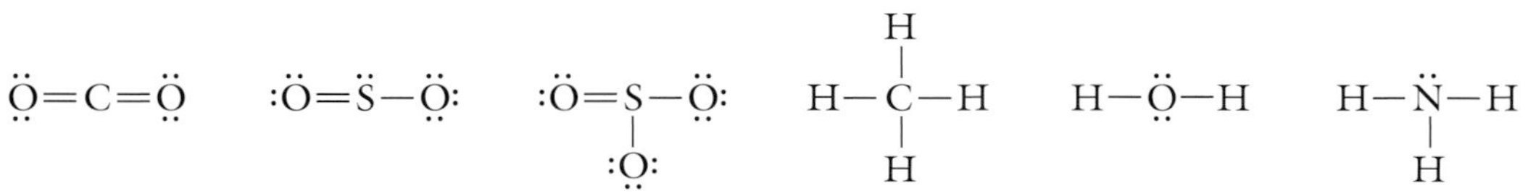

2 regions | 3 regions | 3 regions | 4 regions | 4 regions | 4 regions

Step 3. Determine the electron geometry from the descriptions shown in Figure 8.8. (Note that the lines now represent electron regions, *not* single bonds.)

Electron Regions	Sketch	Electron Geometry
2 regions	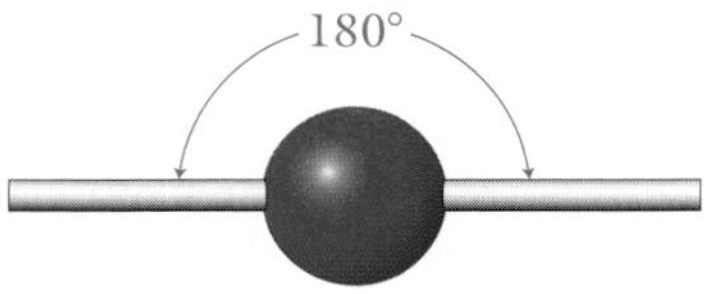	Linear
3 regions	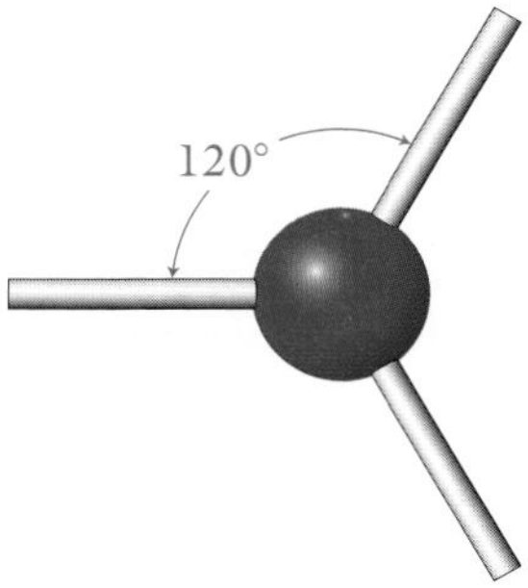	Trigonal planar
4 regions	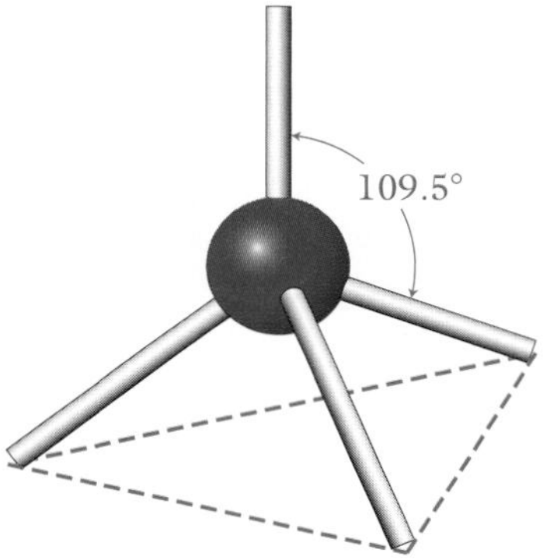	Tetrahedral

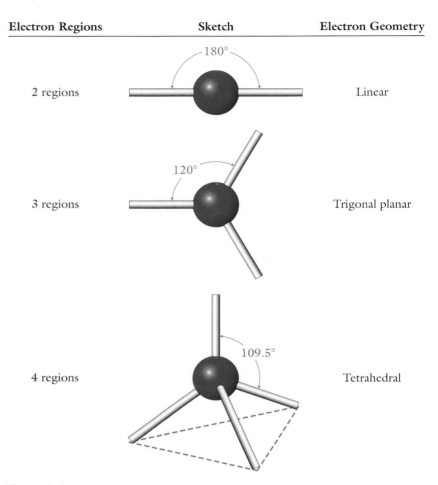

Figure 8.8 Determining electron geometry from the Lewis structure and the number of electron regions.

When two electron regions surround a central atom, they are farthest apart if there is a 180° angle between them. This is called a **linear** arrangement because the two electron regions lie on a line. When a third electron region is added to a central atom, the angles between the regions change to 120° in order to maximize their distance apart. This arrangement is called **trigonal planar** because the three electron regions are located within the same geometrical plane and the tips of each region form a triangle. When four electron regions are located around a central atom, the optimum angle between electron regions is 109.5° with the shape of a tetrahedron. The resulting shape is called **tetrahedral** from the prefix *tetra-* meaning "four." The green dashed lines connecting the bottom three regions of the tetrahedron have been added to help you visualize the three-dimensional shape.

Worked Example 8

Determine the electron geometry of both ions in the potentially explosive ammonium nitrate (NH_4NO_3).

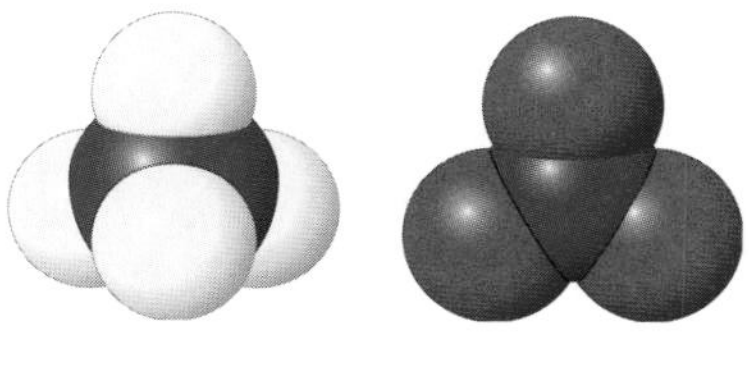

Space-filling models of the ammonium and nitrate ions.

SOLUTION

Step 1. The Lewis structures for the two ions are

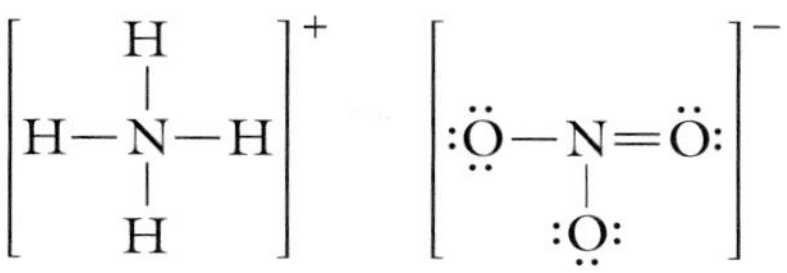

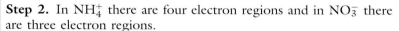

Step 2. In NH_4^+ there are four electron regions and in NO_3^- there are three electron regions.

Step 3. From Figure 8.8, we see that the electron geometry of NH_4^+ is tetrahedral and the electron geometry of NO_3^- is trigonal planar.

Practice 8.8

Determine the electron geometry of nitrogen triiodide, NI_3, a potentially explosive molecule.

ANSWER

There are four electron regions, and the electron geometry is tetrahedral.

Once the electron geometry is known, the next step is to determine the molecular shape by examining the location of the outer atoms in relation to the central atom. The key concept to remember during this part of the process is that lone pair electrons, if present, help to force the atoms into the positions they occupy but are not considered part of the molecular geometry. If no lone pair electrons are present, the molecular geometry is identical to the electron geometry.

Figure 8.9 shows the relationship between the number of electron regions in a molecule and the resulting geometry of the molecule. The geometry of a molecule that has three electron regions around the central atom with three outer atoms is identical to the electron geometry—trigonal planar, because each region terminates with an atom. However, the geometry of a molecule that has three electron regions around the central atom with one of them being lone pair electrons is different. The geometry of this molecule (outlined in green) is called **bent** because of the overall molecular shape.

Similarly, if a molecule has a tetrahedral electron geometry and has four atoms surrounding a central atom, the molecular geometry will also be tetrahedral. However, if one of the four regions is a lone pair of electrons and not an atom, the molecular geometry will change. The new geometry is called **trigonal pyramidal**. If two of the four electron regions are occupied by lone pair electrons, the remaining atoms take on the bent molecular geometry.

Worked Example 9

Determine the molecular geometry for each of the following Lewis structures.

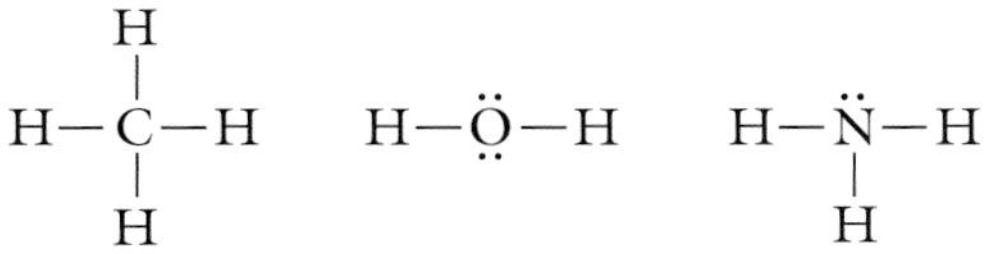

(a) (b) (c)

Electron Regions	Number of Lone Pairs	Sketch	Molecular Geometry
3	0	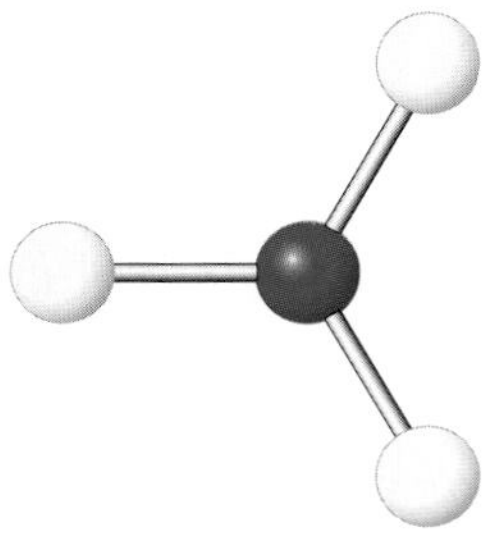	Trigonal planar
3	1	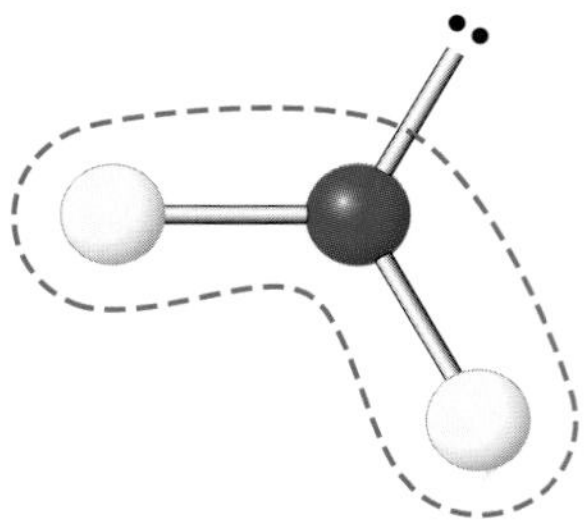	Bent
4	0	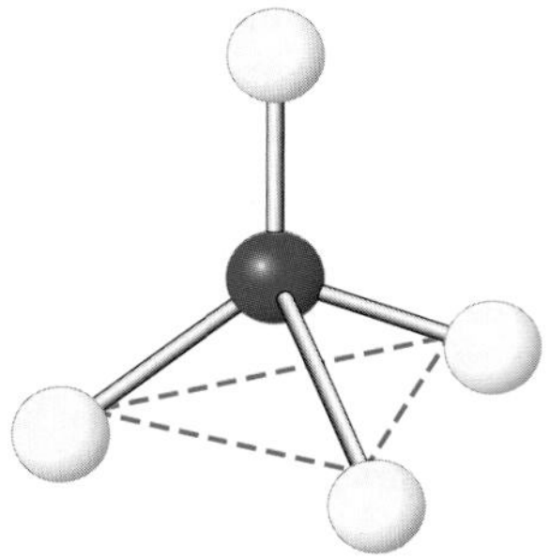	Tetrahedral
4	1	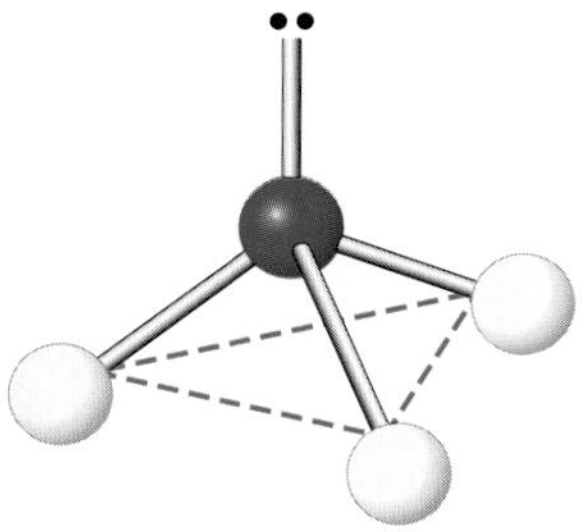	Trigonal pyramidal
4	2	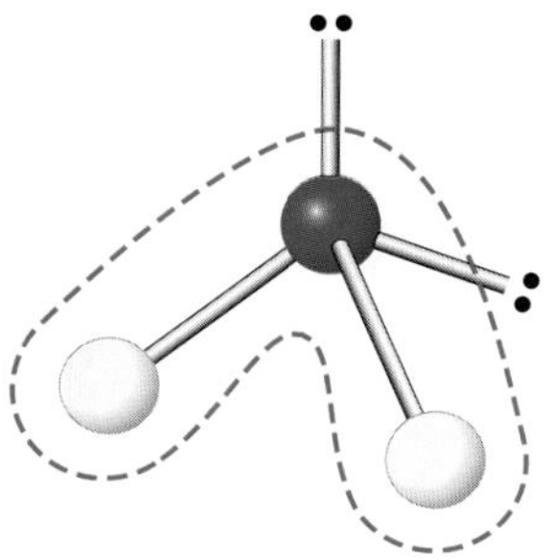	Bent

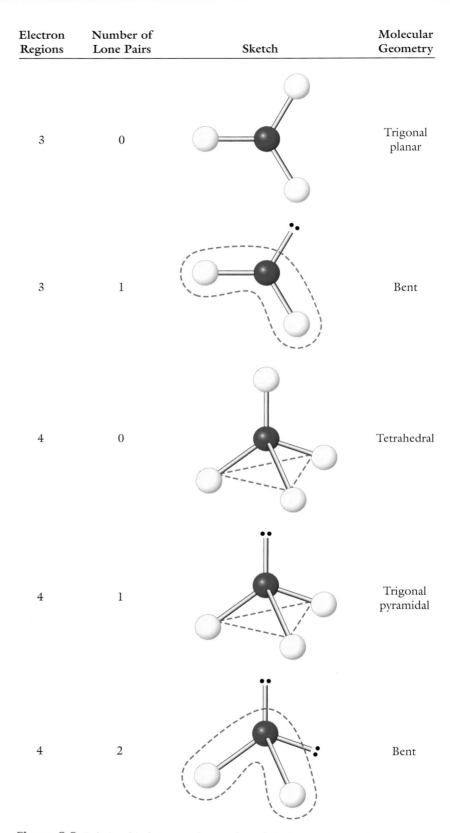

Figure 8.9 Relationship between the number of electron regions and molecular geometry.

SOLUTION

(a) There are four electron regions with no lone pair electrons. This means the molecular geometry is tetrahedral.

(b) There are four electron regions with two lone pairs of electrons. This means the molecular geometry is bent.

(c) There are four electron regions with one lone pair of electrons. This means the molecular geometry is trigonal pyramidal.

Practice 8.9

Determine the molecular geometry for each of the following Lewis structures.

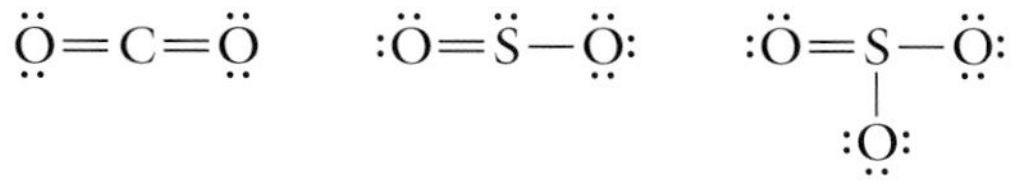

(a) (b) (c)

ANSWER

(a) Linear

(b) Bent

(c) Trigonal planar

8.6 Polarity of Bonds and Molecules

Learning Objective

Determine the polarity and solubility of compounds.

The polarity of a molecule affects many physical variables, such as which solvents it will dissolve in, the melting and boiling points of the pure substance, and how one molecule will interact with another molecule during chromatography, to name just a few. The polarity of a bond is determined by the difference in the electronegativity of the two atoms that are bonded. Recall that electronegativity is the ability of an atom to pull the shared electrons within a covalent bond toward itself. The electronegativity of the elements increases from the bottom to the top of the columns on the periodic table and from the left to the right of the groups. Fluorine has the greatest electronegativity and the noble gases have zero electronegativity. The noble gases have a full shell of valence electrons and do not form covalent bonds. Table 8.1 lists the relative

Table 8.1 Electronegativity of the Nonmetal Elements

Element	Relative Electronegativity
Fluorine	4.0
Oxygen	3.5
Chlorine	3.0
Nitrogen	3.0
Bromine	2.8
Carbon	2.5
Sulfur	2.5
Iodine	2.5
Selenium	2.4
Hydrogen	2.1
Phosphorus	2.1

electronegativity of the nonmetal elements, with fluorine having the maximum value of 4.0.

A **nonpolar covalent bond** occurs between two atoms when the electronegativity of both atoms is equal. The shared electrons are attracted equally to both atoms. Nonpolar covalent bonds occur between diatomic molecules such as nitrogen gas (N_2) and oxygen gas (O_2) and are thought to occur between two atoms of different electronegativity, as long as the difference in electronegativities is between 0.0 and 0.3.

A **polar covalent bond** occurs when the difference in electronegativities is between 0.4 and 1.9. The greater the difference in electronegativities, the more strongly the shared electrons are pulled toward the atom with the highest electronegativity. If the difference in electronegativities is greater than 2.0, the bond is considered to be ionic.

Worked Example 10

Are the following bonds nonpolar covalent or polar covalent?

(a) C—S

(b) S—H

(c) N—Br

(d) O—P

SOLUTION

(a) The difference in electronegativity is $2.5 - 2.5 = 0$, so it is a nonpolar covalent bond.

(b) The difference in electronegativity is $2.5 - 2.1 = 0.4$, so it is a polar covalent bond.

(c) The difference in electronegativity is $3.0 - 2.8 = 0.2$, so it is a nonpolar covalent bond.

(d) The difference in electronegativity is $3.5 - 2.1 = 1.4$, so it is a polar covalent bond.

Practice 8.10

Which element in the bonds from Worked Example 10 will have the partially negative charge?

ANSWER

(a) Neither

(b) Sulfur

(c) Neither

(d) Oxygen

One misconception that occurs with polar and nonpolar covalent bonds is the assumption that nonpolar compounds contain nonpolar covalent bonds and polar compounds contain polar covalent bonds. This is *not* the case. Figure 8.10a shows a polar covalent bond between carbon and chlorine. The electron density is being pulled toward the chlorine atom, as the arrow indicates. When determining the polarity of a molecule such as carbon tetrachloride (CCl_4), it is imperative to examine the

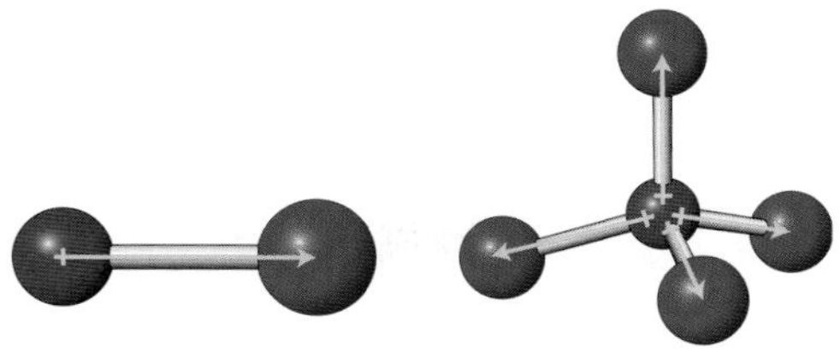

(a) (b)

Figure 8.10 Bond and molecular polarity. The carbon (black) to chlorine (green) bond shown in part (a) is polar in nature due to the higher electronegativity of the chlorine atom, which pulls electron density toward itself. The carbon tetrachloride molecule shown in part (b) is nonpolar because the polarities of the four carbon-chlorine bonds effectively negate one another.

geometry of the molecule. Figure 8.10b shows the electron density being pulled equally in all directions about the tetrahedral geometry. When electron density is pulled equally in all directions, the net result is a nonpolar molecule. If the electron density of the molecule is not symmetrically distributed, the molecule is polar. Any molecule in which the central atom contains lone pair electrons tends to be polar, too.

Worked Example 11

Draw the Lewis structure for ammonia (NH_3), and determine whether or not this is a polar molecule.

SOLUTION The Lewis structure for ammonia: 8 valence electrons, of which 6 are used in forming bonds to hydrogen and the remaining 2 valence electrons are placed on the nitrogen atom as a lone pair of electrons:

$$\begin{array}{c} \text{H}-\ddot{\text{N}}-\text{H} \\ | \\ \text{H} \end{array}$$

The molecular geometry for a molecule with four electron regions and one lone pair of electrons is trigonal pyramidal. The electronegativity of nitrogen is greater than hydrogen, drawing the partial negative charge toward the lone pair electrons, producing a polar molecule as shown at right.

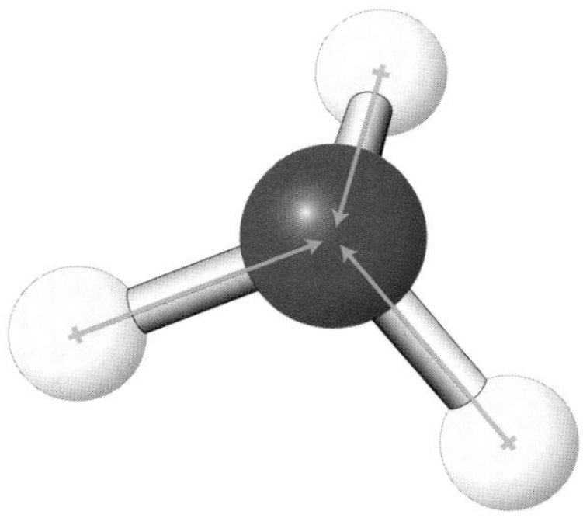

Ammonia has tetrahedral electron geometry with trigonal pyramidal molecular geometry. There is a lone pair of electrons on the nitrogen atom, and the electron density is all being directed to one portion of the molecule. Ammonia is polar.

Practice 8.11

Which molecular geometries will always produce a polar molecule?

ANSWER
Bent and trigonal pyramidal.

Assigning a term such as *polar* or *nonpolar* to a molecule is a dramatic oversimplification of the physical phenomena involved. Some molecules such as water are very polar, while others such as hexane are very nonpolar. There are numerous variables, such as size and the functional groups attached, that determine the polarity of a molecule. Table 8.2 lists many common compounds and gives their polarity on a relative scale, with 0 being nonpolar and 9.0 being polar. The solubility is the maximum amount of the compound that can be dissolved in water.

Table 8.2 Polarity of Common Organic Solvents

Compound	Polarity Index	Solubility in Water (%)
Hexane	0.0	0.001
Pentane	0.0	0.004
Toluene	2.4	0.051
Benzene	2.7	0.18
1-butanol	4.0	0.43
1-propanol	4.4	100
2-butanone	4.7	24
Propanone	5.1	100
Ethanol	5.2	100
Water	9.0	100

From Table 8.2 it is apparent that the alkanes are extremely nonpolar and are virtually insoluble in water. The addition of an alcohol functional group to the carbon chain substantially increases both the polarity and the solubility of the compound in water. Notice, however, that adding a carbon atom to 1-propanol to form 1-butanol causes the polarity to decrease and the solubility to drop dramatically. This pattern repeats itself when a ketone functional group is added to the carbon chain, as illustrated with propanone and 2-butanone.

The polarity of organic compounds decreases with increasing carbon chain length. The polarity is also determined by the functional group contained in the molecule. The relative polarity of the organic compounds can be described as follows:

Alkanes < Aromatics < Ethers < Esters < Aldehydes ≈ Ketones < Amines < Alcohols < Carboxylic acids < Water

Worked Example 12

Place the following compounds in order of increasing polarity: butanal, pentanal, ethanal, and propanal.

SOLUTION Because all of the compounds have the same functional group (aldehyde), the difference in polarity is based on the number of carbon atoms: pentanal < butanal < propanal < ethanal.

Practice 8.12

Place the following compounds in order of increasing polarity: ethanal, ethanoic acid, ethanol, and ethane.

ANSWER

Ethane < ethanal < ethanol < ethanoic acid

Learning Objective

Describe how the molecular geometry at individual carbons can be used to visualize large molecules.

8.7 Molecular Geometry of Drugs

Methamphetamine is an interesting molecule to revisit now that we can add concepts of molecular geometry to our discussion. Examine the two compounds in Figure 8.11 labeled *d*-methamphetamine and

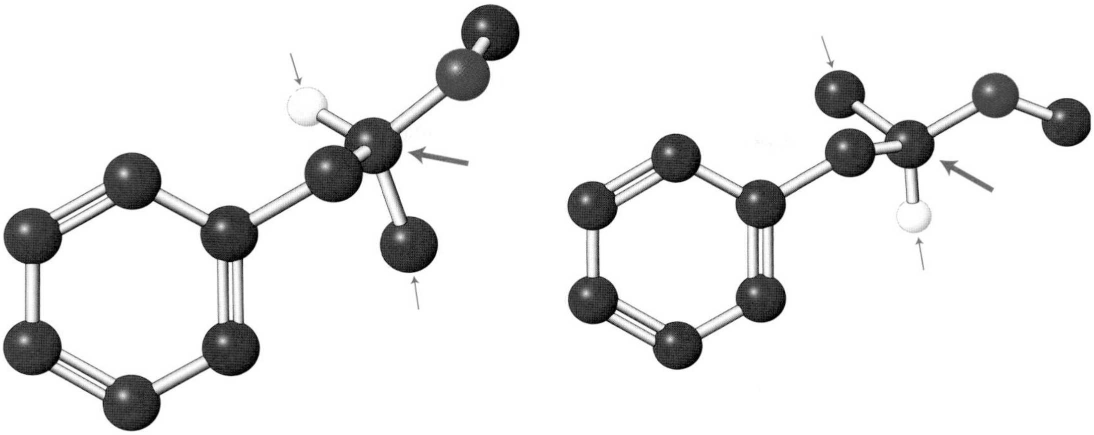

***d*-methamphetamine** ***l*-methamphetamine**

Figure 8.11 Methamphetamine three-dimensional structures. The illegal drug *meth* specifically refers to *d*-methamphetamine, which differs from the isomer *l*-methamphetamine by the special orientation of the four atoms attached to the carbon atom indicated by the large arrow.

l-methamphetamine. Most of the hydrogen atoms are omitted to make the model easier to view. The only hydrogen atom shown is the one attached to the carbon atom indicated by the large arrow in each molecule. The difference between the two structures can be seen by focusing on the position of the hydrogen and carbon atoms attached to the indicated carbon atom. In both molecules, this carbon atom has four single bonds surrounding it and exhibits a tetrahedral geometry. Yet the molecular structures are not identical. Notice that the methyl group and hydrogen group in the two molecules point in different directions.

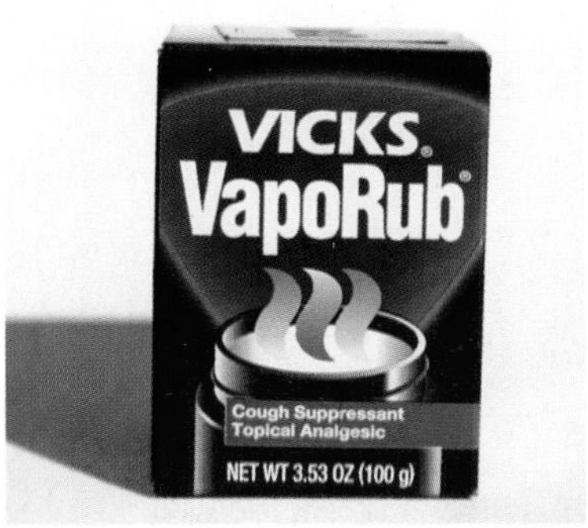

A vapor rub product containing *l*-methamphetamine. (Elyse Rieder)

When four different atoms or groups of atoms are bonded to a tetrahedral carbon, two types of structures are possible. These structures are special isomers called **stereoisomers,** which occur when two compounds share the same chemical formula and the same connections between atoms but exhibit differences in the way their atoms are arranged three-dimensionally, as shown in the figure. Does this small change make a real difference? *d*-Methamphetamine is the extremely addictive illegal drug; *l*-methamphetamine is one of the ingredients in vapor rub, used for relief of nasal congestion, and it has none of the stimulant effects of *d*-methamphetamine! Methods of determining whether a stereoisomer is the *d*- or *l*- form are beyond the scope of this text. However, these stereoisomers illustrate an excellent example of the importance of the shape of drug molecules.

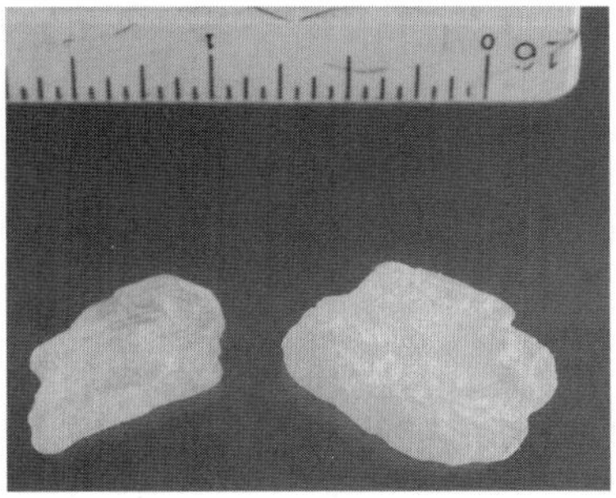

Crystals of *d*-methamphetamine. (U.S. Drug Enforcement Administration)

In the past, convicted methamphetamine dealers sometimes appealed their conviction on a simple legal point: The prosecutors had not proven that the dealers sold *d*-methamphetamine. Dealers would testify that they had actually been selling *l*-methamphetamine, a crime that carried a much less severe sentence. Of course, their proof was usually limited to a claim that they did not experience a sufficient high from the methamphetamine they were selling for it to be the illegal substance. This appeal was seldom successful, and the laws have been changed so that the same sentencing guidelines apply to both stereoisomers of methamphetamine. The

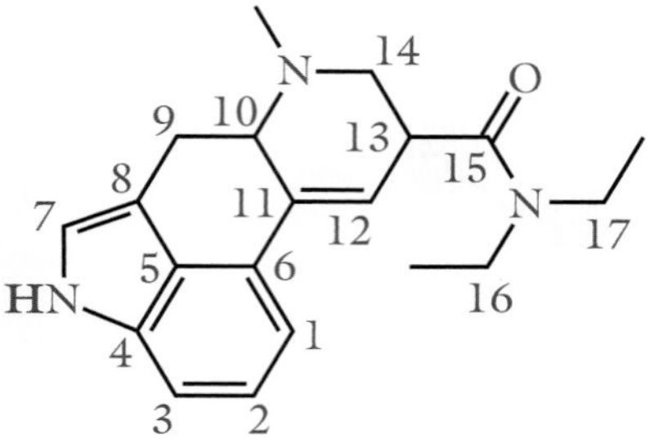

Figure 8.12 Structure of LSD with carbon atoms numbered for the purpose of determining the overall geometry of the drug molecule.

new laws effectively made vapor rubs for colds the only legal method of possessing *l*-methamphetamine.

Understanding the structure of LSD is necessary for comprehending the scientific evidence in the vigilante Jell-O case study. Figure 8.12 shows the chemical structure of LSD. The carbons are numbered as a reference for establishing the shape of the molecule. Recall that all carbon atoms must have four bonds, and any bonds not specifically drawn are assumed to be bonds to hydrogen atoms. The nitrogen atoms each contain a lone pair of electrons that must be taken into consideration when determining molecular geometry.

Consider the carbon atoms C_1 to C_6. The C_1 atom has three bonds drawn, so a fourth bond must be to hydrogen. The three electron regions around C_1 lead to a trigonal planar molecular geometry. The C_2 and C_3 atoms have the same bonding as C_1 and the same trigonal planar geometry. The C_4 atom has four bonds drawn with three electron regions, which indicates that the local geometry about C_4 is also trigonal planar. The C_5 and C_6 atoms have bonding arrangements identical to those of C_4 and are trigonal planar. Because all six carbon atoms are trigonal planar, this part of the molecule is flat, a characteristic of benzene ring structures.

We could continue analyzing the electron geometries surrounding all of the atoms in LSD in this fashion. However, visualizing such a large molecule is challenging for even the most experienced chemists, who usually employ computer-aided software to determine the structure of molecules. Some researchers are now using computer-generated images in virtual reality chambers that allow scientists to explore the molecular structures from the vantage point of standing inside the molecules! After we examine the LSD molecule in a bit more detail, we will view three-dimensional images to get a better sense of the whole structure.

Worked Example 13

Identify each carbon that is arranged in a tetrahedral geometry in the LSD molecule.

SOLUTION A tetrahedral geometry, indicated by a carbon atom with four electron regions, is restricted to those carbon atoms connected by single bonds. (Recall that hydrogen atoms are present but not shown on carbon atoms for which only three single bonds are drawn.) Carbon atoms C_9, C_{10}, C_{13}, and C_{14} have the tetrahedral geometry.

Practice 8.13

What is the local geometry of the bonds surrounding each nitrogen atom present in LSD?

ANSWER
Tetrahedral

The shape of the LSD molecule is an accumulation of the local geometries at each atom. Consider that the four ring structures making up the left side of the molecule (C_1 to C_{14}) have 10 atoms with the trigonal pla-

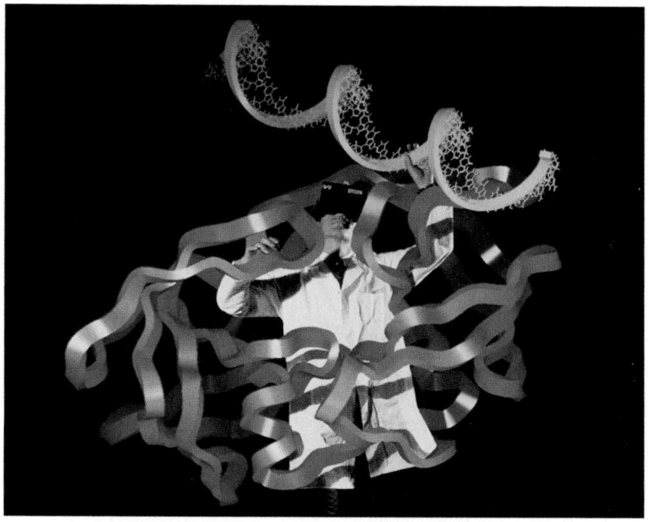

A research scientist is exploring the interaction between a single strand of DNA (blue) and HIV protease (purple) using a virtual reality chamber. (G. Thompkinson/ Photo Researchers, Inc.)

nar geometry. It is fair to assume that the rings make up a primarily flat portion of the molecule. The right portion of the LSD molecule consists of a branched region (C_{15} to C_{17}). The molecular geometry of the individual atoms in this region consists of a single trigonal planar atom (C_{15}), with the remainder (C_{16} and C_{17}) having tetrahedral geometry. Given that the atoms are no longer constrained to form a ring structure, it is reasonable to assume that the branches will not fall in a single geometric plane.

Figure 8.13 shows two different three-dimensional views of LSD. The first presents the molecule in a form that is similar to the structural formula given in Figure 8.12. The second view illustrates the dramatic changes in the geometry going from the ring structures to the branched portion of the molecule.

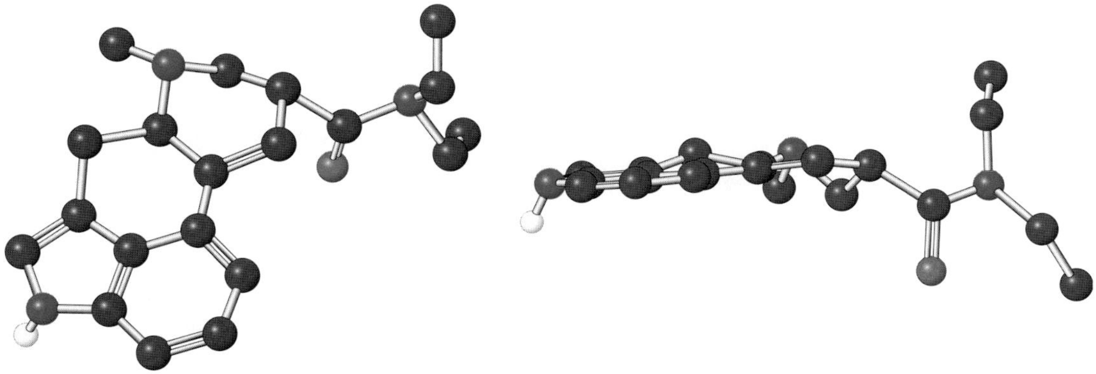

Figure 8.13 LSD models. Three-dimensional models of LSD rotated to highlight the molecular shape of the molecule.

8.8 Drug Receptors and Brain Chemistry

Learning Objective

Describe how drug molecules interact with neurons to produce a high.

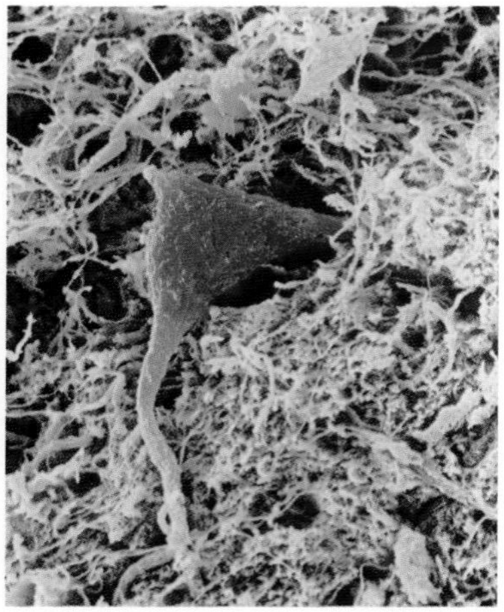

A scanning electron microscope image of a neuron (shown in red). (Quest/Photo Researchers, Inc.)

Many illegal drugs alter the biochemical processes that take place in the brain. One of the dangers of abusing these drugs is that they can cause permanent changes to the normal operation of the brain cells. The three-dimensional shape of drug molecules influences their interaction with the human brain. Before we explore this interaction, we must first discuss how the central nervous system functions.

An estimated 100 billion nerve cells called **neurons** make up the human nervous system. Neurons communicate with one another by sending an electrical signal, called an **action potential,** from one cell to the next. The end of a neuron contains small packets of chemicals called **neurotransmitters,** which are chemical compounds that can travel outside of the neuron, cross a small gap, and arrive at the next neuron. When a neurotransmitter binds with another neuron, it causes the action potential to continue on its path toward a specific location within the human brain. Figure 8.14 illustrates the process. The action potential triggers the release of neurotransmitters from the top neuron. The neurotransmitters travel by diffusion to receptor sites on the second neuron, completing transmission of the action potential. The neurotransmitter molecules then diffuse back to the original neuron through uptake channels and are recycled for use with the next action potential. This entire process occurs on the microsecond time scale.

Illegal drugs create their mind-altering effects by several different mechanisms. Cocaine functions by interfering with the uptake of neurotransmitters, as shown in the right-hand side of Figure 8.14. Cocaine mol-

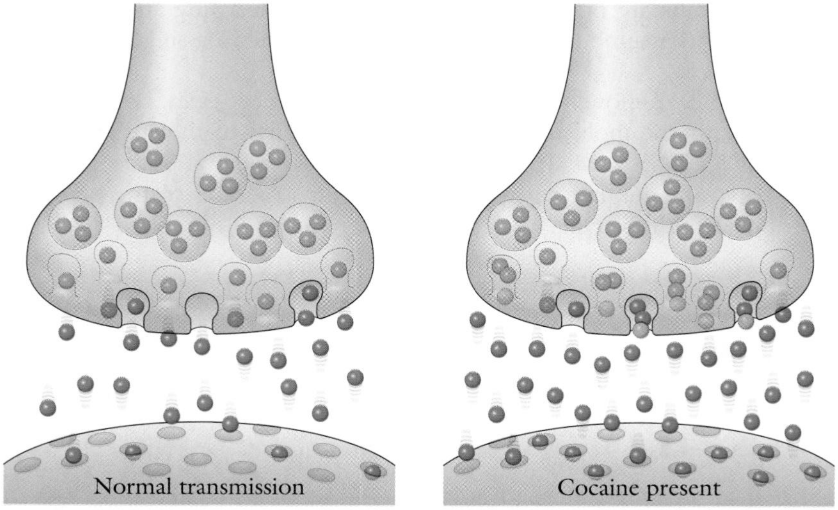

Figure 8.14 Neuron communication. Neurotransmitters (shown in blue) are chemicals released from one neuron that diffuse across a small gap to receptor sites on another neuron. This process allows communication between neurons. Once a neurotransmitter has sent the signal to the receptor, it diffuses back to the original neuron, enters through uptake channels, and is stored for future use. Cocaine molecules (shown in yellow) block the uptake channels, preventing released neurotransmitters from diffusing back into the neuron. This floods the gap with neurotransmitters, which then magnifies the signal being sent.

ecules block the uptake of the neurotransmitter, a process that floods the gap between the cells with an excess of neurotransmitter molecules and causes an amplification of the signal. Because the affected nerve cells are in the pleasure-sensing region of the brain, the person who has taken the cocaine experiences a high.

Although the cocaine molecules are eventually removed, the system does not return completely to normal. People who use illegal drugs build up a tolerance that requires higher doses to continue producing the same pleasurable effect. The user must take in more cocaine to block more uptake channels and produce a large enough neurotransmitter concentration to achieve the same level of sensation. These higher doses have a damaging effect on neurons. When people stop using drugs, the damage to the neurons is not repaired. Such damage can lead to depression because the mood-sensing neurons no longer function properly under the normal release of neurotransmitters.

How exactly does cocaine interfere with the uptake of the neurotransmitters? The uptake channel of the neuron is a portion of the cell membrane that functions as a one-way tunnel for the neurotransmitter. The tunnel walls are strands of protein molecules arranged so that neurotransmitters of a particular shape and size can pass through. Cocaine has the ability to lodge itself into this tunnel, blocking the neurotransmitters from reentering the cell. To block this tunnel, cocaine molecules must interact strongly with portions of the protein molecules through intermolecular forces. The distinct three-dimensional structure of cocaine makes this interaction possible. Figure 8.15 illustrates this interaction. The red coil ribbons represent protein molecules forming the neurotransmitter uptake channel. Key portions of the molecules are drawn in as molecular models. The cocaine molecule is highlighted to help illustrate its location.

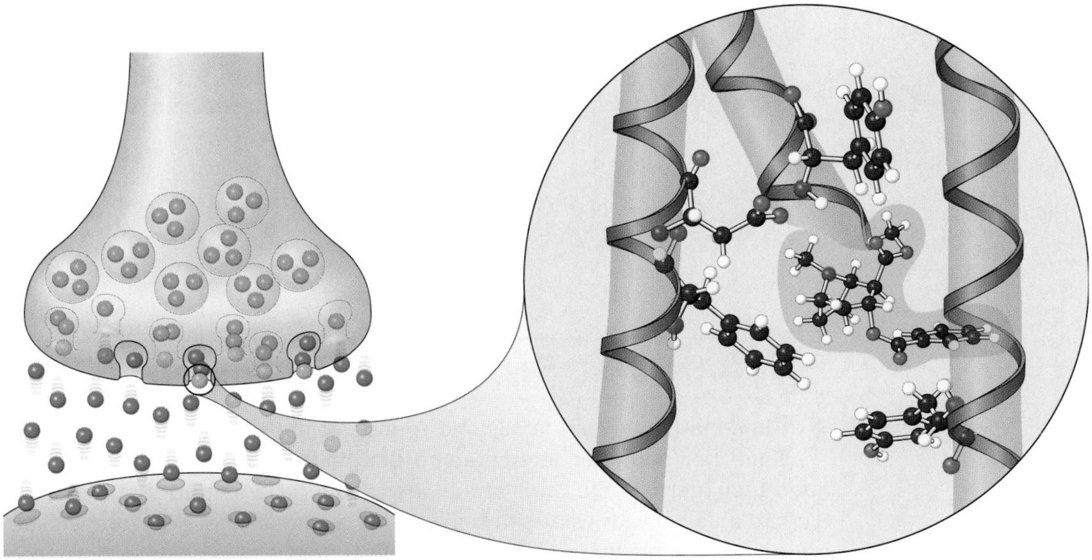

Figure 8.15 Cocaine blocking a neurotransmitter uptake transporter. The cocaine molecule is the correct size, shape, and polarity to wedge itself into the uptake channels of neurotransmitters. Once the uptake of neurotransmitters is blocked, the concentration of the neurotransmitters in the gap between neurons is greatly magnified, which results in the magnification of the signal being sent between neurons. The increased signal produces a pleasurable sensation, or "high." (Adapted from Dahl)

Evidence Analysis | Immunoassay Methods

The case against Christina Martin hinged on the identification of LSD from an immunoassay test. A lock-and-key analogy is often used to explain how an immunoassay test works. The shapes of a lock and key are designed in such a way that only a key with one particular shape can fit into the lock properly. Certain types of molecules have three-dimensional structures that enable them to behave as though they are locks or keys. The molecule that functions as a key is called an **antigen,** and the molecule that functions as the lock is called an **antibody.**

The human body produces antibodies to fight off infections. When a foreign molecule is introduced into the body, antibodies are created to attack the foreign molecules (antigens). The antibody is able to attack the antigen because it has a three-dimensional structure that is specific for binding to the antigen. A bound antigen is unable to disrupt normal cellular functions. Immunoassay techniques depend on similar antigen-antibody reactions that are specific for particular drugs.

There are several methods in which the basic principles of immunoassay are used to detect molecules. Figure 8.16 illustrates the method called **radioimmunoassay (RIA),** a technique in which radioactive isotopes play a part in the analysis. RIA was the specific method used in the Martin case. The first step in the detection of an antigen such as LSD is to mix it with a known amount of LSD that is labeled with a radioactive iodine atom, as shown in step 1 of Figure 8.16.

Recall from Section 3.5 that isotopes are forms of an element that have the same number of protons but different numbers of neutrons. Iodine-131 is an isotope that emits radiation that can be de-

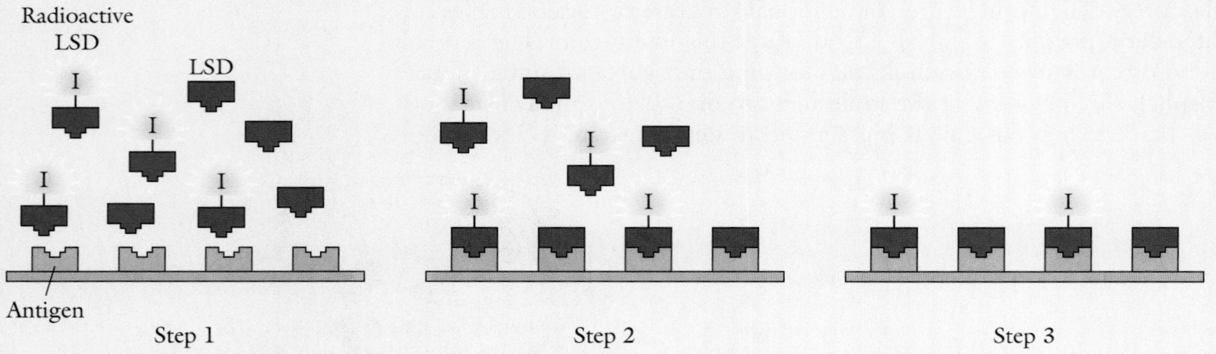

Figure 8.16 The radioimmunoassay procedure.

8.9 Case Study Finale: Vigilante Jell-O

Christina Martin told family and friends of her plan to kill Richard. She and her daughter attempted to purchase mescaline from neighborhood teenagers. In the original trial, her daughter admitted that she laced Richard's Jell-O with what she believed to be mescaline. Two forensic laboratories testified that a radioimmunoassay test showed the presence of LSD, but a judge ruled that Christina Martin's attorney had not questioned the evidence sufficiently and that she deserved a new trial. The state then decided to offer Christina Martin a plea bargain for time served. Although the procedures of the legal case may appear difficult to understand, let's tected by an electronic instrument. When radioactive iodine is incorporated into a molecule such as LSD, the drug molecule is radioactively labeled.

For an RIA analysis, the amount of radioactively labeled LSD is known exactly, but the amount of LSD present in the sample, if any, is unknown. In step 2 of the procedure, the labeled and unlabeled LSD are allowed to react with an antibody that has been physically immobilized on a plate. The labeling of the LSD molecules does not change their ability to bond to the antibody. Therefore, both the labeled and unlabeled LSD have an equal chance of binding to the antibody. If the two types of antigens are present in equal amounts, an equal number of each type of molecule will bond to the antibody. However, if the sample being analyzed contains no LSD, the plate will be filled with only radioactively labeled LSD. After the two antigens are allowed to react with the antibodies, the excess antigens are washed away. The final step is to determine the ratio of labeled to unlabeled antibodies. The greater the radioactivity of the immunoassay test when finished, the less LSD in a sample and vice versa.

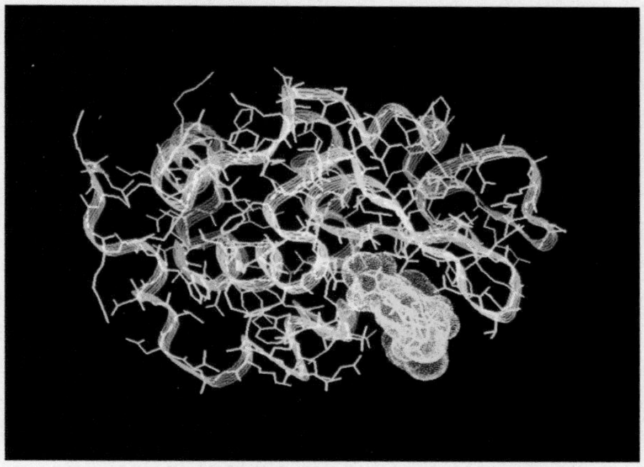

The large protein molecule forms the "lock" with an opening that is specific for the "key" molecule shown in yellow. The ribbon has been superimposed on the large protein molecule to better illustrate the overall shape of the protein molecule. (Science Photo Library/Photo Researchers, Inc.)

Worked Example 14

Is it necessary to separate out the various compounds of blood or urine using chromatography before using the RIA method?

SOLUTION Only those molecules that fit into the antibody will react. Therefore, if other molecules are present, they will simply be washed away between steps 2 and 3, making chromatography unnecessary.

Practice 8.14

Will a radioimmunoassay test that shows positive results for LSD have a larger or smaller signal from the radioactive antigen in step 3 than a radioimmunoassay test that comes back negative for LSD?

ANSWER
The greater the LSD concentration in the sample, the smaller the observed signal from the radioactive isotopes will be in the final step.

take a closer look at the evidence, which turns out to be the key to what happened to Christina Martin in the end.

First, the samples from Richard's body were taken one month after his burial. He had been embalmed, a process that involves removing most of the blood from the body and replacing it with methanal, better known by its common name *formaldehyde,* and other preservatives. Most of the LSD, if it had been present in his blood, would have been removed during the embalming. Furthermore, the possibility of interferences or reactions with the embalming fluid itself would present a risk in the analysis. The coffin had also filled with water at some point after his burial, further compromising the samples.

(Mike Dembeck/Reuters/Corbis)

Considering all of these factors, to say that the samples were not ideal is an understatement. Even so, the positive results obtained from the RIA test had to be considered by the jury. But how valid were the results?

A limitation of immunoassay analysis is that some molecules can trigger false positive results because the molecules share a sufficient portion of the geometry exhibited by the antigen that the test is designed to detect. When a different molecule produces a false positive result, the molecule is said to have *cross-reactivity* with the antigen. Because of the possibility of cross-reactivity, a positive RIA test for LSD means that LSD is *possibly* present in the sample, not that LSD *is* present in the sample. In samples from a body that is not embalmed or waterlogged, very few compounds exhibit cross-reactivity with LSD. Perhaps under those circumstances a greater reliance could have been placed on results of the RIA test. However, ergot fungus, a common component in soil, is cross-reactive with the LSD and could easily have been washed into the coffin. To establish the presence of LSD under such circumstances, a more specific test for LSD should have been used.

In the typical case involving illegal drugs, a screening test such as RIA is used to evaluate the possible components in a sample. After a positive result, the sample is then analyzed with a gas chromatograph mass spectrometer (GC-MS). The GC-MS separates each compound in the sample and determines the molecular weight of each compound. This method gives confirmatory proof that a specific compound such as LSD (and *not* ergot fungus) is present in the sample.

Why were results of GC-MS testing not used against Christina Martin? The forensic chemist in charge of analyzing the evidence conducted five separate experiments with the GC-MS and did not obtain a single positive result. The foolproof method for detecting the presence of drugs did not show any presence of LSD! The negative GC-MS results were minimized and improperly glossed over at the original trial. Perhaps if Christina's attorney had a better understanding of the science behind the RIA tests, he would have done a more effective job in the first trial. Without a confirmatory test for LSD, the prosecutor made a wise decision to offer Christina a plea bargain for time served.

CHAPTER SUMMARY

- The sharing of electrons to produce a covalent bond arises through an overlap of the orbitals in which the bonding electrons have the highest probability of being found.
- The Lewis theory of bonding states that valence electrons are the electrons responsible for the formation of both ionic and covalent bonds. Ionic bonding is achieved by the transfer of electrons between two atoms to produce positively and negatively charged ions. The newly formed ions achieve a completed octet of electrons, a stable state in which each ion is isoelectronic with one of the noble gases. Covalent bonds share electrons in order to achieve a stable octet of electrons.
- Lewis structures for complex molecules can be determined by verifying the total available valence electrons and then distributing them according to the rules for drawing Lewis structures.
- Writing a proper Lewis structure is a critical step for examining both the electron geometry and molecular geometry of molecules. Several common mistakes to avoid include failure to subtract or add electrons to the total valence electron number for cations and anions, placement of more than a single bond on hydrogen, and use of double or triple bonds when the central atom already has a completed octet.
- Resonance structures exist whenever multiple equivalent Lewis structures can represent a compound in which a double bond occurs between equivalent atoms within the structure.
- The VSEPR theory states that electrons around a central atom in bonds or lone pairs will repel one another and will assume positions as far apart from one another as possible. Electron geometry determines molecular geometry, the structural arrangement of the atoms in the molecule.
- Complex molecules have geometries that are dictated by the local geometry of each atom within the molecule. The complex three-dimensional shape of molecules influences the mechanisms by which drug molecules interact with the human body. The molecular geometry of drugs can be exploited for the detection of illegal drugs by employing immunoassay methods.
- Immunoassays are based on the principle that a drug molecule (antigen) of interest has a unique three-dimensional structure that allows it to bind specifically to another molecule called an antibody.
- In radioimmunoassay tests, the drug molecules from an evidence sample compete for the binding sites with a known amount of radioactively labeled drug molecules that have identical bonding properties. If labeled and unlabeled molecules are present in equal numbers, equal amounts will bond to the antibody molecules. The stronger the signal from the radioactive isotope, the lower the amount of the drug present in the evidence sample.

KEY TERMS

immunoassay, p. 238
valence bond theory, p. 239
Lewis theory, p. 240
isoelectronic, p. 240
octet rule, p. 241
single bond, p. 242
double bond, p. 243
triple bond, p. 243
lone pair, p. 244
nonbonding pair, p. 244
valence shell electron pair repulsion (VSEPR) theory, p. 247
electron geometry, p. 247
molecular geometry, p. 247
electron region, p. 247
linear geometry, p. 248
trigonal planar geometry, p. 248
tetrahedral geometry, p. 248
bent geometry, p. 249
trigonal pyramidal geometry, p. 249
nonpolar covalent bond, p. 252
polar covalent bond, p. 252
stereoisomers, p. 255
neurons, p. 258
action potential, p. 258
neurotransmitters, p. 258
antigen, p. 260
antibody, p. 260
radioimmunoassay (RIA), p. 260

CONTINUING THE INVESTIGATION Additional Readings, Resources, and References

The following Web site contains the opinion of a Massachusetts court in *Commonwealth v. Christina Martin*: http://caselaw.lp.findlaw.com/scripts/getcase.pl?court=ma&vol=sjcslip%5C/6789&invol=1

A large number of newspaper articles regarding Christina Martin's appeal can be found in the *Standard-Times* newspaper of New Bedford, MA, during the height of her case in 1998 and 1999.

The following journal articles give specific details relating to several applications provided in the chapter:

Dahl, Svein G. *The Journal of Pharmacology and Experimental Therapeutics*, vol. 307, no. 1, pp. 34–41.

Hirota, O. S., Suzuki, A., Ogawa, T., and Ohtsu, Y. "Application of Semi-Microcolumn Liquid Chromatography to Forensic Analysis," *Analysis Magazine*, 1998, vol. 26, no. 5.

REVIEW QUESTIONS AND PROBLEMS

Questions

1. Discuss the concept of electron sharing between atoms and how this takes place. To illustrate your explanation, sketch a diagram of two atoms sharing a pair of electrons.
2. Which electrons are responsible for the formation of ionic and covalent bonds?
3. What benefit do atoms receive by the exchange or sharing of electrons in reactions?
4. What determines whether an atom will form a cation or an anion?
5. What determines whether an atom will form a covalent bond or an ionic bond?
6. What determines the charge an ion will have?
7. Under what conditions will a compound form a double bond between two atoms?
8. List all of the possible examples of an electron region.
9. Explain why resonance structures do not actually represent accurate bonding within a molecule.
10. What is the basic principle of the valence shell electron pair repulsion (VSEPR) theory?
11. How does the number of electron regions around a central atom influence the electron geometry of a molecule?
12. What are the possible electron geometries? Draw a sketch of each.
13. Does the nature of the electron region (single bond, double bond, triple bond, lone pair) affect the electron geometry of the molecule? Explain.
14. What are the possible molecular geometries? Draw a sketch of each.
15. Does the nature of the electron regions affect molecular geometry? Explain.
16. Discuss how the local geometry of individual atoms contributes to the overall geometry of a larger molecule.
17. What are stereoisomers? Discuss their significance in drug chemistry.
18. Explain how neurotransmitters enable communication between neurons.
19. Describe the role of molecular shape of neurotransmitters in transmission of action potentials from one neuron to another.
20. Discuss the mechanism by which illegal drugs such as cocaine interfere with the process indicated in Question 19.
21. How can illegal drugs such as cocaine cause a permanent change in brain function?
22. Discuss how molecular shape is crucial to immunoassay methods of analysis.
23. Discuss why immunoassays are considered a screening method.

Problems

24. Draw the Lewis structure showing the bonding between the two chlorine atoms in Cl_2.

25. Draw the Lewis structure showing the bonding between the hydrogen and iodine atoms in HI.

26. Draw the Lewis structures for the following elements:
(a) K
(b) F
(c) S
(d) Pb

27. Draw the Lewis structures for the following elements:
(a) As
(b) Be
(c) N
(d) Al

28. Draw the Lewis structures for the following ionic compounds:
(a) $CaCl_2$
(b) $AlBr_3$
(c) SrS
(d) LiF

29. Draw the Lewis structures for the following ionic compounds:
(a) K_3P
(b) BaI_2
(c) CaO
(d) Na_3N

30. Draw the Lewis structures for the following polyatomic ions. Draw resonance structures if applicable.
(a) NO_3^-
(b) NH_4^+
(c) SO_4^{2-}
(d) ClO_3^-

31. Draw the Lewis structures for the following polyatomic ions. Draw resonance structures if applicable.
(a) CN^-
(b) PO_4^{3-}
(c) CO_3^{2-}
(d) OH^-

32. Draw the Lewis structures for the following covalent compounds. Draw resonance structures if applicable.
(a) CO_2
(b) CO
(c) H_2O
(d) CH_4

33. Draw the Lewis structures for the following covalent compounds. Draw resonance structures if applicable.
(a) PH_3
(b) I_2
(c) OF_2
(d) CS_2

34. Determine the electron geometry for the following ions from Problem 30:
(a) NO_3^-
(b) NH_4^+
(c) SO_4^{2-}
(d) ClO_3^-

35. Determine the electron geometry for the following ions from Problem 31:
(a) CN^-
(b) PO_4^{3-}
(c) CO_3^{2-}
(d) OH^-

36. Determine the electron geometry for the following compounds from Problem 32:
(a) CO_2
(b) CO
(c) H_2O
(d) CH_4

37. Determine the electron geometry for the following compounds from Problem 33:
(a) PH_3
(b) I_2
(c) OF_2
(d) CS_2

38. Determine the molecular geometry for the following ions from Problem 30:
(a) NO_3^-
(b) NH_4^+
(c) SO_4^{2-}
(d) ClO_3^-

39. Determine the molecular geometry for the following ions from Problem 31:
(a) CN^-
(b) PO_4^{3-}
(c) CO_3^{2-}
(d) OH^-

40. Determine the molecular geometry for the following compounds from Problem 32:
(a) CO_2
(b) CO
(c) H_2O
(d) CH_4

41. Determine the molecular geometry for the following compounds from Problem 33:
(a) PH_3
(b) I_2
(c) OF_2
(d) CS_2

42. Determine whether the following bonds are polar covalent or nonpolar covalent. If the bond is polar covalent, indicate the polarity of the bond by drawing an arrow indicating the direction of electron density.
(a) F—N
(b) H—I
(c) C—I
(d) H—P

43. Determine whether the following bonds are polar covalent or nonpolar covalent. If the bond is polar covalent, indicate the polarity of the bond by drawing an arrow indicating the direction of electron density.
(a) S—I
(b) C—N
(c) C—F
(d) O—O

44. Determine whether the following compounds from Problem 32 are polar or nonpolar:
(a) CO_2
(b) CO
(c) H_2O
(d) CH_4

45. Determine whether the following compounds from Problem 33 are polar or nonpolar:
(a) PH_3
(b) I_2
(c) OF_2
(d) CS_2

46. Place the following compounds in order of increasing polarity:
(a) Methylamine
(b) Ethylmethylamine
(c) Ethylmethylpropylamine
(d) Ammonia

47. Place the following compounds in order of increasing polarity:
(a) 1-butanol
(b) 1-hexanol
(c) 1-propanol
(d) Ethanol

48. Place the following compounds in order of increasing polarity:
(a) 1-butanone
(b) 1-butanoic acid
(c) Butane
(d) Butanal

49. Place the following compounds in order of increasing polarity:
(a) Propylamine
(b) Propanol
(c) Propanal
(d) Propane

Forensic Chemistry Problems

50. Identify the molecular geometry of each carbon atom (black) in the following structure for the illegal drug GHB.

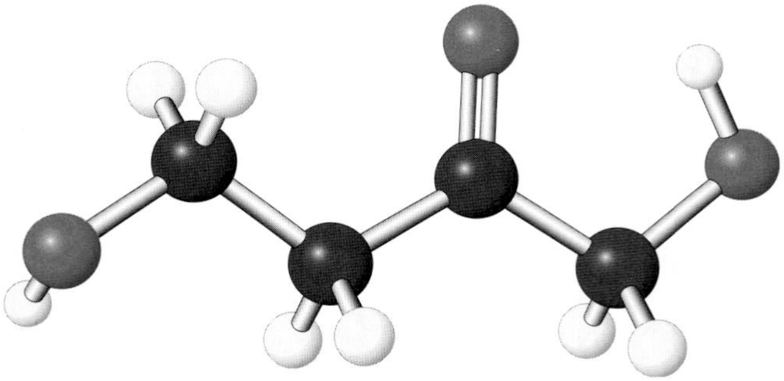

51. Identify the electron geometry of each oxygen atom (red) in the following structure of the illegal drug GBL. Each oxygen atom contains two lone pairs of electrons.

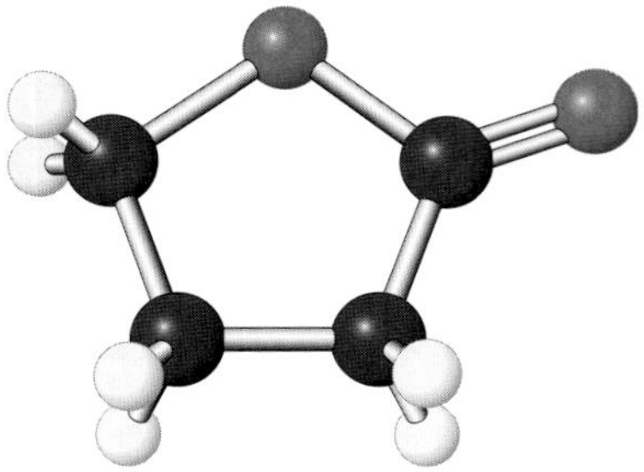

52. Identify the electron geometry and molecular geometry of each atom in the structure of the prescription medicine Ritalin, a drug that is often sold illegally. Each oxygen atom

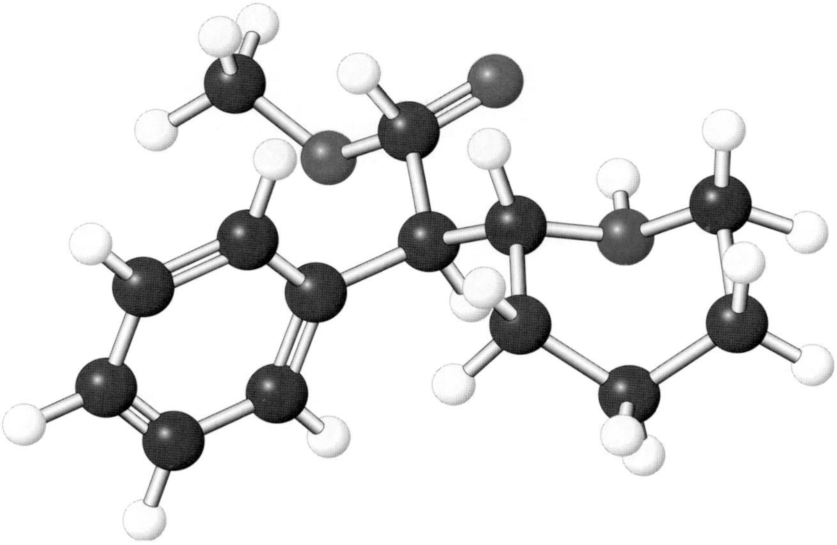

(red) has two lone pairs of electrons, and the nitrogen (blue) has one lone pair of electrons.

Case Study Problems

53. Discuss how a typical blood sample that tests positive for LSD by radioimmunoassay could be considered stronger evidence than the sample taken in the vigilante Jell-O case, even if the same analysis procedure were used.

54. In the trial of Christina Martin, the prosecution relied heavily on the RIA analysis of a questionable sample while downplaying the GC-MS results. The defense failed to challenge the evidence in the first trial, contrary to what any competent defense lawyer would have done. In your opinion, could a conviction have been secured if the evidence had been properly presented? Remember that standard for a jury to convict a defendant is that the person must be guilty "beyond a reasonable doubt."

CHAPTER 9

Arson Investigation

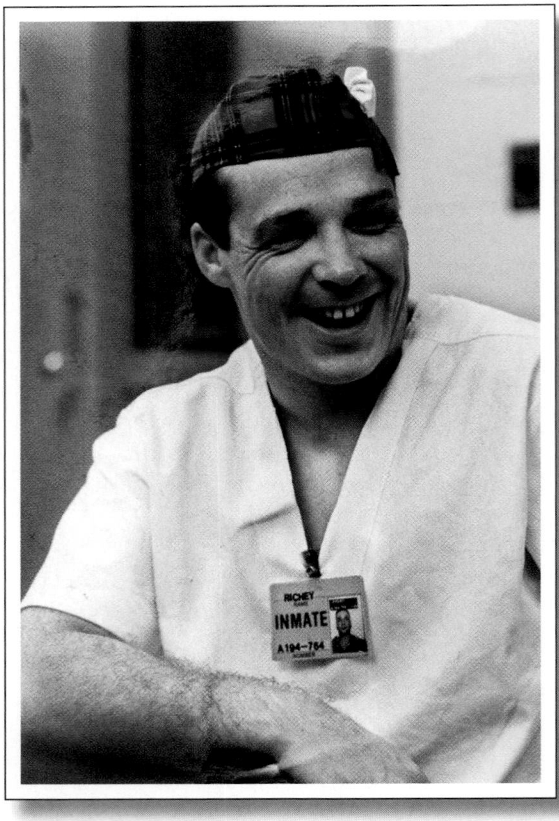

Kenny Richey, on death row since 1987, turned down a plea bargain offer of a 10-year sentence with parole likely after 6 years during his original trial. (AP/Wide World Photos)

CASE STUDY: "False and Unreliable"

Even though this new evidence may establish Mr. Richey's innocence, the Ohio and United States Constitution nonetheless allow him to be executed because the prosecution did not know that the scientific testimony offered at the trial was false and unreliable.

—From the case for the prosecution,
Kenny Richey arson trial (1986)

The quotation above is chilling and contrary to the spirit in which forensic scientists pursue their work. First and foremost, the idea that a person can be executed even though the state realizes serious problems have arisen with the evidence against him clearly violates the rights of the accused. Secondly, it is disturbing that false and unreliable evidence was allowed to be presented at trial without challenge. Yet in spite of these injustices, an appellate judge apparently agreed with the prosecutor's argument and denied the appeal of Kenny Richey, an inmate on death row in Ohio.

This case started on June 30, 1986, in Columbus Grove, Ohio. Kenny Richey was already living a tough life at the age of 18. He readily admits he was prone to binge drinking, fighting, and petty theft. On the evening in question, Kenny was so inebriated that one witness saw him pass out in the bushes for 10 minutes after talking with the occupants of a car. The prosecutor claimed that Kenny, after waking up in the bushes, proceeded to break into a nearby greenhouse storage shed and steal containers of gasoline and paint thinner. According to the prosecutor, Kenny then climbed on top of a storage shed below the deck of an apartment, hoisted himself onto the deck, and gained access to the living room of Hope Collins. Hope had been one of the car's occupants with whom Kenny had been seen speaking earlier. The prosecutor next claimed that Kenny spread the gasoline and paint thinner on the living room floor and deck and set the apartment on fire.

His motive? An ex-girlfriend lived in the apartment below, and he was hoping to burn the apartment building down to kill her and her new boyfriend. Everyone escaped the fire except a two-year-old girl in the apartment where the fire started.

Since Kenny was not only inebriated but also had a broken hand in a cast at the time of the alleged incident, it is unclear how he would have been able to carry out the acts attributed to him. Also, not a single trace of gasoline or paint thinner was found on his hands, clothes, or shoes. In the plea bargain, the prosecutor offered Kenny a 10-year sentence with parole likely after six years, but Kenny refused to plead guilty. The prosecutor apparently used the evidence to make a convincing case to a panel of three judges. The inexperienced public defender representing Kenny offered an ineffective and incompetent defense. The panel of judges found Kenny guilty, and he was sentenced to death.

It is very likely that one day you will sit on a jury and listen to experts present opposing opinions on the evidence against the accused. You will have to decide in a criminal case whether someone is guilty of a crime, or in a civil case whether a person should pay monetary compensation for wrongdoing. The key is to use the critical-thinking skills you are learning throughout your coursework to examine the data and, with the guidance of the expert witnesses, determine whether the evidence fits the conclusion presented in court.

Many details of the physical evidence from the Kenny Richey case will be presented in this chapter to illustrate the chemical principles involved in the investigation of arson cases. Then you can make your own decision about whether the physical evidence supports a guilty verdict.

But first, some background information on the chemistry of fire is necessary to understand the evidence in an arson trial . . .

9.1 The Chemistry of Fire

Learning Objective

Explain the concept of thermal equilibrium and the chemistry of fire.

The ability to fight fires effectively and to investigate the origin of a fire relies on having a firm understanding of the chemistry and physics behind flames. A **fire** is a self-sustaining chemical reaction that releases energy in the form of heat and light. A fire is classified as **arson** if it is deliberately set for the purpose of destroying property.

All fires require the same three ingredients: a source of fuel, a source of heat, and oxygen gas. Together these constitute the **fire triangle**. If any one of these three ingredients is missing, a fire will not start. If any one of these components is removed, the fire will be extinguished. This principle is used by firefighters to combat a fire.

For example, if a natural-gas main is ruptured and the gas bursts into flame, the fire is put out by cutting off the fuel supply. Flame-resistant materials in the home provide another method for preventing the spread of a fire. These materials are flame resistant because as they burn they produce a large amount of char, the black crusty material resulting from partial combustion. Char acts as a barrier to prevent oxygen from reaching the potential fuel (the flammable material) beneath it, and it also acts as an insulator that prevents heat from reaching the fuel.

Water is one of the most important tools used in firefighting to cool the area and remove the heat source. But how does water actually cool a fire? One of the properties of water is its ability to absorb a substantial amount of heat from hotter objects. **Heat** is a form of energy that is transferred from hot objects to cold objects that are in contact with one another.

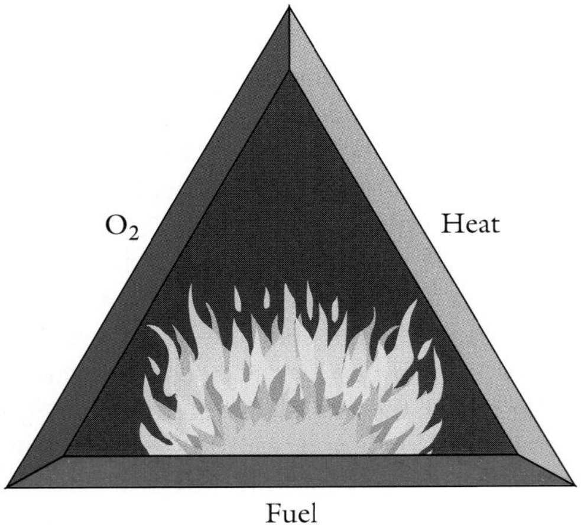

The fire triangle.

(Bill Stormont/Corbis)

Heat energy is measured in the SI units of **joules (J)** that are the parallel unit to the English system's **calorie (cal)**. The calorie is defined as the amount of energy required to raise the temperature of 1 gram of water from 14.5°C to 15.5°C. To convert from calorie to joule or vice versa, use the conversion factor 1 cal = 4.184 J. The nutritional unit we use to express the caloric content of foods is actually a kilocalorie, or Calorie. The conversion factors are 1 kilocalorie = 1 Calorie = 1000 calories.

When an object absorbs energy, the kinetic energy (the energy of motion) of its particles increases; this results in an increase in the temperature of the object. **Temperature** is a measurement of the average kinetic energy of the particles in a system, not a measurement of how much heat is in a system. The difference between a piece of wood at 25°C and 250°C is that the internal motion of the molecules is much greater at 250°C.

When two objects that are at different temperatures come into contact, heat is transferred from the warmer object to the colder object, thereby causing the warmer object to cool. The energy added to the colder object increases the kinetic energy of its particles and warms the colder object. **Thermal equilibrium** is achieved when the temperatures of the two objects become equal and remain constant. At this point, the kinetic energy of the particles in both objects is identical. Figure 9.1 illustrates these processes.

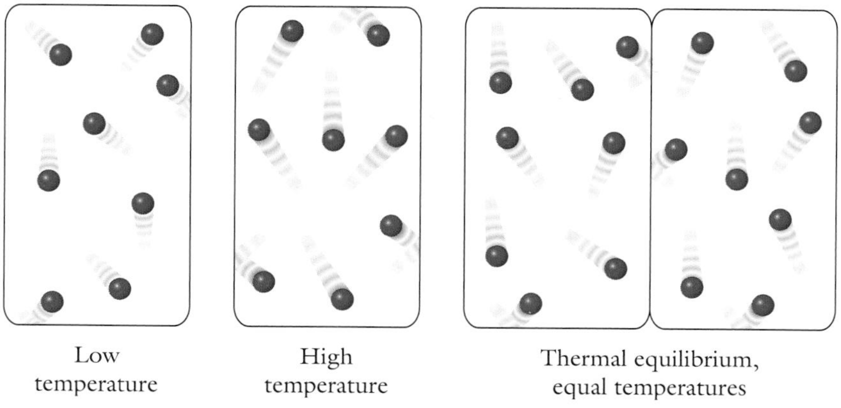

Figure 9.1 Temperature and thermal equilibrium. The temperature of an object is a measure of the kinetic energy of the atoms, molecules, or ions that make up the object. When two objects at different temperatures come in contact, energy in the form of heat is transferred until thermal equilibrium is reached.

When water from a fire hose comes into contact with a burning structure, the water absorbs heat from the structure, thus cooling it and removing the heat component of the fire triangle. If the water brings about sufficient cooling, the fire is extinguished.

Worked Example 1

What is incorrect about the following statement? *A thermometer measures how much heat is in an object.*

SOLUTION A thermometer measures the kinetic energy of the particles in a system. Heat is a form of energy that is transferred from warm objects to cold objects.

Practice 9.1

Using the concept of thermal equilibrium, explain what is incorrect about the following statement. *The tile floor is colder than the carpeted area of the floor.*

ANSWER
Two objects in thermal contact would have reached thermal equilibrium. Therefore, the temperatures of the tile floor and the carpet are identical. However, the tile floor may *seem* colder to the touch because it draws more heat away from your body than the carpet does. But the temperature of the tile and carpet before you touch them is the same.

9.2 Combustion Reactions

Learning Objective

Balance the chemical equations for combustion reactions.

A hidden danger lurks in the fire triangle. When fire heats a poorly ventilated room, the air in the room can fill with flammable compounds created from the incomplete combustion of fuels. The poor ventilation in the room can cause oxygen to be consumed quickly while the heated, flammable vapors build up. Only two of three ingredients for fire (heat and fuel) are present once the oxygen is consumed—until a door is opened or a window is broken and oxygen gas rushes in. Under these circumstances, it is not an ordinary fire that ignites. The presence of such an excess of heated fuel ready to react causes an explosion known as a **backdraft**, or smoke explosion. The safety of firefighters depends on their having a full understanding of the chemical and physical processes that may lead to dangerous conditions during a fire.

A backdraft. (Universal/The Kobal Collection)

In Section 4.9, a *combustion reaction* was first defined as a reaction between an organic compound and oxygen gas to produce water and carbon dioxide. These reactions were presented as a subclass of reduction-oxidation (redox) reactions. However, combustion reactions are not actually confined to organic compounds. We must broaden the definition because combustion reactions can occur with certain metals and nonmetals. Furthermore, even in cases where organic compounds are involved (as in a burning wooden structure), the production of water and carbon dioxide as the only products occurs only under ideal conditions.

(a) (Corbis)

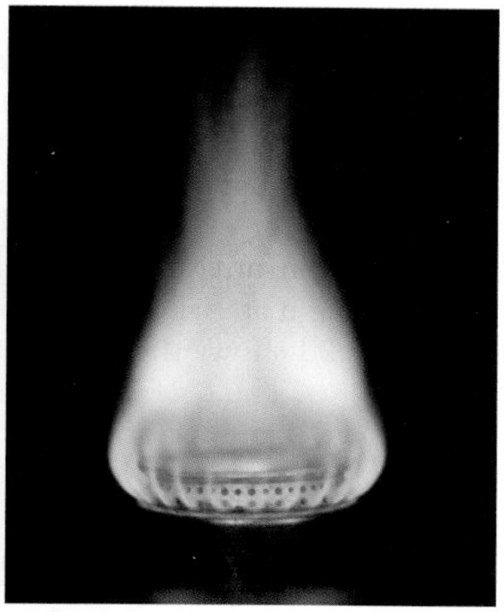

(b) (Corbis)

(c) (Stone/Getty)

(d) (Andrea Comas/Reuters/Corbis)

Figure 9.2 (a) The efficient combustion of natural gas to form carbon dioxide and water. (b) The inefficient combustion of natural gas. (c) A typical forest fire has easy access to atmospheric oxygen. (d) Building fires typically have restricted access to oxygen.

In most structural fires, many other compounds are produced because of incomplete combustion; that is, there is insufficient oxygen present for the fuel to form only carbon dioxide and water. The oxygen gas serves as a limiting reactant. The incomplete combustion of fuels produces compounds such as carbon monoxide and solid carbon particles (soot). The color of the flame and the presence or absence of smoke are actually indicators of how efficient the combustion reaction is. Consider the different fires shown in Figure 9.2.

In the burning of natural gas in a stove, as shown in Figure 9.2a, the flame that is produced is an intense blue with no evidence of smoke. This fire is efficiently converting the fuel into carbon dioxide and water vapor.

In Figure 9.2b, there is insufficient oxygen (or excess fuel) for the reaction, resulting in a flame that has a noticeable orange portion due to the production of small carbon particles. The carbon particles glow when heated by the flame. The smoke in Figure 9.2c, produced in a forest fire, is predominately white in color. This color indicates the condensation of water vapor on small particles of dust and soot, a process similar to the formation of clouds. The white color is indicative of a fire in which the burning material is undergoing nearly complete combustion. Finally, Figure 9.2d shows the typical black smoke and flame associated with building fires. The black color is due to heavy production of solid carbon soot particles from incomplete combustion reactions as the amount of oxygen available is restricted.

Worked Example 2

As illustrated in Figure 9.2, it is common for structural fires to burn with black smoke and for grass or forest fires to have white smoke. Explain why.

SOLUTION The white smoke indicates a more complete combustion reaction, which occurs when sufficient oxygen gas is present. Because outdoor fires have no barriers between the atmospheric oxygen and the fuel, a more complete combustion reaction can occur. The black smoke from a building is a sign of incomplete combustion due to insufficient oxygen gas reaching the fire because the building walls and ceilings restrict air flow.

Practice 9.2

Firefighters often ventilate a structural fire by cutting holes in ceilings. What processes occurring within a fire would be altered by ventilating a room?

ANSWER

The conditions for a backdraft require poor ventilation for the buildup of heated, combustible gases within a room. By ventilating the room, firefighters minimize the likelihood of a backdraft. (It also improves visibility and lowers the temperature within the building.)

Balancing Combustion Reactions

The balancing of chemical equations corresponding to the complete combustion of organic materials can sometimes seem to be a complicated process. The same method for balancing equations that was discussed in Section 4.6 is also used here; however, it is important for the elements to be balanced in the following order: carbon, hydrogen, oxygen. Balancing equations is illustrated below, using as an example the combustion of propane (C_3H_8) with oxygen gas to form water and carbon dioxide.

Step 1. Write the unbalanced reaction and determine the total number of carbon, hydrogen, and oxygen atoms on each side of the reaction:

$$C_3H_8(g) + O_2(g) \rightarrow CO_2(g) + H_2O(g)$$

C: 3	C: 1
H: 8	H: 2
O: 2	O: 3

Step 2. Balance the carbon atoms:

$$C_3H_8(g) + O_2(g) \rightarrow \mathbf{3}CO_2(g) + H_2O(g)$$

C: 3	C: ~~1~~ 3
H: 8	H: 2
O: 2	O: ~~3~~ 7

Step 3. Balance the hydrogen atoms:

$$C_3H_8(g) + O_2(g) \rightarrow 3CO_2(g) + \mathbf{4}H_2O(g)$$

C: 3	C: ~~1~~ 3
H: 8	H: ~~2~~ 8
O: 2	O: ~~3~~ ~~7~~ 10

Step 4. Balance the oxygen atoms:

$$C_3H_8(g) + \mathbf{5}O_2(g) \rightarrow 3CO_2(g) + 4H_2O(g)$$

C: 3	C: ~~1~~ 3
H: 8	H: ~~2~~ 8
O: ~~2~~ 10	O: ~~3~~ ~~7~~ 10

Worked Example 3

Balance the equation for the combustion reaction of 1-butanol (C_4H_9OH) with oxygen gas.

SOLUTION

Step 1. $C_4H_9OH(g) + O_2(g) \rightarrow CO_2(g) + H_2O(g)$

C: 4	C: 1
H: 10	H: 2
O: 3	O: 3

Step 2. $C_4H_9OH(g) + O_2(g) \rightarrow \mathbf{4}CO_2(g) + H_2O(g)$

C: 4	C: ~~1~~ 4
H: 10	H: 2
O: 3	O: ~~3~~ 9

Step 3. $C_4H_9OH(g) + O_2(g) \rightarrow 4CO_2(g) + \mathbf{5}H_2O(g)$

C: 4	C: ~~1~~ 4
H: 10	H: ~~2~~ 10
O: 3	O: ~~3~~ ~~9~~ 13

Step 4. $C_4H_9OH(g) + \mathbf{6}O_2(g) \rightarrow 4CO_2(g) + 5H_2O(g)$

C: 4	C: ~~1~~ 4
H: 10	H: ~~2~~ 10
O: ~~3~~ 13	O: ~~3~~ ~~9~~ 13

Practice 9.3

Balance the equation for the combustion reaction of 1-hexene (C_6H_{12}) with oxygen gas.

ANSWER

$$C_6H_{12}(g) + 9O_2(g) \rightarrow 6CO_2(g) + 6H_2O(g)$$

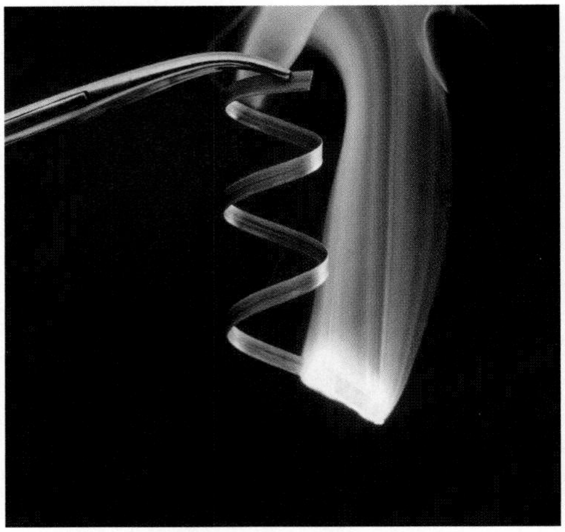

(Both: Richard Megna/Fundamental Photos)

Magnesium (left) and sulfur (right) both undergo combustion reactions, which produce heat and light. Magnesium is often used in mixtures for fireworks displays because of the intense bright light it produces during combustion. Sulfur is one of the three ingredients in gunpowder.

Several metals and nonmetals undergo combustion reactions that produce heat and light when ignited in the presence of oxygen gas. The most common examples are the combustion of magnesium ribbon to form magnesium oxide, and the combustion of sulfur to form sulfur dioxide. Other metals such as copper, when heated, react with oxygen gas to form a metal oxide. However, such a reaction is not considered combustion because heat and light are not produced.

9.3 Redox Reactions

Learning Objective

Assign oxidation numbers and identify reduction and oxidation processes.

The combustion of organic compounds, metals, and nonmetals, and the formation of metal oxides from their elements are all examples of redox reactions involving the gain or loss of electrons. In Section 4.9, *oxidation* was defined as the loss of electrons and *reduction* was defined as the gain of electrons. In this section, we will see how to determine which atoms are oxidized and which are reduced when electrons are not explicitly shown in the reaction.

The concept of **oxidation number** offers a tool for interpreting redox reactions. The oxidation number is merely a value assigned to an atom by the basic rules listed on the facing page and should not be interpreted as a true charge on a particular atom. Some exceptions to the rules exist, but for simplicity's sake, they have been omitted here.

Rule 1. All pure elements have an oxidation number of 0. Examples: Oxygen gas (O_2) and iron (Fe) have an oxidation number of 0.

Rule 2. Monatomic ions have an oxidation number equal to their ionic charge. Examples: Mg^{2+} has an oxidation number of +2, and S^{2-} has an oxidation number of −2.

Rule 3. Hydrogen always has an oxidation number of +1 in a compound.

Rule 4. Oxygen always has an oxidation number of −2 in a compound.

Rule 5. The sum of the oxidation numbers must be 0 for neutral compounds or equal to the ionic charge for polyatomic ions. *First example:* In FeO, the oxygen atom has an oxidation number of −2 according to Rule 4, and there is no rule specific for the iron atom in a compound. Because FeO is a neutral compound, the iron atom must have an oxidation number equal to +2. *Second example:* In NO_3^-, the oxygen atoms each have an oxidation number of −2 by Rule 4. Because there are 3 oxygen atoms, they have a total oxidation number of −6. There is no specific rule for nitrogen atoms in a compound, but the nitrate ion must have an overall −1 charge, so the nitrogen atom must have an oxidation number equal to +5.

Once the oxidation numbers have been assigned, compare the oxidation number of each element from the reactant side with the same element on the product side. You will notice that one of the elements from the compound being *reduced* has gained electrons, and one of the elements from the compound being *oxidized* has lost electrons. For example, the reaction for the combustion of magnesium is shown below.

Lost 2 e^-

$$\overset{0}{2\text{Mg}}(s) + \overset{0}{\text{O}_2}(g) \rightarrow 2\overset{+2}{\text{Mg}}\overset{-2}{\text{O}}$$

Gained 2 e^-

The oxidation numbers of the reactants, Mg and O_2, are both 0 according to Rule 1. The oxidation number of O is −2 in MgO, as indicated by Rule 4. The oxidation number of Mg in MgO is +2 according to Rule 5. Therefore, oxygen gas undergoes a change in oxidation number from 0 to −2 by gaining two electrons and thus must have been reduced. Conversely, magnesium undergoes a change in oxidation number from 0 to +2 by losing two electrons, which fits the definition of oxidation.

Worked Example 4

Assign oxidation numbers to each atom for the following reactions:

(a) $2\text{Cu}(s) + \text{O}_2(g) \rightarrow 2\text{CuO}(s)$

(b) $\text{CH}_4(g) + \text{O}_2(g) \rightarrow \text{CO}_2(g) + 2\text{H}_2\text{O}(g)$

SOLUTION *Remember that losing an electron makes the oxidation number more positive because of the negative charge on the electron!*

(a) $2Cu(s) + O_2(g) \rightarrow 2CuO(s)$

Rule 1: All pure elements have an oxidation number of 0.

$Cu = 0$, $O_2 = 0$

Rule 4: Oxygen in compounds has an oxidation number of −2.

O in $CuO = -2$

Rule 5: The sum of the oxidation numbers must be 0 for neutral compounds.

Cu in $CuO = +2$

(b) $CH_4(g) + O_2(g) \rightarrow CO_2(g) + 2H_2O(g)$

Rule 1: All pure elements have an oxidation number of 0.

$O_2 = 0$

Rule 3: Hydrogen in compounds has an oxidation number of +1.

H in $CH_4 = +1$

H in $H_2O = +1$

Rule 4: Oxygen in compounds has an oxidation number of −2.

O in $CO_2 = -2$

O in $H_2O = -2$

Rule 5: The sum of the oxidation numbers must be 0 for neutral compounds.

C in $CH_4 = -4$

C in $CO_2 = +4$

Practice 9.4

Determine which elements are oxidized and which elements are reduced in the redox reactions from Worked Example 4.

ANSWER

(a) Cu went from −2 to +2, a loss of 4 e^-. Cu is oxidized.
O went from 0 to −2, a gain of 2 e^- per O atom. O is reduced.

(b) C went from −4 to +4, a loss of 8 e^-. C is oxidized.
O went from 0 to −2, a gain of 2 e^- per O atom. O is reduced.

9.4 Thermochemistry of Fire

Learning Objective

Describe how heat affects both physical and chemical processes.

The fire triangle indicates that a fire needs fuel, oxygen, and heat. But if a fire produces heat, why does it require heat to exist? Heat is needed to initiate the combustion reaction for two reasons. First, liquid or solid fuel must be converted into vapors before the combustion reaction can occur. Second, energy is required to start the reaction of the fuel vapors with oxygen gas. Let's look at each of these requirements.

Energy Needed to Produce Gaseous Reactants

For a liquid fuel, there must be sufficient heat to vaporize the liquid to the gaseous state. Solid fuels must undergo a process called **pyrolysis**, a decomposition reaction that produces small, gaseous compounds capable of undergoing the combustion reaction. Many different chemicals are produced as a result of pyrolysis, but some of the more common products are carbon monoxide, oxygen gas, and small alkanes.

A chemical or physical process that must absorb heat to occur is an **endothermic** process. Endothermic processes cease if the source of the heat is removed. For example, when the copper wiring in a building is heated by a fire, the copper atoms increase in energy and, if sufficient heat is added, the solid wire begins to melt. The copper will continue to melt as long as heat is being added to the copper by the fire. When the firefighters extinguish the fire, the source of heat is removed and the copper will stop melting and solidify.

A related example would be the melting of the plastic housing of a smoke detector, which played a role in analyzing the fire scene in the case of Kenny Richey. There are two critical endothermic processes involved in supplying the fuel to a fire: the vaporization of a liquid fuel and the pyrolysis of solid fuels. Both processes require that heat be transferred into the fuel from the surroundings in order for the fire to continue.

Any chemical reaction or physical change that produces or releases heat into the surroundings is called an **exothermic** process. A fire is an excellent example of a process that releases heat to the surroundings. Chemical reactions are exothermic if the final energy of the products is lower than the energy of the reactants. The energy that is released in a chemical reaction was stored in the chemical bonds between atoms. The products of an exothermic chemical reaction have less energy stored in the chemical bonds. The excess energy is released into the surroundings.

For the combustion of a fuel to become a self-sustaining chain reaction, the heat produced in the process must be greater than the heat needed to continually convert the fuel to the gas phase. When water is applied to a fire, it absorbs the heat being produced from the exothermic combustion reaction, which in turn makes less heat available for the endothermic vaporization of the liquid fuel or for pyrolysis of the solid fuel. The fire can thus be extinguished.

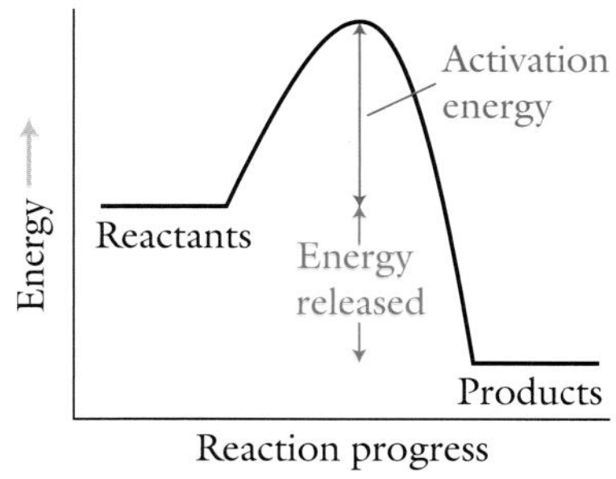

Figure 9.3 Energy diagram for exothermic reaction. The reactant compounds in an exothermic process have a higher energy than the product compounds. The activation energy (green arrow) is the amount of energy the reactants must absorb in order for the reaction to occur. This energy is used to break bonds within the reactant compounds. As new bonds form, energy is released and the amount of energy initially absorbed is recovered. As the reaction continues to release energy, a net release of energy occurs (red arrow).

Activation Energy

What is it about a chemical reaction that makes it endothermic rather than exothermic? In any chemical reaction, the bonds between atoms in the reactants must be broken before the atoms can be rearranged to form new bonds in the products. The breaking of bonds requires that energy be put into the system, whereas the formation of new bonds releases energy. If the energy released by making new bonds is greater than the energy used to break the reactant bonds, the reaction is exothermic. If the energy released by bond formation is less than the energy used to break the reactant bonds, the process is endothermic. The energy changes in an exothermic chemical reaction are graphically depicted in Figure 9.3.

This reaction is exothermic because the final energy of the products is less than the energy of the reactants. Note, however, that even in an exothermic process, energy must first be absorbed by the reactants to initiate the reaction. The energy needed to initiate a reaction is called the **activation energy** and represents a barrier that the reactants must

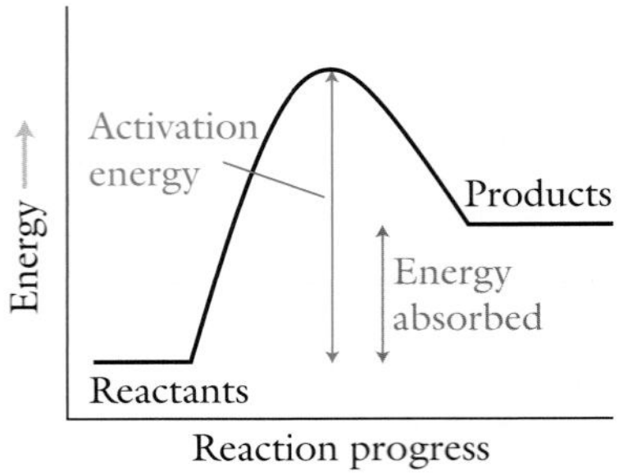

Figure 9.4 Energy diagram for endothermic reaction. The reactant compounds in an endothermic process have a lower energy than the product compounds. The activation energy (green arrow) is the amount of energy required for the reactant molecules to successfully form the product. The amount of energy absorbed (blue arrow) is the amount of energy added into the reactants.

overcome before they can react. The activation energy is the energy needed to break the bonds holding the reactants together.

If methane gas and oxygen gas are mixed together, a combustion reaction does not immediately occur because the molecules do not have sufficient energy to cross the barrier represented by the activation energy. In order for the methane and oxygen to react, energy must be added to the system, usually by an ignition source. The added energy provided gives the reactants sufficient energy to cross the activation energy barrier, allowing the reaction to proceed. But note that just because reactant molecules have sufficient energy to cross the activation energy barrier does not imply that *all* molecules will form product, a topic explored in more depth in Chapter 11.

The energy changes in an endothermic chemical reaction are shown in Figure 9.4. This reaction is endothermic because the products have a greater total energy than the reactant compounds. The activation energy (green arrow) again represents the energy needed for the reaction to occur. However, as the reaction proceeds and new bonds are formed, the energy needed to cross the activation energy barrier is *not* released. Rather, part of the energy added into the system is stored within the bonds of the product compounds. The blue arrow in Figure 9.4 represents the amount of energy stored in the product molecules.

A fire can easily spread from its point of origin to involve the entire contents of a room—a process called **flashover**. In a flashover, the heat from a fire in a small area rises to the ceiling and then spreads out across the room. As the fire progresses, the heated layer of smoke and gases just below the ceiling increases its temperature and thickness. Because the heat is restricted by the ceiling, it radiates downward, heating the remaining combustible furniture that has not yet been ignited.

The radiative heating of the furniture continues until the furniture reaches its **autoignition** temperature, the temperature at which material will ignite without a spark or flame directly contacting it. This can occur because all components of the fire triangle (fuel, oxygen, and heat) are present. The heat radiating down on the furniture is providing the activation energy needed for the combustion reaction. Once the furniture has sufficient energy to overcome the activation barrier, the combustion reaction proceeds as long as sufficient oxygen is present in the room. This entire process can occur in less than two minutes from the start of the fire!

Worked Example 5

How does ventilating a room affect the possibility of a flashover?

SOLUTION Ventilation of a room removes the heated layer of smoke at the ceiling, thereby eliminating the energy source needed for combustible material to reach its autoignition temperature.

Practice 9.5

What happens to the excess energy produced by an exothermic reaction?

ANSWER
The heat is absorbed by the surroundings (air, beaker, and so forth), which increase in temperature due to the increased kinetic energy of the particles absorbing the energy.

9.5 Heat Capacity and Phase Changes

Learning Objective

Describe what happens when heat is added to a substance.

Reconstructing a fire scene to determine the cause of the fire relies on the investigator's ability to interpret physical evidence left behind and to combine this knowledge with an understanding of the physical and chemical properties of substances in a fire. Once the cause of the fire has been established, the investigator can determine whether the fire was started accidentally or intentionally.

One type of physical evidence taken into account is the presence or absence of melted materials. For example, the glass of light bulbs softens in the heat of a fire. Because light bulbs greater than 25 watts are pressurized with inert gases, the bulb bulges outward in the direction of an intense heat source. Plastics and metals used throughout buildings can be indicators of fire temperatures, depending on whether the materials have melted.

Materials can also produce demarcation lines between temperature regions. For example, copper, commonly used for electrical wiring and plumbing, can indicate where along a wall the temperature exceeded 1082°C, the melting point of the metal. Copper that is heated in a fire tends to melt along its surface and flow to lowest points, forming globules and thinned areas.

Not all objects have the same ability to absorb heat. The **specific heat capacity** (C_p), also known as *specific heat,* is a measure of the heat needed to raise the temperature of a 1-gram sample of the substance by 1°C. If a substance has a high specific heat, it takes a large amount of heat to increase the temperature of 1 gram of the substance by 1°C. The same amount of heat could increase the temperature of 1 gram of a different substance by several degrees, depending on the substance.

Water, for example, has a specific heat capacity of 4.184 J/g · °C, indicating that it takes 4.184 joules of heat to raise the temperature of 1 gram of water by 1°C. Water actually has an unusually high specific heat capacity when compared with all other materials. For example, wood construction material has a specific heat capacity of 1.70 J/g · °C. In fire fighting, it takes more energy gram for gram to warm water up than to cool wood down—a tremendous benefit! Table 9.1 provides a representative list of the specific heat capacities associated with materials of interest in fire scene reconstruction.

Table 9.1 Specific Heat Capacities

Material	Specific Heat (J/g · °C)
Aluminum	0.90
Copper	0.38
Gasoline	2.01
PVC (polyvinyl chloride)	1.05
Wood	1.70
Concrete	0.80
Particle board	1.30
Polyethylene	1.90
Silica glass	0.75
Gypsum wallboard	1.05

Worked Example 6

What happens to the individual particles of a cold object as it is being heated?

SOLUTION The particles that make up the object (molecules, atoms, or ions) are absorbing heat, which increases their kinetic energy.

Practice 9.6

The specific heat capacity of water is of 4.184 J/g · °C, which is more than twice the value for gasoline at 2.01 J/g · °C. Explain the role of intermolecular forces that cause the specific heat capacity of water to be almost twice that of gasoline.

ANSWER
The lower specific heat capacity of gasoline compared with water indicates that it is easier to increase the kinetic energy of gasoline particles, which results in an increase in temperature. It is easier to increase the kinetic energy of gasoline particles because the nonpolar organic molecules that make up gasoline have weaker intermolecular forces than the polar water molecules.

The temperature of a typical building fire can easily reach 1300°C or more, well above the boiling point of water (100°C) being used by firefighters to extinguish the fire. Figure 9.5 shows a graph of the temperature of water as heat is added to a system.

Notice that as water absorbs heat, the temperature increases linearly until it reaches a plateau at the boiling point of 100°C. The temperature of the water holds constant at the boiling point until all of the water has been converted to steam (water vapor). Then the temperature of the steam rises above 100°C as more heat is absorbed. The amount of energy required to convert a liquid to a gas is called the **heat of vaporization**. The energy is absorbed by the molecules in the liquid phase, increasing their kinetic energy until the particles can overcome the attractive inter-

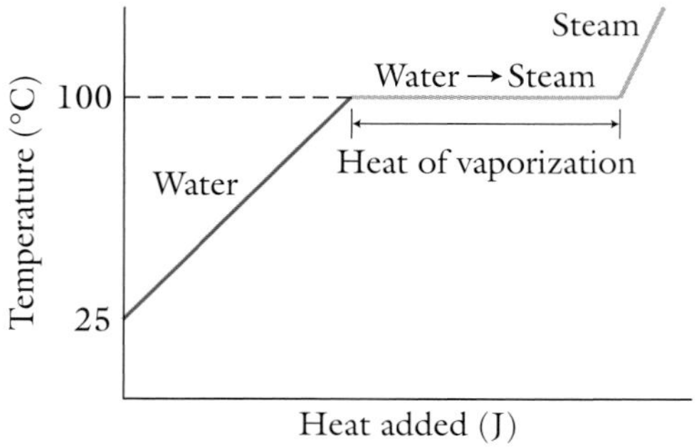

Figure 9.5 Temperature of water versus heat added. Heat is added to water at room temperature, causing the temperature of the water to increase. When the water reaches its boiling-point temperature, the heat energy converts the liquid water into steam. The amount of heat required for this conversion is called the *heat of vaporization*. When all of the water has been converted to steam, the temperature of the steam increases.

Table 9.2 Heat of Vaporization for Ignitable Liquids

Material	Heat of Vaporization (J/g)
Gasoline	349
Diesel fuel	233
Ethanol	921
Turpentine	293
Kerosene	250
Water	2258

molecular forces holding them in the liquid state. For water, the heat of vaporization is 2258 J/g. Table 9.2 is a representative list of the heats of vaporization of ignitable liquids that frequently play a role in an arson investigation.

Worked Example 7

Which process cools a fire more effectively—heating water to 100°C or vaporizing the water to steam? Assume the water is initially at 25°C.

SOLUTION From the heat of vaporization of water, we know that it takes 2258 J to vaporize 1 gram of water. The specific heat of water shows that 1.00 g of water requires 4.184 J to increase the temperature by 1.00°C. The water has to be heated from 25.0°C to 100°C, an increase of 75.0°C. Therefore, the water absorbs

$$4.184 \text{ J/g} \cdot {}^\circ\text{C} \times 1.00 \text{ g} \times 75.0{}^\circ\text{C} = 314 \text{ J}$$

So, more heat is removed by vaporizing the water than by heating the water.

Practice 9.7

Would you expect the heat of vaporization of ethanol (CH_3CH_2OH) to be different from the heat of vaporization of ethane (CH_3CH_3)? Why?

ANSWER

The presence of the alcohol functional group (—OH) in the ethanol molecule increases the heat of vaporization as compared with the simple alkane because the ethanol molecule is polar. The alcohol exhibits dipole-dipole intermolecular forces and has the ability to form hydrogen bonds, neither of which is true for ethane. Overcoming the stronger intermolecular forces in the alcohol requires more energy for the liquid to vaporize.

Arsonists often make the mistake of assuming that a flammable liquid used to start a fire will be completely consumed in the ensuing fire. Yet even when the temperature of a fire reaches over 1000°C, the components of gasoline that have boiling points between 40°C and 200°C can still be found in samples. Flammable liquids are quickly soaked into most objects in a room—such as carpeting, wood flooring, cement, and furniture. Rarely does a fire completely destroy all of these materials, so

Table 9.3 Heats of Fusion for Various Materials

Material	Heat of Fusion (J/g)
Copper	207
Iron	247
Aluminum	397
Low-density polyethylene	98
High-density polyethylene	296

traces of the accelerant can be detected in the debris. Only the vapors of the flammable liquid ignite, and the liquid beneath the flames continues to soak into the absorbent material. Because heat rises, the temperature beneath the flame is considerably lower than the temperature above the flame. Furthermore, as the fire burns upward, debris and ash will fall downward and tend to smother the flames burning the accelerant, thereby preserving it for future analysis. Meanwhile, the fire spreads to other materials and continues burning.

As we saw earlier, fire-scene investigators can estimate the temperature of a fire by determining whether copper wiring in the walls has melted. Other common metals found at a fire scene are aluminum cans (660°C melting point) or steel furniture springs (1100°C to 1600°C melting point, depending on the type of steel).

The amount of heat required to melt a solid completely is referred to as the **heat of fusion** and is analogous to the heat of vaporization discussed at the beginning of this section. When a solid material reaches its melting point, the solid does not melt instantly. It must absorb heat to increase the kinetic energy of the solid particles until they have sufficient energy to overcome the forces holding them in the solid state and break into the liquid phase. Table 9.3 provides a list of heats of fusion for common materials.

Worked Example 8

The following graph represents the heating curve for copper metal. Estimate the melting point and boiling point of copper.

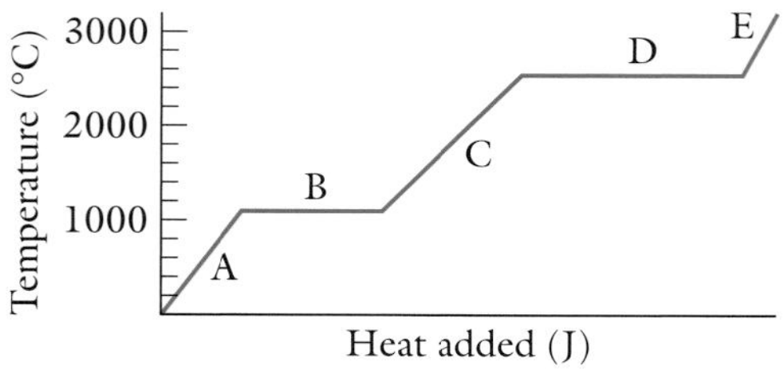

SOLUTION The melting point corresponds to the temperature of the plateau region B. Extrapolating the line back to the *y* axis, an estimate of 1050°C is reasonable. (The actual value is 1083°C.) The boiling point corresponds to plateau region D with an estimated value of 2600°C (the actual value is 2595°C).

Practice 9.8

For each distinct region on the heating curve of copper, provide a description of the physical process occurring.

ANSWER

In region A, the temperature of solid copper increases. In region B, the solid copper changes to liquid copper; the temperature remains constant. In region C, the temperature of liquid (molten) copper increases. In region D, the liquid copper boils and changes to copper in the gaseous state; the temperature remains constant. In region E, the temperature of gaseous copper increases.

9.6 Mathematics of Heat Capacity

Learning Objective

Calculate the heat needed for phase changes and temperature changes.

The specific heat capacity of a material is most commonly used in calculating the amount of energy required to raise the temperature of an object made of that material. For example, how much energy in joules is needed to heat a 10.0-g copper (Cu) wire from room temperature at 25.0°C to its melting point at 1082°C? The equation for calculating the heat required is:

$$\underbrace{q}_{\substack{\text{Heat}\\(\text{J})}} = \underbrace{C_p}_{\substack{\text{Specific heat}\\\left(\frac{\text{J}}{\text{g}\cdot{}^\circ\text{C}}\right)}} \times \underbrace{M}_{\substack{\text{Mass}\\(\text{g})}} \times \underbrace{\Delta T}_{\substack{\text{Temperature}\\\text{change}\\({}^\circ\text{C})}}$$

Look up the specific heat of copper in Table 9.1 and examine the units carefully:

$$0.38 \overbrace{\underbrace{\frac{\text{J}}{\cancel{\text{g}}\cdot\cancel{{}^\circ\text{C}}}}_{\text{Cancel units}}}^{\substack{\text{Desired}\\\text{final units}}} \times 10.0\,\cancel{\text{g}} \times \overbrace{1057\cancel{{}^\circ\text{C}}}^{1082\,-\,25} = 4.0 \times 10^3\ \text{J}$$

Worked Example 9

Will it take more or less energy than that used in the previous example to heat a 10.0-g aluminum (Al) wire from 25°C to its melting point at 660°C, given the specific heat of Al is 0.90 J/g · °C?

SOLUTION

$$q = C \times m \times \Delta T$$
$$C = 0.90\ \text{J/g}\cdot{}^\circ\text{C}$$
$$m = 10.0\ \text{g}$$
$$\Delta T = T_{\text{final}} - T_{\text{initial}} = 660 - 25 = 635^\circ\text{C}$$

The heat required is

$$0.90\ \frac{\text{J}}{\cancel{\text{g}}\cdot\cancel{{}^\circ\text{C}}} \times 10.0\,\cancel{\text{g}} \times 635\cancel{{}^\circ\text{C}} = 5.7 \times 10^3\ \text{J}$$

More energy is needed to heat 10.0 g of Al to 660°C than to heat 10.0 g of Cu to 1082°C.

Practice 9.9

If 3.50×10^3 J of energy is absorbed by a 25.3-g metal object, and the temperature increases from 25°C to 235°C, what is the specific heat of the metal?

ANSWER

0.659 J/g · °C

Worked Example 10

Calculate the amount of heat necessary to vaporize 100.0 mL of gasoline. The density of gasoline is 0.730 g/mL, and the heat of vaporization of gasoline is 349 J/g.

SOLUTION

Step 1. Determine the mass of gasoline present (100.0 mL does *not* equal 100.0 g unless you are dealing with water, which has a density of 1 g/mL) using $D = m/V$ or $m = D \times V$:

$$0.730 \frac{\text{g}}{\text{mL}} \times 100.0 \text{ mL} = 73.0 \text{ g}$$

Step 2. The heat required to vaporize 73.0 g of gasoline is

$$73.0 \cancel{\text{g}} \times \frac{349 \text{ J}}{\cancel{\text{g}}} = 2.54 \times 10^4 \text{ J}$$

Practice 9.10

How many grams of Cu can be melted by heat energy of 2.50×10^3 J?

ANSWER

12.1 g

9.7 The First Law of Thermodynamics and Calorimetry

Learning Objective

Describe how to measure the amount of heat produced in combustion reactions.

As discussed in previous sections of this chapter, combustion reactions require not only activation energy but also energy for phase changes and pyrolysis to convert fuel materials into gaseous reactants. However, once a combustion reaction becomes self-sustaining, more energy is produced by the fire than is required to sustain the reaction. What happens to this excess energy?

The **first law of thermodynamics**, also known as the *law of conservation of energy,* is based on the principle that energy cannot be created or destroyed. The amount of energy released by a process must be equal to the amount of energy absorbed by the rest of the system. For example, the energy released in a fire is absorbed by the air surrounding the fire or by any object in the vicinity of the reaction. The total amount of energy before and after the fire has remained constant.

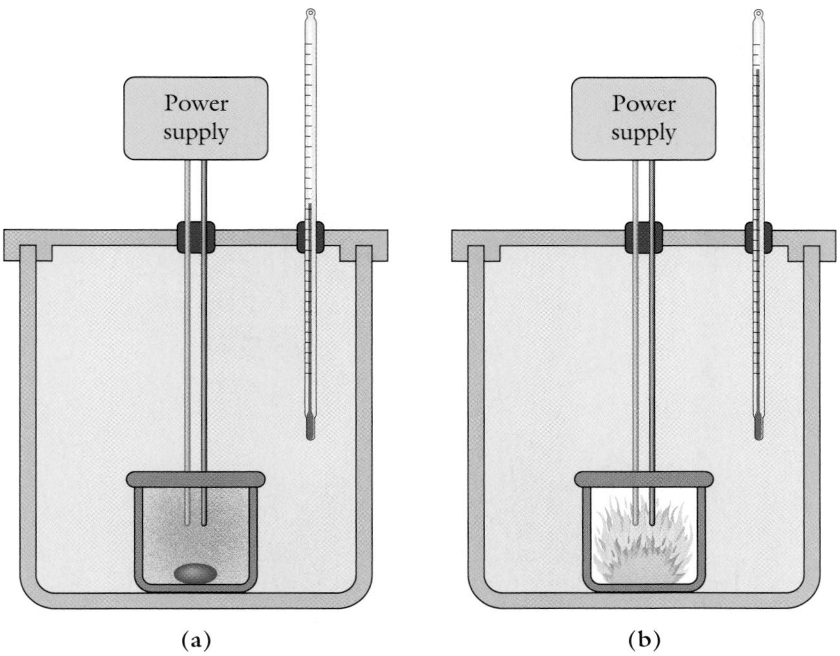

Figure 9.6 Calorimeter for measuring energy content. (a) The inner chamber of a calorimeter contains a sample for analysis (fuel), a pure oxygen environment, and an ignition source. (b) The heat produced by the combustion reaction is transferred to the water that completely surrounds the reaction vessel. The heat produced can be measured from the increase in the temperature of the water.

The law of conservation of energy is used experimentally in the science of **calorimetry**. Calorimetric methods determine the heat released in a combustion reaction by measuring the increase in the temperature of water that surrounds the reaction. A typical calorimeter is shown in Figure 9.6.

The calorimeter consists of a sealed chamber that is completely submerged in a reservoir of water. The chamber contains a fuel source and oxygen gas. The fuel is ignited by an electrical spark from an external power supply. Energy from the combustion reaction is fully absorbed by the water surrounding the chamber, causing the temperature of the water to increase. The heat released can then be determined from the specific heat of water, the temperature change of the water sample, and the volume of water.

This instrument is often called a **bomb calorimeter** because the fuel is converted to gaseous products that are confined within a sealed container at high pressures. Care is taken to use small amounts of fuel, and the container is typically constructed out of thick stainless steel to withstand the pressure.

Calorimeters are also used to study the amount of heat produced by burning furniture, upholstery, carpet, and various construction materials. Because fire investigators are interested not only in how much heat is produced when such materials burn, but also in how long it takes for these materials to produce their maximum heat output, a slightly modified calorimeter, called a **cone calorimeter**, is used. A cone calorimeter is de-

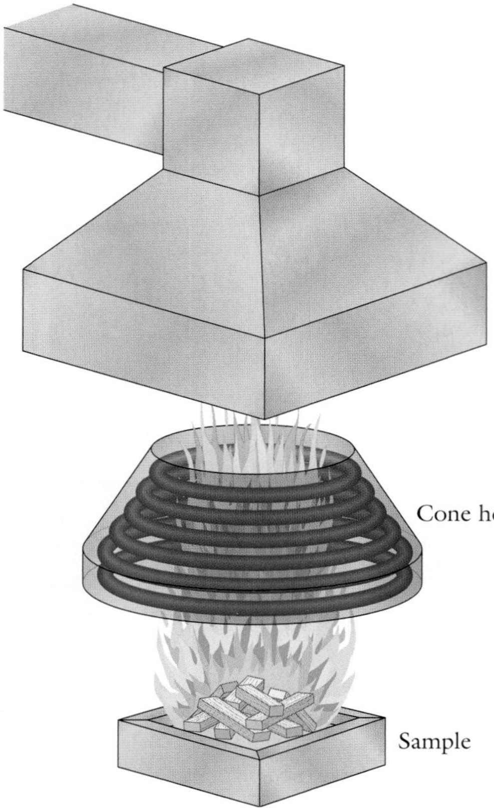

Figure 9.7 A cone calorimeter. A cone calorimeter heats a sample to the point of ignition while measuring the concentration of oxygen in the air above the burning object. The amount of oxygen consumed is used to calculate the amount of heat released by the burning sample. The heat released is recorded as a function of time to provide an accurate depiction of how the sample material would behave in a real fire.

signed to measure the heat of combustion of ignitable materials by monitoring the decrease in oxygen in the air collected above a burning sample. Figure 9.7 shows a typical cone calorimeter.

A cone calorimeter functions by heating a sample until it ignites, capturing the gases above the flames in a hood and passing the gases by an oxygen sensor. It is known that in a combustion reaction, approximately 13 kJ of heat are produced per gram of oxygen gas consumed. The decrease in oxygen gas concentration above the fire corresponds to greater consumption of oxygen gas and, therefore, to a greater release of heat. Cone calorimeters are used because they are based on using atmospheric oxygen to supply the reaction, whereas a bomb calorimeter uses a pure oxygen environment. There are also large-scale calorimeters made for studying the combustion of entire pieces of furniture.

Cone calorimeters have changed the way fire investigators evaluate the cause of a fire. Over the last 50 years, our homes have been increasingly filled with foam materials, especially in furniture and mattresses. Because foams are made from petroleum products, they tend to make excellent fuel sources and often burn very easily, releasing a large amount of heat back into the fires. Heat is one of the three necessary parts of the fire triangle and one of the key causes for flashovers. Cone calorimeters have helped to show that fires in homes today burn faster and reach higher temperatures sooner than they did 50 years ago.

In the Kenny Richey case, an expert witness testified that the speed at which the fire engulfed the building was corroborating evidence of arson. However, due to the amount of petroleum-based products in a modern home, this is not an accurate method for determining whether a fire was in fact arson.

9.8 Mathematics of Calorimetry

Learning Objective

Perform calculations in calorimetry.

The heat produced (or absorbed) by a reaction or process must be absorbed (or supplied) by the surroundings. Mathematically, it is necessary to use a sign convention to describe the transfer of heat. If the system under study releases heat through an exothermic process, the amount of heat is indicated by a negative sign; if heat is absorbed in an endothermic process, the amount of heat is indicated by a positive sign.

For example, if a heated metal object is submerged in cold water, heat is transferred from the hot metal to the cold water. The amount of heat leaving the metal is given a negative sign; the amount of heat gained by the water is given a positive sign. According to the law of conservation of energy, the amount of energy lost by the metal must equal the amount of energy gained by the water. The equation for heat transfer becomes:

$$q_{\text{lost}} = q_{\text{gained}}$$
$$q_{\text{lost}} + q_{\text{gained}} = 0$$

When the amount of heat lost is added to the amount of heat gained, they must sum to zero. Otherwise, there would be a violation of the law of conservation of energy. Exothermic reactions can also serve as the source of energy in calorimetry problems. In this situation, the energy released by the chemical reaction is absorbed by the water in the calorimeter. For calculations, it is often convenient to summarize the information.

Worked Example 11

An unknown amount of copper is heated to 975°C and then plunged into a calorimeter containing 75.0 mL of water at 24.0°C. The temperature of the water increases to 31.3°C. Determine the amount of copper heated. The specific heats are provided in Table 9.1.

SOLUTION Look up the specific heats of copper and water in Table 9.1 and use the known density of water:

Variable	Copper	Water
Mass	X	$75.0\ \text{mL} \times \dfrac{1.0\ \text{g}}{1\ \text{mL}}$
Specific heat	0.38 J/g · °C	4.184 J/g · °C
Initial temperature	975°C	24.0°C
Final temperature	31.3°C	31.3°C

Calculation 1: The heat gained by the water is calculated using $q = C \times m \times \Delta T$:

$$4.184 \frac{\text{J}}{\text{g} \cdot {}^\circ\text{C}} \times 75.0 \text{ g} \times (31.3^\circ\text{C} - 24.0^\circ\text{C}) = 2.29 \times 10^3 \text{ J}$$

Calculation 2: The heat released by the copper is calculated similarly:

$$0.38 \frac{\text{J}}{\text{g} \cdot {}^\circ\text{C}} \times X \times (31.3^\circ\text{C} - 975^\circ\text{C}) = \left(-359 \frac{\text{J}}{\text{g}}\right) X$$

Calculation 3: Since the heat lost plus the heat gained is zero,

$$\left(-359 \frac{\text{J}}{\text{g}}\right) X + 2.29 \times 10^3 \text{ J} = 0$$

$$X = \frac{-2.29 \times 10^3 \text{ J}}{-359 \frac{\text{J}}{\text{g}}} = 6.4 \text{ g}$$

Practice 9.11

The heat of combustion of gasoline is -4.40×10^4 J/g. How many grams of gasoline are burned in a calorimeter containing 1000.0 mL of water if the temperature increases from 25.5°C to 46.5°C?

ANSWER

2.00 g of gasoline

9.9 Petroleum Refinement

Learning Objective

Describe the physical processes involved in refining petroleum.

Gasoline and paint thinners, the accelerants allegedly used by Kenny Richey, are the most common accelerants used in arson cases. These liquids and a few others, such as kerosene and lighter fluid, comprise a set of highly ignitable and easily obtained liquids that could be used to accelerate a fire in a home or business.

What interests forensic scientists who perform accelerant analysis of these ignitable liquids is that they are homogeneous mixtures made during the processing of crude oil from many different compounds. To understand how these mixtures are analyzed and identified, it is necessary to understand the basics of oil refining. The Evidence Analysis box on page 292 goes into detail about how the flammable petroleum products are analyzed using gas chromatography.

Crude oil contains thousands of compounds. To refine the crude oil into useable products, it is first heated at the base of a tower, as shown in Figure 9.8. The temperature within the tower is lowest at the top and highest at the bottom. As crude oil is heated, compounds with low boiling points, such as butane and propane, enter the gas phase and proceed to travel upward toward the top of the tower. When a compound reaches a point in the tower where the temperature is lower than its boiling point, the compound condenses back into liquid form. The compounds are in

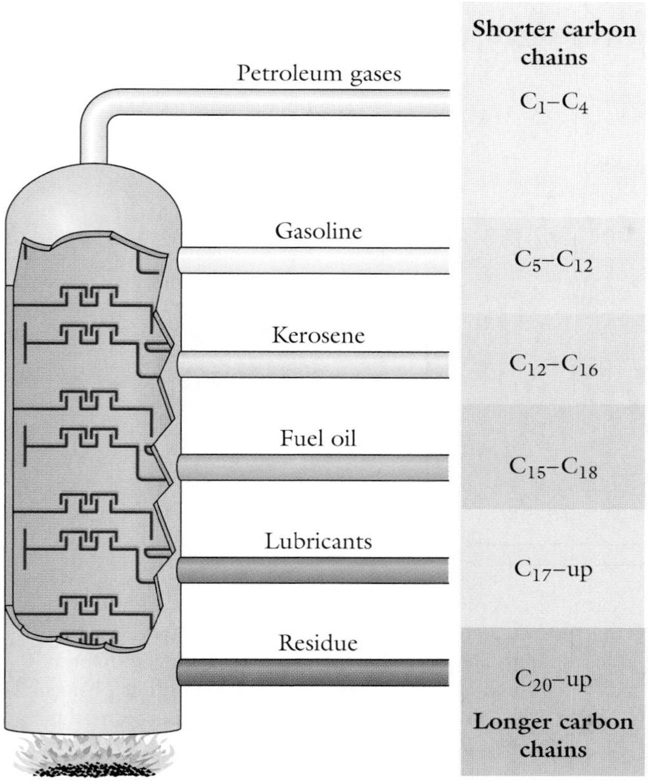

Figure 9.8 Crude-oil refinement. (Adapted from www.schoolscience.co.uk/content/4/chemistry/fossils/p8.html)

this way grouped by boiling points and removed from different heights along the tower. Gasoline has components with boiling points from approximately 40°C to 220°C. The boiling points of kerosene's components range from 175°C to 270°C.

There are two categories of ignitable liquids—flammable liquids and combustible liquids. The difference between the two is based on their volatility. A **flammable liquid** is defined as a liquid that produces sufficient vapors to support combustion at a temperature lower than 37.8°C. A **combustible liquid** is one that must reach a temperature higher than 37.8°C for sufficient vapors to be produced to support combustion. Gasoline is an example of a flammable liquid; kerosene is an example of a combustible liquid. If you are wondering why such an odd temperature was chosen, the definition is based on the Fahrenheit scale—37.8°C is the same as 100°F.

Each product made in the refinement of crude oil has a unique set of compounds. The mixture of compounds that make up gasoline is different from the mixture of compounds that make up kerosene. In order to identify whether an accelerant is kerosene or gasoline, it is necessary to determine what components make up the unknown accelerant and compare those components with standard samples of each accelerant. Chromatography provides an experimental method for separating and detecting multiple compounds within a mixture.

Evidence Analysis | Gas Chromatography

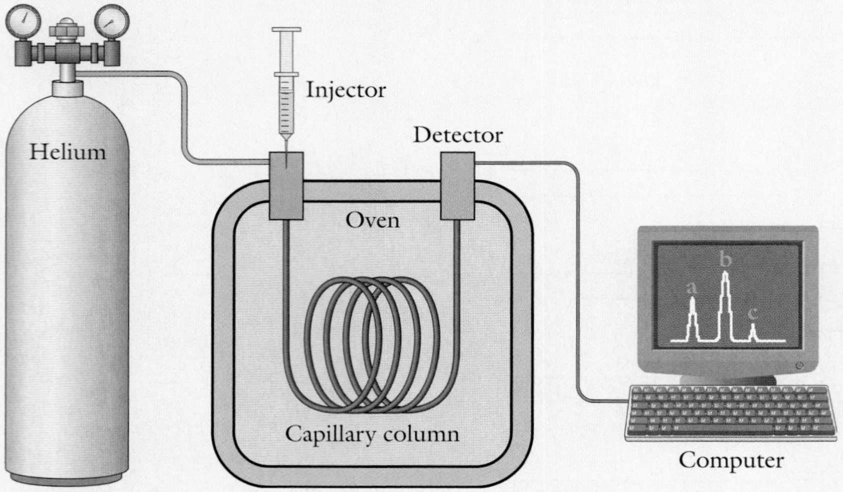

Figure 9.9 A typical gas-chromatography system. A mixture of compounds is placed onto a stationary phase, and a mobile phase is passed through the system. Samples injected into the system are vaporized and pass over the stationary phase in a temperature-controlled oven. Each compound spends a different amount of time in the mobile and stationary phase as it passes through the system, based on its physical attraction for the stationary phase and mobile phase. The separation of the compounds is based on the polarity and boiling point of each compound.

If an accelerant is used to start a fire, trace amounts of liquid recovered after the fire can be used to identify the type of ignitable liquid used. The compounds present in a sample of gasoline are different from the compounds present in a sample of kerosene. To identify the accelerant, the identity and types of compounds present in the sample must be determined. **Gas chromatography (GC)** is the ideal method for analyzing volatile liquids because it separates compounds based on their boiling points and their polarity. A sketch of a typical gas chromatography system is shown in Figure 9.9.

The sample to be analyzed is loaded into a syringe and injected into a heated injector port. The amount of sample needed is very small, as little as several microliters (μL). The heated injector serves two purposes: It vaporizes the liquid sample and loads the sample onto the column that contains the stationary phase. The mobile phase, usually helium gas, carries the sample through a long, narrow glass tube, called a *capillary column*. The capillary column is coated on the inside surface with a thin layer of a nonvolatile liquid that is in the stationary phase. The capillary column is kept inside an oven so that the temperature can be controlled. The most volatile compounds in the sample, those with the lowest boiling point, are quickly swept through the column by the helium gas. The compounds that have similar polarities to the stationary-phase coating will take longer to reach the detector. The temperature of the column is typically increased throughout the experiment so that those compounds with higher boiling points will be able to travel through the column to the detector in a reasonable amount of time.

9.10 Case Study Finale: "False and Unreliable"

The scientific evidence against Kenny Richey included gas chromatography analysis of fire debris, analysis of the fire scene, and a witness who claimed Richey admitted to torching the apartment. What is the new

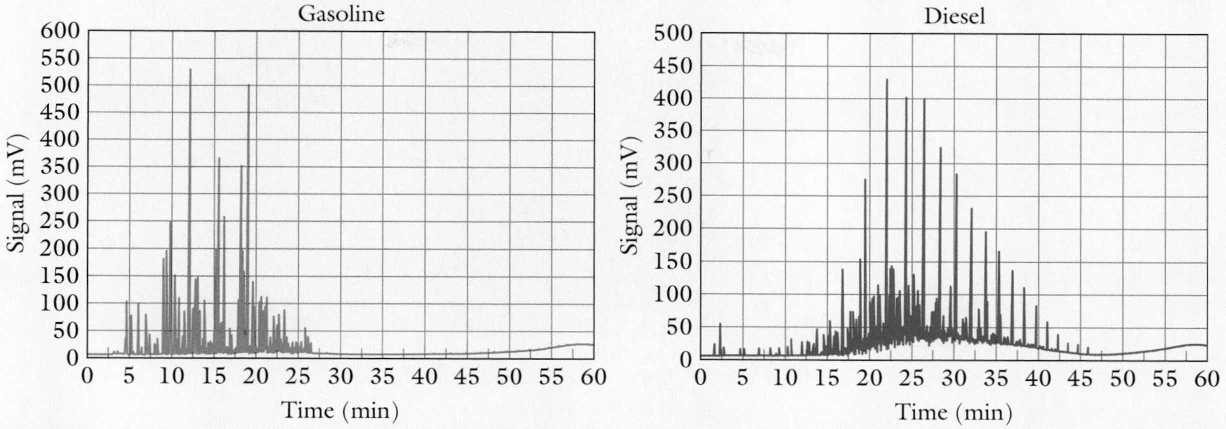

Figure 9.10 Each line on the chromatogram represents a unique compound present in gasoline or diesel fuel. Analysis of an arson sample attempts to match the major peaks found in the sample under investigation to peaks obtained from a known sample of accelerant. (www.eti-geochemistry.com/pdf/HRGC-standards.pdf)

When each compound reaches the detector, it triggers a response in which the peak height is proportional to the compound's concentration in the sample. The time from injection of the sample to the time at which the peak response of the compound is detected is called the **retention time**. As long as all the samples are analyzed on the same chromatography system using the same method, the retention time of each compound will not change. A graph of the data, called a **chromatogram**, is then compared against standards made up from known accelerants. Representative chromatograms of gasoline and diesel fuel are shown in Figure 9.10.

If an arson investigator believes that an accelerant was poured on carpeting, the investigator will collect a sample of the carpeting far away from the suspected point of origin of the fire. This is called a *blank sample* because it is used to determine the background levels of organic compounds. Some of the same compounds that make up gasoline and kerosene can be found in household cleaners, insecticides, and adhesives used for carpeting or laminate flooring. These compounds may also be produced as a result of pyrolysis of plastics and polymers. One could get a false positive reading for an accelerant if a blank sample is not evaluated.

Suspected arson samples usually consist of charred wood, clothing, or carpeting taken from the fire scene. To isolate the possible accelerants, the samples are sealed in air-tight vials. If an accelerant such as gasoline or kerosene is present, vapors from these volatile compounds build up inside the sealed container. The air above the charred sample is then sampled with a gas-tight syringe and injected into the gas chromatography system for analysis.

A more sophisticated method, called *gas chromatography-mass spectroscopy* (GC-MS), is used to measure not only the retention time but also the molecular mass of each compound. This ensures that peaks appearing at a common retention time on two different chromatograms actually represent the same compound. Mass spectroscopy is covered in the Evidence Analysis box in Chapter 10.

evidence that is referred to in the quotation by the prosecutor in the case study? The new evidence was actually the old evidence as evaluated by some of the leading experts in the world. In Kenny's first trial, his lawyer hired an inept and unqualified expert witness who ended up testifying for the prosecution.

Figure 9.11 Improper collection and storage of possible arson evidence. (Courtesy of Goodwin Procter, LLP)

Dr. Andrew Armstrong was the expert responsible for analyzing the debris from the fire that killed 71 people at the Branch Davidian Compound in Waco, Texas. In the Kenny Richey case, Dr. Armstrong testified about the gas chromatography evidence analysis and determined that not only were the samples improperly taken, they were improperly stored. He had serious doubts about the integrity of the samples because of the way the evidence was handled. The fire marshal initially ruled the fire was an accident and allowed the property owner to remove all of the apartment's contents for disposal at the local landfill. When the police decided to investigate the fire as a crime, officers were sent to the landfill to retrieve the carpeting from the apartment! Most likely because of the carpet's large size, odor, and damaged condition, the police decided to leave the carpet outside on the asphalt next to the gas pump, as shown in the upper right corner of Figure 9.11.

Armstrong also evaluated the methods of analysis used by the original forensic scientists and determined that the methods were not accepted by the general scientific community. Furthermore, he concluded that the data were misinterpreted.

According to Armstrong, the evidence actually showed no traces of ignitable fluids at all! Were some petroleum products found on the samples? Yes. But were they the same compounds that would be found in the suspected accelerants? Armstrong testified that the samples were not even close to what would be expected if accelerants were present.

The new evidence also consisted of testimony by a nationally respected expert in the analysis of fire scenes, Richard L. P. Custer. He debunked the initial analysis of the fire scene, which the prosecutors had claimed to show evidence of arson. During the original trial, evidence of wood charring between the edges of wood planks on the deck flooring was presented as proof that a liquid accelerant had been poured and leaked in between the boards. Custer explained that this is in fact a very common pattern caused by air (oxygen) rushing up between the wood decking to feed the fire.

Another piece of evidence used against Kenny Richey in the original trial was that there were pooling marks near the furniture, which were explained as marks left by pools of burning accelerant. Again, an alternative explanation is that modern furniture contains foam made from petroleum ingredients. The foam melts into a pool when ignited, leaving marks identical to those created by accelerants.

Finally, the witness who testified against Richey almost immediately recanted her testimony after the trial, insisting that she was nervous and felt pressured to say what she thought the prosecutor wanted her to say. Yet the prosecutors chose to stay with their original view of the testimony and interpretations given by the laboratory personnel who analyzed the samples.

Through the appeals process, Richey was seeking a new trial at which to present the new evidence. During that time, the prosecutor did not challenge any of the arguments debunking the original physical evidence. The prosecution instead decided to focus on the circumstantial evidence and testimony of witnesses to justify the verdict, despite relying heavily on the physical evidence during the original trial.

The Sixth Circuit Court of Appeals recently reversed Kenny Richey's conviction and ordered either his release or a new trial. As of the writing of this book, the Supreme Court of the United States stayed that order while it decides whether to hear the state's appeal of the reversal. The county prosecutor has indicated that if the conviction is reversed, the state will seek an indictment from a grand jury and take this case to trial once again.

CHAPTER SUMMARY

- A fire is a self-sustaining chemical reaction that produces heat and light. Oxygen, heat, and a fuel must simultaneously be present for a fire to exist. If one of these three ingredients is removed, the fire will not burn.
- Heat is a form of energy that is transferred between objects at different temperatures. Temperature is a measure of the average kinetic energy of an object. When heat is transferred from one object to another, the heat will flow from the warmer to the colder object until thermal equilibrium is achieved.
- Solid fuels must undergo pyrolysis to form smaller gaseous molecules. Liquid fuels must be vaporized before they will combust. Pyrolysis and the vaporization process require that energy be consumed (endothermic process). The overall combustion reaction releases energy (exothermic process) because more heat is produced than consumed in the reaction.
- The activation energy for a reaction is the energy required to initiate the breaking of bonds in the reactant molecules.
- The specific heat capacity of a substance is the amount of energy required to increase the temperature of 1 gram of the material by 1°C and is determined by the intermolecular forces present in the substance.
- Energy is required by a material to overcome intermolecular forces during a phase change from solid to liquid (the heat of fusion) and from liquid to gas (the heat of vaporization).
- The first law of thermodynamics, also called the law of conservation of energy, states that energy is conserved in any physical or chemical process. This law is the basis of calorimetry experiments.
- Gas chromatography is an instrumental method that exploits differences in the boiling points and polarities of compounds to separate the components of a mixture. As the compounds travel the length of the chromatographic column, they interact with a stationary-phase material. Compounds with polarities similar to those of the column's stationary phase tend to be slowed. This results in a unique retention time, the time from injection until the compound is detected at the end of the column, for each compound.

KEY TERMS

fire, p. 270
arson, p. 270
fire triangle, p. 270
heat, p. 270
joule (J), p. 271
calorie (cal), p. 271
temperature, p. 271
thermal equilibrium, p. 271
backdraft, p. 272
oxidation number, p. 276
pyrolysis, p. 278
endothermic, p. 279
exothermic, p. 279
activation energy, p. 279
flashover, p. 280
autoignition, p. 280
specific heat capacity, p. 281
heat of vaporization, p. 282
heat of fusion, p. 284
first law of thermodynamics, p. 286
calorimetry, p. 287
bomb calorimeter, p. 287
cone calorimeter, p. 287
flammable liquid, p. 291
combustible liquid, p. 291
gas chromatography (GC), p. 292
retention time, p. 293
chromatogram, p. 293

CONTINUING THE INVESTIGATION Additional Readings, Resources, and References

National Fire Protection Association, *NFPA 921 Guide for Fire and Explosion Investigations*, Quincy, MA: National Fire Protection Association, 2004.

Richey, Tom. *Death Row Scot—My Brother Kenny's Fight for Justice*, Edinburgh: Black & White Publishing, 2005.

The following Web sites contain information about the Kenny Richey case: www.kennyrichey.org www.tcforensic.com.au/features/richey.html

For information about the composition of gasoline and standards in gas chromatography: www.eti-geochemistry.com/pdf.

REVIEW QUESTIONS AND PROBLEMS

Questions

1. Draw a molecular view, similar to the one shown in Figure 9.1, for equal masses of Cu at 50°C and at 500°C that come into thermal contact. Explain what would happen if the mass of the Cu at 500°C were twice that of the Cu at 50°C.
2. Draw a molecular view, similar to the one shown in Figure 9.1, for equal masses of H_2O at 50°C and at 100°C that are mixed together. Explain what would happen if the mass of the H_2O at 50°C were twice that of the H_2O at 100°C.
3. What is the difference between heat and temperature?
4. What are the common units for energy and how are they defined?
5. Why does a thermometer placed in a solution alter the temperature it is attempting to measure?
6. What is the difference between mercury atoms at 20°C and mercury atoms at 50°C? How is this behavior exploited in a mercury-filled thermometer for measuring temperature?
7. What combustion conditions can lead to a backdraft? How can a backdraft be prevented?
8. How does the absence or presence of smoke, and the color of smoke when it is present, indicate the efficiency of a combustion reaction? Give the main products of the reaction in each example you provide.
9. What types of compounds can undergo combustion reactions? Give an example of a specific compound from each type and write a balanced equation for its combustion reaction.
10. Does the oxidation number of an atom represent a real charge on the atom?
11. How can you determine whether an element within a compound is oxidized or reduced in a reaction? What precaution has to be followed in determining which compounds are oxidized and which are reduced?

12. What is the nature of the activation energy barrier in a chemical reaction?

13. What determines whether a reaction is endothermic or exothermic?

14. In which type of reaction, endothermic or exothermic, do the products have more energy than the reactants?

15. Explain how the process of a flashover is related to the activation energy of a combustion reaction. How can flashover conditions be prevented?

16. Despite the addition of energy to a system during a phase transition, an increase in the temperature does not occur. Explain this observation.

17. How is specific heat capacity influenced by intermolecular forces? What types of compounds would tend to have large specific heat capacities?

18. How are the heat of fusion and heat of vaporization influenced by intermolecular forces? What types of compounds require very little heat to melt or vaporize?

19. Tile floor will feel colder than carpeting despite being in the same room and at the same temperature. Explain the nature of this sensation by discussing relative heat capacities and the flow of heat from one object to another.

20. Sketch a temperature-versus-heat curve for an ice cube at −5°C that is converted to steam at 115°C. Label each region with the phases present and the physical processes that are happening.

21. Why are accelerants typically not completely consumed in a fire? What process involving the liquid fuel can actually serve to cool it?

22. Explain how the first law of thermodynamics is used in calorimetry. When a heated object cools down and releases heat, what happens to that energy?

23. Explain how crude oil is refined, and describe the chemical makeup of the resulting petroleum products.

24. Describe how a mixture of volatile compounds in gasoline is separated by gas chromatography. What would the difference be between a kerosene sample and a gasoline sample?

Problems

25. Convert the following amounts of energy to units of joules.
(a) 25.0 calories
(b) 110.0 Calories
(c) 50.0 kilojoules
(d) 73.5 kilocalories

26. Convert the following amounts of energy to units of joules.
(a) 5.05×10^4 Calories
(b) 1.00 Calorie
(c) 8.87×10^{-3} kilojoules
(d) 1.00×10^3 calories

27. Heats of combustion for many fuels are reported in units of Cal/g. Convert the following heats of combustion to values of J/g.
(a) Kerosene: 11.0 Cal/g
(b) Gasoline: 11.5 Cal/g
(c) Benzene: 10.0 Cal/g
(d) Paraffin wax: 9.92 Cal/g

28. Heats of combustion for many fuels are reported in units of Cal/g. Convert the following heats of combustion to values of J/g.
(a) Octane: 11.4 Cal/g
(b) Alcohol: 7.4 Cal/g
(c) Fuel oil: 10.7 Cal/g
(d) Methanol: 5.42 Cal/g

29. Heats of combustion for many fuels are reported in units of megajoules per kilogram (MJ/kg), *mega* being the prefix for one million. Convert the following heats of combustion to values of J/g.
(a) Kerosene: 43.3 MJ/kg
(b) Jet fuel: 43.5 MJ/kg
(c) Gasoline: 43.7 MJ/kg
(d) Fuel oil: 44.0 MJ/kg

30. Heats of combustion for many fuels are reported in units of megajoules per kilogram (MJ/kg), *mega* being the prefix for one million. Convert the following heats of combustion to values of J/g.
(a) Propane: 50.0 MJ/kg
(b) Coal: 30.5 MJ/kg
(c) Octane: 47.7 MJ/kg
(d) Wood (average): 20.0 MJ/kg

31. Write and balance equations for combustion of the following compounds. Assume complete combustion to form carbon dioxide and water.

(a) CH_4 (c) $C_6H_{14}O$
(b) C_5H_{12} (d) $C_8H_{16}O$

32. Write and balance equations for combustion of the following compounds. Assume complete combustion to form carbon dioxide and water.
(a) C_3H_8 (c) C_7H_{12}
(b) C_4H_8 (d) C_2H_6O

33. Write and balance equations for combustion of the following compounds. Assume complete combustion to form carbon dioxide and water.
(a) Propyne (c) Decane
(b) 2-butene (d) 1-octene

34. Write and balance equations for combustion of the following compounds. Assume complete combustion to form carbon dioxide and water.
(a) 1-butanol (c) Propanone
(b) Ethene (d) 2-pentyne

35. Assign oxidation numbers to each element in the following compounds or ions.
(a) Mg^{2+} (c) SO_2
(b) CuO (d) NH_4OH

36. Assign oxidation numbers to each element in the following compounds.
(a) Al_2O_3 (c) PH_3
(b) Fe_2O_3 (d) CO_2

37. In the following reactions, determine which element in the reactants is oxidized and which is reduced.
(a) $6CO_2(g) + 6H_2O(l) \rightarrow C_6H_{12}O_6(s) + 6O_2(g)$
(b) $CH_4(s) + 2O_2(g) \rightarrow CO_2(g) + 2H_2O(l)$

38. In the following reactions, determine which element in the reactants is oxidized and which is reduced.
(a) $Fe(s) + O_2(g) \rightarrow Fe_2O_3(s)$
(b) $CuCl_2(aq) + Zn(s) \rightarrow Cu(s) + ZnCl_2(aq)$

39. In the following reactions, determine which element in the reactants is oxidized and which is reduced.
(a) $2CO(g) + O_2(g) \rightarrow 2CO_2(g)$
(b) $2Na(s) + 2H_2O(l) \rightarrow 2NaOH(aq) + H_2(g)$

40. In the following reactions, determine which element in the reactants is oxidized and which is reduced.
(a) $Cu(s) + O_2(g) \rightarrow CuO(s)$
(b) $C_2H_5OH + 3O_2(g) \rightarrow 2CO_2(g) + 3H_2O(l)$

41. Calculate the specific heat capacity of the metal in a 445-g block that requires 1.00×10^4 J of heat to raise the temperature from 24.7°C to 49.7°C.

42. Calculate the specific heat capacity of a concrete cinder block with a mass of 1.09×10^4 g that requires 1.96×10^6 J of heat to raise the temperature from 17°C to 242°C.

43. Determine how much heat is required to increase the temperature of PVC (polyvinyl chloride, a common type of plastic) from 25°C to its autoignition temperature of 507°C, given that the specific heat capacity of PVC is approximately 1.2 J/g · °C.

44. Determine how much heat is required to increase the temperature of No. 2 diesel fuel from 25.0°C to its autoignition temperature of 600.0°C, given that the specific heat capacity of the fuel is 1.8 J/g · °C.

45. Determine the mass of copper that would release 438 J of heat while cooling from 284°C to 25°C.

46. Determine the mass of water that would release 1.50×10^3 J of heat while cooling from 100°C to 25°C.

47. Determine the amount of energy absorbed by 3.785 L of water as it is converted to steam at its boiling point.

48. Determine the amount of energy absorbed by 3.785 L of gasoline as it is converted to the vapor phase at its boiling point.

49. Calculate the amount of water at its boiling point that would be vaporized by 115.0 Calories.

50. Calculate the amount of gasoline at its boiling point that would be vaporized by 115.0 Calories.

51. How much heat is required to melt 75.8 g of copper starting at a temperature of 25°C? The melting point of copper is 1082°C.

52. How much heat is required to convert 15.0 g of ice at 0°C to steam at 100°C?

53. Draw a temperature-versus-heat curve depicting the processes indicated in Problem 51. Label all regions, list all phases present, and indicate the melting point.

54. Draw a temperature-versus-heat curve depicting the processes indicated in Problem 52. Label all regions, list all phases present,

and indicate the melting point and boiling point.

55. If the heat of combustion of Teflon is 5×10^3 J/g, how many grams of Teflon are consumed in a bomb calorimeter if 500.0 mL of water increases in temperature from 25.5°C to 31.2°C?

56. If the heat of combustion of newspaper is 1.97×10^4 J/g, what is the final temperature that 500.0 mL of water, initially at 25.0°C, will reach in a bomb calorimeter if 5.00 g of newspaper is burned?

57. If 2.00 g of a flammable fuel is ignited in a bomb calorimeter and the temperature of 1.00 L of water increases from 26.8°C to 31.2°C, what is the heat of combustion (J/g) of the fuel?

Forensic Chemistry Problems

58. Would the heat of combustion calculated in a bomb calorimeter be the same as that observed in a cone calorimeter? Explain why or why not.

59. Fire sprinklers were first patented in 1872 and operate on a simple principle: Heat from a fire triggers a physical change in a material that is plugging a hole in the water pipe. The plug can either be a piece of metal or a liquid-filled glass vial. What are the desirable properties for (a) a solid plug and (b) a liquid-filled vial?

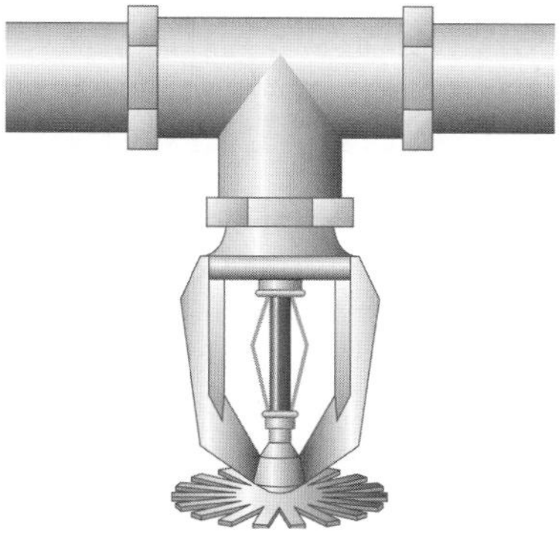

60. The optimal use of water at a fire is to efficiently remove the heat from the fire by completely vaporizing the water to steam. An added benefit is that 4.0 L of water will produce approximately 6.0 m^3 of steam capable of displacing oxygen and preventing it from reaching the fuel source. Can too much water be added to a fire to fight it effectively?

61. If benzene is found in a blank sample of carpeting that has been analyzed by gas chromatography, could the presence of benzene in the suspected sample still be used to determine whether an accelerant was used to ignite the fire?

62. If benzene is not present in the blank sample of carpeting analyzed by gas chromatography but *is* present in the suspected sample, does this constitute proof that a benzene-containing fuel was used to start the fire?

63. If signs of electrical arcing are present in an electrical box, is this proof that a structural fire was caused by the electric arcing?

64. What thermal properties are desirable in the equipment a firefighter uses and carries into fires?

65. Sometimes flames appear to be forming out of the smoke that is pouring out of a structural fire. Is this possible?

66. In a structural fire, what is most commonly the limiting reagent? When is it advisable for firefighters to ventilate a room?

Case Study Problems

67. In the Kenny Richey case, a plastic fire detector in the apartment building was found melted and dangling from the ceiling by wires. Is this proof of tampering or is there a reasonable alternative explanation based on the physical properties of the components and basic principles of fire science?

68. In the Kenny Richey case, the fact that the apartment building became engulfed so quickly was used as evidence that the fire must have been set with an accelerant. Is this a scientifically sound view?

69. In the Kenny Richey case, the deck boards were found to have charring in between the boards, a fact presented as proof that an accelerant had been poured across the deck's surface and seeped down between the cracks in the boards. Is there another explanation, based on the three requirements of fire, that would explain this observation?

70. What problems exist in relying on any information obtained from the analysis of the carpet of the apartment fire in the Kenny Richey case?

CHAPTER 10

Chemistry of Explosions

Nathan Allen was murdered on May 10, 1979, when a stick of dynamite exploded as he started his truck. *(Alexander Tsiaras/Photo Researchers, Inc.)*

CASE STUDY: Tracing Explosives

On May 10, 1979, Nathan Allen left his job at Bethlehem Steel in Sparrows Point, Maryland, and got into his 1977 Dodge truck. As Nathan started the engine, an explosion ripped through the cab of his truck. Although he survived the initial blast at 10:40 P.M., he passed away at 1:17 A.M. in the trauma unit of the local hospital. The size and scope of the blast led investigators to believe that a powerful explosive material had been used in the blast that took Mr. Allen's life.

Local authorities called on state and federal agencies to assist in the investigation of this explosion. Evidence from the bomb scene in parking lot 7 of the steel mill was sent to a national laboratory in Rockville, Maryland. The laboratory, operated by the Bureau of Alcohol, Tobacco, Firearms and Explosives, was able to identify the explosive material as Tovex 220, a brand of dynamite. This information was important. Further laboratory investigation found that the dynamite had been manufactured the previous December and that part of that shipment went to Jenkins Explosives in Martinsburg, West Virginia.

Investigators traveled to Martinsburg to interview Lawrence Jenkins, owner of the store that sold the Tovex 220 dynamite. There they discovered that on March 10, 1979, James McFillin bought two sticks of dynamite, reportedly for removing tree stumps. The owner was able to make a positive identification of McFillin as the person who purchased the dynamite, and he was able to describe the vehicle McFillin had driven.

The next step of the investigation was to explore any connection between McFillin and his victim, Nathan Allan. Investigators were shocked to learn that McFillin was Nathan's next-door neighbor—and uncle! After interviewing family members, it soon became apparent that McFillin had never accepted Nathan's friendship with McFillin's wife, Sandra, and accused them of having an affair. He had made these accusations before but had never resorted to violence. However, earlier on the afternoon of May 10, 1979, James McFillin reportedly had seen Sandra and Nathan sitting on the front step talking. After he confronted them with his usual accusations, Nathan simply returned home. Sandra, on the other hand, informed James that she was intending to leave him as soon as their children were grown. That night, James disappeared for over an hour before supper with no explanation of where he had been. Investigators now knew that McFillin had the means, the motive, and the opportunity to kill Nathan. James was found guilty and sent to prison for the murder of his nephew.

How could residue from the explosion provide investigators with the brand name, manufacturing, and shipping information for the dynamite? . . .

10.1 Explosives 101

Learning Objective

Differentiate between high and low explosives.

What are the requirements for a compound or mixture to be explosive? Certainly, the explosive material must be capable of releasing a large amount of energy. However, the combustion of coal actually releases much more energy per gram than nitroglycerin does, yet we do not consider coal to be explosive. The difference between the reactions of coal and of nitroglycerin is that the combustion of coal is a slow process whereas nitroglycerin reacts very rapidly.

In an explosion, substantial amounts of energy and gaseous products are released almost instantaneously. The energy produced in the reaction is released as thermal energy and the kinetic energy of flying debris. There are different mechanisms for initiating explosive reactions and different uses for explosions based on the chemical and physical properties of the explosives. Explosives are separated into two broad classes: low explosives and high explosives. A **low explosive** is a compound that must be confined within a container to produce an explosion and can be triggered by a flame that ignites the explosive compound. The combustion reaction of a low explosive occurs in milliseconds (thousandths of a second) and produces a large volume of gaseous products. Gunpowder is an excellent example of a low explosive. If ignited in the open, it will simply burn and exhaust both heat and gases into the surroundings with little effect. However, when gunpowder is confined within the chamber of a gun, it becomes an explosive compound. Most low explosives serve as propellants for bullets in guns and shells in military artillery.

High explosives produce a violent, shattering effect without having to be confined like low explosives. The detonation is initiated either by heat or by a shockwave, and the explosion occurs in microseconds (millionths of a second). The explosion produces a tremendous change in pressure that shatters any material near the explosive.

High explosive materials must be relatively stable compounds to ensure the safe handling of the explosive until it is needed. However, if the explosive material is too stable, the initiation of the reaction will be difficult. It is common to use several types of explosives with varying degrees of stability in commercial or military applications. The bulk of the explosive is made up of a stable explosive compound. A second explosive is housed in a detonator, which serves to trigger the main charge. The explosive in the detonator is usually very reactive and easily triggered, such as lead azide, $Pb(N_3)_2$. Because of its reactivity, only very small amounts are used, usually less than 0.5 gram. The detonator is never brought in contact with the main explosive charge until just before use. Without the detonator, the bulk explosive material is much safer to handle.

Figure 10.1 shows the chemical structures of several common explosive molecules. Note that the structures are those of organic compounds that contain nitro ($-NO_2$) groups. When these compounds undergo combustion reactions, the products include carbon dioxide, carbon monoxide, nitrogen gas, hydrogen gas, and water vapor. The detonation of these compounds occurs in microseconds, converting nearly all of the

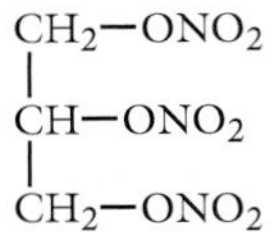

Nitroglycerin (NG)

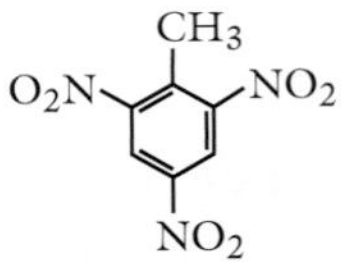

2,4,6-trinitrotoluene (TNT)

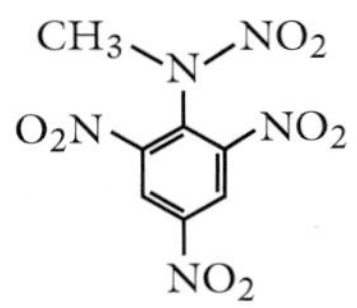

2,4,6-trinitrophenyl-n-methylnitramine (tetryl)

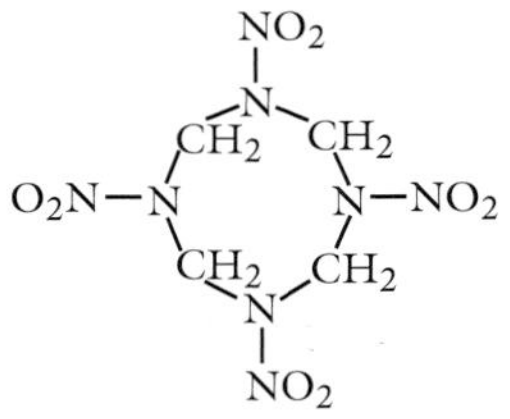

Cyclotetramethylenetetranitramine (HMX)

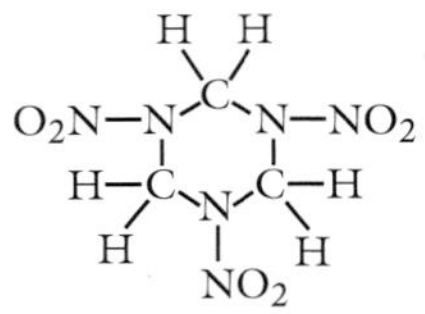

Cyclotrimethylenetrinitramine (RDX)

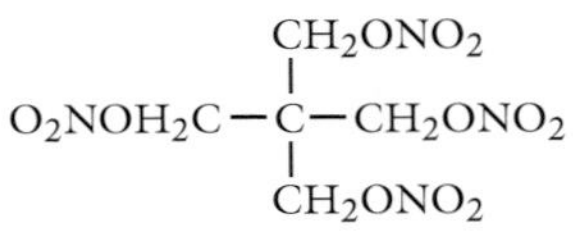

Pentaerythrytoltetranitrate (PETN)

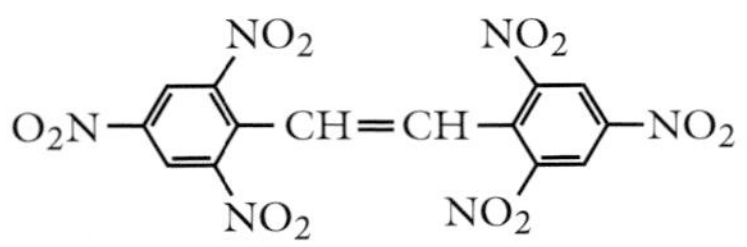

Hexanitrostilbene (HNS)

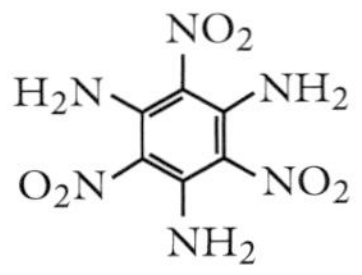

Trinitrotriaminobenzene (TATB)

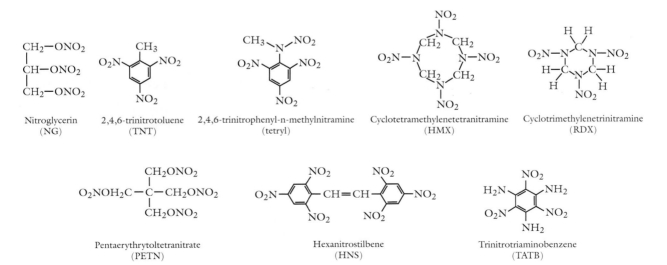

Figure 10.1 Structures of explosive molecules.

solid mass into highly compressed gaseous compounds that expand outward rapidly, creating a blast wave. Consider the detonation of nitroglycerin, as shown in the following reaction:

$$4C_3H_5N_3O_9(l) \rightarrow 12CO_2(g) + 10H_2O(g) + 6N_2(g) + O_2(g)$$

On a molecular level, the reaction of four liquid nitroglycerin molecules will produce 29 total molecules of gas. On the much larger scale of bulk explosives, this reaction produces a tremendous volume of gas. The typical explosion will produce 750 to 1000 liters of gas per kilogram of explosive detonated.

Plastic explosives such as C4 are mixtures of a traditional explosive (RDX) with materials such as rubber or oils that make the mixture moldable. While there are plastics that are true explosives, most plastic explosives are simply moldable mixtures.

10.2 Redox Chemistry of Explosives

Learning Objective

Determine an explosive molecule's oxygen balance.

In Section 9.1 on the chemistry of fire, we discussed combustion reactions and their requirements for fuel and oxygen. Explosions are nothing more than specialized combustion reactions. However, they typically do not rely on oxygen gas from the air. Consider, for example, the reaction of gunpowder:

$$4KNO_3(s) + 7C(s) + S(s) \rightarrow 3CO_2(g) + 3CO(g) + 2N_2(g) + K_2CO_3(s) + K_2S(s)$$

The carbon component of the reaction is the fuel, and the potassium nitrate serves as the source of oxygen in the reaction. Assigning oxidation numbers to the components of the reaction reveals that carbon is oxidized (loses electrons) and that nitrogen is reduced (gains electrons). Because the fuel source and oxygen source are both solids and must be physically in contact, it is critical for the gunpowder to be a homogenous mixture of fine powders.

The molecules in Figure 10.1 provide a fuel source (carbon and hydrogen atoms) and an oxygen source ($-NO_2$ groups) built into a single molecule. This combination of fuel and oxygen source increases the efficiency of the explosion and eliminates the need to mix the explosive compounds from Figure 10.1 with other compounds. Some explosive compounds are found in mixtures, but the purpose is to alter the physical or chemical properties of the explosive, as occurs when making a plastic explosive.

The reactions of both nitroglycerin and 2,4,6-trinitrotoluene (TNT) are shown below. Assigning oxidation numbers to the components of each reaction reveals that carbon is oxidized (loses electrons) and that nitrogen is reduced (gains electrons).

Nitroglycerin: $4C_3H_5N_3O_9(l) \rightarrow 12CO_2(g) + 10H_2O(g) + 6N_2(g) + O_2(g)$
2,4,6-trinitrotoluene: $2C_7H_5N_3O_6(l) \rightarrow 7CO(g) + 7C(s) + 5H_2O(g) + 3N_2(g)$

Notice that the detonation of nitroglycerin produces all gaseous compounds, whereas the detonation of TNT produces carbon particles and

The muzzle flash seen when a gun is fired results from the combustion of hot carbon monoxide gas as it reacts with oxygen gas from the atmosphere. (Scott Doyle/Firearms.com)

gaseous compounds. To understand the reason for this difference, we must consider a concept called the **oxygen balance** of the explosive.

If there is sufficient oxygen within the molecule to completely oxidize the carbon and hydrogen atoms, the compound is said to have a *neutral* oxygen balance. A *negative* oxygen balance exists when the molecule contains insufficient oxygen atoms for a complete oxidation reaction; a *positive* oxygen balance exists when there is a surplus of oxygen. Nitroglycerin, an example of a compound with a positive oxygen balance, produces a small surplus of oxygen gas. TNT, however, lacks sufficient oxygen to react completely with the carbon present in the explosive and, therefore, has a negative oxygen balance.

The use of explosives for mining or in confined spaces must be done carefully, because carbon monoxide poisoning can occur if explosives with negative oxygen balances are used. Propellants used in firing guns usually have a negative oxygen balance. When the heated, flammable carbon monoxide gas is released at the end of a barrel, it ignites in the muzzle flash.

Worked Example 1

Determine whether the explosive compound RDX from Figure 10.1 has a positive, negative, or neutral oxygen balance. Assume an ideal reaction in which all carbon is converted to carbon dioxide, all hydrogen is converted to water vapor, and all nitrogen is converted to nitrogen gas. Compare the oxygen needed with that present in the molecule.

SOLUTION The formula of RDX is $C_3H_6N_6O_6$. The complete oxidation of RDX would be:

3 C atoms form 3 CO_2 molecules → requires 6 O atoms

6 H atoms form 3 H_2O molecules → requires 3 O atoms

6 N atoms form 3 N_2 molecules → requires 0 O atoms

Total O atoms needed: 9

RDX contains 6 oxygen atoms, but 9 oxygen atoms are needed. RDX therefore has a negative oxygen balance and would produce CO, not CO_2.

Practice 10.1

Ammonium nitrate (NH_4NO_3) is not typically considered an explosive material. However, when mixed with certain organic compounds, it can form an explosive mixture. This was the explosive used in the Oklahoma City bombing of the Alfred P. Murrah Federal Building. Determine whether pure ammonium nitrate has a negative or positive oxygen balance upon combustion.

ANSWER

Ammonium nitrate has a positive oxygen balance.

Explosives are often manufactured as mixtures that benefit from properties of other explosive and nonexplosive compounds. For example, TNT has a negative oxygen balance and ammonium nitrate has a positive oxygen balance. The mixture of the two is called *amatol* and has a neutral oxygen balance. Amatol, first used in World War I, was critical in the war effort, because TNT manufacturing plants could not keep up with the demand for explosives. The amatol mixture reduced the amount of TNT needed in the bombs and provided superior performance.

10.3 Kinetic-Molecular Theory of Gases

Learning Objective

Explain the properties and behaviors of gases using kinetic-molecular theory.

Changes in the temperature, volume, and pressure of gases produced in an explosion occur rapidly. Before discussing how these variables are related, we must first look at a molecular model of gases. In the following description of the properties of gases, the term *gas particle* refers to either atoms or molecules.

Kinetic-Molecular Theory of Gases

1. Gas particles are extremely small and have relatively large distances between them. *Gases can be compressed easily, but liquids and solids cannot be compressed to any appreciable extent because the particles that make them up are already very close to one another.*
2. Gas particles act independently of one another because there are no significant attractive or repulsive forces between gas particles. *Compared with the energy of the gas particles, the intermolecular forces between gas particles are so weak that they do not have any real effect. Liquid and solid particles are affected by intermolecular forces, as discussed in Chapter 6.*
3. Gas particles are in continuous random, straight-line motion as they collide with one another and with the container walls. *Gas particles move until a collision alters their course, which is why gases fill the entire volume of their container.*
4. The average kinetic energy of gas particles is proportional to the temperature of the gas. *Energy added into the system increases the kinetic energy of the particles, which translates to an increase in temperature of the gas.*

The **kinetic-molecular theory** of gases provides a basis for understanding the observable behaviors of gases. For example, gases can be compressed in containers because of the large distances between gas particles. The particles in liquids and solids cannot be compressed to any great extent because there is very little space between particles in the condensed phases. Collisions of the gas particles against the walls of the container create the pressure of the gas within the container. When a gas is heated, the velocity of the gas particles increases and so does their kinetic energy. Because the temperature of a gas is a measure of the average kinetic energy of all the gas particles present, the temperature of the gas increases.

To understand what happens in explosions, we will use the insights into gas behavior provided by the kinetic-molecular theory and consider how explosive compounds react. For example, if we have a mixture of fuel vapors and oxygen gas from the air, the kinetic-molecular theory tells us that molecules of fuel and oxygen move randomly and collide. Is the energy of the collision sufficient to overcome the activation energy barrier? Detonators or, in the case of fuel-air explosions, a simple flame provides the required activation energy. Yet whether or not the collisions result in a fuel-air explosion depends on several additional factors.

Fuels have a **lower explosive limit (LEL)**, the lowest ratio of fuel to air at which the mixture can propagate a flame. Below this level, no explosion will occur because there is not enough fuel present in the mixture for a sufficient number of collisions to sustain a continued reaction. Natural gas, which is mostly methane gas, has an LEL of 3.9%. If the concentration is any lower, natural gas in air will not produce an explosion.

Perhaps surprisingly, fuels also have an **upper explosive limit (UEL)**, the highest ratio of fuel to air at which the mixture will support a combustion reaction. Natural gas has a UEL of 15%. At this concentration, the mixture becomes too fuel-rich for the natural gas to react explosively, because there is an insufficient quantity of oxygen present to sustain a reaction. Although such a mixture will not explode, it presents a dangerous situation because gases diffuse quickly. The concentration of fuel vapors may start above the UEL, but the gas will be diffusing at its edges and may quickly fall within the limit that supports combustion and, perhaps, an explosion.

Fuel-air explosions are used in military bombs and can also be improvised by terrorists. At the inauguration of President George W. Bush for his second term, for instance, the U.S. Secret Service watched for limousines that could be carrying propane tanks to create a fuel-air explosion. Terrorists had used an improvised explosive device (IED) of this type in the bombing of a Tunisian synagogue in April 2002. Therefore, when terrorists' plans for a limo bomb were discovered before the presidential inauguration, the threat was taken seriously.

10.4 Gas Laws

Learning Objective

Using the gas laws, describe how gases behave.

Whether an explosion is from a harmless firecracker or an IED built by terrorists, the power of the explosion is due to the almost instantaneous release of energy. In this highly energetic environment, large volumes of gas molecules are produced. Immediately following the explosion, the temperature of the gases reaches several thousand degrees Celsius, the volume of the gases expands at a tremendous rate, and the pressure exerted by the gases can propel solid objects—whether bullets from a gun or fragments from a bomb—at high speeds.

Although explosions are extremely complex, we can comprehend some aspects of what happens by studying a group of principles called the *gas laws*. These laws describe how the temperature, pressure, volume, and number of moles of a gas are related. The gas laws apply to all gases, no matter what their identity.

Avogadro's law states that equal volumes of gases at the same temperature and pressure contain the same number of gas particles. This law means that the volume of a gas is directly proportional to the number of gas particles (and, therefore, the number of moles of gas) in the sample, as long as the pressure and temperature of the gas remain constant. Figure 10.2 illustrates this concept. The gas-filled containers in Figures 10.2a and 10.2b hold different gases but have the same number of gas particles because the containers are equal in volume and have the same temperature and pressure. Avogadro's law is useful because the coefficients in balanced equations that describe chemical reactions of gases can be interpreted as volumes of reactants and products at identical temperatures and pressures.

Figure 10.2 Avogadro's law. Containers of gases at equal temperatures and pressures have the same number of moles.

Worked Example 2

Hydrogen and oxygen gas can form an explosive mixture. Sketch what the missing gas container in the following diagram looks like when 2 L of hydrogen gas react with 1 L of oxygen gas to form water vapor:

$$2H_2(g) \quad + \quad O_2(g) \quad \longrightarrow \quad 2H_2O(g)$$

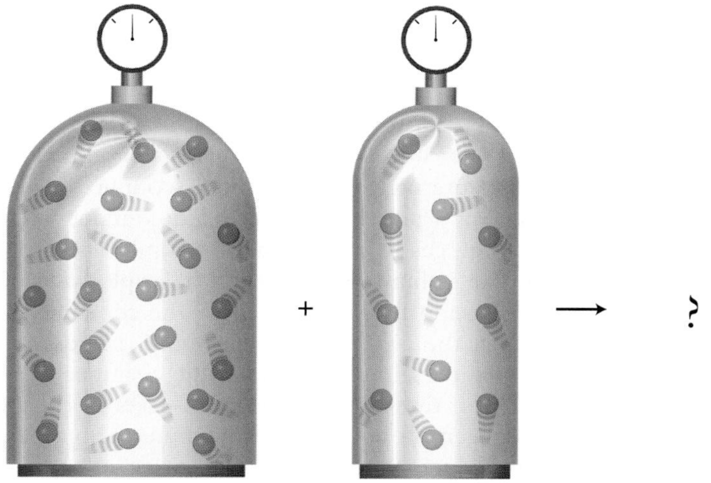

SOLUTION The coefficients of the balanced chemical equation can be interpreted as representing the volumes of the gases that react. Therefore,

2 L of hydrogen gas react with 1 L of oxygen to produce 2 L of water vapor, as shown below:

Practice 10.2

If two large, unlabeled industrial gas cylinders of equal volume are filled to the same pressure and temperature, one with hydrogen gas and one with argon gas, is there any way to determine which container has hydrogen and which has argon? The information provided on the periodic table about hydrogen and argon may be useful when thinking about your answer.

ANSWER

The two containers have exactly the same number of H_2 molecules as Ar atoms. However, because H_2 has a molecular mass of 2.02 amu and Ar has an atomic mass of 39.95 amu, the mass of the argon gas is approximately 20 times greater than the mass of hydrogen gas. Thus, the Ar-filled container would have a much greater mass than the H_2-filled container.

For low explosives, the creation of a large amount of gaseous products results in a large pressure increase if the gas is confined in a small space—such as the center of a pipe bomb. The rigid container walls do not allow the volume of the gas to increase as the number of gas particles increases. Because the volume is constant, the explosive reaction causes a tremendous increase in the pressure inside the container. Thus, the container (in this case, the pipe) ruptures, flinging dangerous fragments in all directions. If the explosion occurs in the chamber of a gun, the bullet is propelled out of the barrel to allow the gases to expand.

The relationship between the pressure and volume of a gas is summarized in **Boyle's law**, which states that the pressure of a gas is inversely proportional to its volume. The inverse relationship indicates that if one variable increases, the other must decrease, provided the temperature and number of gas particles stay constant. Figure 10.3 illustrates Boyle's law.

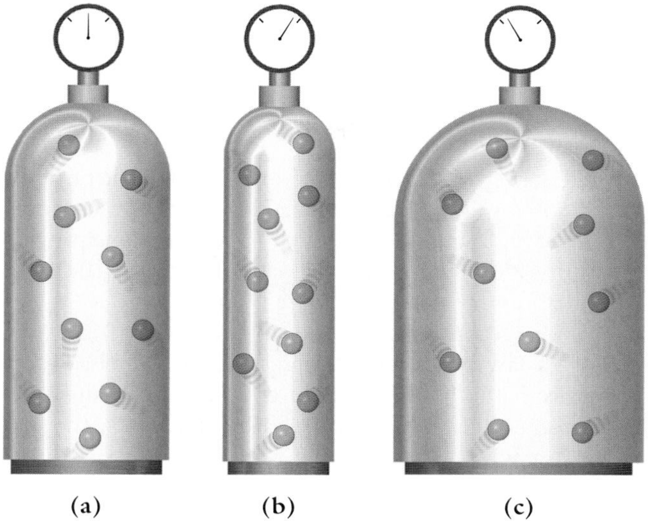

Figure 10.3 Boyle's law states that if the volume of a gas decreases, as from (a) to (b), the pressure will increase. Conversely, if the volume increases, as from (a) to (c), the pressure will decrease.

The container in Figure 10.3a holds a fixed number of gas particles. If the volume of the gas is decreased to that shown in Figure 10.3b, the gas particles will not have to travel as far to collide with the container walls and therefore will undergo more collisions. The greater number of collisions results in an increase in pressure. If the volume of the container increases as in Figure 10.3c, the gas particles will have a greater distance to travel between collisions with the container wall, resulting in a decrease in pressure.

Worked Example 3

At an altitude of 3000 m, the air pressure is approximately 0.35 atm. Most airplane cabins are pressurized to 0.70 atm. If the pressurized cabin of an airplane is breached at an altitude of 3000 m, describe what will happen to the volume of air (doubled, halved, no change) in the lungs of an airplane passenger.

SOLUTION Boyle's law states that if the pressure decreases, the volume must increase. Because the pressure drops from 0.70 atm to 0.35 atm, the volume must increase by an equal ratio (0.70/0.35). Therefore, the volume of air in a passenger's lungs will double.

Practice 10.3

Does the density of air change as the altitude increases? Explain your answer.

ANSWER

Air pressure decreases as the height above sea level increases. As the pressure decreases for gases, the volume must increase. The density of a gas is the mass of gas divided by the volume the gas occupies. The equation

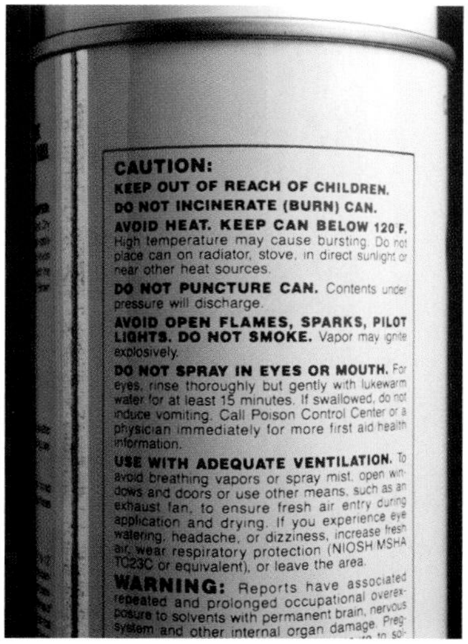

Aerosol cans have temperature warnings for storage because increasing the temperature will increase the internal pressure to potentially dangerous levels. (Michael Dalton/Fundamental Photographs)

for density is $D = m/V$, as explained in Chapter 2 (page 42). Because the volume of air increases at higher altitudes, there is less mass within a given volume, and the density of air decreases.

Compressed gas cylinders are used in many different businesses, from welding shops to hospitals, florists to laboratories. By compressing the gases at extremely high pressures, a much larger supply is provided to the consumer within a tank of manageable size. However, the storage temperature of high-pressure gas tanks must be monitored so that the tanks do not become dangerous at elevated temperatures.

Gay-Lussac's law states that the pressure of a gas is directly related to its temperature if the volume is held constant, as in an enclosed container. Therefore, as the temperature of a gas increases, the pressure increases proportionately.

The kinetic-molecular theory provides an insight into how temperature and pressure are related because the temperature of a gas is a measure of the kinetic energy of the gas molecules. Consider the illustrations in Figure 10.4. The kinetic energy of the gas particles shown in Figure 10.4a is related to the velocity of the gas particles. As the container is heated, the gas particles increase their kinetic energy, which corresponds to a higher average velocity for the gas particles, as shown in Figure 10.4b. The increased velocity of the gas particles increases the pressure because the particles undergo collisions more often with the container walls. The opposite condition occurs if the temperature of the gas particles is decreased, as illustrated in Figure 10.4c.

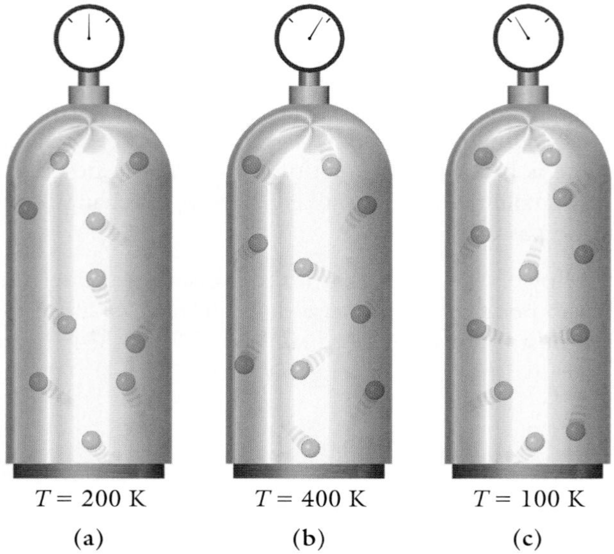

Figure 10.4 Gay-Lussac's law states that if the temperature of a gas increases, as from (a) to (b), the pressure will increase. Conversely, if the temperature decreases, as from (a) to (c), the pressure will decrease.

One of the many dangers that firefighters must confront is explosions that result from the heating of compressed gases in aerosol cans and compressed gas cylinders. If the contents are flammable, they may further react after the explosion. Most compressed gas cylinders are equipped with a safety valve designed to vent the contents if the internal pressure gets too high.

Compressed hydrogen gas is a nonliquefied gas because it will not form a liquid at room temperatures even when subjected to high pressures. Other gases such as propane will form a liquid at high pressures and are stored in liquefied form. Liquid propane evaporates as the propane is removed from the cylinder to provide a steady flow of gas. When a liquefied gas cylinder is heated, the number of particles in the gas state increases, thus causing the gas pressure to increase until the container walls fail in an explosion.

Worked Example 4

Car tires are heated by friction with the road surface while driving. Is this related to the recommendation by tire manufacturers that the air pressure in the tires be checked before driving a long distance?

SOLUTION The amount of air in the tire and the volume of the tire are fixed. However, as a result of friction between the tire and the road surface as a car is driven, the temperature of the tires will increase. As the temperature of the air inside the tire rises, so does the pressure. If the tires happen to be overinflated at the outset, the added increase in pressure while driving can cause a tire to rupture.

Practice 10.4

Why do the manufacturers of automobile tires recommend that tire pressures be measured several times per year?

ANSWER

The pressure of air in the tires changes with seasonal temperature variations. During the winter months, it is necessary to add air to the tires to maintain the proper tire pressure. As the weather warms up, the air pressure in the tire increases and the excess pressure must be released.

Thus far we have considered gas laws that express the relationship between volume and number of moles (Avogadro's law), between pressure and volume (Boyle's law), and between pressure and temperature (Gay-Lussac's law). The last of the gas laws we will discuss that relate two variables is **Charles's law**, which states that the volume of a gas is directly proportional to its temperature, provided that the pressure and amount of gas remain constant. By increasing the temperature of a gas particle, the kinetic energy and, therefore, the velocity of the gas particle will increase. If the pressure is to remain constant, the volume of the gas must increase. The gases produced in an explosion are initially heated by the exothermic reaction, but as they cool to the temperature of the surroundings, the volume of gas must decrease.

The final volume (the volume after expansion) of the gas produced in an explosion is dictated by the final temperature of the gas, which will be the same as the ambient temperature, or the temperature of the surroundings. A single stick of dynamite has a volume of approximately 400 cm^3 (slightly more than a can of cola), which will upon explosion be converted nearly instantaneously to gaseous products with a final volume of 462 L at a temperature of 25°C. This represents more than a thousandfold increase in volume.

Worked Example 5

Does the density of a gas depend on the temperature of the gas? Explain.

SOLUTION Yes, the density depends on the temperature of the gas. It is calculated by dividing the mass by the volume of the gas: $D = m/V$. Because the volume of a gas increases with an increase in temperature, the density of a hot gas is lower than the density of a cold gas.

Practice 10.5

In Section 9.4, the process of a flashover was discussed. Explain this process in terms of Charles's law.

ANSWER
A flashover happens when a fire located in one region of the room produces hot gases that become trapped at the ceiling, radiating heat to the remaining contents of a room until the autoignition temperature of the contents is reached. The hot gases produced in the combustion reaction have a lower density than cold gases, due to their increased volume. This is why the less dense gas rises to the ceiling of the room.

The individual gas laws are summarized in Table 10.1. Remember that the individual gas laws assume that all variables, other than those undergoing change according to the law, are held constant. Figure 10.5 focuses on what occurs in a system when the volume changes. The increase in volume can be accomplished by either increasing the temperature or the number of gas particles in a system. In order for the volume of a system to be decreased, either the temperature of the system or the number of gas particles in the system must be decreased.

Figure 10.6 illustrates the variable that must change in order for the pressure of a system to change. If the pressure increases, the volume of the system must be decreased or the temperature of the system must be increased. If the pressure of a system decreases, either the number of gas particles decreased or the temperature of the system decreased.

Table 10.1 Summary of Gas Laws

Law	Variables	Constants	Relationship
Avogadro's	n and V	T and P	Directly proportional
Boyle's	P and V	n and T	Inversely proportional
Gay-Lussac's	P and T	n and V	Directly proportional
Charles's	T and V	n and P	Directly proportional

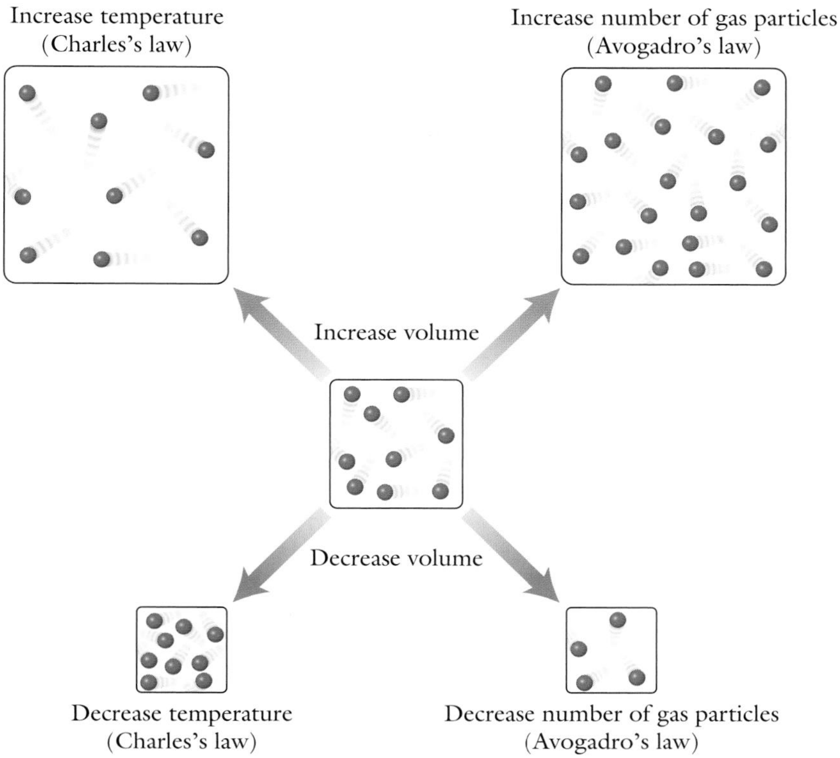

Figure 10.5 Summary of gas laws related to volume changes (with pressure held constant).

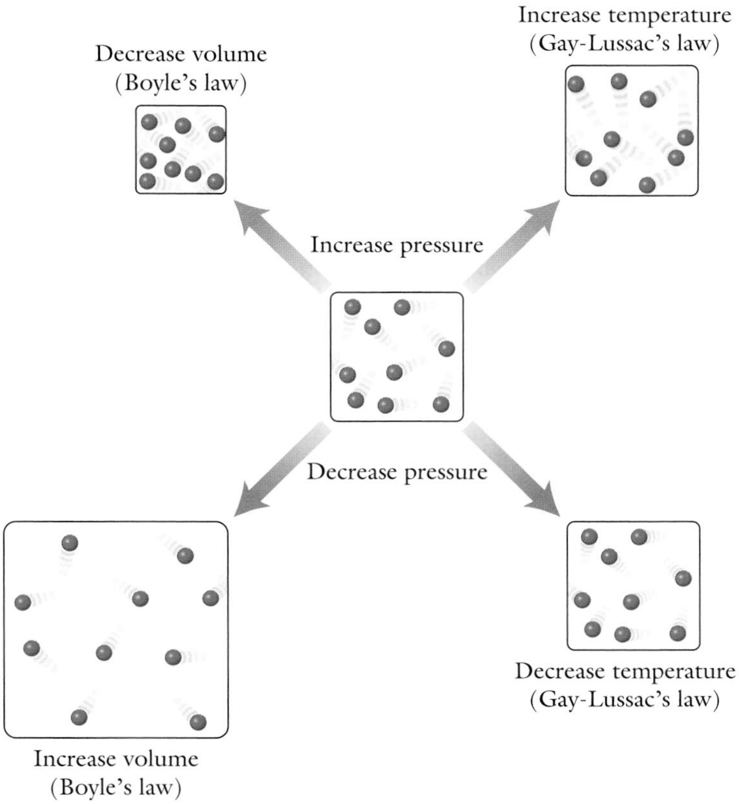

Figure 10.6 Summary of gas laws related to pressure changes (with number of moles held constant).

10.5 Mathematics of the Gas Laws

Learning Objective

Perform calculations using the gas laws to illustrate the behavior of gases.

Avogadro's Law

Avogadro's law states that the volume of a gas is proportional to the number of gas particles present at constant temperature and pressure. If additional gas particles are added to a container, the volume must increase proportionately. The new volume can be calculated using the equation shown below. The initial and final conditions for the gases are noted by the subscripts 1 and 2, respectively.

$$\frac{V_1}{n_1} = \frac{V_2}{n_2}$$

The volume occupied by one mole of gas is 22.4 L when the temperature of the gas is 0°C and its pressure is 1 atm. These conditions of temperature and pressure are often referred to as **standard temperature and pressure (STP).** We can express Avogadro's law at STP as follows:

$$\frac{22.4\ \text{L}}{1\ \text{mol}} = \frac{V_2}{n_2}$$

Worked Example 6

Automobile airbags inflate when sodium azide (NaN_3) contained in the bags explodes during a collision. Calculate the volume of gas produced at STP conditions if 32.5 g of sodium azide are used to inflate a side-impact airbag according to the following equation:

$$2NaN_3(s) \rightarrow 2Na(s) + 3N_2(g)$$

SOLUTION The first step is to calculate moles of sodium azide reacted:

$$32.5\ \cancel{\text{g}} \times \underbrace{\frac{1\ \text{mol}}{65.0\ \cancel{\text{g}}}}_{\text{Molar mass of NaN}_3} = 0.500\ \text{mol}$$

The second step is to calculate how many moles of nitrogen gas are produced from 0.500 mol of NaN_3:

$$0.500\ \cancel{\text{mol NaN}_3} \times \underbrace{\frac{3\ \text{mol N}_2}{2\ \cancel{\text{mol NaN}_3}}}_{\text{Stoichiometric coefficients from balanced equation}} = 0.750\ \text{mol N}_2$$

The final step is to calculate the volume of 0.750 mol of N_2 at STP:

$$\frac{22.4\ \text{L}}{1\ \text{mol N}_2} = \frac{V_2}{0.750\ \text{mol N}_2} \Rightarrow V_2 = \frac{22.4\ \text{L}}{1\ \cancel{\text{mol N}_2}} \times 0.750\ \cancel{\text{mol N}_2} = 16.8\ \text{L}$$

Practice 10.6

Calculate the number of moles of nitrogen gas produced and grams of sodium azide consumed if 115 L of N_2 results from a sodium azide explosion.

ANSWER

5.13 mol of N_2 and 222 g of NaN_3

Boyle's Law

Boyle's law states that the pressure of a gas is inversely proportional to the volume of the gas at constant temperature. Compressed gas cylinders are designed with Boyle's law taken into consideration because a large amount of gas is stored in a small volume. This compression leads to extremely high pressures within the cylinder. As the pressure is reduced, the volume of the gases increases. Boyle's law calculations use the following equation:

$$P_1 V_1 = P_2 V_2$$

Compressed gases have been used to increase the damage done by improvised explosive devices because of the massive increase in volume of gas that occurs when a tank is physically ruptured.

Worked Example 7

If a compressed gas cylinder has a pressure of 135 atm and a volume of 15.0 L, calculate the volume the gas will occupy at 1.00 atm of pressure.

SOLUTION The first step is to summarize the variables of the problem:

$P_1 = 135$ atm

$V_1 = 15.0$ L

$V_2 = ?$

$P_2 = 1.00$ atm

The next step is to use the Boyle equation to solve for V_2:

$$P_1 V_1 = P_2 V_2 \Rightarrow V_2 = \frac{P_1 V_1}{P_2} = \frac{(135 \text{ atm})(15.0 \text{ L})}{1.00 \text{ atm}} = 2.03 \times 10^3 \text{ L}$$

Practice 10.7

Calculate the pressure inside a pipe bomb before it ruptures, given that the internal volume of the pipe is 475 cm^3 and the contents of the bomb generate exactly 1 mol of gas at STP.

ANSWER

47.2 atm

Gay-Lussac's Law

Gay-Lussac's law states that the pressure of a gas is directly proportional to the temperature when the volume of the gas is constant. The equation expressing this relationship between pressure and temperature is given below:

$$\frac{P_1}{T_1} = \frac{P_2}{T_2}$$

An important point about this equation is that the temperatures must be expressed using the Kelvin scale (K) rather than the Celsius scale (°C). The Kelvin temperature of a gas is calculated by adding 273.15 to the Celsius temperature (K = °C + 273.15). For most calculations, in which temperature is expressed only to the nearest degree (as in 25°C), 273.15 is rounded off to 273. Also note that the degree sign is *not* used in the Kelvin units.

Worked Example 8

The pressure inside a compressed gas cylinder is 134 atm at 25°C. Calculate the new pressure inside the cylinder if it is heated to 48°C.

SOLUTION The first step is to summarize the variables of the problem:

$P_1 = 134$ atm

$T_1 = 25 + 273 = 298$ K

$P_2 = ?$

$T_2 = 48 + 273 = 321$ K

The next step is to use the Gay-Lussac equation to solve for V_2:

$$\frac{P_1}{T_1} = \frac{P_2}{T_2} \Rightarrow P_2 = \frac{P_1 T_2}{T_1} = \frac{(134\ \text{atm})(321\ \cancel{K})}{298\ \cancel{K}} = 144\ \text{atm}$$

Practice 10.8

The pressure of CO_2 inside a cola bottle is approximately 1.35 atm at 25°C. What will be the pressure inside the bottle if it is chilled to 0°C?

ANSWER

1.24 atm

Charles's Law

Charles's law states that the volume of a gas is directly proportional to the temperature of the gas at constant pressure. We know that the volume of one mole of gas at 0°C is 22.4 L. But most fires and explosions occur at temperatures much greater than 0°C. Charles's law, shown below in mathematical form, provides a method for calculating the volume of the gas at other temperatures. Temperatures must be expressed in kelvin, as above.

$$\frac{V_1}{T_1} = \frac{V_2}{T_2}$$

Worked Example 9

Calculate the density of 1.00 mol of carbon dioxide at STP. Then calculate the density of the gas if the temperature is increased to 500°C.

SOLUTION To calculate the density of CO_2, substitute into the formula for density:

$$D = \frac{m}{V} = \frac{\overbrace{44.0\text{ g}}^{\text{Molar mass of } CO_2}}{\underbrace{22.4\text{ L}}_{\text{Volume of 1 mol at STP}}} = 1.96\text{ g/L}$$

Before calculating the density of CO_2 at 500°C, we need the volume at that temperature. Summarize the variables in the problem:

$V_1 = 22.4$ L
$T_1 = 0°\text{C} + 273 = 273$ K
$V_2 = ?$
$T_2 = 500°\text{C} + 273 = 773$ K

Use the Charles equation to solve for V_2:

$$\frac{V_1}{T_1} = \frac{V_2}{T_2} \Rightarrow \frac{22.4\text{ L}}{273\text{ K}} = \frac{V_2}{773\text{ K}} \Rightarrow V_2 = \frac{(773\text{ K})(22.4\text{ L})}{273\text{ K}} = 63.4\text{ L}$$

Calculate the density at 500°C:

$$D = \frac{m}{V} = \frac{\overbrace{44.0\text{ g}}^{\text{Molar mass of } CO_2}}{\underbrace{63.4\text{ L}}_{\text{Volume of 1 mol at 500°C}}} = 0.694\text{ g/L}$$

Practice 10.9

The explosion of 2.00 mol of TNT at 0°C produces a total of 15 mol of gas-state products. Calculate the volume of the gas produced when the final temperature of the gas reaches 25°C. (*Hint:* Use the molar volume of gases at STP to calculate the initial condition for volume and initial temperature.) The balanced equation for the reaction is:

$$2C_7H_5N_3O_6(l) \rightarrow 7CO(g) + 7C(s) + 5H_2O(g) + 3N_2(g)$$

ANSWER
367 L

10.6 The Combined and Ideal Gas Laws

Learning Objective

Show how the pressure, volume, temperature, and quantity of a gas are interrelated.

On Christmas Eve of 2004, the family of Graham Foster received horrible news. Graham had been in a high-speed car accident—he was thrown

Investigating a car accident in which airbags were deployed. (Age Fotostock/SuperStock)

from the car and suffered fatal injuries. Police interviewed his close friend, David Munn, who said that Graham had dropped him off at his home before the accident occurred. Police would soon learn, however, that the events of that night had not transpired as David described them.

The police evidence that contradicted David's story was found through an investigation of the automobile's airbags, which had inflated during the accident. The collision of the car triggered a mechanism for exploding the sodium azide (NaN_3) contained in the airbags; this produces nitrogen gas. As the nitrogen gas filled the airbags, the occupants' forward momentum was slowed by their impact with the airbags. Investigators looked on the surface of the airbags for skin cells, saliva, blood, and cosmetics from the occupants.

The DNA evidence from the airbag on the driver's side matched that of David Munn, not Graham Foster. When Munn was presented with the evidence, he admitted to being behind the wheel, under the influence of methamphetamine, when he lost control while attempting to navigate a curve at a speed in excess of 100 mph. The explosion that had saved his life provided the evidence needed to prove his guilt. The science behind the airbag mechanism is based on knowledge of the gas laws.

The detonation of an explosive compound such as sodium azide produces gas-phase products at high pressures that are expanding in volume and simultaneously changing temperature. The gas laws described in the previous section explain how one variable will respond to a change in another, with the assumption that the remaining variables hold constant. In an explosion, however, the pressure, volume, and temperature of the gas *all* are changing. The **combined gas law** is a combination of Boyle's, Charles's, and Gay-Lussac's laws and is written:

$$\frac{P_1 V_1}{T_1} = \frac{P_2 V_2}{T_2}$$

The combined gas law is used when multiple variables of a gas are simultaneously changed.

The **ideal gas law** is based on measurements of the pressure, volume, temperature, and moles of a gas and is given by the equation:

$$PV = nRT$$

R is the universal gas constant and has the value

$$R = 0.08206 \frac{\text{L} \cdot \text{atm}}{\text{mol} \cdot \text{K}}$$

The ideal gas law allows calculations of the properties of gases under a specific set of conditions. For example, the ideal gas law can be used to determine the moles of nitrogen gas needed to properly inflate a vehicle airbag by detonation of sodium azide. The combined gas law can be used to calculate the change in volume of the airbag during various phases of deployment.

A typical airbag on the passenger's side can require up to 70 L of gas to inflate completely, whereas an airbag on the driver's side requires only 36 L of gas for full inflation because the airbag is closer to the driver. The side-impact airbags designed to cushion the head are typically 12 L in size. The amount of sodium azide required for each type of airbag differs greatly. The equation for the reaction of sodium azide that fills an airbag is given below. The sodium metal produced in the reaction undergoes further reactions (not shown below) with oxygen gas from the atmosphere to form sodium oxide (Na_2O).

$$2NaN_3(s) \rightarrow 2Na(s) + 3N_2(g)$$

The equation for the explosive decomposition of sodium azide shows that 2 mol of sodium azide will produce 3 mol of nitrogen gas. From the ideal gas law, the number of moles of nitrogen needed to fill a 70-L airbag at 1 atm and 25°C can easily be calculated. Once the amount of nitrogen is determined, the stoichiometry of the reaction is used to determine the amount of sodium azide. Approximately a 130-g quantity of sodium azide is needed for the airbags on the passenger's side. The airbags on the driver's side contain approximately half this amount, and side-impact bags contain about one-fourth as much sodium azide.

Sodium azide is an extremely toxic substance. Environmentalists are greatly concerned about the release of this substance into the environment from automobile junkyards. Research is being conducted to find an alternative design for the deployment of airbags.

10.7 Mathematics of the Combined and Ideal Gas Laws

Learning Objective

Use the gas laws to calculate the variables of a gaseous system.

How do you determine which one of the gas laws to use when solving a problem? Carefully examine the conditions set forth in the problem. If only two variables are involved, one that is changing and one that is unknown, use Avogadro's, Boyle's, Charles's, or Gay-Lussac's law. If the problem involves temperature, pressure, and volume, then use the combined gas law. Finally, if the problem involves all four variables, use the ideal gas law to calculate the unknown variable.

When using the ideal gas law, the unit of the universal gas constant (R) is a combination of atmospheres, liters, kelvin, and moles. For the calculation to be accurate, the unit of each variable in the gas must match those used in R. If the variable units in a problem do not match those needed for the calculation, it is best to convert them immediately to the proper units when summarizing the data.

Worked Example 10

A 0.500-L gas container is heated to 95°C and has an internal pressure of 65.0 atm when the container walls undergo a critical failure, releasing the contents. Calculate the final volume the gas will occupy at 25°C and 1.00 atm.

SOLUTION This problem involves a change in the volume, pressure, and temperature of a compressed gas. Therefore, the combined gas law should be used.

First, summarize the provided data:

$P_1 = 65.0$ atm and $P_2 = 1.00$ atm

$V_1 = 0.500$ L and $V_2 = ?$

$T_1 = 95°C + 273 = 368$ K and $T_2 = 25°C + 273 = 298$ K

Always convert temperature to the Kelvin scale immediately!

Now solve for the missing variable using the combined gas law:

$$\frac{P_1 V_1}{T_1} = \frac{P_2 V_2}{T_2} \Rightarrow \frac{65.0 \text{ atm} \times 0.500 \text{ L}}{368 \text{ K}} = \frac{1.00 \text{ atm} \times V_2}{298 \text{ K}}$$

$$V_2 = \frac{65.0 \cancel{\text{atm}} \times 0.500 \text{ L} \times 298 \cancel{\text{K}}}{368 \cancel{\text{K}} \times 1.00 \cancel{\text{atm}}} = 26.3 \text{ L}$$

Practice 10.10

If a 1.00-L container was rated to withstand pressures up to 25.0 atm, calculate what the temperature must have been when it ruptured. The final volume of the gas at 1.00 atm and 25°C was 20.0 L.

ANSWER

373 K, or 100°C

Worked Example 11

Mercury fulminate, $Hg(ONC)_2(s)$, is a high explosive compound that historically was used in detonators. Calculate the final volume of gases produced when 0.500 mol of mercury fulminate is detonated at a pressure of 1.05 atm and a temperature of 28°C. The reaction of mercury fulminate is:

$$Hg(ONC)_2(s) \rightarrow Hg(l) + N_2(g) + 2CO(g)$$

SOLUTION List the variables:

$P = 1.05$ atm

$V = ?$

$T = 28°C + 273 = 301$ K

n = The detonation of 1 mol of $Hg(ONC)_2$ produces 1 mol of $N_2(g)$ and 2 mol of $CO(g)$, for a total of 3 mol of gas. Thus, when 0.500 mol of $Hg(ONC)_2$ is detonated, 1.5 mol of gas will be produced. Because there are four variables, use the ideal gas law to calculate V:

$$PV = nRT \Rightarrow V = \frac{nRT}{P} = \frac{(1.5\text{ mol})(0.8206\text{ L}\cdot\text{atm/mol}\cdot\text{K})(301\text{ K})}{1.05\text{ atm}} = 35.3\text{ L}$$

Practice 10.11

Lead azide, $Pb(N_3)_2$, is a high explosive used in detonators. It decomposes to form lead and nitrogen gas. Determine how many liters of nitrogen gas have to be produced for the resulting pressure of nitrogen in a 15.0-mL container to be 74.5 atm at a temperature of 32°C.

ANSWER
0.0446 L

10.8 Detection of Explosives: Dalton's Law of Partial Pressures

Learning Objective
Explain how Dalton's law of partial pressures relates to mixtures of gases.

The tragic events surrounding the destruction of Pan Am Flight 103 in 1988 over Lockerbie, Scotland, prompted the U.S. Congress to enact a new law called the *Aviation Security Improvement Act* of 1990 (H.R. 5732). Section 107 of this act specifically directed the Federal Aviation Administration (FAA) to determine the types and amounts of explosive materials that posed the greatest risk for use by terrorists. The FAA was further directed to determine whether current technology could be used to screen passengers for explosives and, if not, then to invest in the development of such technology. If you have traveled by air, you are certainly aware of the X-ray machines used to screen luggage and the metal detectors to screen passengers. However, plastic explosives can easily avoid detection by these means and so require other detection methods.

One method developed in response to this need is based on the detection of very small amounts of gas-phase molecules emitted by solid

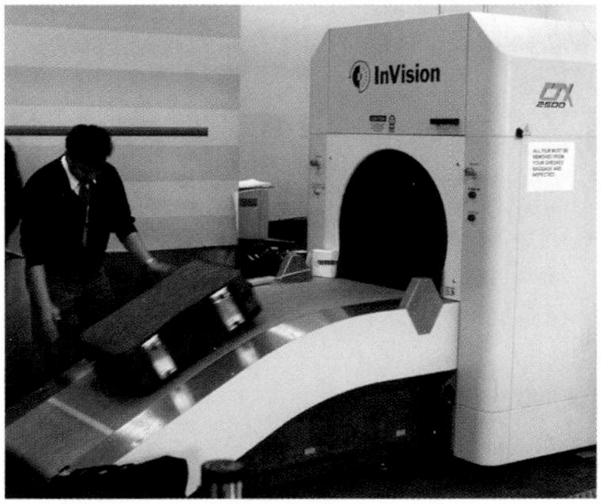

Luggage being screened for explosives. (Reuters/Corbis)

Evidence Analysis | Mass Spectroscopy

GC-MS (All: Courtesy of Varian, Inc.)

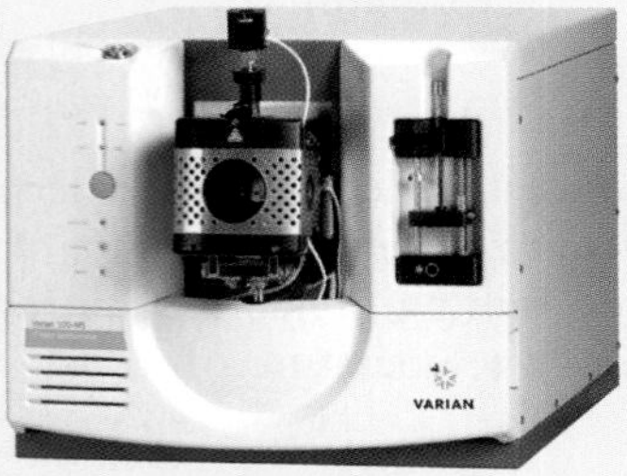

LC-MS

ICP-MS

One of the single most powerful instruments a forensic chemist has in the laboratory to analyze evidence is a **mass spectrometer,** which has the ability to determine the exact molecular mass of an unknown substance. Some uses of a mass spectrometer in the forensic crime laboratory include the confirmation of illegal and prescription drugs, and identification of arson accelerants and explosives.

The mass spectrometer can be found coupled to the other methods discussed earlier in this book, such as gas chromatography-mass spectroscopy (GC-MS), liquid chromatography-mass spectroscopy (LC-MS), and inductively coupled plasma-mass spectroscopy (ICP-MS).

Mass spectroscopy is based on the ability of a magnetic field to deflect a moving ion. Figure 10.7 is an illustration of the major components of a mass spectrometer. The ion source allows a very small number of gas-phase ions created from the unknown compound(s) to enter into the vacuum chamber. It is important for the mass spectrometer to be inside a vacuum so the ions do not collide with atmospheric gases that would divert the ions from their paths. Once the ions enter the mass spec-

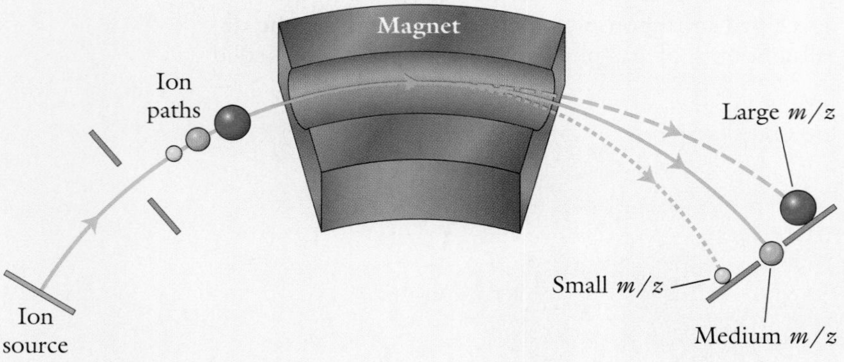

Figure 10.7 Mass spectrometer.

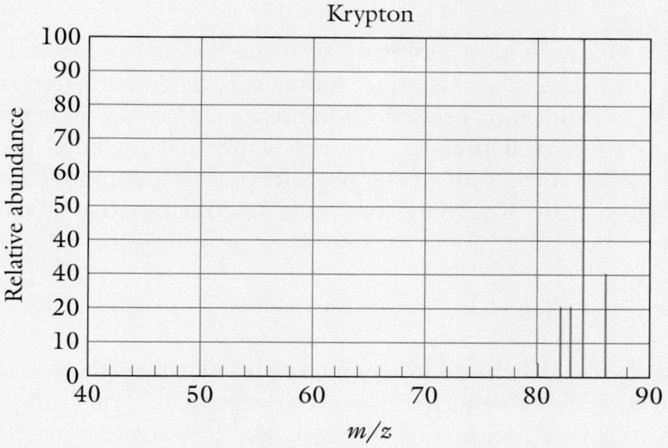

Figure 10.8 Mass spectroscopy data for krypton. (Webbook.nist.gov/chemistry)

trometer, they pass metal plates that have voltages applied such that the ions are repelled by one plate and attracted toward the next. This process increases the velocity of the ions.

The ions next enter a region of the spectrometer where a magnetic field is applied to the ions. The ions are deflected by the magnetic field, much the same way as was first described in the cathode ray experiment from Section 3.5, involving the deflection of electrons. In a mass spectrometer, the ions travel in a curved path, as shown in Figure 10.7. Not all ions will be able to reach the detector, which is on the other side of the narrow slit.

The degree to which each ion is deflected by the magnetic field depends on two variables: the mass of the ion (m) and the charge of the ion (z). The magnetic field of the mass spectrometer is adjusted so that only ions corresponding to a specific mass-to-charge (m/z) ratio can reach the detector. The ions with a larger mass-to-charge ratio are not deflected sufficiently, whereas the ions with a smaller mass-to-charge ratio are deflected too much. The magnetic field of the mass spectrometer is adjusted sequentially so that all possible ions are allowed to reach the detector.

The number of ions at each mass-to-charge ratio is measured for each sample. The relatively simple data obtained for krypton gas are shown in Figure 10.8 as an example. The graph shows that krypton has five isotopes (Kr-80, Kr-82, Kr-83, Kr-84, and Kr-86). For mass spectral data, the peak from the most abundant isotope (in this case, Kr-84) is assigned a relative abundance of 100%, and the ions from all other isotopes are shown as a percentage of that most abundant ion.

The data obtained can be very complex, as shown for cocaine in Figure 10.9. Isotopes alone cannot explain the complexity of the cocaine spectrum. The additional complexity comes from creating ions out of large organic molecules that can break apart into smaller pieces, each forming ions. When chemists analyze an unknown, they compare the experimental results with a known sample of cocaine, because cocaine will always form the same pattern, as shown in Figure 10.9, provided similar experimental conditions are used.

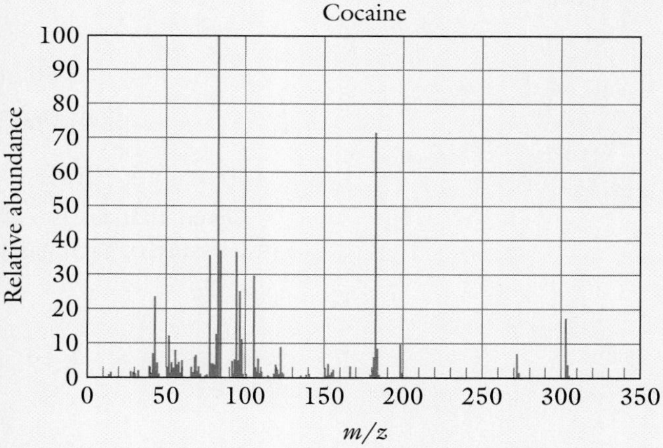

Figure 10.9 Mass spectroscopy data for cocaine. (Webbook.nist.gov/chemistry)

explosives. A sample of air taken from an area near the explosive material will contain molecules of the material, which can be detected by an electronic instrument designed for such analysis or by the nose of a trained bomb-sniffing dog.

Explosives that have a high vapor pressure will have more molecules that enter the gas phase and are therefore more easily detected than those compounds with a low vapor pressure. Mixtures of gases are governed by **Dalton's law of partial pressures**, which states that the total pressure of a gas mixture is the sum of the pressures of each component in the mixture. Dalton's law of partial pressures is described mathematically as follows:

$$P_{\text{total}} = P_1 + P_2 + P_3 + \ . \ . \ .$$

Simply put, each gas behaves independently of all other gases in the mixture and acts as if the other gases were not present. For example, the air we breathe is a mixture of nitrogen at 78%, oxygen at 21%, argon at 0.9%, with all other gases making up the remainder. If the total atmospheric pressure is 1.0 atm, then the pressure of nitrogen is 0.78 atm, the pressure of oxygen is 0.21 atm, and the pressure of argon is 0.009 atm.

Worked Example 12

The pressure of a methane-oxygen-nitrogen gas mixture is 2.50 atm. If the partial pressure of oxygen gas is 0.25 atm and of nitrogen gas is 1.45 atm, what is the partial pressure due to methane?

SOLUTION Substitute into Dalton's law of partial pressures:

$$P_{\text{total}} = P_{O_2} + P_{N_2} + P_{CH_4}$$
$$2.50 \text{ atm} = 0.25 \text{ atm} + 1.45 \text{ atm} + P_{CH_4}$$
$$P_{CH_4} = 2.50 \text{ atm} - 0.25 - 1.45 \text{ atm} = 0.80 \text{ atm}$$

Practice 10.12

Given that air is 78% N_2, 21% O_2, and 0.9% Ar, calculate the partial pressure of each gas at an elevation with an atmospheric pressure of 0.90 atm.

ANSWER
0.70 atm N_2, 0.19 atm O_2, and 0.008 atm Ar

Table 10.2 lists the vapor pressures of some common explosive compounds in air. The first five compounds listed are commonly used high explosives. The vapor pressures of high explosives range widely, but even those like nitroglycerin on the higher end of the range are not easily detected. RDX and PETN have the lowest vapor pressures of the list, and these two compounds are often used to make plastic explosives powerful enough that only a small amount is required to endanger the safety of

Table 10.2 Explosive Compounds and Vapor Pressure

Name	Vapor Pressure at 25°C (1 atm)
Ammonium nitrate (AN)	6.6×10^{-9}
Nitroglycerin (NG)	3.2×10^{-8}
Pentaerythritol tetranitrate (PETN)	5.0×10^{-13}
1,3,5-trinitro-1,3,5-triazacyclohexane (RDX)	1.8×10^{-12}
2,4,6-trinitrotoluene (TNT)	3.9×10^{-9}
2,3-dimethyl-2,3-dinitrobutane (DMNB)	2.8×10^{-6}
Ethylene glycol dinitrate (EGDN)	3.7×10^{-5}
Para-mononitrotoluene (p-MNT)	5.4×10^{-5}
Ortho-mononitrotoluene (o-MNT)	2.4×10^{-4}

Table adapted from *Containing the Threat from Illegal Bombings*, National Research Council, National Academy Press, Washington D.C., 1998.

those on an airplane. Therefore, the detection of plastic explosives has become a priority.

In 1991 the United Nations Council of the International Civil Aviation Organization agreed on a standard for the marking of plastic explosives for the purpose of detection. This convention requires that plastic explosives be manufactured with small amounts of **marker compounds** (0.1% to 1.0%). The markers are explosive compounds that have large vapor pressures that ease the detection of the plastic explosives. The bottom four compounds on Table 10.2 are the marker compounds designated for addition to plastic explosives. The chemical markers DMNB, EGDN, p-MNT, and o-MNT have vapor pressures 10,000 to 1 billion times higher than the primary explosive compounds. The high vapor pressures produce a sufficiently large partial pressure in the air samples found near the explosive that they can be detected by either canine or electronic detection systems. Although the marker compounds do not provide any information on the origin of plastic explosives, they do increase the ability of aviation and law enforcement officials to detect the explosives before they can be used.

10.9 Case Study Finale: Tracing Explosives

The investigation into the death of Nathan Allen was dramatically accelerated because the manufacturer added to the dynamite various micron-sized polymer chips, called **taggants.** The taggant is layered with multiple colors of plastics. It acts as a bar code and can identify the manufacturer, date of manufacturing, lot number, and even the intended distribution sites of the explosive. The average stick of dynamite contains several thousand taggants, many of which are destroyed in an explosion.

Explosive taggants contain information for identifying the lot number and manufacturer. (Microtrace, LLC, Minneapolis, MN)

However, because the taggants do not contain a source of oxygen (as explosive molecules do), they can be destroyed only by combustion with atmospheric oxygen. Even gunpowder is not completely consumed when used, so it is reasonable to expect that enough taggants will survive for a positive identification. The detection and recovery of taggants in blast residue is made easier if the taggants are made of magnetic or fluorescent chips.

The batch of explosives used to manufacture the Tovex 220 purchased by James McFillin was part of a one-time government-sponsored trial to test whether labeling explosives with taggants would be practical. The information provided by the taggant normally is not readily available to investigators because the government has never instituted a regulation requiring manufacturers to add a specific identifying marker to explosives. Switzerland, however, has required since 1980 that all explosives contain taggants—the lone country with such a requirement. Arguments against requiring manufacturers to incorporate identification taggants center on increased costs to industry, possible contamination with other taggants at the bomb site, the use by terrorists of homemade bombs that would not have taggants, and reduced performance of explosives containing the taggants.

While the addition of taggants can aid in the identification of the manufacturer, lack of taggants does not mean that tracing the source of an explosive is impossible. A method that is currently being developed, researched, and evaluated relies on the "fingerprint" of trace components, such as isomers of the explosive compound present in the explosive mixture.

Consider the example of TNT, which is made by the nitration of toluene using nitric acid. Specifically, the compound produced is 2,4,6-trinitrotoluene, also known as 2,4,6-TNT. In most chemical reactions, unintended compounds are produced, and these include isomers of the main product. The structure of 2,4,6-TNT is illustrated in Figure 10.10. The carbon atoms of the benzene ring are numbered 1 through 6, with number 1 always being the carbon bonded to the methyl group.

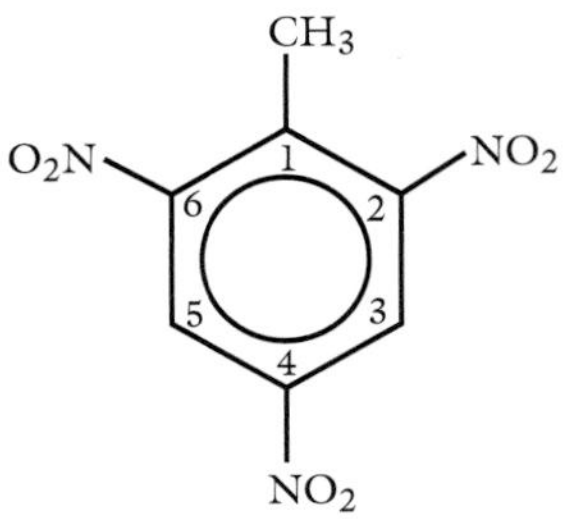

Figure 10.10 2,4,6-trinitrotoluene, also known as TNT.

Worked Example 13

There are five isomers of 2,4,6-TNT that occur by placing the three nitro ($-NO_2$) groups on different carbon atoms. Draw and name the five isomers.

SOLUTION

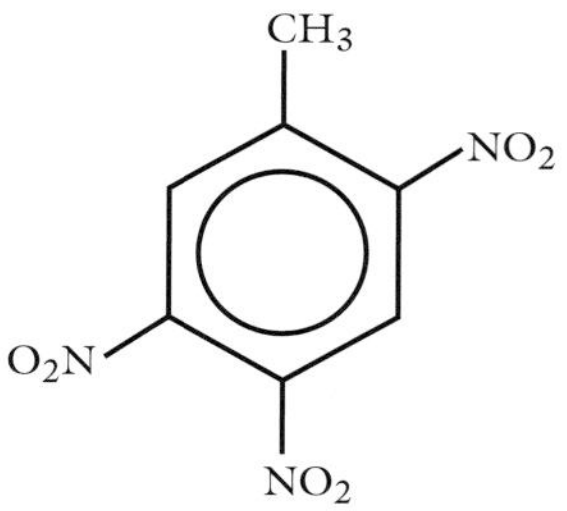

2,4,5-TNT

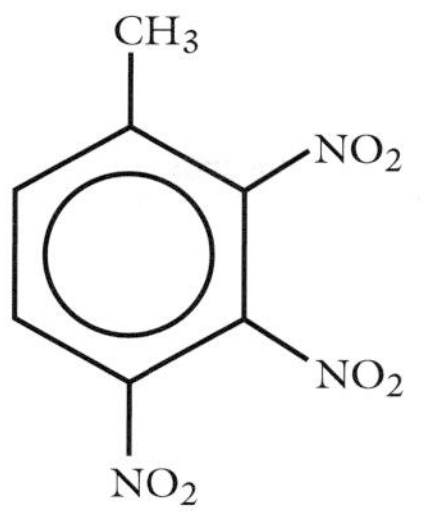

2,3,4-TNT

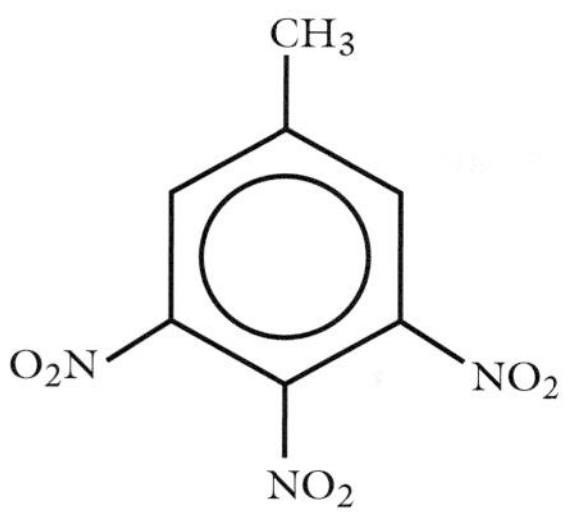

3,4,5-TNT

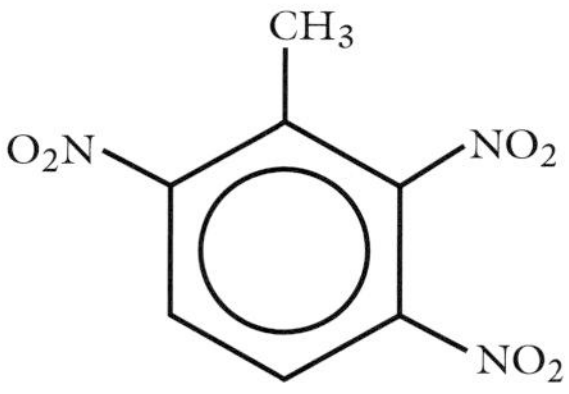

2,3,6-TNT

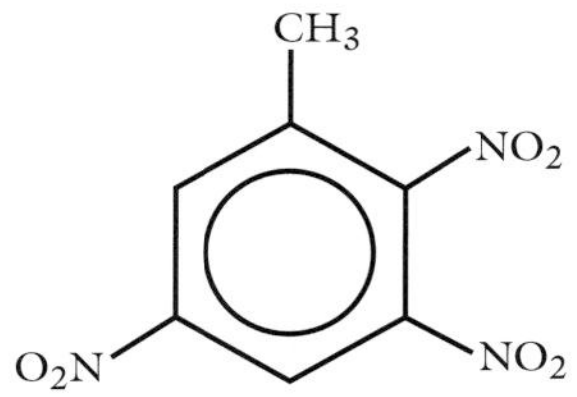

2,3,5-TNT

Practice 10.13

Explain why 3,4,6-TNT is not one of the isomers listed in Worked Example 13.

ANSWER

The 3,4,6-TNT isomer is actually the 2,4,5-TNT isomer, as shown below. 2,4,5-TNT is the correct name for this structure.

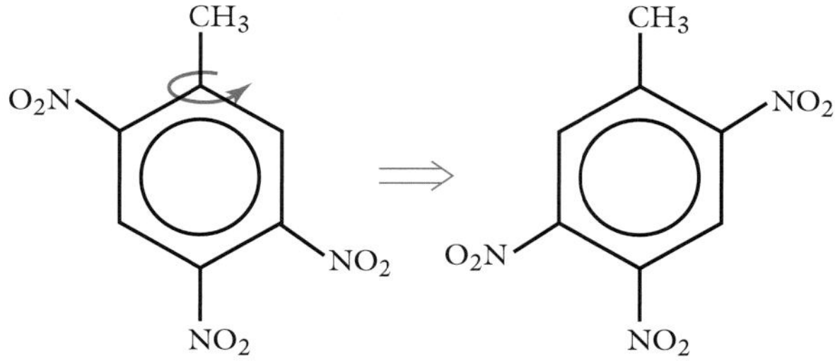

3,4,6-TNT 2,4,5-TNT

The five isomers of 2,4,6-TNT are minor components that are produced in the manufacturing process and are present in the final product mix at very low concentrations. These minor components can provide clues to where the TNT was produced because every manufacturer uses slightly different reaction procedures, equipment, and reaction temperatures that result in varying amounts of each isomer in the final product. If law enforcement officials collect a suspect sample of TNT, the levels of each isomer can be measured and matched to standards obtained from each factory. The isomers act as a bar code or fingerprint for the factory that manufactured the explosive.

LC-MS is an effective tool for the analysis of 2,4,6-TNT and its isomers, as it is for many mixtures. The liquid chromatography method separates

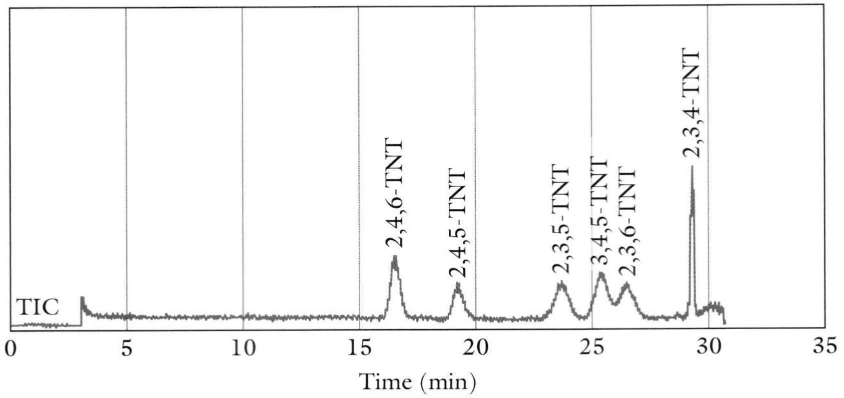

Figure 10.11 Separation of TNT isomers through liquid chromatography-mass spectrometry. (Zhao, X., and Yinon, J. "Characterization and Origin Identification of 2,4,6-TNT through its By-product Isomers. . ." *Journal of Chromatography A,* 946, 125–132, 2002)

each isomer of the mixture, and the mass spectrometer is able to identify each of the TNT isomers within the sample. A representative chromatogram of the separation and detection of a TNT isomer standard is shown in Figure 10.11. It should be noted that in a real TNT sample (as opposed to a laboratory mixture of its isomers used as a standard), the size of the signal for the isomers other than 2,4,6-TNT would be extremely small because they would be present in trace amounts.

CHAPTER SUMMARY

- The chemistry of explosives is characterized by reactions that release a large amount of energy in a short period of time and form substantial quantities of gaseous products. Low explosives are compounds capable of undergoing a rapid combustion reaction and must be physically confined to a small area to become explosive. High explosives are compounds that detonate whether confined or not, and they undergo rapid decomposition to produce a shock wave that will shatter any material near the explosive.
- The kinetic-molecular theory provides an intellectual model for understanding how gases behave and why their properties differ considerably from those of solids or liquids. The effects of temperature, pressure, and volume changes on gas behavior can be understood by using the principles of the kinetic-molecular theory.
- The gas laws allow us to examine the interrelationships of the pressure, volume, temperature, and amount of gas present in a system. We can calculate the effect that a change in one variable has on another. The combination of the individual gas laws leads to the ideal gas law, $PV = nRT$, which shows the relationship of pressure, volume, temperature, and number of moles.
- Dalton's law of partial pressure states that the partial pressures of all gases present in a mixture contribute to the total pressure. One method for the detection of explosives is based on spiking low-vapor-pressure explosives with compounds that have a larger vapor pressure. The vapors from the high-vapor-pressure compounds will produce a measurable partial pressure in the vicinity of the explosive, making them easier to detect.
- One way to identify the source of an explosive is for the manufacturer to add a multilayered, colored plastic chip called a taggant, which acts like a bar code. However, not all manufacturers use taggants. Another method that has been employed is based on measuring trace isomers of the explosive molecules to "fingerprint" the manufacturing facility. The slight variations in the levels of trace isomers reflect varying conditions under which the explosives were manufactured.

KEY TERMS

low explosives, p. 301
high explosives, p. 302
oxygen balance, p. 304
kinetic-molecular theory, p. 306
lower explosive limit (LEL), p. 306
upper explosive limit (UEL), p. 306
Avogadro's law, p. 307
Boyle's law, p. 308
Gay-Lussac's law, p. 310
Charles's law, p. 311
standard temperature and pressure (STP), p. 314
combined gas law, p. 318
ideal gas law, p. 318
mass spectrometer, p. 322
Dalton's law of partial pressures, p. 324
marker compounds, p. 325
taggants, p. 325

CONTINUING THE INVESTIGATION Additional Readings, Resources, and References

"Killer trapped by his DNA," *The Journal* (Newcastle, UK), September 18, 2004.

Semple, Erin. "Invisible Taggants Ensure Authenticity," *Government Security,* August 1, 2005.

Witkin, Gorden. "The Debate about Invisible Detectives," *U.S. News & World Report,* September 16, 1996.

Zhao, X., and Yinon, J. "Characterization and Origin Identification of 2,4,6-Trinitrotoluene through Its By-product Isomers by Liquid Chromatography-Atmospheric Pressure Chemical Ionization Mass Spectrometry," *Journal of Chromatography A,* vol. 946, pp. 125–132 (2002).

For information about the *Aviation Security Improvement Act of 1990:*

www4.law.cornell.edu/usc-cgi/get_external.cgi?type=pubL&target=101-604

For information about bombing incidents in the United States:

http://www.pbs.org/wgbh/nova/transcripts/2310tbomb.html

For an article about taggants published in the *Chicago Sun-Times* on May 13, 1995: home.speedsite.com/ccohen/taggants.htm.

REVIEW QUESTIONS AND PROBLEMS

Questions

1. What are the requirements for a compound to be explosive?
2. What is the difference between a low explosive and a high explosive?
3. What is the main use of a low explosive?
4. Why is a detonator used to trigger the main charge of a high explosive?
5. Explosive molecules contain carbon, hydrogen, and oxygen. Why is this important in the detonation process?
6. What is the oxygen balance of an explosive? Why is this important information?
7. What are the four premises of the kinetic-molecular theory of gases?
8. What is the nature of gas pressure within a container? Illustrate your answer with a drawing showing two containers, one with a gas at high pressure and the other with a gas at low pressure.
9. Explain why gases can be compressed to a much greater extent than liquids and solids can. Illustrate your answer with a drawing showing a particle-size view of each physical state.
10. Explain why fuel-air mixtures below the LEL will not explode.
11. Explain why fuel-air mixtures above the UEL will not explode.
12. Explain the physical phenomenon that creates the pressure of a gas.
13. State Avogadro's law and explain how the variables are related. Illustrate your answer with a drawing.
14. State Boyle's law and explain how the variables are related. Illustrate your answer with a drawing.

15. State Gay-Lussac's law and explain how the variables are related. Illustrate your answer with a drawing.
16. State Charles's law and explain how the variables are related. Illustrate your answer with a drawing.
17. What does STP represent?
18. How can you tell which gas law to use in a calculation?
19. Explain how the kinetic-molecular theory of gases can be used to understand Dalton's law of partial pressures.
20. Discuss the principles behind using taggants in explosive materials.
21. Discuss the principles behind using chemical markers in plastic explosives.
22. Discuss the use of isomers in tracing the origin of explosives.
23. How do taggants, markers, and isomers help security officials?
24. What limitations are there on the use of taggants, markers, and isomers?

Problems

25. Determine how many moles of oxygen would be necessary for the complete oxidation of one mole of each of the following compounds to CO_2 and H_2O.
    (a) CH_3CH_3
    (b) $CH_3CH_2CH_2CH_3$
    (c) $CH_2{=}CH_2$
    (d) CH_3CH_2OH
26. Determine how many moles of oxygen would be necessary for the complete oxidation of one mole of each of the following compounds to CO_2 and H_2O.
    (a) CH_3OH
    (b) $CH_2{=}CHCH_3$
    (c) CH_3OCH_3
    (d) $CH{\equiv}CH$
27. Determine how many moles of oxygen would be necessary for the complete oxidation of one mole of each of the following compounds to CO_2 and H_2O.
    (a) Methane
    (b) Propane
    (c) 1-butyne
    (d) Benzene
28. Determine how many moles of oxygen would be necessary for the complete oxidation of one mole of each of the following compounds to CO_2 and H_2O.
    (a) Hexane
    (b) 2-methyloctane
    (c) Pentane
    (d) Octane
29. Determine what happens to each variable below (increases, decreases, no change) under the stated conditions. Assume that all other variables not specifically mentioned are constant values.
    (a) What happens to the pressure when the volume decreases?
    (b) What happens to the volume when the temperature increases?
    (c) What happens to the pressure when the temperature decreases?
    (d) What happens to the pressure when the amount of gas decreases?
30. Determine what happens to each variable below (increases, decreases, no change) under the stated conditions. Assume that all other variables not specifically mentioned are constant values.
    (a) What happens to the volume if the amount of gas doubles?
    (b) What happens to the temperature when the pressure decreases?
    (c) What happens to the temperature when the volume increases?
    (d) What happens to the volume when the pressure increases?
31. Calculate the missing variables in each experiment below using Avogadro's law.
    (a) $V_1 = 2.00$ L, $n_1 = 0.553$ mol, $V_2 = 0.575$ L, $n_2 = ?$
    (b) $V_1 = ?$, $n_1 = 1.00$ mol, $V_2 = 750.0$ mL, $n_2 = 1.25$ mol
    (c) $V_1 = 0.334$ L, $n_1 = 0.872$ mol, $V_2 = ?$, $n_2 = 6.77$ mol
    (d) $V_1 = 17.2$ L, $n_1 = 3.14$ mol, $V_2 = 5.86$ L, $n_2 = ?$
32. Calculate the missing variables in each experiment below using Avogadro's law.
    (a) $V_1 = 9.15$ L, $n_1 = ?$, $V_2 = 4.65$ L, $n_2 = 3.56$ mol
    (b) $V_1 = ?$, $n_1 = 0.661$ mol, $V_2 = 1.37$ L, $n_2 = 1.45$ mol

(c) $V_1 = 612$ mL, $n_1 = 9.11$ mol, $V_2 = 575$ mL, $n_2 = ?$
(d) $V_1 = 5.58$ L, $n_1 = 0.330$ mol, $V_2 = 9.98$ L, $n_2 = ?$

33. Calculate the missing variables in each experiment below using Boyle's law.
(a) $P_1 = 6.97$ atm, $V_1 = 3.74$ L, $P_2 = ?$, $V_2 = 6.64$ L
(b) $P_1 = 0.245$ atm, $V_1 = 565$ mL, $P_2 = 4.71$ atm, $V_2 = ?$
(c) $P_1 = 6.03$ atm, $V_1 = ?$, $P_2 = 8.66$ atm, $V_2 = 0.904$ L
(d) $P_1 = ?$, $V_1 = 7.43$ L, $P_2 = 3.38$ atm, $V_2 = 2.85$ L

34. Calculate the missing variables in each experiment below using Boyle's law.
(a) $P_1 = 8.51$ atm, $V_1 = 5.92$ L, $P_2 = 0.773$ atm, $V_2 = ?$
(b) $P_1 = ?$, $V_1 = 2.91$ L, $P_2 = 9.76$ atm, $V_2 = 4.02$ L
(c) $P_1 = 6.99$ atm, $V_1 = ?$, $P_2 = 1.24$ atm, $V_2 = 12.6$ L
(d) $P_1 = 3.87$ atm, $V_1 = 5.17$ L, $P_2 = ?$, $V_2 = 25.6$ L

35. Calculate the missing variables in each experiment below using Gay-Lussac's law.
(a) $P_1 = 1.02$ atm, $T_1 = 300.0$ K, $P_2 = 8.15$ atm, $T_2 = ?$
(b) $P_1 = ?$, $T_1 = 266$ K, $P_2 = 4.97$ atm, $T_2 = 607$ K
(c) $P_1 = 8.80$ atm, $T_1 = ?$, $P_2 = 3.47$ atm, $T_2 = 279$ K
(d) $P_1 = 9.98$ atm, $T_1 = 287$ K, $P_2 = ?$, $T_2 = 484$ K

36. Calculate the missing variables in each experiment below using Gay-Lussac's law.
(a) $P_1 = 9.91$ atm, $T_1 = 103$ K, $P_2 = 0.216$ atm, $T_2 = ?$
(b) $P_1 = 5.89$ atm, $T_1 = 334$ K, $P_2 = ?$, $T_2 = 210$ K
(c) $P_1 = 3.66$ atm, $T_1 = ?$, $P_2 = 7.37$ atm, $T_2 = 797$ K
(d) $P_1 = ?$, $T_1 = 977$ K, $P_2 = 5.57$ atm, $T_2 = 244$ K

37. Calculate the missing variables in each experiment below using Charles's law.
(a) $V_1 = 2.55$ L, $T_1 = 379$ K, $V_2 = 9.92$ L, $T_2 = ?$
(b) $V_1 = ?$, $T_1 = 118$ K, $V_2 = 14.0$ L, $T_2 = 612$ K
(c) $V_1 = ?$, $T_1 = 298$ K, $V_2 = 3.33$ L, $T_2 = 828$ K
(d) $V_1 = 4.31$ L, $T_1 = 318$ K, $V_2 = 6.83$ L, $T_2 = ?$

38. Calculate the missing variables in each experiment below using Charles's law.
(a) $V_1 = 3.99$ L, $T_1 = 901$ K, $V_2 = 6.98$ L, $T_2 = ?$
(b) $V_1 = ?$, $T_1 = 659$ K, $V_2 = 17.1$ L, $T_2 = 280$ K
(c) $V_1 = ?$, $T_1 = 258$ K, $V_2 = 16.0$ L, $T_2 = 818$ K
(d) $V_1 = 5.76$ L, $T_1 = 408$ K, $V_2 = ?$, $T_2 = 277$ K

39. Calculate the volume of the following amounts of gases at STP.
(a) 7.66 mol of carbon dioxide
(b) 33.1 mol of hydrogen
(c) 0.148 mol of sulfur dioxide
(d) 0.996 mol of hydrogen sulfide

40. Calculate the volume of the following amounts of gases at STP.
(a) 0.166 mol of helium
(b) 4.29 mol of fluorine
(c) 64.4 mol of nitrogen
(d) 7.05 mol of sulfur trioxide

41. Calculate the volume of the following amounts of gases at STP.
(a) 3.30 g of methane
(b) 1.96 g of oxygen
(c) 78.2 g of carbon monoxide
(d) 30.3 g of dinitrogen monoxide

42. Calculate the volume of the following amounts of gases at STP.
(a) 23.4 g of nitrogen
(b) 42.7 g of argon
(c) 171 g of propane
(d) 96.5 g of butane

43. Calculate the density of the gases (g/L) listed in Problem 41.

44. Calculate the density of the gases (g/L) listed in Problem 42.

45. A gas initially occupies 2.00 L at 25°C and 3.00 atm. Calculate its volume after the temperature is increased to 125°C and the pressure is decreased to 1.00 atm.

46. A gas initially occupies 25.0 L at 25°C and 3.00 atm. Calculate its volume after the temperature is increased to 125°C and the pressure is decreased to 1.00 atm.

47. A gas initially occupies 10.00 L at 55°C and 25.0 atm. Calculate its volume after the temperature is decreased to 25°C and the pressure is reduced to 1.50 atm.

48. Calculate the pressure of a gas if 25.0 L of the gas at 25°C and 1.00 atm is compressed into a volume of 2.50 L at a pressure of 5.0 atm.

49. What is the pressure of a 25.5-L compressed gas tank that holds 32.4 mol of nitrogen gas at 357 K?

50. What is the amount in moles of nitrogen gas found in a 79.4-L compressed gas tank that has a pressure of 89.3 atm at 357 K?

51. What is the volume of a compressed gas tank that contains 2983 g of butane gas at a pressure of 56.4 atm and a temperature of 22.7°C?

52. What is the pressure of a 10.8-L compressed gas tank that contains 716 g of nitrogen gas at a temperature of 57.1°C?

53. What is the temperature of a 926-L compressed gas tank that contains 64.1 mol of carbon monoxide at a pressure of 38.1 atm?

54. What is the pressure of 118 g of hydrogen gas stored in a 63.5-L compressed gas cylinder at 276 K?

55. Using both the ideal gas law and Dalton's law of partial pressures, calculate the total pressure of a 1.00-L container at 25°C that contains 32.0 g of O_2 and 14.0 g of N_2.

56. Using both the ideal gas law and Dalton's law of partial pressures, calculate the total pressure of a 5.00-L container at 20°C that contains 120.0 g of argon and 22.0 g of carbon dioxide gas.

Forensic Chemistry Problems

57. Determine whether the explosive HMX ($C_4H_8N_8O_8$) has a positive, negative, or neutral oxygen balance.

58. Determine whether the explosive nitroglycerin ($C_3H_5N_3O_9$) has a positive, negative, or neutral oxygen balance.

59. Would the color of the smoke produced from an explosive differ if it had a positive oxygen balance versus a negative oxygen balance?

60. A fuel-air explosive is most powerful when there is an exact stoichiometric amount of oxygen present at detonation. Determine how many grams of oxygen gas would be needed to combine exactly with 13.2 kg of propane gas.

61. Combustible gases exhibit a concentration range, from the lower explosive limit (LEL) to the upper explosive limit (UEL), within which a mixture becomes explosive. Use the kinetic-molecular theory of gases and Dalton's law of partial pressures to explain why.

62. In conjunction with Figure 10.1 and the organic functional groups covered throughout Chapter 7, identify compounds that contain the following modified organic functional groups that are present in explosive molecules.
    (a) Nitroester
    (b) Nitroether
    (c) Nitroamine
    (d) Nitroaromatics

63. Using Boyle's law, explain why a person who has a penetrating chest wound struggles to breathe and risks having a collapsed lung.

64. If the density of TNT is 1.65 g/cm^3, calculate the volume of a 454-g quantity of the explosive before and after the explosion. Assume that the volume of the C(*s*) in the product of the following reaction is negligible, and assume STP conditions.

$$2C_7H_5N_3O_6(l) \rightarrow 7CO(g) + 7C(s) + 5H_2O(g) + 3N_2(g)$$

65. During the synthesis of 2,4,6-trinitrotoluene, trace amounts of the dinitrotoluene (DNT) isomers are formed. Draw and name all of the isomers of DNT that could form.

Case Study Problems

66. Cesare Lombroso, a famed criminologist, described a primitive lie detector called a volumetric glove in his 1876 book *Criminal Man*: "The glove is filled with air, and the greater or smaller the pressure exercised on the air by the pulsation of the blood in the veins of the hand acts on an aerial column. . . . [T]his chamber supports a lever carrying an indicator which rises and falls with the greater or slighter flow of blood in the hand." Explain the science (gas laws) of how the "volumetric gloves" function as a lie detector.

67. After reading the following quotes, comment on the scientific and investigative challenges that might occur if physical taggants were used in high explosives.

> The taggants in materials derived from explosives will live forever, so they will be in all the building blocks, and they'll be in all the roads, and they'll be in all the foodstuffs and things around, and for agriculture where explosives are used. It wouldn't be too many years before they would be so common that when you did a crime scene almost anywhere, you would be likely to find a taggant from contamination. But very seldom is it going to be from the bomb.
>
> —Chris Ronway (industry representative)

> The average stick of dynamite in the taggant program had more than two thousand taggants in it. So when you go to a scene and you're investigating, you may find stray taggants that were in the construction material. But the overwhelming concentration of taggants will be from the device itself.
>
> —Reynold Hoover (ATF agent)

CHAPTER 11

Estimating the Time of Death

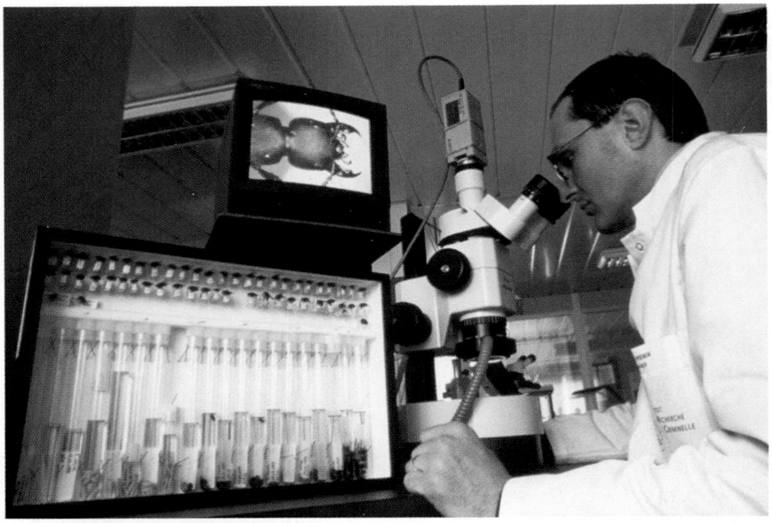

Forensic entomologist identifying a rove beetle (family Staphylinidae) found on a corpse. *(Pascal Goetgheluck/Photo Researchers, Inc.)*

CASE STUDY: Cold-Blooded Evidence

There is a segment of society that lives in the shadows. They are not hardened criminals, but they have had their run-ins with the law. When a person from these edges of society is missing, he or she will be lucky if anyone takes the time to file a missing person report, let alone highlight the case on local, state, or even national news, as might be done for individuals from more prominent or affluent stations in life.

Kari lived in the shadows on the Hawaiian island of Oahu. When she disappeared, no one was surprised. This was not the first time she had dropped below the radar of those who cared about her. After several weeks, though, her family reported her missing to the police. A few weeks after the report was filed, human skeletal remains were discovered in a sugar cane field.

The medical examiner had very little evidence on which to base an identification of the body because insects had already consumed the soft tissue of the body. A forensic anthropologist was called in from the military's Central Identification Laboratory (CIL) to consult. The CIL's main mission is to identify the remains of U.S. military personnel recovered from wars and conflicts dating back as far as the Civil War. The anthropologist examined the skeleton and determined that the victim was female, Caucasian, and 45 to 48 years old, with evidence of stab wounds to the torso.

The investigators used this information and missing person reports to tentatively identify the victim as Kari. Kari's dental records were obtained for the forensic odontologist, and a positive identification was made. The investigators of the murder had a suspect, but to evaluate his alibi, they needed to know more precisely when she died.

A dead body on the Hawaiian islands can be completely skeletalized in about 18 days. Kari had been missing for well over a month. This created a window of perhaps 16 days within which she could have perished. To close this window, the investigators turned to a group of witnesses who had been with the body at the crime scene for quite some time: the flies, beetles, and larvae that had been devouring the remains.

Dr. Lee Goff has been studying insects for several decades. When he is not surfing or teaching courses at Chaminade University in Honolulu, he is using his knowledge of insect development and the factors that affect it to provide investigators with estimated times of death for cases such as Kari's. To make an estimate, he must know in great detail how various species of insects mature, how temperature (and, therefore, the weather) affects the process, and the order in which particular species are attracted to a dead body.

But before we can understand how this information helps to provide a time of death, we must know something about factors that affect the rates of chemical reactions . . .

11.1 Introduction to Chemical Kinetics

Chemical kinetics is the study of the rate of chemical reactions. A forensic entomologist is concerned with how quickly insects grow, which is regulated by the rate of biochemical reactions that take days or weeks to occur. Other reactions such as explosions occur in microseconds, whereas some reactions such as corrosion can take years. Knowing the rate of a

Learning Objective

Explain why chemical kinetics is important in understanding reactions.

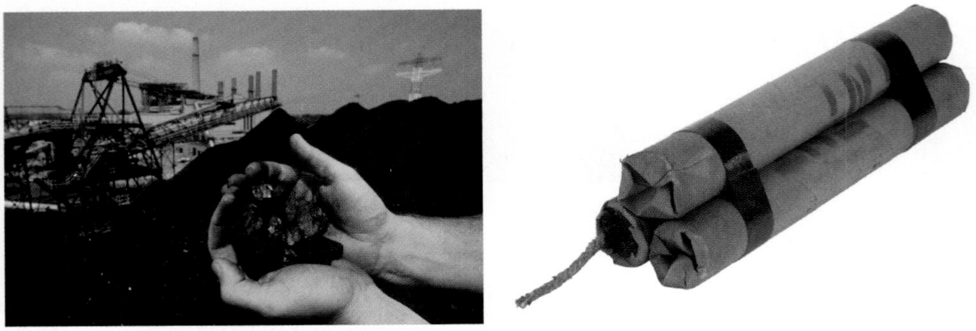

Coal contains more energy gram for gram than nitroglycerin. The rate at which the energy is released, though, is significantly faster for nitroglycerin. (Left: Tim Wright/ Corbis; right: Ingram Publishing/Alamy)

reaction is very important to understanding what occurs during a chemical reaction.

For example, explosions are characterized by the release of large amounts of energy and the formation of gaseous products. But these are not the only requirements for an explosion. The combustion of a piece of coal yields more energy per gram than nitroglycerin and produces gases, yet there is no explosion. The other factor for an explosion is how quickly the reaction occurs. The danger presented by nitroglycerin lies in the speed of its reaction.

The reaction rate is determined by measuring either the decrease of reactants or the increase of products during a given time period. Consider the formation of sulfur dioxide from the combustion of sulfur with oxygen, as shown below. At the start of the reaction, the S and O_2 are combined and start to form SO_2. The rate of this reaction is determined by measuring how fast the amount of S and O_2 decreases or how fast SO_2 is created.

$$S + O_2 \rightarrow SO_2$$

$$\text{Reaction rate} = \frac{\text{amount of } SO_2 \text{ formed}}{\text{change in time}} = \frac{\text{amount of S reacted}}{\text{change in time}} = \frac{\text{amount of } O_2 \text{ reacted}}{\text{change in time}}$$

Worked Example 1

Consider the oxidation of carbon monoxide by oxygen:

$$2CO + O_2 \rightarrow 2CO_2$$

Is the rate at which the amount of carbon monoxide decreases the same as the rate at which oxygen decreases? Consider the coefficients of the balanced equation in your explanation.

SOLUTION No, the rates are not the same. For every 2 mol of CO that react, 1 mol of O_2 is required. Therefore, in a given amount of time, twice as many moles of CO will react as will O_2.

Practice 11.1

Compare the rate of carbon dioxide formation with the rates of loss of both reactants.

ANSWER
Carbon monoxide will be lost at the same rate at which carbon dioxide forms. The rate of carbon dioxide formation is twice as fast as the rate of oxygen consumption.

11.2 Collision Theory

Learning Objective

Explain what happens on a molecular scale during a reaction.

The observation of many chemical reactions led scientists to some interesting questions about reaction rates. Why do two solids that are mixed together usually react very slowly, whereas if the solids are dissolved in water and mixed, the reaction proceeds quickly? Why is the reaction rate in many instances fastest at the start of a reaction but slower as the reaction continues? Why do most reactions proceed faster at warmer temperatures? To explain these observations about chemical kinetics, scientists use **collision theory,** a model of how chemical reactions occur at a molecular level.

Collision Theory

1. The molecules of the reactants must collide for a reaction to take place. *However, not all collisions will result in the formation of products.*
2. The collisions must have a high energy. *However, not all high-energy collisions will result in the formation of products.*
3. The colliding particles must be properly oriented for a reaction to occur. *If the collision has sufficient energy and proper orientation of the colliding particles, the reaction will proceed.*

The first point of collision theory—that the molecules of the reactants must collide for a reaction to occur—seems obvious enough. When a reaction in a solution first starts, collisions between reactant molecules happen quite frequently because there is a large amount of reactants in the solution. Having a large number of collisions is one of the prerequisites for a reaction to occur rapidly. However, as the reaction progresses, the reactants are consumed as the products are made. This decreases the amount

(a)

(b)

Lead(II) iodide, a bright yellow compound, is formed from the reaction of potassium iodide and lead(II) nitrate. Why is the reaction occurring much faster when the reactants are dissolved in solution, as shown in (b), compared with mixing the two solids, as shown in (a)? (Both: Richard Megna/Fundamental Photographs)

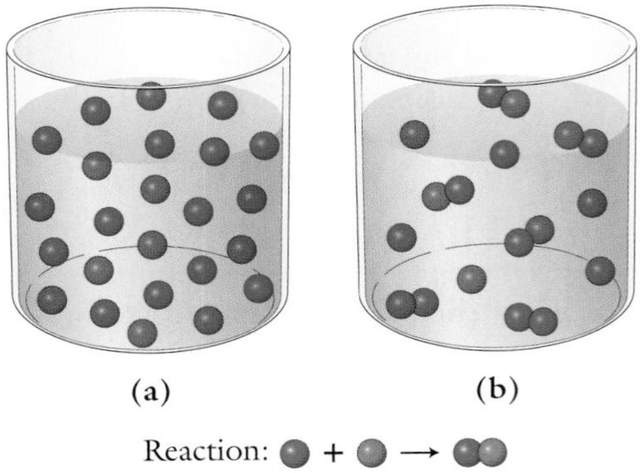

Figure 11.1 Reaction rates versus concentration. The reaction rate is greatest at the start of the reaction—as is the number of collisions between reactant molecules—because there is a high concentration of reactant molecules, as shown in part (a). As the reaction continues, increasing numbers of reactant molecules are consumed in the reaction. The remaining reactants are less likely to collide because fewer molecules are present, as shown in part (b), so the reaction rate decreases.

of reactant available to undergo collisions, so fewer collisions will occur. As the number of collisions decreases, so does the rate of the reaction.

Consider Figure 11.1, which illustrates how the concentration of reactants influences the frequency of collisions as the reaction proceeds. At the beginning of a reaction, as shown in Figure 11.1a, the reactants are present at a high concentration and undergo many collisions. As the reaction continues, the concentration of reactants decreases, as shown in Figure 11.1b. This decrease in concentration causes the reaction rate to slow because of fewer collisions.

Worked Example 2

An experimental procedure calls for mixing reactants that have a concentration of 0.1 M. What will happen to the reaction rate if, by mistake, 1.0 M solutions are mixed? Explain your answer.

SOLUTION According to collision theory, reactants must collide for a reaction to occur. If the reactants have a higher concentration, they will undergo more collisions and start out with a higher reaction rate.

Practice 11.2

Is it always desirable to have a reaction proceed as quickly as possible? Explain your answer.

ANSWER

No, there are situations in which a fast reaction is not desirable. For example, explosions are reactions that occur extremely fast, and it is not desirable for all reactions to proceed so quickly. Another example in which a fast reaction rate is not desirable is the corrosion of metal in vehicles or buildings.

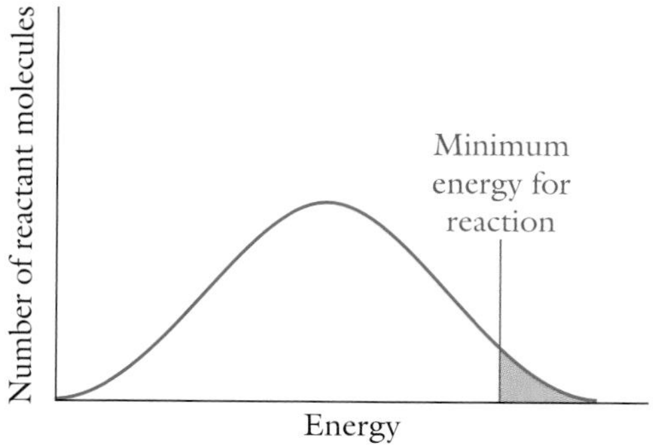

Figure 11.2 Reactant energy distribution. The energy of reactant molecules is distributed in the shape of a bell curve. Only those molecules with sufficiently high energy are capable of undergoing the reaction. The activation energy barrier determines the minimum energy necessary for a reaction, and the shaded area of the bell curve represents the fraction of molecules with sufficient energy.

The requirement that a collision must occur also explains why reactions with solid reactants tend to have very slow reaction rates, whereas reactions in aqueous solutions are fast in comparison. Not all collisions of reactant molecules result in the formation of products. Only high-energy collisions have the ability to bring about a reaction. The kinetic energy of the reactant molecules can vary dramatically, as shown in Figure 11.2. Most molecules do not have the minimum amount of energy needed to react. The shaded area in Figure 11.2 represents the small fraction of reactant molecules that have sufficient energy to react.

Why is energy needed for a reaction to start? Recall Section 9.4, in which the concept of an activation energy for a reaction was first discussed. The activation energy is the minimum energy needed in a reaction to break the bonds in the reactants. As new bonds in the products are formed and energy is released, the energy used to overcome the barrier to activation comes from collisions between the reactant molecules. We might expect that all high-energy collisions will lead to formation of reaction products, but this is not the case. An additional factor, the orientation of the colliding molecules, is also critical.

Consider the reaction illustrated in Figure 11.3, the hydrogenation of the double bond in propene ($CH_3CH{=}CH_2$). The energy of the collision needs to be focused on the region of the molecule with the double bond, as that is where the hydrogen atoms are going to react. If a collision between hydrogen and propene occurs at another location of the molecule, a reaction will not take place even if there is sufficient energy to supply the activation energy. Figure 11.3a shows a collision that does not have the proper orientation for the needed bonds to break, whereas Figure 11.3b has both adequate energy and the proper orientation.

Worked Example 3

Consider the reaction of ethane ($CH_2{=}CH_2$) with H_2. Will this reaction have a greater rate than the reaction illustrated in Figure 11.3, assuming the same conditions of concentration and temperature?

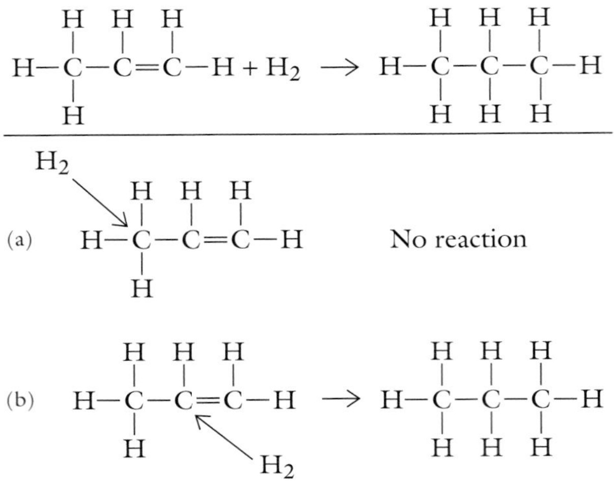

Figure 11.3 Reaction collisions. The overall reaction written at the top shows the addition of the hydrogen gas to the double bond of propene to form propane. For this reaction to occur, hydrogen gas must collide with sufficient energy to rearrange the necessary bonds. It must also strike the region of propene that actually contains the double bond, as shown in part (b), rather than a different region of the molecule, as in part (a).

SOLUTION The reaction of $CH_2{=}CH_2$ will occur at approximately the same rate as the reaction in Figure 11.3 because the only successful collisions will be those in which hydrogen has a direct hit on the double bond. One might argue that the additional $CH_3{-}$ group in propene will reduce the number of successful collisions, but the difference is minimal.

Practice 11.3

In the reaction of hydrogen gas with oxygen gas to form water, will the orientation of the collision be a limiting factor?

ANSWER

No, there will be no difference in collisions of hydrogen and oxygen gas because of orientation. Every collision will bring the necessary components of the reaction together.

11.3 Kinetics and Temperature

Learning Objective

Relate temperature to the rate of reaction.

Insects are cold-blooded creatures, and their level of activity depends on the temperature of the surrounding environment. Not only is an insect's activity level dependent on the temperature, but so is its growth rate. Insect growth is governed by a complex set of biochemical reactions that occur very slowly at cold temperatures but can occur quite rapidly at elevated temperatures. The temperature-dependent growth rate of insects has several implications for the study of insects at a crime scene.

A forensic entomologist will collect several specimens of each species of insect found at the crime scene for later identification at a laboratory. However, there is a challenge in collecting samples of insects in the larva form: The larva of one insect species is almost identical to the larva of another. Live samples of the larvae are therefore collected and taken to the laboratory to be raised in a temperature-controlled chamber until the insects reach adulthood, when the exact species identification can be made.

To determine how long the larvae have been growing at the time a body has been discovered, larvae are collected and preserved. It is important to collect and preserve the largest and smallest larvae present, as they represent the range of oldest and youngest larvae. Forensic entomologists have studied the length of time it takes an insect species to reach various sizes and growth stages under highly controlled temperature conditions. For example, in a laboratory setting, it takes 8 hours at 25°C for a certain species' deposited egg to transform to the larva stage.

As the temperature increases or decreases, the chemical reactions governing the growth of the insect increase or decrease correspondingly. The total amount of heat to get to a specific growth stage is the same whether it happens quickly at high temperatures or slowly at cold temperatures. The amount of heat needed for a species to develop is measured in units called **accumulated degree hours (ADH)** that simply express the number of hours the insect spends developing times the temperature of each hour. For example, if it takes 8 hours at 25°C for an egg to hatch into a larva, the heat required for an egg of this particular species to mature is

$$8 \text{ hr} \times 25°\text{C} = 200 \text{ ADH}$$

For this particular species, if the temperature outside were 20°C, it would take 10 hours to reach the total of 200 ADH. At 30°C, the species would mature in 6.7 hours.

The larvae of many different fly species look much alike. It is necessary to raise to maturity the larvae collected at a crime scene to properly identify the insect species present on a corpse. (Perennou Nuridsany/Photo Researchers, Inc.)

Worked Example 4

A body was discovered on November 12 at 1:00 P.M. and was found to have larvae of the blow fly species *Phormia regina* present. In a laboratory it was determined that this species requires 16 hours at 27°C to develop from the egg stage to the larva stage. Given that the average temperature on November 12 was 15°C and on November 11 was 17°C, when did the blow fly first arrive at the scene?

SOLUTION The heat required by the blow fly to go from the egg stage to the larva stage is

$$16 \text{ hr} \times 27°\text{C} = 432 \text{ ADH}$$

On November 12, from midnight to 1:00 P.M., a period of 13 hours, the blow fly received

$$13 \text{ hr} \times 15°\text{C} = 195 \text{ ADH}$$

This means that on November 11, the blow fly received 432 ADH − 195 ADH = 237 ADH. The average temperature on this date was 17°C, which means the blow fly was getting 17 ADH per hour. The blow fly was there on November 11 for

$$\frac{237 \text{ ADH}}{17 \text{ ADH per hr}} = 14 \text{ hr}$$

Working backwards from midnight, 14 hours would put the arrival of the eggs on the body at 10:00 A.M. on the morning of November 11.

Practice 10.4

The development of a larva actually has three stages, called the *first instar, second instar,* and *third instar*. For the blow fly species *Phormia regina*, the first instar stage requires 18 hours at 27°C to develop. If larvae collected from a body had just entered into the second instar stage, how long had they been growing if the body was inside a house where the thermostat was set at 20°C?

ANSWER
46 hr

What is occurring on a chemical level to slow or increase reaction rates as the temperature changes? Recall that temperature is a measurement of the average kinetic energy of the particles within a system. When heat is added to a system, the energy is absorbed by the particles and their kinetic energy increases, as shown in Figure 11.4. When the kinetic energy increases, the particles travel a greater distance per unit of time and undergo a greater number of collisions per unit of time. Because there are more collisions, there are more opportunities to react successfully.

In addition to increasing the number of collisions, the number of reactant molecules that have the minimum energy to overcome the activation energy barrier increases. Figure 11.5 shows the energy of the molecules of the reactants at two different temperatures. The number of molecules with sufficient energy to react is indicated by the shaded areas.

The faster rate of a reaction is due to both an increase in the number of molecules with sufficient energy to react and an increase in the number of collisions. The rule of thumb is that for every 10°C increase in temperature, the rate of a reaction will double.

Worked Example 5

Explain why samples taken by a medical examiner performing an autopsy are routinely refrigerated after collection.

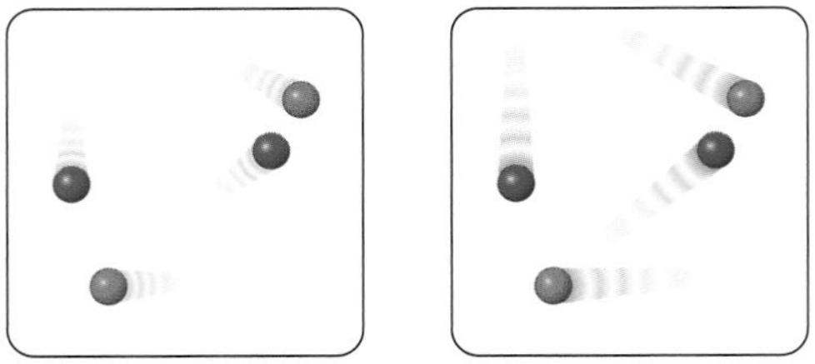

(a) (b)

Figure 11.4 Kinetic energy and collisions. Particles represented in (a) have a lower kinetic energy than those represented in (b). An increase in the kinetic energy of the reactant particles causes them to move a greater distance in the same amount of time, resulting in more collisions and a faster reaction rate.

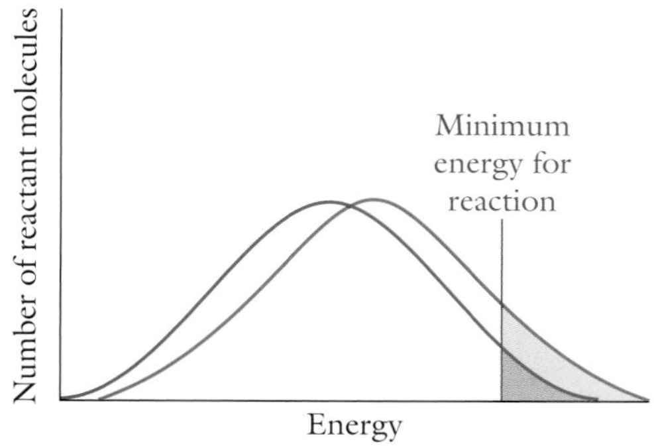

Figure 11.5 Energy diagram for two temperatures. Increasing the average kinetic energy of the reactant molecules increases the total number of molecules that have sufficient energy to overcome the activation energy necessary for the reaction. The rate of a reaction is approximately doubled for every 10°C increase in temperature or, conversely, halved for every 10°C decrease in temperature.

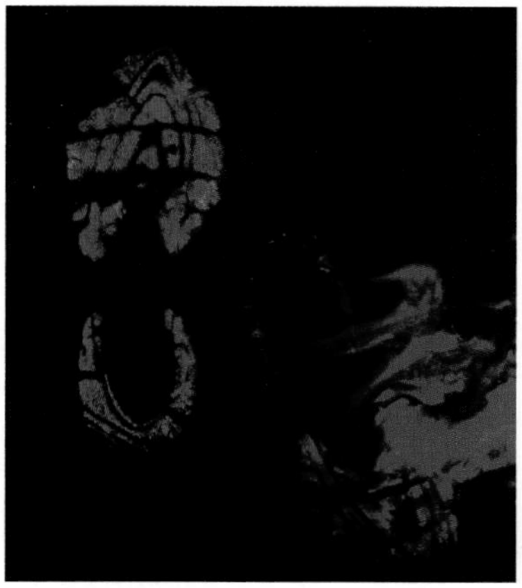

Luminol lights up the traces of blood left after the commission of a crime even after attempts are made to clean up the evidence. (Courtesy of Michael E. Stapleton/Stapleton Associates, LLC)

SOLUTION It is important to preserve the evidence in the same condition in which it was collected. By refrigerating a sample, the rate of any reaction is decreased dramatically.

Practice 11.5

If the temperature of the human body is approximately 37°C and a sample taken from the body is stored at 7°C, how much has the reaction rate changed?

ANSWER

The rate of the reaction is decreased to 1/8 the original rate.

11.4 Kinetics and Catalysts

Learning Objective

Describe how a catalyst increases the rate of a chemical reaction.

The forensic entomologist generally processes what is called a *secondary crime scene,* that is, a location that contains evidence of the crime, such as a remote field where a body is dumped to avoid detection. Where the crime actually occurred is the *primary crime scene.* When someone commits a violent crime, blood leaves a trail that is extremely hard to erase. Trace amounts of blood, invisible to the naked eye, remain behind despite the best efforts of the guilty to clean up the evidence. The two ingredients needed to reveal this evidence are a darkened room and a luminol solution.

Luminol is a compound that releases energy in the form of blue-green light when the luminol is oxidized by a reaction with hydrogen peroxide. The emission of light in a chemical reaction is called **chemiluminescence.** During the oxidation of luminol, an energetically excited molecule is produced. The excess energy in the molecule is released as a photon of light in the visible region of the electromagnetic spectrum.

When luminol and hydrogen peroxide are combined, the reaction rate is extremely slow. Fortunately, for crime-scene investigators, the reaction rate for the oxidation of luminol is increased by iron, which is present in the hemoglobin molecules in blood.

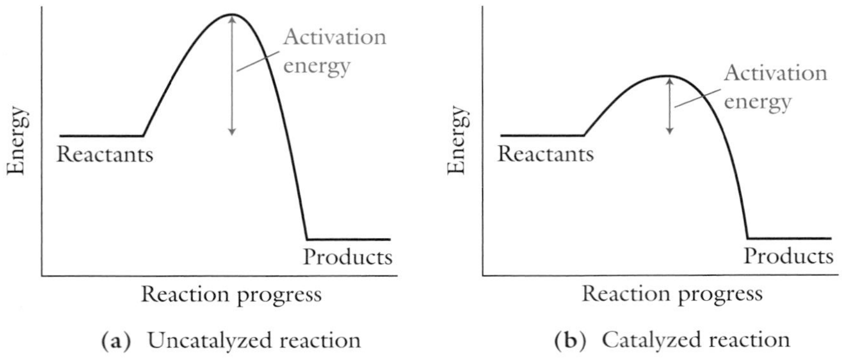

Figure 11.6 Activation energy changes with catalysts. (a) All reactions must overcome the activation energy barrier. The energy for doing this comes from the collisions of reactant molecules. (b) A catalyst increases the rate of a reaction by providing an alternate path for the reaction, which lowers the required activation energy. A larger number of molecules will have sufficient energy to react in the catalyzed system.

Substances that increase the rate of a chemical reaction but are not actually consumed as part of the reaction are called **catalysts.** The iron-containing portion of hemoglobin catalyzes a luminol oxidation reaction and reveals the location of blood, however faint the stain, in a room.

For a better understanding of how a catalyst works, examine the activation energy of a reaction, as illustrated in Figure 11.6. The catalyst in Figure 11.6b lowers the activation-energy requirement as compared with the uncatalyzed reaction by providing an alternate pathway for the reactants to yield the products. The reaction rate increases because more reactants have sufficient energy in their collisions to overcome the lower activation energy barrier.

How exactly does a catalyst provide an alternate pathway? There are two different classes of catalysts, each with its own unique properties and characteristics. Consider the decomposition reaction of hydrogen peroxide to oxygen gas and water:

$$2H_2O_2 \rightarrow 2H_2O + O_2$$

Because the rate of the reaction is extremely slow, H_2O_2 can easily be stored without significant decomposition. However, when hydrogen peroxide is placed on a laceration, the reaction rate increases dramatically. Oxygen gas bubbles can be seen forming because the blood in the wound acts as a catalyst.

When the catalyst is in the same physical state as the reactant, it is called a **homogeneous catalyst.** In the case of hydrogen peroxide, a homogeneous catalyst would be another aqueous compound. Figure 11.7 shows the decomposition of hydrogen peroxide by the iodide ion in aqueous solution, which makes the iodide ion a homogeneous catalyst. In the first step of the catalyzed pathway, the hydrogen peroxide reacts with the iodide ion to form the hypoiodite ion (IO^{-1}) and water. In the second step, the hypoiodite ion reacts with another hydrogen peroxide molecule to produce oxygen gas and water. The hypoiodite ion is a **reaction intermediate,** which is a compound formed in the first step of a reaction

$$\begin{array}{ll} 1. & H_2O_2 + \underbrace{I^-}_{\text{Catalyst}} \rightarrow \underbrace{OI^-}_{\text{Intermediate product}} + H_2O \\ 2. & OI^- + H_2O_2 \rightarrow O_2 + H_2O + I^- \} \text{ Catalyst reformed} \\ \hline 1+2: & 2H_2O_2 \rightarrow 2H_2O + O_2 \end{array}$$

Figure 11.7 Decomposition of hydrogen peroxide with iodide ion as a catalyst. The hydrogen peroxide reacts with the iodine to form the hypoiodite ion, a reaction intermediate. The hypoiodite ion reacts with another hydrogen peroxide molecule to form the final reactants. Bubbles in the solution result from formation of oxygen gas in the reaction. (Richard Megna/Fundamental Photographs)

mechanism but immediately consumed in the following step so that it does not appear in the overall reaction.

Notice that the first step of the catalysis mechanism is the reaction of the iodide ion with hydrogen peroxide. It is a common misconception that catalysts do not undergo chemical reactions themselves. The iodide ion is regenerated in step 2 of the mechanism, so that it appears not to have reacted because there is not a net consumption or production of the ion.

A **heterogeneous catalyst** is typically a solid substance that provides a surface to which liquid or gaseous reactants adhere. The interaction between the reactant and the catalyst weakens the bonds holding the reactants together, thus lowering the activation energy, which in turn increases the reaction rate. Pictured in Figure 11.8 is the catalyzed decomposition

(a) (b)

Figure 11.8 Hydrogen peroxide reacting with manganese(IV) oxide in (a) pellet and (b) powder form. The reaction takes place more rapidly when there is greater surface contact between the hydrogen peroxide and the catalyst, as in (b). (Both: Richard Megna/Fundamental Photographs) of hydrogen peroxide by solid manganese(IV) oxide. A limiting factor for a heterogeneous catalyst can be the accessibility of the reactants to the surface of the catalyst. Figure 11.8 shows two identical reactions using the same amount of hydrogen peroxide and the same mass of the catalyst MnO_2. The only difference is that MnO_2 is in pellet form in Figure 11.8a and in powdered form in Figure 11.8b.

Worked Example 6

Sketch the energy diagram for a catalyzed and uncatalyzed reaction.

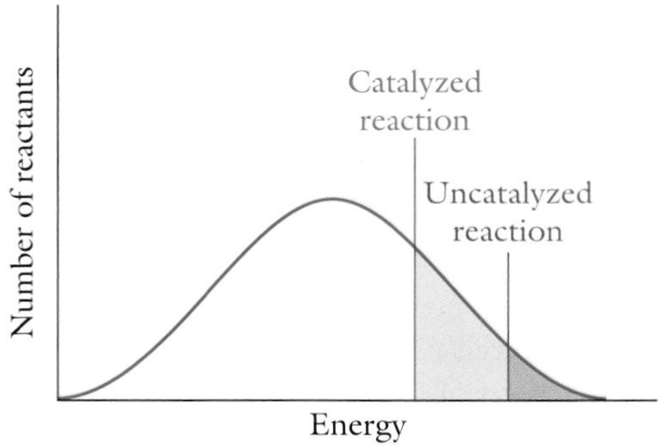

SOLUTION A catalyst lowers the activation energy barrier, which is the minimum amount of energy required during a collision. A greater number of reactant molecules then possesses sufficient energy to react during collisions.

Practice 11.6

In Figure 11.8, the rate for the decomposition of hydrogen peroxide by MnO_2 is much faster for the powdered MnO_2 than for the pellet MnO_2. Using collision theory, explain why.

ANSWER
The heterogeneous catalyst lowers the activation energy barrier, which allows more collisions to be successful. The powdered MnO_2 is providing more sites for the reactants to adhere to and, therefore, more collision sites and ultimately more successful collisions.

The use of heterogeneous and homogeneous catalysts is commonplace in industry, with each type of catalyst having benefits and drawbacks. For example, homogeneous catalysts will mix thoroughly with the reactant solution, which is desirable. However, recovery of the catalyst involves separating the components of a mixture using chromatography or perhaps distillation. Heterogeneous catalysts have the benefit of simply being filtered from solution after use, but the mixing of the catalyst with the reactants is not nearly as efficient.

Learning Objective

Explain how enzymes catalyze reactions.

11.5 Zero-Order Reactions

The police are charged with pulling over and evaluating any driver suspected of violating the drunk-driving laws. Individuals who fail a field so-

(Bill Fritsch/Brand X Pictures/Alamy)

briety test are often transported to a hospital or testing laboratory for further determination of their blood alcohol content (BAC) beyond that of any breath test administered by the arresting officer. The analysis of blood alcohol can be done hours after a person has been detained, and an accurate value for the BAC can be determined at the time of arrest. The method for this analysis relies on knowing the rate of oxidation of ethanol to carbon dioxide in the body.

According to the collision theory, if the concentration of a reactant is increased, the rate of the chemical reaction should also increase. However, there is a class of chemical reactions in which the rates will not increase, despite an increase in the reactant concentration. These particular reactions are known as **zero-order reactions**—the concentration of a reactant does not influence the reaction rate. The oxidation of ethanol in the human body is an example of a zero-order reaction. The rate of elimination of alcohol does not slow down as the concentration of alcohol decreases, nor does it increase if more alcohol is consumed. The only way for alcohol to be eliminated from the body is to wait until sufficient time passes for the completion of the metabolic processes that remove alcohol.

Figure 11.9 shows a typical graph of alcohol levels found in blood after a person consumes alcohol. Once the alcohol is fully absorbed, it is eliminated at a constant rate from the body. On average, a person eliminates from the blood 15 mg/dL of blood per hour.

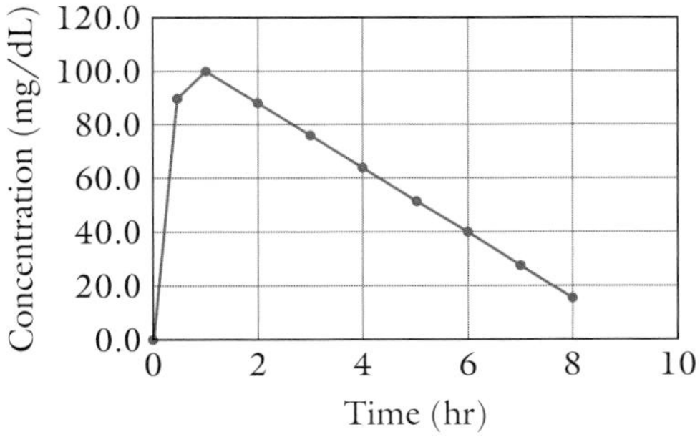

Figure 11.9 Graph of blood concentration versus time for alcohol. After the consumption of an alcoholic beverage, the resulting concentration of ethanol in the blood increases dramatically as it is absorbed through the stomach. The concentration then decreases at a constant rate of 15 mg/dL, showing a linear decline.

Section 7.8 discussed the German practice of using alcohol congener analysis to combat drunk driving. The body oxidizes ethanol at a constant rate until the ethanol has completely reacted. The alcohol congeners build up in a person's system until the ethanol concentration decreases sufficiently to the point at which the congeners will be oxidized. By measuring the concentration of congeners in a person's blood and comparing that result with the amount found in alcoholic beverages, the quantity and time of an individual's alcohol consumption can be determined using calculations based on the kinetics of ethanol oxidation.

As stated earlier, it is unusual for a reaction rate to be independent of the concentration, as is the case for zero-order reactions. This unusual behavior can be explained by taking a closer look at how alcohol is oxidized.

An enzyme called *alcohol dehydrogenase* (ADH) catalyzes the oxidation of ethanol in the body. **Enzymes** are large molecules designed to speed up or allow a specific chemical reaction in a living system. In the case of ADH, the consumption of a single alcoholic beverage swamps the capacity of the enzyme to oxidize ethanol. The excess amounts of ethanol cannot reach the limited amount of ADH in the body. Hence, the only way for the body to eliminate the alcohol is to cease intake and wait for the ADH to oxidize the ethanol. The amount of enzyme effectively acts like the mouth of a funnel that limits how much liquid can flow through the funnel. If more liquid is added, it just builds up in the funnel reservoir waiting to pass through. Enzyme-catalyzed reactions are the most common form of zero-order reactions.

11.6 First-Order Reactions

Learning Objective

Explain why collision theory applies to first-order reactions.

Collision theory predicts that increasing the reactant concentration produces an increase in the reaction rate, and in fact this does occur most of the time. A reaction is said to be a **first-order reaction** when the rate of the reaction mimics changes in the reactant concentration. If the reactant concentration is doubled, so is the reaction rate; if the reactant concentration is halved, the rate of the reaction is also halved.

Consider the elimination of cocaine from the body, as illustrated in Figure 11.10. The cocaine concentration decreases much faster in the be-

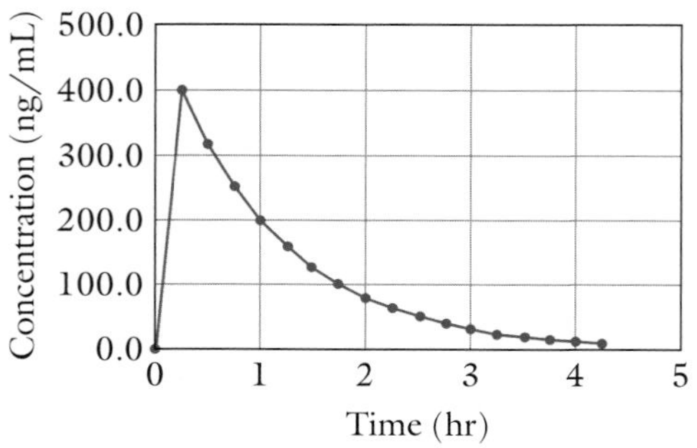

Figure 11.10 Graph of blood concentration versus time for cocaine. The concentration of cocaine in blood shows a dramatic increase at the onset of use. The rate at which cocaine is then eliminated from the body is greatest at the beginning of the reaction and decreases as the reaction progresses.

ginning of the reaction, as evidenced by the large drop in concentration values in the first hour compared with the decreases in concentration levels later in the reaction. Most common drugs, both legal and illegal, follow first-order reaction rates for elimination.

Worked Example 7

What is the decrease in concentration of cocaine during the first hour, according to Figure 11.10?

SOLUTION The cocaine level peaks at 400 ng/mL and then decreases to approximately 200 ng/mL during the first hour, for a total decrease of 200 ng/mL.

Practice 11.7

What is the decrease in concentration of cocaine between the first and second hours, according to Figure 11.10?

ANSWER
110 ng/mL

A toxicology screening is ordered for any suspicious death. Investigators search the victim's home for signs that the person was taking prescription medications. This information can often provide potential investigative leads. For example, the person might have accidentally overdosed, might have had a deadly reaction to combining illegal drugs or alcohol with the prescription medication, or might have stopped taking a prescribed medicine.

Investigators try to determine what levels of a prescription drug should be in the bloodstream if a person had been taking the prescriptions as directed. Most prescription drugs are partially metabolized by the body while also being excreted in urine. Furthermore, each time a new dose is taken, the pharmaceutical levels in the body increase. It turns out that most pharmaceuticals follow a first-order reaction in their elimination from the body, as seen in Figure 11.10. Still, to determine the proper therapeutic levels that should be in a blood sample, more information—such as the dosage amount and the dosage interval—is necessary.

(Michael Keller/Corbis)

11.7 Half-Life

Learning Objective

Explain how the half-life of a drug corresponds to the dosage.

A blood sample is taken from a murder victim and analyzed for both prescription and illegal drugs. The results come back positive for aspirin at a level in the blood that correlates to a total amount of 8 mg of aspirin in the body. If the aspirin bottle recovered from the person's home contained 500-mg tablets, how much time elapsed since the person took the aspirin, assuming only one tablet was taken?

An approximate time can be calculated using the half-life of aspirin. The **half-life** ($t_{1/2}$) is the amount of time required for one-half of a substance to react. In pharmaceuticals, the half-life is how long it takes for one-half of the drug to be eliminated from the body. The elimination of the drug is the result of its metabolism into other compounds or its physical removal through bodily wastes. Since the half-life of aspirin is approximately three hours, the following concentration profile can be written for the above case:

$$500 \xrightarrow{3\text{ hr}} 250 \xrightarrow{3\text{ hr}} 125 \xrightarrow{3\text{ hr}} 62.5 \xrightarrow{3\text{ hr}} 31.25 \xrightarrow{3\text{ hr}} 15.63 \xrightarrow{3\text{ hr}} 7.81$$

Therefore, the aspirin underwent six half-lives: 6×3 hr = 18 hr.

Worked Example 8

Determine the amount of time required for 97% of aspirin to leave a person's system, provided that the person took a single dose of 800 mg and the half-life of aspirin is 3 hours.

SOLUTION Since 97% of the aspirin has been eliminated,

$$0.97 \times 800 \text{ mg} = 776 \text{ mg eliminated}$$

This corresponds to 24 mg remaining in the person's system. Starting with 800 mg to calculate the number of half-lives,

$$800 \text{ mg} \rightarrow 400 \text{ mg} \rightarrow 200 \text{ mg} \rightarrow 100 \text{ mg} \rightarrow 50 \text{ mg} \rightarrow 25 \text{ mg}$$

This means it would take approximately five half-lives to eliminate 97% of the aspirin, which corresponds to 5×3 hr = 15 hr.

Practice 11.8

Determine the half-life (in hours) of the pain reliever ibuprofen, given that a person takes a 600-mg dose at 10:00 A.M. and by 4:00 P.M. only 75 mg remain.

ANSWER
$t_{1/2} = 2$ hr

Pharmaceutical half-lives are also important in determining whether individuals have been taking their prescription medicine regularly, a factor that may be of investigative use. Many medications are prescribed so that the dosage interval corresponds to the half-life of the particular drug. This dosage interval leads to a *steady state* in the concentration of the pharmaceutical that is equal to twice the dosage, as illustrated in Figure 11.11.

When an individual starts taking a prescription drug, the maximum level in the body usually does not occur for two to three days, depending on the interval between doses. When the second dose is taken, only half of the original dose remains in the body. This process continues as each dose is taken and gradually increases the concentration in the body. By the seventh dose, the body is within 99% of the final, maximum value that will be attained. It is important to realize that as the drug concentration is being increased with each of these doses, the drug is also being metabolized and excreted. Eventually, a steady state is reached in which the drug concentration stays fairly constant. The values in Figure 11.11 are maximum values.

Dose Number:	1	2	3	4	5	6	7	8	9
	400	200	100	50	25	12.5	6.3	3.1	1.6
		400	200	100	50	25	12.5	6.3	3.1
			400	200	100	50	25	12.5	6.3
				400	200	100	50	25	12.5
					400	200	100	50	25
						400	200	100	50
							500	200	100
								400	200
									400
Total	400	600	700	750	775	788	794	797	799

Figure 11.11 Steady-state pharmaceutical dosage. This chart gives the total dosage after a dose of 400 mg is administered in intervals equal to the half-life of the drug within a person's body. After a time interval of one half-life, the total dosage is 400 mg plus what remains of the initial dose, and so on. The bottom total line shows that the total concentration of a pharmaceutical taken regularly reaches a near steady-state value after about seven doses.

11.8 Case Study Finale: Cold-Blooded Evidence

In the remains of Kari, Dr. Goff identified a fly of the species *Piophila casei,* commonly known as the cheese skipper. Under normal circumstances in Hawaii, this fly will typically arrive at a decomposing body 15 days after death, and larvae from eggs laid in the remains mature and leave 21 days later—at 36 days after death. Therefore, Kari could not have been dead for more than 36 days. The second insect species identified by Dr. Goff was a type of rove beetle called *Philonthus longicornis,* which typically arrives 25 days after death and remains until day 53. This information narrowed the time of death estimate to the range 25 to 36 days, which is still too large a timeframe for investigators.

Luckily, a third species, *Hermetia illucens,* also known as the black soldier fly, was present. The black soldier fly arrives on day 20 and, based on the size of the most mature larva present, had been growing for 14 days. This information excludes days 25 to 33 from the time of death. Dr. Goff was able to tell investigators that Kari had fallen victim in a time period 34 to 36 days before the discovery of her remains.

The investigators had been interviewing a suspect named Cory about the disappearance of Kari, because he had been seen with her around the time of her disappearance. Cory had given police a statement indicating that he gave Kari a ride home from the bars one evening 32 days earlier and had not seen her since he dropped her off. According to Dr. Goff's estimate, Kari had probably already been dead for two to four days by that time.

Based on the discrepancy between Cory's statement and Dr. Goff's estimate, investigators obtained a search warrant for Cory's home. Once inside, investigators immediately became suspicious. The home of this single, middle-aged man had been scrupulously cleaned, and not a single item was out of order. The house looked too clean for normal use. Investigators

Larva and adult of the species *Piophila casei,* commonly known as the cheese skipper. (Top: The Cleveland Museum of Natural History; bottom: Tom Myers/Micropics)

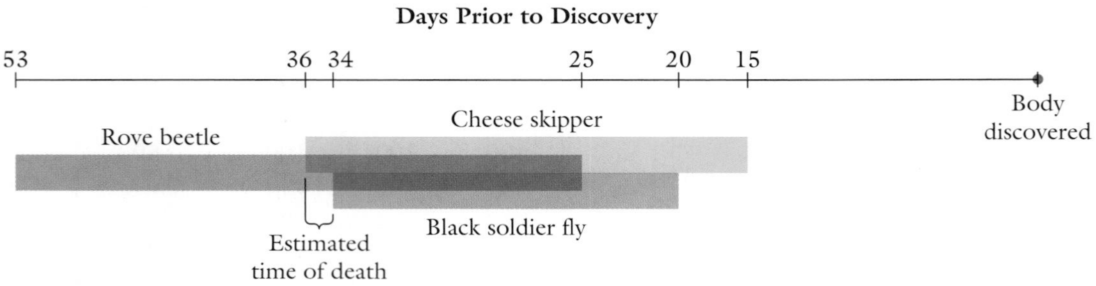

The presence of the cheese skipper was evidence that the body had been at the location of its discovery between 15 and 36 days. The rove beetle does not appear until day 25 and leaves by day 53. This narrowed the estimated time of death to 25 to 36 days before discovery (brown region). The black soldier fly will not arrive at a corpse until day 20, at which time it will lay eggs. Dr. Goff calculated, based on the temperature-dependent growth rate of the black soldier fly, that the oldest larva was 14 days old. These data meant that the body had been left in the cane field at least 34 days before discovery, but no longer than 36 days.

believed he had scrubbed the place down to remove all traces of evidence of his crime.

The crime-scene investigators opened up their processing kits and pulled out luminol, the organic compound introduced earlier in this chapter as a tool for blood detection. The reaction of luminol with hydrogen peroxide exhibits chemiluminescence, a process in which a reaction produces a compound that "glows" by emitting excess energy in the form of photons of visible light. However, the kinetics of the reaction are very slow unless iron present in hemoglobin acts as a catalyst for the reaction.

When Cory's home was sprayed with luminol, a blue glowing outline of a body was found on the living room floor. Investigators could also follow a blue trail of blood toward the door, and the trunk of Cory's car glowed brightly with traces of blood. Cory was arrested for the murder of Kari and placed on trial.

During the trial, the defense tried to question the validity of the entomological evidence because it was the justification for the search warrant. When they failed in this attempt, they presented a counter-theory for the presence of the blood in Cory's home. Because this case predated DNA analysis, and there was insufficient blood left to do standard blood typing, the blood could not be positively identified as Kari's. Therefore, Cory's attorneys presented the parrot defense. According to their version of events, Cory had a parrot that needed its claws trimmed. While trimming the parrot's claws, Cory cut too deeply. The injured and hysterical parrot proceeded to hop around the living room spreading blood everywhere. Cory finally caught his parrot and attempted to take it to the veterinarian. He could not drive while restraining his frantic parrot. His solution was to place the parrot into the trunk of his car . . .

Cory is serving a life sentence for the murder of Kari.

(Comstock Images/Alamy)

CHAPTER SUMMARY

- Kinetics is the study of the rate at which chemical reactions occur. Reaction rates can vary from microseconds to years. Variables that affect the rate of a chemical reaction include temperature, reactant concentration, physical state of the reactants, and presence or absence of catalysts.

- Collision theory describes the requirements for two reactant molecules to successfully come together and form the new product(s). The three main principles of collision theory are: (1) Reactant molecules must collide in order to react; (2) they must collide with sufficient energy to overcome the activation energy barrier; and (3) the molecules that collide must have the proper orientation for the reactant bonds to break and the new chemical bonds to form in the products.
- The temperature of a system directly affects the kinetic energy of the reactants, increasing the frequency and energy of collisions.
- A catalyst increases the rate of a chemical reaction by providing an alternative reaction mechanism that has a lower activation energy. Catalysts can be either homogeneous with the reaction solution or a heterogeneous solid material. Enzymes are large protein molecules that serve as biological catalysts designed to speed up a specific reaction in a living system.
- The reaction rate of a zero-order reaction is independent of concentration. Enzyme-catalyzed reactions tend to be zero-order reactions. First-order reactions have reaction rates that mimic changes in the reactant concentrations. The rate will double when the concentration of a reactant doubles, and the rate is halved when the concentration is halved. In the elimination of most pharmaceuticals, reactions in the body follow first-order reaction rates.
- The time required for the elimination of one-half of a drug in the body is the half-life. Half-life calculations are commonly used to determine the proper dosage interval for pharmaceutical drugs to maintain a therapeutic level of the drug in the body.

KEY TERMS

chemical kinetics, p. 335
collision theory, p. 337
accumulated degree hours (ADH), p. 341
chemiluminescence, p. 343
catalyst, p. 344
homogeneous catalyst, p. 344
reaction intermediate, p. 344
heterogeneous catalyst, p. 345
zero-order reaction, p. 347
enzyme, p. 348
first-order reaction, p. 348
half-life, p. 349

CONTINUING THE INVESTIGATION Additional Readings, Resources, and References

Goff, M. Lee. *A Fly for the Prosecution: How Insect Evidence Helps Solve Crimes,* Cambridge: Harvard University Press, 2001.

Javaid, J. I., Musa, M. N., Fischman, M., Schuster, C. R., and Davis, J. M. "Kinetics of Cocaine in Humans After Intravenous and Intranasal Administration." *Journal of Biopharmaceutical Drug Dispositions,* 4(1):9–18, 1983.

Siegel, Jay A., Saukko, Pekka J., Knupfer, Geoffrey C., ed. *Encyclopedia of Forensic Sciences,* San Diego: Academic Press, 2000.

REVIEW QUESTIONS AND PROBLEMS

Questions

1. Discuss why it is important to understand the rates of chemical reactions.
2. What are the factors that can influence the rate of a chemical reaction?
3. What are the three principles of the collision theory? What are the restrictions on each principle?
4. Why do many gas-phase reactions occur with rapid kinetics, whereas solid-state reactions occur slowly?
5. Why does the initial concentration of the reactants affect most reaction rates?
6. Why do the rates of most chemical reactions slow down over time?
7. If bloody clothes are recovered from a crime scene, one of the first things done before

storing the evidence is to dry the items out. Why?

8. Why do reactants require high-energy collisions to form products?
9. Does increasing the temperature of a reaction system change the amount of energy needed for reactants to cross the activation energy barrier?
10. When the temperature of a system is increased, what two factors are changed that lead to an increase in the reaction rate?
11. What is the only method of increasing the reaction rate that works by lowering the activation energy?
12. Explain the similarities and differences between homogeneous and heterogeneous catalysts.
13. Explain why the surface area of a heterogeneous catalyst can affect the rate of a chemical reaction.
14. Sketch the activation energy diagram for a reaction with and without a catalyst present.
15. Sketch an energy distribution diagram for a reaction indicating the fraction of reactant molecules that have sufficient energy to react. Is the number of molecules with sufficient energy the same as the number of reactant molecules that *will* react? Explain your answer.
16. Carbon monoxide is the flammable gas that is partially responsible for the muzzle flash seen from a firearm. It is also one of the gases that can cause a backdraft to happen when firefighters open up poorly ventilated rooms. Automobiles produce carbon monoxide as a result of the negative oxygen balance of the fuel-air explosion that powers the engines. Why does the carbon monoxide generated in a gun barrel or in a backdraft ignite, whereas there is no such igniting in the muffler of a car?
17. Carbon monoxide produced in automobile engines cools at it passes through the exhaust system and, before exiting the vehicle, passes through the catalytic converter for conversion to carbon dioxide. Is the catalytic converter an example of a homogeneous or heterogeneous catalyst?
18. What is the rule of thumb that relates a temperature change to the kinetics of a reaction? How does that rule relate to preventing food from spoiling by placing it in a refrigerator?
19. Explain how the concentration of alcohol in blood, when measured hours after the ingestion of the alcohol, can be extrapolated backwards to the time a suspected drunk driver was pulled over by police.
20. Does the rate of an enzyme-catalyzed, zero-order reaction change as a function of time? Explain your answer.
21. What would happen to the rate of an enzyme-catalyzed reaction if the concentration of the enzyme were doubled? Explain your answer.
22. Does the rate of a first-order reaction change as a function of time? Explain your answer.

Forensic Chemistry Problems

23. Explain what investigative information can be provided by determining blood serum levels of both legal and illegal drugs and combining this information with the half-lives of those drugs.
24. How much aspirin will remain after 12 hours if a person took a single dose of 400 mg?
25. What is the half-life of a pharmaceutical if after 20 hours a patient had 0.469 mg remaining out of a 30-mg dose?
26. When forensic entomologists arrive at the location of a crime scene, they determine the closest weather station. How can the location of the weather station affect the determination of the time of death?
27. The injured victim in a pedestrian-vehicle accident states that he walked out into the street because he did not see any headlights. The driver of the car insists the headlights were turned on at the time of the accident. How can a broken headlight conclusively prove whether or not the headlights were turned on? (*Hint:* The temperature of a filament is approximately 3000 K and the headlight is filled with inert gases such as xenon to prevent oxidation of the filament.)
28. The production of nitroglycerin is an exothermic reaction. Explain why scaling up the production of nitroglycerin could lead to an especially dangerous situation.

Case Study Problems

29. An individual was reported missing on February 25 from a home in northern California. Local law enforcement, FBI investigators, and community members failed to locate the missing person. On March 21, a neighbor investigating a foul odor located the remains inside an abandoned ceramic kiln. The forensic entomologist received maggots collected from the scene and estimated that onset of the insect activity corresponded approximately to March 14. The condition of the body was such that the individual had been dead for more than seven days. What factors could be responsible for the delay of insect activity between February 25 and March 14? Consider the location and time of year in your answer.

30. It requires approximately 2700 accumulated degree hours for the species *Chrysomya megacephala*, a type of blow fly, to become a 14-mm-long larva. If this species of fly is one of the first to deposit eggs on a body, determine the estimated time of death, given that the body was discovered on January 3 at 9:45 A.M. in an apartment where the thermostat was set at 20°C.

CHAPTER 12

Dirty Bombs and Nuclear Terrorism

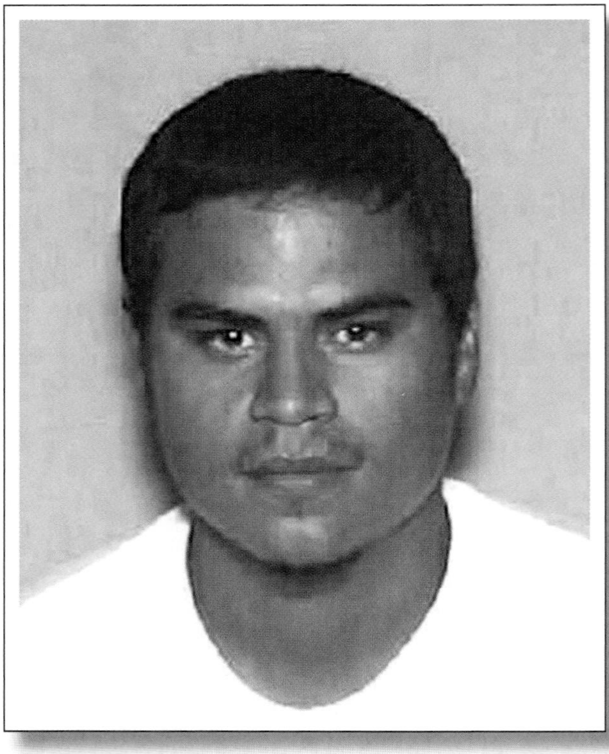

José Padilla was arrested on June 10, 2002, as a suspected al-Qaeda operative allegedly trained to create dirty bombs. (Reuters/Landov)

CASE STUDY: Gangster Turned Terrorist?

On June 10, 2002, U.S. Attorney General John Ashcroft announced the capture of a suspected al-Qaeda operative who was returning to Chicago from a trip to Pakistan. The suspected terrorist is an American citizen named José Padilla, who also goes by the name Abdullah al-Muhajir. Padilla was born in New York City in 1970, but he moved to Chicago at the age of five. As a teenager, he was involved in a local gang and was arrested five times. In one case, he spent three years in a juvenile detention center for armed robbery and assault.

Padilla converted to Islam after marrying a Muslim woman in 1996. He spent most of the late 1990s living in the Middle East. U.S. officials say that Padilla was identified by multiple independent sources as an al-Qaeda operative who was studying how to make explosive devices, particularly a type known as a **radiological dispersion device.** Such a device is also called a *dirty bomb* and is made from a conventional bomb that has radioactive materials added to the explosive.

At first, Padilla was taken into custody as a material witness in the September 11, 2001, attacks on the World Trade Center. In July 2002, President George W. Bush declared Padilla an "enemy combatant," and he was transferred into military custody. In November 2005, Padilla was indicted in federal court in the Southern District of Florida and charged with providing material support to terrorists and conspiring to murder individuals overseas.

The priority given to Padilla's detention as a suspect highlights the importance officials place on avoiding a possible terrorist incident involving a dirty bomb. One question you might ask is: If a dirty bomb has not been used, how can this be considered a forensic science case study? The answer is that we must be prepared to handle the situation if it does occur, understand the risks involved, and have a plan for how the investigation should proceed.

If Padilla was in fact going to build a dirty bomb, where would he obtain the radioactive materials? Two possible sources investigators would examine closely are hospitals, which use radioactive materials to treat and diagnose diseases, and nuclear power plants, which use radioactive materials to generate electricity. Is there any difference between the radioactive material used in a hospital and in a nuclear power plant? What risks do these materials pose if used in a dirty bomb?

To understand the potential short-term and long-term hazards of a dirty bomb, you must first have a better understanding of nuclear chemistry . . .

12.1 The Discovery of Natural Radioactivity

Learning Objective

Trace Becquerel's experiments that led to the discovery of radioactivity.

Nuclear medicine, radiation therapy, atomic weapons, nuclear power plants, and dirty bombs all trace their origins back to the discovery of natural radioactivity one cloudy day in 1896. Henri Becquerel, a French

The discovery of natural radioactivity. Henri Becquerel discovered rays of energy being emitted spontaneously from uranium-containing minerals as he was investigating phosphorescence. (Time Life Pictures/Getty Images)

Glow-in-the-dark toys contain a phosphorescent substance. (Doug Steley/Alamy)

scientist, was studying the phosphorescence of minerals after exposure to sunlight. **Phosphorescence** is the spontaneous emission of light from a substance after it has been activated by exposure to light. This is the process by which glow-in-the-dark toys function.

Becquerel hypothesized that the process of phosphorescence was related to X-rays, which had only recently been discovered. Although Becquerel was incorrect about a link between X-rays and phosphorescence, his experiments set in motion a chain of events that would usher in the nuclear age.

Becquerel's experiment consisted of wrapping a photographic plate in black paper to prevent its exposure to light and placing it next to a mineral crystal made of salts of uranium. The next step in the experiment was to expose the mineral-photographic plate combination to sunlight in order to induce phosphorescence in the mineral. When the photographic plate was developed, the resulting image corresponded to the silhouette of the mineral. Becquerel incorrectly assumed that the image was due to the emission of X-rays during phosphorescence.

Becquerel intended to repeat this experiment but was delayed several days by cloudy weather. He placed his mineral-photographic plate combination in a desk drawer. When sunshine returned later that week, he decided he should start over with a fresh photographic plate. Becquerel removed the mineral sample and unused plate from his desk drawer and, by chance, developed it. To his surprise, the photographic plate showed exactly the same pattern of the mineral's shape as had been obtained from the previous experiments. Because there was no phosphorescence in this case, it was clear that something else caused the pattern to appear on the photographic plate in his desk drawer.

Becquerel continued his experiments with the mineral and soon determined that the uranium present in the mineral sample was causing the exposure of the photographic plate. He postulated that the uranium was spontaneously emitting rays of some sort.

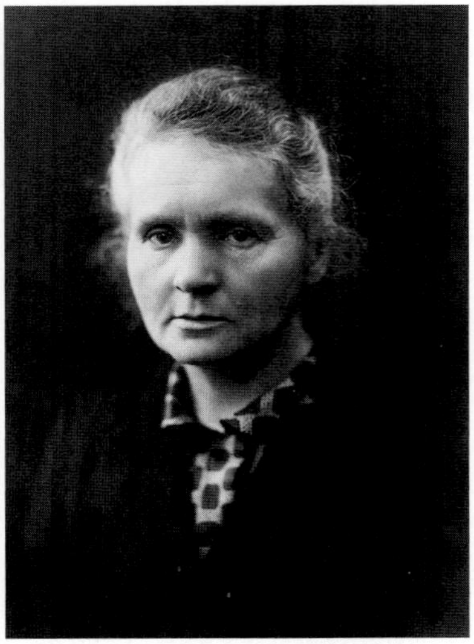

Marie Curie, winner of two Nobel Prizes—one in chemistry and one in physics. (Hulton-Deutsch Collection/Corbis)

A young student by the name of Maria Sklodowska, who was looking for a thesis project to earn her doctorate, took up the study of these "uranic rays." Better known today by her married name, Madame Marie Curie, she is responsible for coining the term **radioactivity** to describe the spontaneous emission of high-energy rays and particles from an atomic nucleus.

Marie Curie is also responsible for the discovery of two previously unknown radioactive elements, radium and polonium. Her isolation of radium required the processing of eight tons of ore to obtain a single gram of radium! During her distinguished career, she became the first person ever to win two Nobel Prizes—one in physics and one in chemistry. The element with the atomic number 96 is named *curium,* in honor of her work.

12.2 Radiation Types and Hazards

Learning Objective

Explain the risks in handling radioactive material.

A dirty bomb is not a nuclear weapon—rather, it is a bomb designed to scatter and contaminate a large area with radioactive materials. If Padilla was planning to build a bomb, why would he bother adding radioactive material? What hazards would result from the detonation of a dirty bomb? The answer depends on what kind of radioactive material was used in the bomb.

The danger of radioactivity is that the rays and particles emitted from the radioactive source collide with molecules in body cells. The collision ionizes the molecules by breaking their internal bonds. The ionized molecules can react with other molecules in the cell and adversely affect bodily processes. Extended exposure to elevated levels of ionizing radiation can increase the odds of developing cancer; exposure to high levels of ionizing radiation can cause severe illness or even death.

We are all exposed to low levels of ionizing radiation, called **background radiation,** that occurs naturally in the environment from a variety of sources. Figure 12.1 illustrates the various sources of background radiation. The most prevalent is radon gas, a product of the radioactive decay of trace amounts of uranium naturally found in the soil. The U.S. Environmental Protection Agency estimates that one out of 15 homes has elevated radon gas levels. This is of concern because the second leading cause of lung cancer is elevated radon levels.

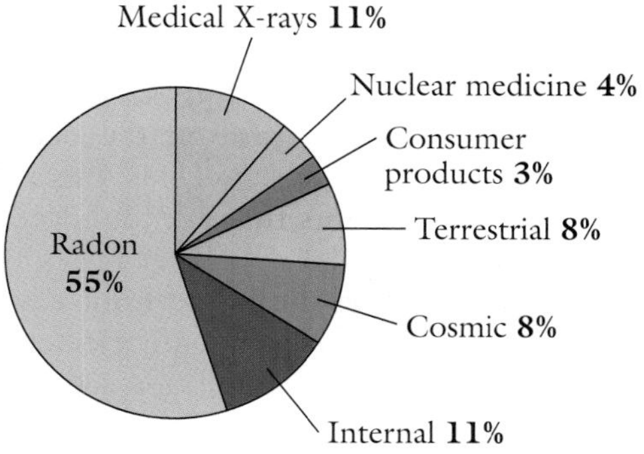

Figure 12.1 Sources of background radiation. The major source is radon gas that collects in homes from the natural decay of trace amounts of uranium occurring naturally in soil. Ionizing radiation from medical X-rays and nuclear medicine is used to diagnose and treat medical disorders. Cosmic radiation is produced by the sun, and a small percentage passes through the earth's atmosphere to reach the surface of the earth. Terrestrial radiation comes from the earth's natural radioactive isotopes. Some of the naturally occurring radioactive isotopes are also present in our bodies and serve as internal sources of radioactivity. (www.nrc.gov/reading-rm/basic-ref/glossary/exposure.html)

Medical X-rays and nuclear medicine are another source of background radiation, but the benefit in diagnosing and treating medical conditions outweighs the risk involved in exposure to the radiation. Yet another form of background radiation comes from outer space and passes through our atmosphere; the sun is the main producer of the cosmic radiation that reaches earth. Terrestrial radiation, on the other hand, refers to the naturally occurring radioactive elements found in the earth's crust. Radioactive materials are also intentionally placed in certain consumer products such as smoke detectors and static eliminators.

The final source of background radiation is our own bodies—radioactive carbon, lead, and potassium are among the most prominent radioactive isotopes that are natural components of our body and are present even at birth. Even though we are continually bombarded with background radiation, it is prudent to limit unnecessary exposure to ionizing radiation.

Many elements have naturally occurring radioactive isotopes. Uranium is probably the best known, but we will encounter other examples in this chapter. What makes some isotopes radioactive and others stable? Hydrogen consists of three isotopes—${}^{1}_{1}H$, ${}^{2}_{1}H$, ${}^{3}_{1}H$—of which the first two are stable and the third is radioactive. To understand why a particular isotope may be unstable, we have to look more in depth at the nucleus of an atom, which was originally discussed in Chapter 3.

The nucleus makes up a minute fraction of the atom's total volume. Although it is approximately 1/100,000 the size of the hydrogen atom, the nucleus contains all of the positively charged protons of the atom. Recall that like charges repel each other. The repulsive forces between the protons within the nucleus are stabilized by the neutrons, which can be thought of as the glue holding the nucleus together.

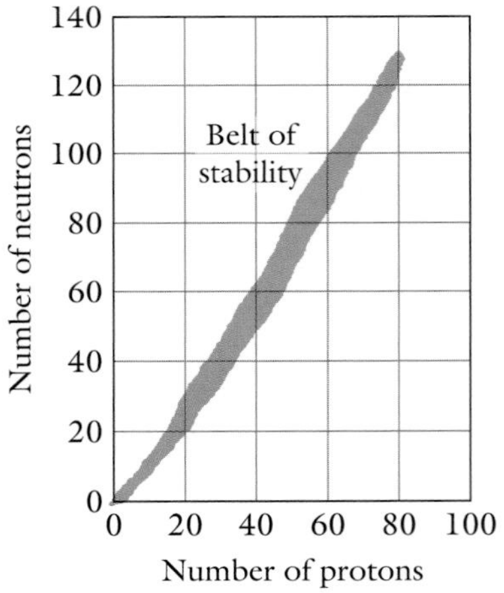

Figure 12.2 Isotope stability. The belt of stability represents isotopes that have a stable ratio of neutrons to protons within the nucleus. Radioactive isotopes have a ratio with either too many neutrons as compared with protons (above the belt of stability), or too many protons as compared with neutrons (below the belt of stability).

Figure 12.2 shows a graph that compares the number of neutrons with the number of protons of all stable (nonradioactive) isotopes. For lighter elements, stable isotopes have one neutron for every proton. Heavier elements, however, require more neutrons to stabilize the increased number of protons present in the nucleus. The shaded area of the graph in Figure 12.2 represents the location of stable isotopes, which is referred to as the **belt of stability.** Radioactive isotopes have neutron-to-proton ratios that fall outside of the belt of stability. By emitting radioactive particles from their nucleus, unstable isotopes will ultimately form isotopes or elements that are found within the belt of stability.

Consider the periodic table on the inside front cover of this book. Most elements have radioactive isotopes, but any element with an atomic number greater than 83 consists entirely of radioactive isotopes. Notice that the atomic masses listed for most of these elements are in parentheses; this number is the mass of the most stable isotope. Thorium (Th), protactinium (Pa), and uranium (U), however, have normal atomic masses because their most stable isotopes have an extremely long half-life that lasts millions to billions of years. Two elements with atomic numbers lower than 83 also consist of only radioactive isotopes, technetium (Tc) and promethium (Pm).

Uranium, with atomic number 92, is the heaviest known naturally occurring element. Most of the elements with an atomic number greater than 92 are artificially created in nuclear research facilities.

Worked Example 1

What changes will occur to a 1-mol sample of a radioactive element over time?

SOLUTION The decay process will generate new isotopes over time. The mass of the sample is constantly changing due to the radioactive decay of the atomic nuclei to produce lighter elements.

Practice 12.1

How many elements do not have a single stable isotope and consist of only radioactive isotopes?

ANSWER
30

How does the emission of radioactive particles change the number of protons and neutrons within the nucleus to create completely different elements? Three types of radiation are emitted, and the emission of each type has a different effect on the original nucleus.

Alpha particles (α), the heaviest of the three particles, consist of two protons and two neutrons. Because alpha particles are identical to the helium +2 ion, the symbol for the alpha particle is $^{4}_{2}He$. The alpha particle is emitted to decrease both the number of protons and the number of neutrons in the nucleus. Note that the ionic charge is generally not shown in the symbol for the alpha particle because the charge is not necessary for balancing nuclear equations.

The danger presented by alpha particles must be gauged on two different scales. The **ionizing power** of a radioactive particle refers to its ability to ionize another molecule. The ionizing power of the alpha particle is extremely high because it is massive in size (compared with the other two types of radioactive particle) and has a +2 charge, which easily attracts electrons away from other molecules.

The **penetrating power** of ionizing radiation is a measure of how far the particles can penetrate into a material and cause damage. Alpha particles have very weak penetrating power because of their large size. A sheet of paper or a layer of clothing is sufficient to stop alpha particles. The location of the source of the alpha radiation is therefore important. An alpha emitter within a person's body does tremendous damage, yet an alpha emitter outside the body doesn't pose as much of an immediate threat.

An isotope undergoes alpha decay when it emits an alpha particle from the nucleus of an atom. The new isotope formed has two fewer protons and two fewer neutrons and will be shifted closer to the belt of stability. Often, a radioactive isotope undergoes many decay processes before achieving a stable isotope. A uranium nucleus, for example, undergoes 14 separate decay processes before it finally forms a stable isotope of lead ($^{206}_{82}Pb$). In the transformation of uranium to lead, not all of the decay steps involve alpha particles. A second type of decay occurs when a neutron is converted to a proton by emitting a **beta particle (β)**.

A beta particle, often written as $^{0}_{-1}e$, is identical to an electron. Beta particles tend to have less ionizing power than alpha particles because the mass of the beta particle is much smaller and transfers less kinetic energy in a collision. However, the penetrating power of beta particles is greater than that of alpha particles. Several millimeters of metal sheeting or very dense wood is required to stop the flow of beta particles. When a beta particle is emitted during a process of beta decay, the atomic number (number of protons) increases by 1. However, the total number of neutrons and protons remains unchanged because the proton was created from a neutron already present in the nucleus. The mass lost due to the beta particle is too small to be noticeable. When a beta particle is emitted, the newly created nucleus is closer to the belt of stability, because the net change is a gain of one proton and a loss of one neutron.

The third type of radiation, the **gamma ray (γ)**, is emitted from a nucleus during the emission of alpha and beta particles. A gamma ray has no mass and no charge, but it is an extremely high-energy photon—a discrete quantity of electromagnetic energy, or a packet of light. Gamma rays have low ionizing power but very high penetrating power. To shield personnel from gamma rays requires either several centimeters of lead or thick cement walls.

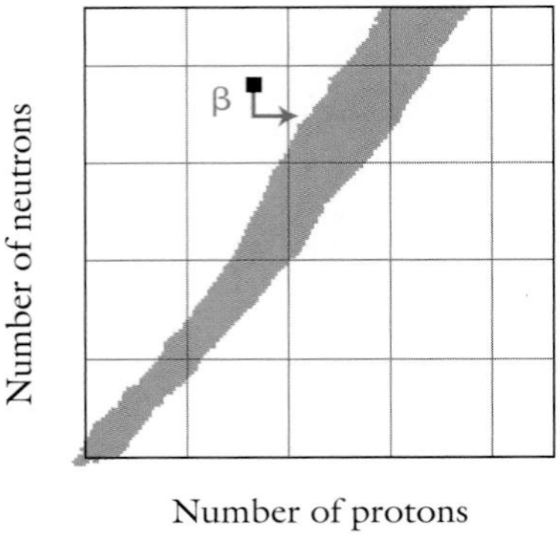

A distinct pattern can be seen when the number of neutrons is plotted against the number of protons for stable isotopes. Radioactive isotopes have ratios of neutrons to protons that fall outside of this stable region (belt of stability).

Worked Example 2

If an unstable isotope situated just above the belt of stability undergoes beta decay, will it be closer to or farther away from the belt of stability?

SOLUTION Beta decay involves a nucleus converting a neutron into a proton and giving off a beta particle. Beta decay decreases the number of neutrons by 1 and increases the number of protons by 1, moving the nucleus down and to the right, as shown in the figure at left. Therefore, the nucleus located just above the belt of stability will shift toward the stable region.

Practice 12.2

If a radioactive isotope undergoes alpha decay, which location will the new isotope have with respect to the original isotope on the belt of stability graph?

ANSWER
Alpha decay involves a nucleus decreasing the total number of protons and neutrons by 2 each, which would move the nuclei down and to the left.

12.3 Balancing Nuclear Equations

Learning Objective

Write balanced reactions for the radioactive decay of an isotope.

What becomes of an atom of uranium-238 ($^{238}_{92}U$) once it has emitted an alpha particle? We can analyze this change by writing and balancing a nuclear equation—a process that is different from balancing a traditional chemical equation. In a traditional chemical equation, we make sure to account for the total number of atoms of each element on both sides of the equation. If two atoms of magnesium appear on the left side of an equation, two atoms of magnesium must appear on the right.

In contrast, a nuclear equation involves an atom emitting a radioactive particle that can change the identity of the element. The result of emitting a radioactive particle is that an atom of a radioactive element such as uranium can appear on the left side of the equation, but not on the right. Nuclear equations are balanced by making sure that the *total number* of neutrons and protons on both sides of the equation is equal. The directions for balancing a nuclear equation are listed below and include as an example U-238 emitting an alpha particle.

Balancing Nuclear Equations

Step 1. Write all elements and particles using isotope notation, with the mass numbers written as superscripts and the atomic numbers written as subscripts.

$$^{238}_{92}\text{U} \rightarrow {}^{4}_{2}\text{He} + ?$$

Step 2. Since the total number of neutrons and protons does not change, the mass numbers must add up to the same value on both sides of the equation.

$$238 = 4 + \underline{234}$$

Step 3. The atomic numbers must also add up to the same value on both sides of the equation.

$$92 = 2 + \underline{90}$$

Step 4. The identity of the missing element or particle is determined by the atomic number calculated in the previous step. In the example, the element with the atomic number of 90 is thorium.

Step 5. Using information from steps 1 to 4, construct the balanced nuclear equation:

$$^{238}_{92}\text{U} \rightarrow {}^{4}_{2}\text{He} + {}^{234}_{90}\text{Th}$$

Worked Example 3

One possible source of radioactive materials for dirty bombs is medical waste, because radioactive isotopes are used to diagnose and treat various disorders. Balance the following nuclear equations for (a) alpha emission of americium-241 and (b) beta emission of cobalt-60.

SOLUTION

(a) Step 1. The atomic number of americium is 95, and the problem states that the mass number is 241. Therefore, the symbol is $^{241}_{95}\text{Am}$. The alpha particle is $^{4}_{2}\text{He}$. The unbalanced equation is:

$$^{241}_{95}\text{Am} \rightarrow {}^{4}_{2}\text{He} + ?$$

Step 2. Determine the mass number: $241 = 4 + \underline{237}$

Step 3. Determine the atomic number: $95 = 2 + \underline{93}$

Step 4. Identify the missing element by its atomic number: 93 = neptunium (Np).

Step 5. Write the balanced nuclear equation:

$$^{241}_{95}\text{Am} \rightarrow {}^{4}_{2}\text{He} + {}^{237}_{93}\text{Np}$$

(b) Step 1. The atomic number of cobalt is 27, and the problem states that the mass number is 60. Therefore, the symbol is $^{60}_{27}\text{Co}$. The symbol for the beta particle is $^{0}_{-1}\text{e}$. The unbalanced equation is:

$$^{60}_{27}\text{Co} \rightarrow {}^{0}_{-1}\text{e} + ?$$

Step 2. Determine the mass number: 60 = 0 + <u>60</u>

Step 3. Determine the atomic number: 27 = −1 + <u>28</u>

Step 4. Identify the missing element by its atomic number: 28 = nickel (Ni).

Step 5. Write the balanced nuclear equation:

$$^{60}_{27}\text{Co} \rightarrow {}^{0}_{-1}\text{e} + {}^{60}_{28}\text{Ni}$$

Practice 12.3

Determine the initial isotope that would produce (a) cobalt-59 and a beta particle, and (b) radon-222 and an alpha particle.

ANSWER
(a) $^{59}_{26}\text{Fe}$ (b) $^{226}_{88}\text{Ra}$

12.4 Half-Lives and Risk Assessment

Learning Objective

Use half-lives of isotopes to determine the risk level of different sources of radiation.

In 1987, the second largest accidental release of radioactivity in history occurred at a junkyard in Goiânia, Brazil. Salvagers who had targeted an abandoned hospital clinic for scrap metal chanced upon a radiation therapy machine and took it for scrap. Several days later, an employee pried open the machine, found an unusual lead container, and forcibly opened it to reveal a powdery material that glowed blue. The workers shared this novel but deadly discovery with friends, family, and neighborhood children. The powdery material was cesium-137.

By the time the first person was diagnosed with radiation sickness nearly a week later, 244 people had been exposed to the cesium-137 and several city blocks contaminated. Radiation sickness consists of nausea, diarrhea, skin burns, hair loss, and bleeding from the mouth, nose, and gums. The severity of radiation sickness depends on the amount of radiation that enters the body.

Entire homes and layers of soil had to be removed and placed in cement-lined storage barrels for disposal at a nuclear landfill. Within the next week, four persons died from their acute exposure. The long-term chronic effects of the exposure are still being determined. Because cesium-137 has a half-life of 30.0 years, trace amounts that could not be removed are still present in the area.

Table 12.1 Half-Lives of Selected Radioisotopes

Symbol	Half-Life ($t_{1/2}$)	Radiation
$^{243}_{99}Es$	21 s	α
$^{106}_{45}Rh$	29.8 s	β
$^{259}_{102}No$	1.0 h	α
$^{24}_{11}Na$	14.7 h	β
$^{240}_{95}Am$	2.1 days	α
$^{131}_{53}I$	8.0 days	β
$^{32}_{15}P$	14.3 days	β
$^{243}_{96}Cm$	28.5 yrs	α
$^{202}_{82}Pb$	5.3×10^4 yrs	α
$^{60}_{26}Fe$	1×10^5 yrs	β
$^{235}_{92}U$	7.6×10^6 yrs	α
$^{238}_{92}U$	4.5×10^9 yrs	α

Which source of exposure is more harmful—short-term exposure to high levels of radiation or long-term exposure to low levels of radiation? In Section 11.7, the concept of a half-life was introduced to measure the length of time a pharmaceutical or illicit drug remains in a person's body. The half-life of radioactive isotopes is the time required for one-half the original amount of isotope to undergo radioactive decay.

One might think that radioisotopes with long half-lives are safer than those with short half-lives (see Table 12.1). For example, uranium-238 has a half-life of 4.5 billion years, whereas iodine-131 has a half life of eight days. If a person is exposed to equal molar amounts of each isotope for the same period of time, fewer total uranium-238 atoms will have spontaneously undergone radioactive decay in that time than iodine-131. But iodine-131 decays by beta emission and uranium-238 decays by alpha emission. The danger presented by radioactive isotopes has to be gauged by the length of time of exposure, the type of radiation, whether the radiation source is external or internal, and the dose of radiation received.

The half-life of a radioactive isotope determines the type of storage and disposal methods required. It is very common for radioactive medical waste to be collected and buried in fairly shallow landfills. The landfills are carefully monitored, but as long as the radioisotopes have short half-lives, the natural decay process will eliminate low-level radioactive materials in a short time. However, the same cannot be said for all radioisotopes used in medicine. Some, like cesium-137, have long half-lives and require special disposal procedures.

The disposal of waste from nuclear power facilities is another problem, especially the spent fuel rods from nuclear reactors. These materials have long half-lives and emit dangerous levels of radiation. For example, the half-life for uranium-235 is 710 million years and the half-life for uranium-238 is 4.5 billion years, as mentioned earlier. Most radioactive nuclear waste materials are stored on site at a power plant.

12.5 Medical Applications of Nuclear Isotopes

Learning Objective

Explain how the properties of radioisotopes are used for medical diagnosis and treatment of disease.

Since the discovery of radioactivity, the field of nuclear medicine has provided for the diagnosis and treatment of a vast array of diseases and disorders. From the earliest days of the discovery of radioactivity, Marie Curie knew that radiation could be used to treat cancer. This application alone could have made the Curie family very wealthy, had they patented the processes.

> Physicists always publish their results completely. If our discovery has a commercial future, that is an accident by which we must not profit. And radium is going to be of use in treating disease. . . . It seems to me impossible to take advantage of that.
>
> —*Madame Curie: A Biography,* by Eve Curie

In general, nuclear medicine has two roles. The first role is to diagnose diseases and disorders. The second role is to treat cancerous tumors by exposing the tumors to radiation. The isotopes chosen for diagnostic purposes have an affinity for a particular organ or target a specific system within the body.

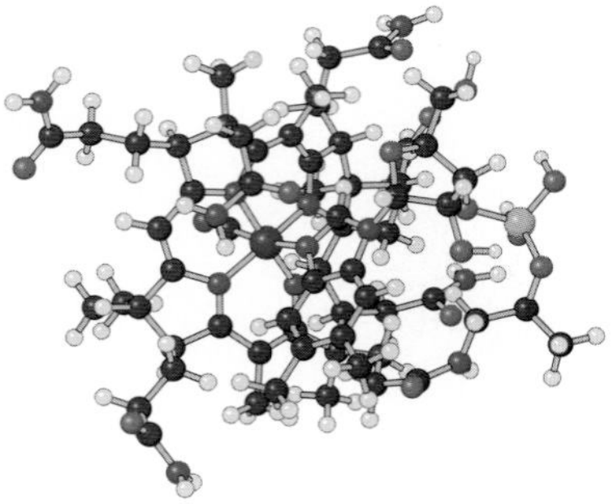

Figure 12.3 Vitamin B_{12} ($C_{63}H_{88}CoN_{14}O_{14}P$). Vitamin B_{12} contains a single cobalt atom (purple) located within the center of the structures. By substituting a radioactive isotope for the normal cobalt atom, the efficiency of the body's absorption of vitamin B_{12} can be determined.

Suppose, for example, that a patient is having problems absorbing vitamin B_{12}. The structure of vitamin B_{12}, shown in Figure 12.3, consists of a large organic molecule with a cobalt atom near the center. Radioactive vitamin B_{12} is made with cobalt-60 and is ingested by the patient. The radioactive vitamin B_{12} will act the same as normal vitamin B_{12} in the body because the chemistry of molecules is unaffected by the identity of the isotope. However, the radioactive cobalt-60 atom will emit gamma rays that are easily detected by medical sensors. If the radioactive vitamin B_{12} is being evenly absorbed by the intestines, the image produced by the radiation will be even and bright. However, if a portion of the intestines is not absorbing vitamin B_{12}, a corresponding dark spot on the image will result from a lack of gamma rays originating from that region.

Most isotopes intended for diagnosing disorders emit gamma rays. Recall that gamma rays have a large penetrating power that allows the gamma rays to escape through a person's body to reach the detector. Secondly, they produce the lowest amount of ionization of the three radioactive particles, minimizing damage to the body's tissues.

The second major use of radioisotopes in nuclear medicine is to destroy tumors. Radiation interferes with a cell's ability to reproduce and affects both healthy and cancerous cells. However, because the cancer cells are growing very rapidly, they are affected to a greater extent. Healthy cells that are damaged can often repair themselves.

Because the ionizing nature of radiation is being exploited to cause damage to the cancer cells, all forms of radioactive emission can be used. Gamma rays are generally used when the radiation is provided externally by focusing the emitted gamma rays onto the location of the tumor. Since beta rays do not have the penetration power of gamma rays, beta emitters are implanted into a tumor to directly expose the cancer cells to the radioisotope. Alpha emitters are also employed in a similar way. Table 12.2

Table 12.2 Medical Uses of Radioactive Isotopes

Isotope	Half-Life	Decay	Form	Use
^{60}Co	5.271 yrs	β, γ	Radioactive vitamin B_{12}	Diagnostic for defects of intestinal vitamin B_{12} absorption
^{113}In	1.658 hrs	γ	Indium-labeled red blood cells	Determination of blood volume
^{131}I	8.040 days	β, γ	Sodium iodide	Imaging and function studies of the thyroid gland
^{32}P	14.282 days	β	Sodium phosphate	Localization of eye, brain, and skin tumors
^{42}K	12.360 hrs	β, γ	Potassium chloride	Tumor detection; determination of total exchangeable potassium
^{87}Sr	2.795 hrs	γ	Strontium nitrate	Bone imaging
^{133}Xe	5.245 days	β, γ	Xenon	Lung ventilation studies
^{241}Am	432.7 yrs	α	Americium metal	Treatment of malignancies

Adapted from *Merck Index*, 13th ed.

lists a representative sampling of the many radioisotopes used in nuclear medicine. The second column of the table lists each radioisotope's half-life—the time required for one-half the original amount of isotope to undergo radioactive decay.

The half-lives of radioactive isotopes used for imaging range from hours to days, which serves to minimize a patient's total exposure to radiation because the source is quickly eliminated. Isotopes that are used for the treatment of tumors have a longer half-life and emit large doses of radiation.

Worked Example 4

Why is Sr-87 useful for bone imaging? Why is the short half-life of Sr-87 desirable? (*Hint:* Consider what bones are made of, and use the periodic table!)

SOLUTION Recall that all elements found within a group in the periodic table undergo similar reactions. Because strontium is among the alkaline earth metals, it will undergo reactions similar to those of calcium, the main elemental ingredient in bone. The radioactive strontium replaces some of the calcium atoms that make up the bone material and becomes a permanent part of the skeleton. It would be undesirable to have elements with a long half-life become a permanent source of internal radioactive exposure. The short half-life ensures that the radiation does not last long in a person's body.

Practice 12.4

Why does xenon-133 work well for studying the function of the lungs?

ANSWER

Xenon is a noble gas and, as such, is inert. The xenon gas will fill the lungs upon inhalation but will not react with any components of the respiratory system. The xenon gas can be expelled from the lungs, reducing exposure to the radioisotope.

Forensic anthropologists specialize in analyzing skeletal human remains to provide investigators with useful information such as the sex, age, race, and manner of death of a victim. Forensic anthropologists use their extensive knowledge of human anatomy to determine such information. Estimating a firm time of death is usually beyond a forensic anthropologist's ability, however, because a skeleton may be from a victim murdered several weeks before or the skeleton may belong to a tribal burial site of an indigenous people from hundreds of years ago.

That uncertainty may change with the radioisotopic analysis of human remains. You may have heard of carbon-14 dating. Radioactive carbon-14 is generated in the upper atmosphere by cosmic rays and is then absorbed by plants in the form of radioactive carbon dioxide. The plants convert the radioactive carbon dioxide into the organic molecules that make up the plant. Animals ingest the plants, and the radioactive carbon-14 passes into the food chain of humans by the ingestion of meat and plants. Carbon dating measures the amount of carbon-14 remaining in an object. If a bone sample has 50% of the carbon-14 it had when the organism died, the bone has been decaying for a time equal to the half-life of carbon-14, or 5730 years. This method is useful for archeologists but not for identifying skeletal remains from recent victims, because the half-life of the radioisotope is so long. An insufficient number of carbon-14 atoms would have undergone decay to make a useful measurement.

Current research is underway to use two different radioactive isotopes to measure the time of death. The lead-210 isotope has a half-life of 22.3 years, making this measure useful for determining whether a discovered skeleton is of modern or ancient origins. The shorter half-life would also be helpful in linking skeletal remains to decades-old crimes. The other isotope being studied is polonium-210, which has a half-life of 138 days, making it useful for identifying skeletons of people who died within the previous several months. Both lead-210 and polonium-210 are found in trace levels in food because they are two of the natural sources contributing to background radiation, and they are continually incorporated into human bodies until death.

12.6 Nuclear Power Plants

Learning Objective

Describe how nuclear power plants generate electricity.

The security of nuclear power plants is a major priority for the Department of Homeland Security for several reasons. A terrorist attack on such a plant would seriously disrupt the power grid and could release radioac-

The skyline of a nuclear power plant is dominated by large cooling towers that dissipate heat generated in the production of electricity. The large white plumes coming out of the cooling towers are created by water vapor condensation and do not contain smoke. (W. Cody/Corbis)

tive isotopes into the environment, causing widespread panic. Because nuclear power plants are always located next to rivers or lakes, any catastrophic release of radioactive materials would spread not only through the atmosphere but also through surface waters. Finally, nuclear power plants store highly radioactive materials that could be used by terrorists if the security of the storage areas were breached. While nuclear power plants in the United States are highly secure facilities, this is not necessarily the case in other countries around the world.

To better assess these risks, it is necessary to understand how nuclear power plants generate electricity. Figure 12.4 shows a sketch of a nuclear power plant, with photos featuring some of the areas important in the generation of electricity.

A nuclear reactor has three independent heat transfer systems: the reactor core system (shown in orange), the steam generation system (shown in blue), and the cooling tower system (shown in gray). Each system is linked to the others in much the same way as three links in a chain are attached—each system crosses into another, but the contents of each system never mix. The reactor core system generates the heat that is transferred to the steam generation system, which converts water into steam. The pressure of the steam is used to spin the turbine fan blades on an electrical generator to produce electricity. Once steam passes through the generators, it is condensed back into water.

The cooling of the steam is achieved by pumping water into the cooling tower system, which cools the water before it flows back to the river. The cooling towers are by far the largest structures in a nuclear power plant. Common misconceptions are that the reactor core is located in this structure and that the substance coming out of the top of the cooling towers is smoke. What is actually coming out of the cooling towers is steam.

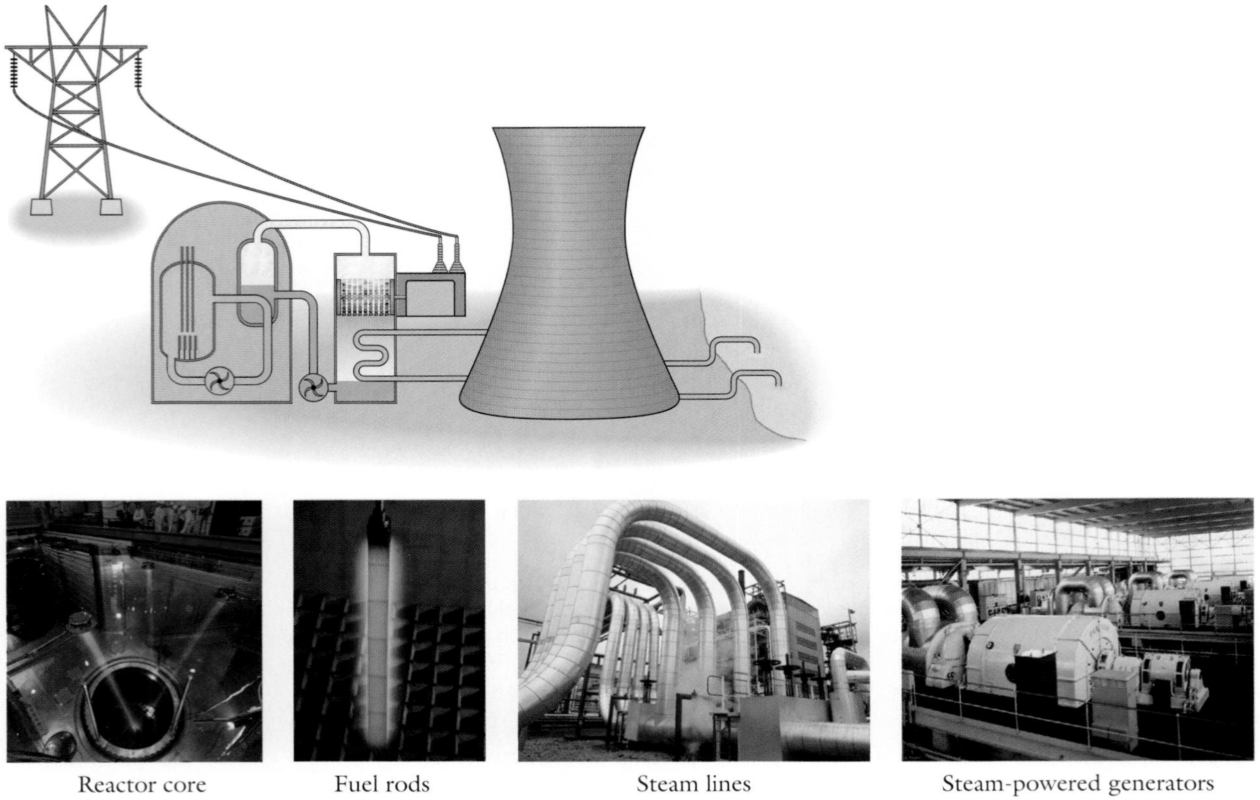

Figure 12.4 Electrical power generated from a nuclear power plant. Uranium fuel rods release large amounts of heat while undergoing fission reactions in the reactor core. The heat from the reactor core is used to convert water into steam, which is used to turn a steam turbine that generates the electricity. The steam is then cooled and converted back into liquid water by fresh water pumped in from a lake or river. The fresh water is then cooled in the cooling towers before it is returned to the lake or river. (Illustration adapted from Howstuffworks.com; photos, left to right: Yann Arthus-Bertrand/Corbis; Tim Wright/Corbis; Corbis; Charles E. Rotkin/Corbis)

The reactor core generates heat by nuclear **fission,** a reaction in which a large nucleus is split into smaller components. The reaction that powers most nuclear power plants is the fission of uranium-235 ($^{235}_{92}U$), as shown in Figure 12.5. A neutron striking the uranium-235 nucleus destabilizes it, causing it to split into barium-91 ($^{91}_{36}Ba$), krypton-142 ($^{142}_{56}Kr$), and three additional neutrons. The reaction produces a large amount of heat.

The fission of a single uranium-235 atom produces three neutrons. Notice that each of the three neutrons created in the first fission reaction strikes another uranium-235 atom, which in turn releases three neutrons per atom. If the concentration of uranium-235 atoms is great enough, the reaction can become a self-sustaining **chain reaction** within the reactor core.

The fission reaction inside a nuclear power plant generates heat. If left unregulated, the amount of heat generated by a chain reaction would be too great for the reactor core to handle—resulting in a **meltdown.** To prevent this, the number of uranium atoms undergoing fission must be regulated. **Control rods** within a nuclear reactor are made out of boron

$$^{1}_{0}n + ^{235}_{92}U \longrightarrow ^{91}_{36}Ba + ^{142}_{56}Kr + 3^{1}_{0}n$$

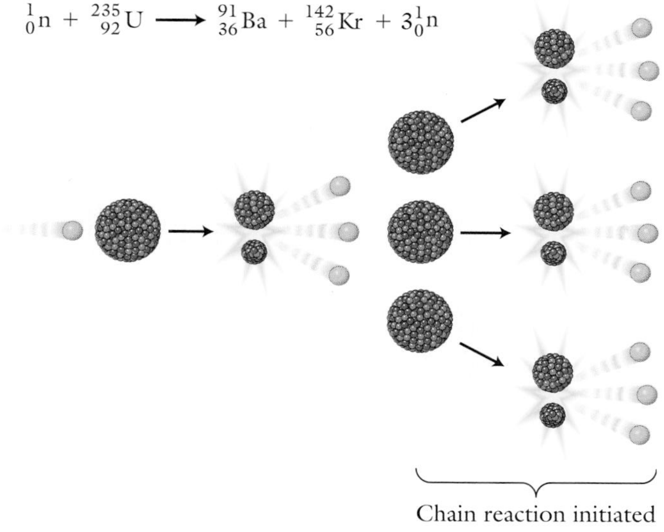

Figure 12.5 Fission of uranium-235.

or hafnium, both of which will absorb the neutrons produced during fission. If the control rods are lowered into the reactor, they absorb the neutrons and less uranium undergoes fission. If more heat is desired in the reactor core, the control rods are raised.

The actual products of uranium-235 fission are more numerous than those illustrated in Figure 12.5. There are hundreds of radioactive isotopes that can be produced. All of the waste products from the nuclear reactor are radioactive and present a long-term storage problem for nuclear facilities. Most facilities have highly secured waste storage on site.

The security of nuclear wastes, both domestically and internationally, is a high priority to the United States because terrorist organizations are known to be actively searching for sources of radioactive materials. The federal government is currently constructing a centralized storage facility in Yucca Mountain, Nevada, where the radioactive waste will be buried over 500 meters deep in the mountain. The site was chosen because it is not located near any population center and has little likelihood of being struck by earthquakes. The site is also in an extremely dry area, which inhibits the corrosion of storage containers.

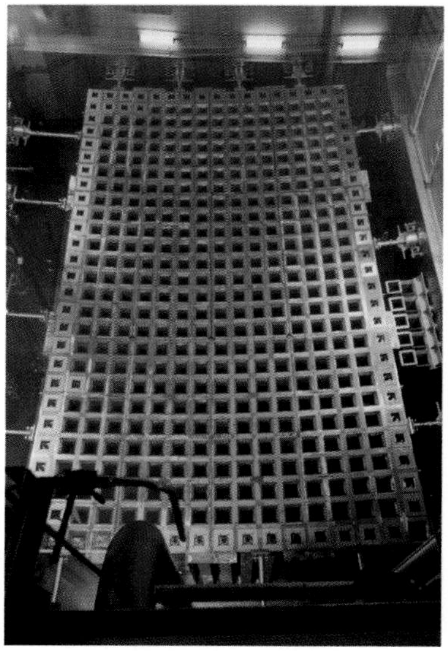

On-site storage of spent fuel rods. (Roger Ressmeyer/Corbis)

Research into a new type of nuclear reactor that may one day provide electrical power without the production of highly radioactive wastes is currently being studied. A nuclear **fusion** reactor produces energy by joining two light nuclei to form a heavier nucleus. The amount of energy released by fusion reactions is much greater than that produced by fission reactions. However, temperatures in excess of 10 million degrees Celsius are needed to initiate a fusion reaction. Since no material can withstand such a high temperature, it is easy to see the problem of constructing a fusion reactor. Current research is directed toward using magnetic fields as the "container" for a fusion reaction and triggering the reaction with lasers.

The most notable fusion reactors are stars, our sun being the closest example. The energy of the sun comes from the fusion of hydrogen nuclei to form helium atoms as follows:

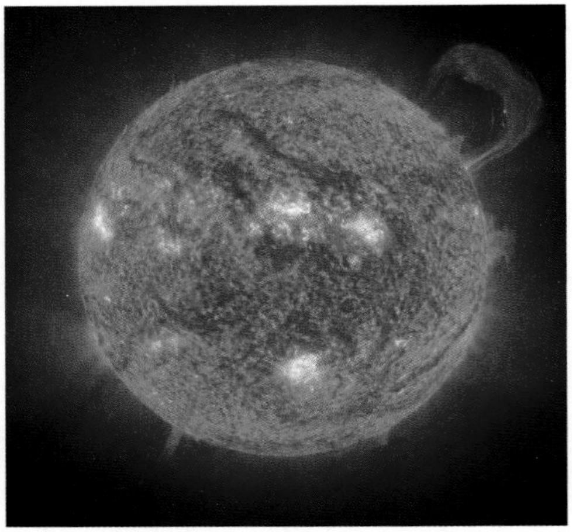

The energy of the sun is generated by the fusion of hydrogen atoms to form helium atoms. (NASA/JPL-Caltech/Corbis)

$$
\begin{aligned}
{}^{1}_{1}\text{H} + {}^{1}_{1}\text{H} &\rightarrow {}^{2}_{1}\text{H} + {}^{0}_{+1}\text{e} \\
{}^{2}_{1}\text{H} + {}^{1}_{1}\text{H} &\rightarrow {}^{3}_{2}\text{He} + \gamma \\
{}^{3}_{2}\text{He} + {}^{3}_{2}\text{He} &\rightarrow {}^{4}_{2}\text{He} + 2\,{}^{1}_{1}\text{H}
\end{aligned}
$$

The first reaction is the fusion of two hydrogen nuclei together to form the deuterium isotope of hydrogen and a positron (${}^{0}_{+1}\text{e}$). A positron is a particle with identical properties to an electron, except the charge is positive, not negative. The second step of the reaction is the fusion of a deuterium nucleus with a hydrogen nucleus to form the helium-3 isotope and a gamma particle. The final reaction is the combination of two helium-3 isotopes to form a helium-4 isotope and two protons.

12.7 Military Uses of Nuclear Isotopes

> Today, the physicists who participate in watching the most formidable and dangerous weapon of all time . . . cannot desist from warning and warning again: we cannot and should not slacken in our efforts to make the nations of the world and especially their governments aware of the unspeakable disaster they are certain to provoke unless they change their attitude towards each other and towards the task of shaping the future. We helped in creating this new weapon in order to prevent the enemies of mankind from achieving it ahead of us. Which, given the mentality of the Nazis, would have meant inconceivable destruction, and the enslavement of the rest of the world . . .
>
> —Albert Einstein

Nuclear weapons were first developed in World War II as part of the top-secret Manhattan Project. Initiated at the urging of Albert Einstein out of fear that Nazi Germany was working toward the deployment of such weapons, Einstein later opposed the use of nuclear weapons and warned of the potential chaos that would ensue if all nations pursued nuclear

(Corbis)

weapons programs. The Germans never built a fully operational nuclear weapon and surrendered before the United States had readied its nuclear weapons for use. However, the United States unleashed nuclear weapons on Japan to end the war in the Pacific.

Nuclear weapons proliferation is still a matter of international concern that challenges the stability and peace of the world. One reason is that the technology used for the production of electricity in nuclear power plants can be used to produce material for nuclear weapons. The existence of such facilities in countries that support the efforts of international terrorists is a cause for concern. Could terrorists obtain radioactive material overseas and transport it into the United States?

Uranium has three isotopes, all of which are radioactive: uranium-238 (99.2745%), uranium-235 (0.720%), and uranium-234 (0.0055%). Uranium in its natural composition is not usable for either weapons or power plants. Only uranium-235 will undergo the fission reaction. Uranium metal is processed to increase the concentration of uranium-235 for use in either nuclear power plants or nuclear weapons. The resulting uranium is referred to as *enriched* uranium, and the level of enrichment determines its use. If the uranium is enriched to contain 3 to 5% uranium-235, it can be used in a nuclear reactor. If the concentration is 20%, the uranium becomes suitable for a crude nuclear weapon, although modern nuclear weapons use uranium enriched to contain 85% uranium-235. Note that the concentration of uranium-235 in a fuel rod is too low to support a nuclear detonation, so nuclear power plants will not under any circumstance produce a nuclear explosion.

The amount of uranium-235 needed to produce a chain reaction is called the **critical mass.** The basic nuclear weapon, illustrated in Figure 12.6, functions by combining two subcritical masses of uranium-235 at the time of detonation to create a critical mass. At the same moment the two subcritical masses are joined, a neutron emitter initiates the fission of

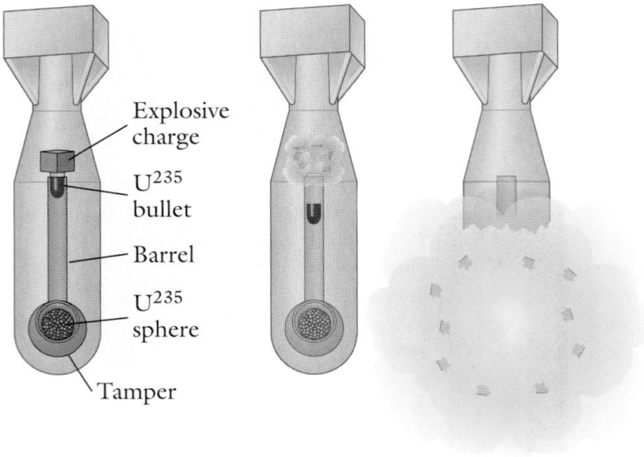

Figure 12.6 How nuclear weapons operate. (Adapted from Howstuffworks.com)

the uranium-235 nuclei. The reaction is allowed to accelerate with no constraints, and a nuclear explosion results. The actual mass of uranium needed to achieve a critical state depends on the level of enrichment.

Armor-piercing missiles are routinely made from ***depleted*** uranium metal, which is uranium that has had the uranium-235 isotope extracted from it. The depleted uranium has a density of 19.0 g/cm^3, which is much greater than the density of steel at 7.85 g/cm^3. This high density is significant because it enables depleted uranium to easily penetrate steel armor when used as a projectile for antitank weapons. The depleted uranium is also used as armor plating to make some tanks resistant to the use of antitank weapons made of depleted uranium.

There is some controversy about using depleted uranium because the uranium metal, now mostly uranium-238, still is a radioactive alpha emitter. Recall that the alpha particles produce high levels of ionization but have low penetrating power. An alpha emitter is most dangerous when it can enter a person's body. When a depleted uranium shell strikes an armored vehicle, it pierces a hole through the target but is vaporized in the process. The greatest risk to personnel would be near a battlefield in which a large amount of depleted uranium was used, because the depleted uranium might be breathed in with the air and dust.

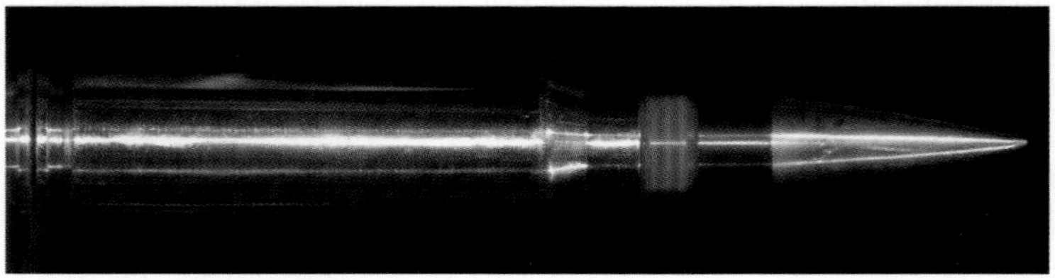

Standard 30-mm antitank shell tipped with depleted uranium. (Reuters/Corbis)

Learning Objective

Determine the isotope formed during a nuclear transmutation reaction.

12.8 Nuclear Transmutations

In a nuclear **transmutation** reaction, a stable isotope is converted into a different element or isotope through a nuclear reaction. Ernest Rutherford conducted the first nuclear transmutation experiment in 1919 when he used alpha particles emitted from a radioactive source to bombard a sample of nitrogen-14. The two nuclei fused together and formed one oxygen-17 atom plus one hydrogen atom, as shown below.

$$^{14}_{7}\text{N} + ^{4}_{2}\text{He} \rightarrow ^{17}_{8}\text{O} + ^{1}_{1}\text{H}$$

Creation of both stable and radioactive isotopes, also known as **radioisotopes,** through transmutation reactions is now a common procedure for making the isotopes used in the field of medicine for the diagnosis and treatment of various diseases. Major hospitals have the capability to create radioisotopes and use them directly at the facility. Can the same technology used in hospitals across the world be used to create the material for a dirty bomb? More information on nuclear chemistry is required before that question can be answered.

Worked Example 5

Balance the reaction that occurs when sodium-23 is bombarded with alpha particles to produce a proton ($^{1}_{1}\text{H}$) and another isotope. Identify the other isotope formed.

SOLUTION

Step 1. Write the equation:

$$^{23}_{11}\text{Na} + ^{4}_{2}\text{He} \rightarrow ^{1}_{1}\text{H} + ?$$

Step 2. Determine the mass number of the unknown:

$$23 + 4 = 1 + \underline{26}$$

Step 3. Determine the atomic number of the unknown:

$$11 + 2 = 1 + \underline{12}$$

Step 4. Determine the unknown element from the atomic number:

$$12 = \text{magnesium (Mg)}.$$

Step 5. Write the balanced equation:

$$^{23}_{11}\text{Na} + ^{4}_{2}\text{He} \rightarrow ^{1}_{1}\text{H} + \underline{^{26}_{12}\text{Mg}}$$

Practice 12.5

Balance the following nuclear transmutation reactions:

(a) Fluorine-19 bombarded with neutrons to produce an alpha particle and another radioisotope.

(b) Uranium-238 bombarded with neutrons to produce a beta particle and another radioisotope.

ANSWER

(a) $^{16}_{7}\text{N}$ **(b)** $^{239}_{93}\text{Np}$

Evidence Analysis | Neutron Activation Analysis

Neutron activation analysis (NAA) is an experimental method based on the nuclear transmutation reactions of a sample that is placed inside a nuclear reactor and bombarded by neutrons.

NAA has been used to gather data in some interesting forensic cases. Bullet fragments from the assassination of President John F. Kennedy on November 22, 1963, were analyzed by NAA, as were hair and nail samples from President Zachary Taylor's remains to determine whether he might have died of arsenic poisoning. A brass plaque purported to have been left by Sir Francis Drake in 1579, commemorating his landing on the California coastline, was analyzed to see if the plaque was genuine or a forgery. Even though the samples in these cases were obtained from 40 to 425 years ago, NAA provided useful information.

President John F. Kennedy (Ted Spiegel/Corbis)

Neutron activation analysis is based on the creation of isotopes of an element by bombarding the element with neutrons. A neutron combines with the nucleus of the sample atom to create an energetically excited isotope, as shown in Reaction 12.1 for the bombardment of arsenic-75.

$$^{75}_{33}\text{As} + ^{1}_{0}\text{n} \rightarrow ^{76}_{33}\text{As}^* \quad \textbf{(Reaction 12.1)}$$

In Reaction 12.2, the energetically excited arsenic atom (arsenic-76) releases excess energy as a gamma ray. Recall that gamma rays are photons—packets of light. The emission of the gamma particle enables a forensic scientist to determine what the excited isotope is because the energy is unique to the isotope that generated it.

$$^{76}_{33}\text{As}^* \rightarrow ^{0}_{0}\gamma + ^{76}_{33}\text{As} \quad \textbf{(Reaction 12.2)}$$

The first gamma ray is released almost instantly, and the gamma ray detector must be in the nuclear reactor with the sample. It is common for a third,

President Zachary Taylor (Bettmann/Corbis)

Sir Francis Drake (Hulton Archive/Getty Images)

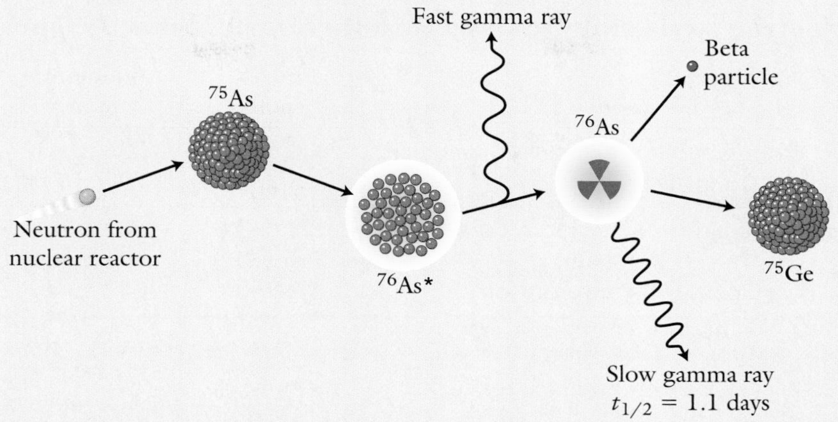

Figure 12.7 Neutron activation analysis. The atoms are bombarded by neutrons in the center of a nuclear reactor. The neutrons will combine with the nucleus of an atom to create an energetically excited isotope of the same element. The excited nucleus will relax by emitting a gamma ray photon with an energy unique to the element of origin. The relaxed nucleus may still undergo further radioactive decay through beta emission, which is also accompanied by gamma emission. (Adapted from www.missouri.edu/~glascock/naa_over.htm)

slower reaction to occur that releases a second gamma ray during a beta decay of the nucleus. The beta decay process can take place anywhere from minutes to days later, depending on the half-life of the newly created isotope. The half-life for arsenic-76 is 1.1 days. Therefore, the sample can be removed from the nuclear reactor, and the energy of the second gamma ray measured, as shown in Reaction 12.3.

$$^{76}_{33}As + {}^{0}_{-1}e \rightarrow {}^{0}_{0}\gamma + {}^{76}_{32}Ge \quad \textbf{(Reaction 12.3)}$$

This reaction is a form of beta decay called *electron capture,* since an electron outside of the nucleus combines with a proton to form a neutron.

The bullet fragments recovered from the assassination of President Kennedy were analyzed by NAA to measure the amounts of the silver and antimony impurities in the lead fragments. Impurity levels that are close in value would indicate that the fragments *could* have come from the same bullet. Levels of trace impurities that vary greatly would suggest that the fragments most likely came from different bullets, assuming that the impurities are distributed homogeneously within the bullet.

Critics of the Warren Commission Report of the assassination have long questioned the number of bullets that struck President Kennedy and surmised that a second gunman was involved. The NAA results showed that two bullets struck the presidential motorcade that day—there was no evidence to support the existence of a third bullet. This analysis does not contradict the Warren Commission Report of the assassination, which stated that *only* two bullets struck President Kennedy. The experimental results of the NAA are summarized in Table 12.3. When examining the results, recall that distribution of trace metals within the lead bullet will vary slightly if they come from the same bullet but differ greatly if they are from different sources.

Was President Zachary Taylor assassinated by arsenic poisoning? Some historians believed that the symptoms associated with President Taylor's illness matched those of arsenic poisoning. In 1991, the descendants of President Taylor agreed to allow his body to be exhumed and to have samples taken for NAA at the Oak Ridge National Laboratory. The samples proved that President Taylor did not have arsenic levels in his body beyond the natural trace levels.

(*continued*)

Evidence Analysis | Neutron Activation Analysis (*continued*)

Table 12.3 Neutron Activation Analysis Results for the Bullet Fragments

Sample	Description	Silver ppm	Antimony ppm	Conclusion
Q1	Bullet from Governor Connally's stretcher	8.8 ± 0.5	833 ± 9	Bullet No. 1
Q9	Governor Connally's wrist	9.8 ± 0.5	797 ± 7	
Q2	Large fragment recovered from car	8.1 ± 0.6	602 ± 4	
Q4	Fragment from JFK's skull	7.9 ± 0.3	621 ± 4	Bullet No. 2
Q14	Small fragment recovered from car	8.2 ± 0.4	642 ± 6	

Guinn, Vincent P. "JFK Assassination: Bullet Analyses," *Analytical Chemistry,* vol. 51, pp. 484A–493A, 1979.

The World Encompassed, published in 1628, describes a plaque that Sir Francis Drake allegedly left on the coastline of California to commemorate his landing there in 1579. Because the plaque had never been located, it was of interest to many historians. When it was discovered in 1936, it soon became the most significant archeological find in the history of California. However, it was proven conclusively in 1977 that the plaque was a forgery constructed at the time of its discovery.

NAA of the brass, conducted by the Lawrence Berkeley National Laboratory, revealed a high purity of the metal alloy and an elemental composition that would not have been possible using technology from 1579. The found plaque was created as a practical joke among a group of historians. Before they had a chance to reveal the joke, a friend and unwitting victim had publicly declared the plaque's authenticity, and a media frenzy ensued. The perpetrators of the joke decided to avoid publicly humiliating their friend and colleague and let the hoax go unchallenged.

Worked Example 6

Does the NAA evidence prove that President Taylor died of natural causes? Explain your answer.

SOLUTION No, the NAA results simply prove that President Taylor did not die from arsenic poisoning. He may or may not have died from another poisonous substance that was not detected by the NAA procedure.

Practice 12.6

Does the NAA evidence in the Kennedy assassination case prove definitively that only two shots struck the President? Explain your answer.

ANSWER
The evidence suggests that two bullets struck President Kennedy. It eliminates the possibility that only one bullet struck the president. If a third bullet struck President Kennedy, either it had an elemental composition identical to one of the other bullets or it was never recovered and analyzed.

12.9 Case Study Finale: Gangster Turned Terrorist?

Recall from the opening of our case study that a dirty bomb, or radiological dispersion device, is made from conventional explosives to which radioactive materials have been added. The U.S. government has focused a great deal of attention on the possibility that a terrorist cell might obtain and detonate a dirty bomb. It is important to note that a nuclear explosion does not occur when a dirty bomb detonates. But radioactive substances are widely dispersed by the explosion, potentially exposing many individuals.

What purposes might a terrorist organization have for exploding a dirty bomb? And what would be the real effects of the detonation of such a device? Government officials must address these issues in developing response plans in case of an emergency. Below is a summary of questions that will arise if there is a dirty bomb incident.

Dirty Bomb Issues

- What type of radioactive material was used and where did it come from?
- What are the casualties and destruction from the explosion?
- What is the exposure risk to the first responders on the crime scene?
- How will the public respond to the explosion?
- How will the area be decontaminated and individuals evacuated?
- What will be the long-range economic damage?

By all estimates, the initial deaths from the detonation of a dirty bomb will be relatively small, particularly when compared with the impact of a nuclear weapon. Casualties might number in the hundreds. The type of radioactive material contained in the bomb would be of great importance in assessing the hazards at the scene. For example, most medical radioisotopes will not present the kind of long-term contamination issues that would result if uranium radioisotopes from a nuclear reactor were used. Risk estimates have shown that the first responders to the scene, who would not be aware of the radiological nature of the explosion, will most likely not be exposed to life-threatening levels of radiation. The amounts and types of radiological materials that could be used will not present an imminent danger after detonation.

Even though a dirty bomb is not likely to create a large number of casualties, it can, nevertheless, wreak havoc in other ways. A potent psychological weapon, it might demoralize and exhaust the resources of the citizens who must deal with a large-scale decontamination and evacuation of a region of their community. The United States Nuclear Regulatory Commission refers to a dirty bomb as a "weapon of mass disruption." Terrorists undoubtedly hope that the public's first response, when news of a radioactive bomb detonation is announced, will be mass panic and chaos.

Terrorists will be likely to choose a target that would cause economic hardship or make uninhabitable an area of real and symbolic importance. Imagine if the White House, the U.S. Capitol, or the Wall Street financial district were evacuated for long-term decontamination. The costs of the cleanup and decontamination could quickly escalate, especially since many radioactive isotopes would be permanently absorbed into concrete or asphalt, requiring the complete removal of those materials.

In the planning and preparation by state and federal government agencies for the response to a dirty bomb incident, the purchase and stockpiling of **radiation pills** has been recommended. However, it is important to understand the limitations of these pills. The radiation pill is simply a dose of potassium iodide that serves to protect the thyroid gland, and only the thyroid gland, from absorbing radioactive iodine ($^{131}_{53}I$). The thyroid gland serves as a regulator of bodily functions and growth, which is why protecting it is desirable.

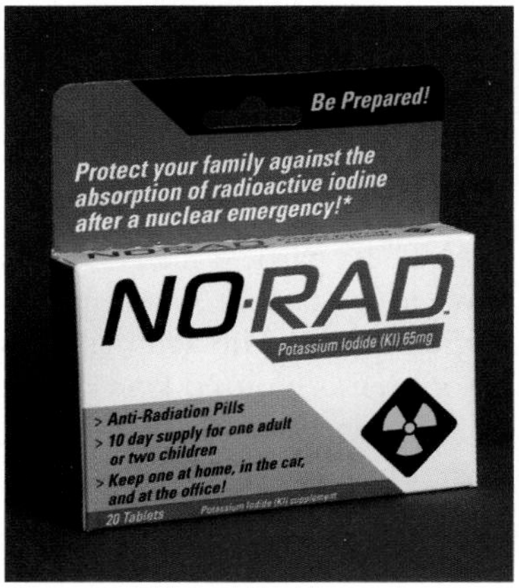

Radiation pills used for thyroid protection from radioactive iodine-131. (Leonard Lessin/Photo Researchers, Inc.)

Iodine-131 has a half-life of 8.0 days. Therefore, it is doubtful that terrorists would use iodine-131 in a dirty bomb because it would be transformed into stable isotopes before use or soon after. The only real danger for exposure to iodine-131 comes from two sources: fallout from the explosion of a nuclear weapon and a catastrophic failure at a nuclear power plant. If either of these two events should happen, countless radioactive isotopes as well as iodine-131 would be present, and there are no pills to counteract the other isotopes. Potassium iodide was used in Poland following the Chernobyl disaster, when a large amount of iodine-131 was released into the atmosphere. The pills proved to be a safe and effective method for protecting the thyroid gland until the radioactive iodine naturally decayed.

If José Padilla had the capability to create and detonate a dirty bomb, it would have delivered a costly blow to the U.S. economy. As the defense attorneys and the White House debate the legality of his detention, you should now be able to better understand why the government decided to detain him. This is not to suggest whether you should or should not agree with the actions of the federal government or debate the constitutionality of José Padilla's detention.

CHAPTER SUMMARY

- The discovery of radioactivity by Henri Becquerel and the subsequent development of the field of nuclear chemistry by Marie Curie have created a new world in which radioisotopes have cured countless numbers of people from disease, provided the allies with a victory in World War II, and generated electricity to power millions of homes.
- The radioactivity of an isotope depends on the ratio of protons to neutrons in the nucleus of its atoms. Atoms that lie outside of the belt of stability try to adjust the ratio by releasing one or more of the three main forms of radiation: the alpha particle, beta particle, or gamma ray.
- The alpha particle consists of two protons and two neutrons and is identical to the helium nucleus. The beta particle is an energetic electron produced when a neutron is converted to a proton within the nucleus. The gamma ray is a high-energy photon that is emitted by a nucleus in an elevated energy state.
- The danger presented by radiation comes in two forms: the power of the radiation to penetrate matter and the ionizing power of the radiation. Ionizing power is the ability to ionize the matter through which the radiation passes. The alpha particle has the greatest ionizing power but the lowest penetrating power. The gamma ray is at the other extreme, requiring several centimeters of lead to be absorbed, but it has the lowest ionizing power. The properties of beta particles are intermediate between the alpha and gamma rays.
- Balancing a nuclear equation is based on the law of conservation of mass. The total mass of protons and neutrons on each side of the equation is conserved, but the identity of the elements changes.
- The fission reaction of uranium-235 provides the fuel source for both atomic weapons and nuclear power plants, although the fuel in a power plant cannot detonate like a nuclear weapon. Nuclear power plants generate electricity by using the heat of the fission reaction to convert water into steam and turn a traditional steam turbine.
- The dangers of nuclear power lie in the extremely long half-life of the waste products, which need to be stored indefinitely. The waste produced by most medical isotopes generally does not pose a long-term storage problem because of the relatively short half-life of those isotopes.
- A radiological dispersion device, or dirty bomb, is actually geared to inflict economic and psychological damage rather than mass casualties.

KEY TERMS

radiological dispersion device, p. 357
phosphorescence, p. 358
radioactivity, p. 358
background radiation, p. 359
belt of stability, p. 360
alpha particle (α), p. 361
ionizing power, p. 361
penetrating power, p. 361
beta particle (β), p. 361
gamma ray (γ), p. 362
fission, p. 370
chain reaction, p. 370
meltdown, p. 370
control rod, p. 370
fusion, p. 371
critical mass, p. 373
transmutation, p. 374
radioisotope, p. 375
neutron activation analysis (NAA), p. 376
radiation pill, p. 379

CONTINUING THE INVESTIGATION Additional Readings, Resources, and References

Curie, Eve. *Madame Curie,* New York: Doubleday, 1939.

Guinn, Vincent P. "JFK Assassination: Bullet Analyses," *Anaytical Chemistry* **51**, 1979: 484A–493A.

For more information on the capture and detention of José Padilla, search any major news agency such as:
archives.cnn.com/2002/US/06/10/dirty.bomb.suspect/index.html
archives.cnn.com/2002/LAW/10/12/padilla.detention/index.html
www.cnn.com/2003/LAW/01/29/padilla.hearing/index.html
www.cnn.com/2004/LAW/03/18/padilla/index.html
www.cnn.com/2005/LAW/11/22/padilla.case/index.html
www.cnn.com/2006/LAW/01/12/padilla.plea.ap/index.html

For more information on Drake's plaque: www.lbl.gov/Publications/Currents/Archive/Mar-07-2003.html#Drake

For a comprehensive article on nuclear forensic science: www.llnl.gov/str/March05/Hutcheon.html

U.S. Environmental Protection Agency, *A Citizen's Guide to Radon: The Guide to Protecting Yourself and Your Family From Radon,* Washington D.C.: U. S. Environmental Protection Agency, 2005. www.epa.gov/radon/pubs/citguide.html

For information on which consumer products contain radioisotopes: www.orau.org/ptp/collection/consumer%20products/consumer.htm.

For use of half-lives in forensic anthropology: news.bbc.co.uk/2/hi/science/nature/3632770.stm

For more on neutron activation analysis and the cases mentioned: www.ornl.gov/info/ornlreview/rev27-12/text/ansside6.html

A list of commonly found medical radioisotopes can be found in the *miscellaneous tables* section of the Merck Index.

REVIEW QUESTIONS AND PROBLEMS

Questions

1. Discuss Henri Becquerel's original hypothesis about phosphorescence and how, using the scientific method, he disproved it.
2. What is phosphorescence?
3. How would history have changed had Becquerel been using a mineral ore other than one that contained uranium?
4. Would history have significantly changed if Becquerel had had sunny weather throughout his experiments?
5. Explain the role of the neutron in the nucleus.
6. Explain why some isotopes are stable and others are radioactive.
7. What happens to the total number of neutrons and protons of a nucleus during alpha, beta, and gamma decays?

8. What happens to the total charge of a nucleus during alpha, beta, and gamma decays?

9. Which radioactive decay particle has the greatest ionizing power? Which has the least ionizing power?

10. Which radioactive decay particle has the greatest penetrating power? Which has the least penetrating power?

11. Which radioactive decay particle poses the greatest risk if present inside the body? Why?

12. Which radioactive decay particle poses the greatest risk if it is external to the body? Why?

13. Explain how the conservation of the total number of neutrons and protons applies to balancing nuclear equations.

14. What is a nuclear transmutation reaction and how does it differ from radioactive decay?

15. One advantage of neutron activation analysis is that it does not destroy the sample being tested. Why is this desirable for forensic analysis of evidence?

16. What is one limitation on the widespread use of neutron activation analysis?

17. Sketch a nuclear power plant and describe how the three independent heat exchange systems generate electricity.

18. What is depleted uranium and what is it used for? How is enriched uranium used after it is processed from natural uranium?

19. How does a fission reaction differ from a fusion reaction? Which is more desirable for generating electricity and why?

20. What are the limitations on creating a fusion reactor?

21. Describe how a nuclear chain reaction occurs. Include a sketch of U-235 undergoing a chain reaction. What is the role of control rods in a nuclear power plant in relation to a chain reaction?

22. Describe two military uses of radioisotopes.

23. What are some desirable characteristics of radioisotopes for medical diagnosis? For cancer treatment?

24. Because mass casualties from the explosion of a radiological dispersion device are highly unlikely, what are the true goals of a terrorist organization using such a device? What are the scientific issues that would have to be addressed in case of such an incident?

25. Historically, the testing of nuclear weapons technology moved from above-ground or underwater detonations to underground testing. Why would underground testing of nuclear weapons be considered a safer alternative?

Problems

26. Write a balanced equation for the following radioactive decay systems.
(a) Uranium-238 undergoing alpha decay
(b) Carbon-14 undergoing beta decay
(c) Krypton-85 undergoing beta decay
(d) Radon-222 undergoing alpha decay

27. Write a balanced equation for the following radioactive decay systems.
(a) Iodine-131 undergoing beta decay
(b) Francium-220 undergoing alpha decay
(c) Cobalt-60 undergoing beta decay
(d) Radium-226 undergoing alpha decay

28. Provide the missing information to balance the following radioactive decay systems.
(a) ${}^{233}_{92}\mathrm{U} \rightarrow$ ______ $+ {}^{229}_{90}\mathrm{Th}$
(b) ${}^{137}_{55}\mathrm{Cs} \rightarrow \beta +$ ______
(c) ${}^{232}_{90}\mathrm{Th} \rightarrow {}^{4}_{2}\mathrm{He} +$ ______
(d) ${}^{42}_{19}\mathrm{K} \rightarrow$ ______ $+ {}^{42}_{20}\mathrm{Ca}$

29. Provide the missing information to balance the following radioactive decay systems.
(a) ${}^{198}_{79}\mathrm{Au} \rightarrow$ ______ $+ {}^{198}_{80}\mathrm{Hg}$
(b) ${}^{239}_{94}\mathrm{Pu} \rightarrow \alpha +$ ______
(c) ${}^{3}_{1}\mathrm{H} \rightarrow {}^{0}_{-1}\mathrm{e} +$ ______
(d) ${}^{235}_{92}\mathrm{U} \rightarrow {}^{4}_{2}\mathrm{He} +$ ______

30. Provide the missing information to balance the following nuclear transmutation reactions.
(a) ${}^{14}_{7}\mathrm{N} +$ ______ $\rightarrow {}^{17}_{8}\mathrm{O} + {}^{1}_{1}\mathrm{H}$
(b) ${}^{238}_{92}\mathrm{U} + {}^{4}_{2}\mathrm{He} \rightarrow 3{}^{1}_{0}\mathrm{n} +$ ______
(c) ______ $+ {}^{239}_{94}\mathrm{Pu} \rightarrow {}^{243}_{96}\mathrm{Cm}$
(d) ${}^{238}_{92}\mathrm{U} + {}^{1}_{0}\mathrm{n} \rightarrow$ ______ $+ {}^{0}_{-1}\mathrm{e}$

31. Provide the missing information to balance the following nuclear transmutation reactions.
(a) ${}^{57}_{27}\mathrm{Co} + {}^{0}_{-1}\mathrm{e} \rightarrow$ ______ $+ \gamma$
(b) ${}^{59}_{27}\mathrm{Co} + {}^{1}_{0}\mathrm{n} \rightarrow$ ______ $+ \gamma$
(c) ______ $+ {}^{111}_{49}\mathrm{In} \rightarrow {}^{111}_{50}\mathrm{Sn}$
(d) ${}^{32}_{16}\mathrm{S} + {}^{4}_{2}\mathrm{He} \rightarrow$ ______

Forensic Chemistry Problems

32. One suggested method for tagging explosives involves using varying amounts of isotopes from rare elements. Why wouldn't radioactive isotopes be used?

33. Neutron activation analysis of a hair sample can be used to determine a person's history of drug use. Explain how hair analysis could provide data on a person's drug use six months earlier.

34. If analysis of skeletal remains shows that lead-210 has undergone 2.5 half-lives, how old is the skeleton?

35. If analysis of skeletal remains shows that polonium-210 has undergone 5.25 half-lives, how old is the skeleton?

Case Study Problems

36. Mark Hofmann was an expert forger who sold hundreds of fake documents. He is best known for duping the Mormon Church into purchasing many fake writings from its early history. As his web of lies started to unravel, Hofmann turned to murder to cover his crimes, killing two individuals in two separate incidents. One of his last sales was to the Library of Congress of the only known copy of the Oath of a Freeman, a loyalty oath taken by the original settlers of the Massachusetts Bay colony. Carbon-14 dating of the document verified both the age of the paper and the ink, despite its modern origin. How could this be?

CHAPTER 13

Poisons

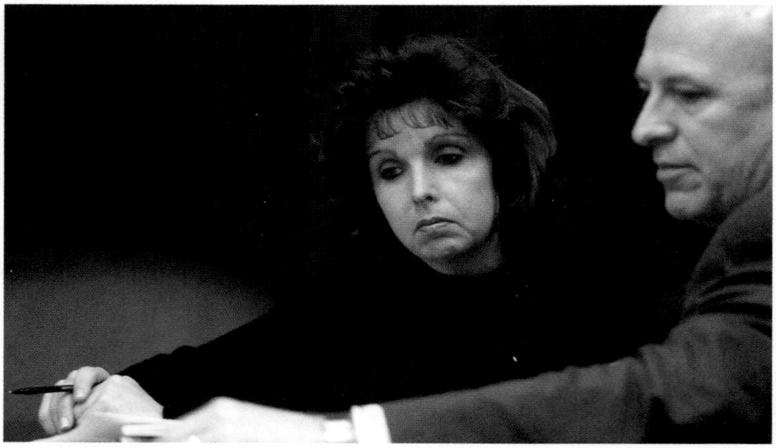

The mothers of Randy Thompson and Glenn Turner became suspicious that their sons' deaths were more than tragic accidents and that Julia Turner, pictured above, might be responsible. (Ric Feld/AP Photo)

CASE STUDY: A Mother-in-Law's Justice

Nita Thompson had to do what every parent fears most—bury one of her own children. Her son Randy died suddenly and unexpectedly at the age of 32. Although Randy had a dangerous job as a firefighter with the Forsyth County Fire Department in Georgia, his death was not work related. He had gone to the emergency room on January 20, 2001, with intense nausea and vomiting. After being given intravenous fluids and a prescription for the nausea, Randy was sent home. He was found dead on January 22. Because the medical examiner ruled the cause of death to be heart failure and the manner of death to be natural, no investigation into Randy's death was initiated.

Kathy Turner had experienced a similar tragedy six years earlier, with the death of her 31-year-old son. Glenn Turner had worked as a police officer with the Cobb County Police Department in Georgia, but his death was also not work related. He had gone to the emergency room on March 2, 1995, with intense nausea and vomiting, was given intravenous fluids and a prescription for the nausea, and sent home. Glenn was found dead the next day. In his case, as in the Randy Thompson case, the medical examiner ruled the cause of death to be heart failure and the manner of death to be natural. Thus, no investigation into his death was initiated.

Kathy Turner and Nita Thompson shared more than just the same set of tragic events surrounding the deaths of their sons. Both women had family connections to each other through Julia Lynn Turner. Julia had been Glenn Turner's wife for one year when she began having an affair with Randy Thompson. Her unhappy marriage to Glenn was headed for divorce in the spring of 1995 when Glenn died, leaving Julia with more than $150,000 in life insurance money. Within four days of Glenn's death, Julia had moved in with Randy. Although Randy and Julia never married, they had two children. Their relationship was rocky, and after a time, Randy moved out but was still seeing Julia and the children. One of the last times he would see Julia was for dinner at a restaurant, shortly before he became ill.

Kathy Turner, the former mother-in-law of Julia Turner, and Nita Thompson, grandmother of Julia's two children by Randy, realized that the similarities in the symptoms and circumstances of their sons' deaths seemed beyond belief.

The women could come to only one horrifying conclusion: Their sons had been murdered, and now they needed to find proof . . .

13.1 Introduction to Equilibrium

Alle Dinge sind Gift, und nichts ist ohne Gift.
Allein die Dosis macht, dass ein Ding kein Gift ist.
All things are poison, and nothing is without poison.
The dosage alone determines that a thing isn't poison.
—Paracelsus (1493–1541)

Learning Objective
Describe chemical equilibrium and show how the concept relates to poisoning.

A **poison** is defined as any compound that injures or harms a living organism. As Paracelsus suggested, all things have the potential to be a poison—only the dose dictates whether a compound is a poison or not. The term **toxic compound** is used to indicate that a compound is dangerous even in small amounts.

Many poisons work by interfering with an organism's normal cell functions in a way that results in cell death. Poisons can be in the form of pure elements (mercury), inorganic compounds (carbon monoxide), or organic compounds (strychnine). The term **toxin** usually refers to a naturally occurring poisonous substance produced by a living organism.

Worked Example 1

Can a person suffer from water poisoning?

SOLUTION Yes. If excessive amounts of water are consumed, it can act as a poison by seriously disrupting the body's balance of electrolytes. News reports indicate that several college students in the last few years have died from hazing incidents that involve drinking excessive amounts of water.

Practice 13.1

Iron is a common element that is included in many over-the-counter multivitamins. Are iron supplements a toxic compound? Search the Centers for Disease Control's Web site at www.cdc.gov for information regarding the safety of iron supplements.

ANSWER

As with other substances that are usually benign when ingested at recommended levels, iron supplements are safe for adults. However, iron can be fatal if ingested by children.

> [Ghost:] With juice of cursed hebona in a vial,
> And in the porches of my ears did pour
> The leperous distilment; whose effect
> Holds such an enmity with blood of man
> That swift as quicksilver it courses through
> The natural gates and alleys of the body,
> And with a sudden vigour it doth posset
> And curd, like eager droppings into milk,
> The thin and wholesome blood. So did it mine;
>
> —*Hamlet,* Act I, Scene 5, by William Shakespeare

William Shakespeare weaved several methods of poisoning into his telling of *Hamlet:* The ghost of Hamlet's father reveals that a poison was poured into his ear; Hamlet's mother dies from drinking poisoned wine; and Hamlet himself dies from being stabbed with the poison-coated tip of a foil during a duel. As illustrated by the play, the means by which an individual is exposed to a poison is a crucial factor in the danger presented by the compound.

Some poisons are extremely hazardous by inhalation, but contact with skin is not particularly dangerous. Other poisons may have multiple routes of exposure and entry. If a poison is discovered during an autopsy, the medical examiner will attempt to determine how the poison entered the victim's system; this information can help investigators determine whether the case is a homicide or an accidental death.

When a poison is introduced into the human body—via inhalation, physical contact, ingestion, or injection—it passes into the circulatory system and is transported throughout the body. The ability of the toxic compound to pass out of the blood and into the cells of the tissues is largely determined by whether or not it will bind with albumin, one of the main proteins found in blood plasma. When a toxic compound

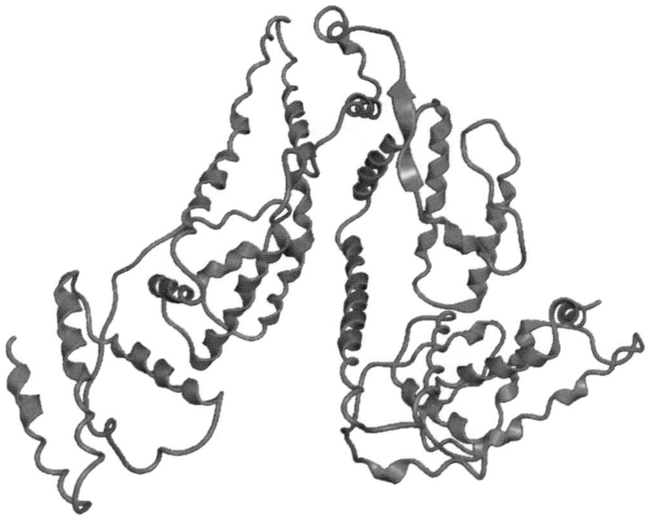

The protein albumin is shown as a ribbon structure, which is used to help visualize the overall three-dimensional shape rather than each individual atom.

binds to albumin, it is no longer able to pass into the cells and disrupt their activity. However, compounds differ in the extent to which they bind to albumin. The unbound form of the poison is available to cells and can disrupt cell function. Determining the relative amounts of bound and unbound toxic substance requires an understanding of *chemical equilibrium.*

Indeed, familiarity with *any* chemical reaction requires a true understanding of **chemical equilibrium.** This discussion will bring together concepts from earlier chapters such as stoichiometry, limiting reagents, excess reagents, kinetics, catalysts, collision theory, and activation energy.

It would be prudent to review these topics if you need to before reading the material on equilibrium.

A common misconception about chemical reactions is that the reactants are completely changed to the products. This notion is conveyed by the simplified description of chemical reactions in which a single arrow ($\rightarrow$) is used to indicate that a reaction takes place. The reaction arrow is translated as "to form," without any qualifier. In reality, most reactants do not completely react to form the product even if sufficient quantities of the reactants are present in the exact molar proportions needed for a complete reaction. One must consider the possibility that the products of the reaction may react together to re-form the reactants. Consider the following generic reaction:

$$A + B \rightarrow C + D$$

Recall from Section 11.2 the requirements for a chemical reaction according to collision theory: Molecular collisions of sufficient energy and orientation will be successful. Since there will be successful collisions between the products, the **reverse reaction** of products to form reactants can occur:

$$C + D \rightarrow A + B$$

Rather than writing two equations for each reaction, a double arrow notation is used to represent both reactions:

$$A + B \rightleftarrows C + D$$

If the reverse reaction of C and D coming together to remake the reactants A and B occurs, can the reaction ever be finished? As soon as A and B are formed, they can react together to form the products C and D once again. To understand the extent to which the forward and reverse reactions take place, and what it means for a reaction to be "finished," we need to have a more complete understanding of collision theory.

The magnitude of both the forward and reverse reactions is controlled by the restrictions of collision theory and especially the activation energy barrier. Consider the reaction progress in relation to the activation energy, as shown in Figure 13.1.

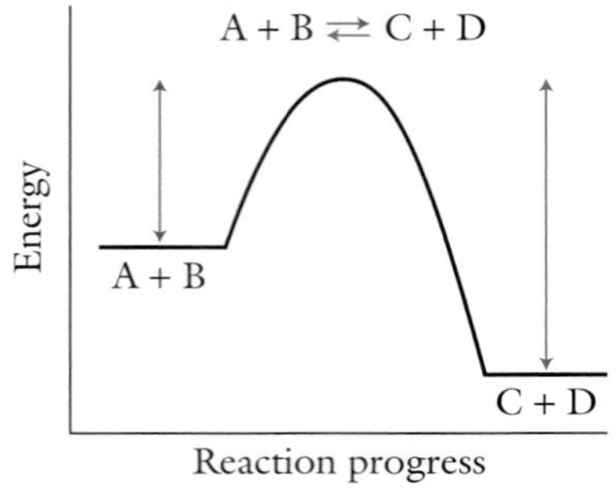

Figure 13.1 Activation energy barrier. The energy required for the forward reaction (A + B → C + D) is much less than the energy required for the reverse reaction (C + D → A + B).

The forward reaction is limited to those molecules that have sufficient energy to overcome the activation energy barrier illustrated by the blue arrow in Figure 13.1. The reverse reaction occurs to a lesser extent because its activation energy, illustrated by the green arrow, is much greater than that of the forward reaction.

13.2 Dynamic Equilibrium

Learning Objective

Explain how dynamic equilibrium is established and the effect of adding a catalyst.

A chemical reaction is at equilibrium when the rate of the forward reaction is equal to the rate of the reverse reaction. Under these circumstances, for every set of reactants that form products, a set of products will undergo the reverse reaction to re-form the reactants. The energy diagram in Figure 13.1 might easily be misinterpreted to mean that the rate of the forward reaction will always be greater than the rate of the reverse reaction due to the differences in activation energy. Recall, though, that activation energy is not the only factor influencing a reaction—the frequency of collisions and the orientation of the molecules are also important.

When compounds A and B first react, as shown in Figure 13.2a, no reverse reaction takes place. As compounds C and D are formed, they are initially far apart and seldom collide, as illustrated in Figure 13.2b. Because compounds A and B are present in high concentrations at the start of the reaction, the forward reaction occurs rapidly. As the reaction continues, the concentrations of A and B decrease, causing fewer collisions between A and B and a slower rate in the forward direction. Simultaneously, the concentrations of C and D increase, causing more collisions between C and D and a higher rate in the reverse direction, as shown in Figure 13.2c.

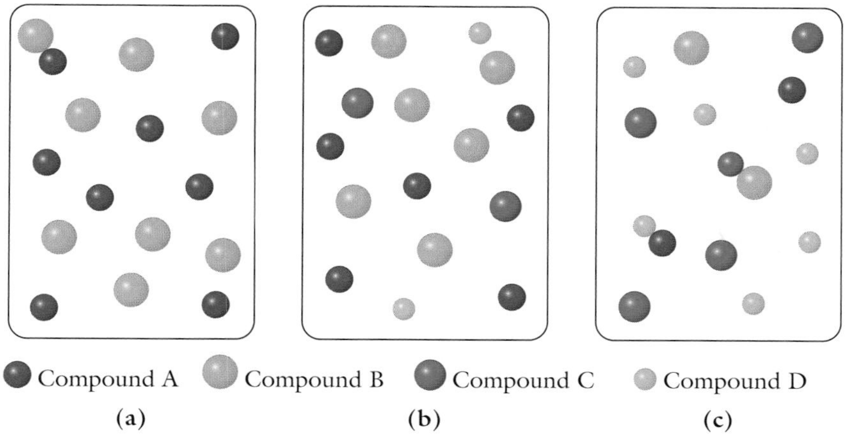

Figure 13.2 Establishment of equilibrium for A + B ⇄ C + D. (a) At the start of the reaction, only the forward reaction is possible because only compounds A and B collide. (b) As the reaction progresses, the number of collisions between compounds A and B decreases because their concentrations are lower. The number of collisions between compounds C and D increases because they are now both present in solution. (c) When the reaction reaches equilibrium, the number of successful collisions between reactants in the forward reaction is equal to the number of successful collisions between products in the reverse reaction.

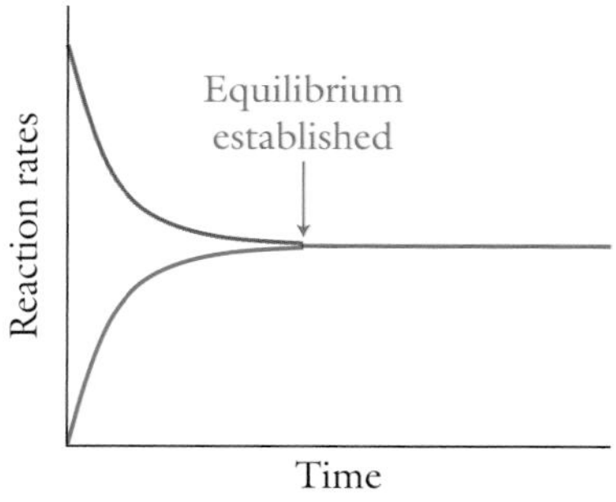

Figure 13.3 Rate changes in equilibrium systems. The rate of the forward reaction (blue) decreases as the reaction consumes the reactant molecules, and fewer collisions occur. The rate of the reverse reaction (green) increases as the reaction progresses because the reaction increases the concentration of the product molecules. Equilibrium is established when the rates of the forward and reverse reactions are the same.

Figure 13.3 illustrates the changes in reaction rates until the point at which equilibrium is established. The blue line indicates the rate of the forward reaction, and the green line represents the rate of the reverse reaction. At equilibrium, both the forward and reverse reactions occur at the same rate, and the concentrations of reactants and products do not change over time. We often refer to this as *dynamic equilibrium* because the reactions continue but do not produce any overall change in the concentrations of reactants and products. It is important to note that just because the concentrations are no longer changing does not mean that the concentrations are equal at equilibrium.

Figure 13.4 shows a concentration profile for the generic reaction of

$$A + B \rightleftarrows C + D$$

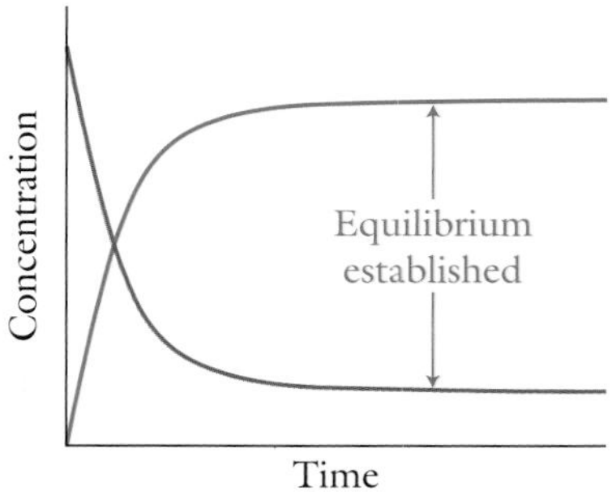

Figure 13.4 Equilibrium concentration profile. The concentration of the reactants (blue) decreases and the concentration of the products (green) increases as the reaction progresses. Equilibrium is established when the concentrations of reactants and products become constant. Note that both the forward and reverse reactions occur at the same rate.

The concentration profile differs from the kinetic graphs shown in Figure 13.3 because the concentrations become constant but not equal to one another. Figure 13.4 shows that the final concentration of reactants (in blue) is less than the final concentration of products (in green). This is not always the case, because some reactions do not create a significant amount of the products.

Worked Example 2

Does the following sketch of a reaction vessel represent a reaction at equilibrium, given that the vessel in Figure 13.2c represents the same reaction at equilibrium? Explain your answer.

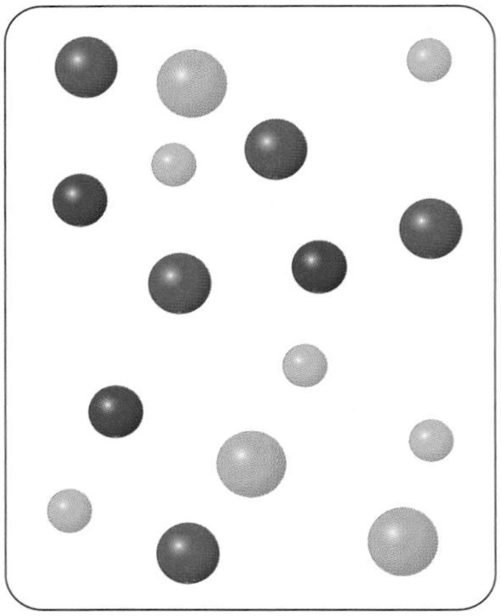

SOLUTION This system is not in equilibrium with the one shown in Figure 13.2c. The concentration of the reactants in this vessel is higher, and the concentration of the products is lower, than shown for the vessel in Figure 13.2c.

Practice 13.2

Draw a reaction vessel that is in equilibrium with that shown in Figure 13.2c.

ANSWER

The vessel should have six units each of compounds C and D, and two units each of compounds A and B.

The presence of a catalyst in a reaction increases the rate of a reaction by providing an alternate mechanism that lowers the activation energy required by the reactants. How will this affect equilibrium? Figure 13.5 is an energy diagram for a reaction, with the catalyzed reaction path shown in red. Notice that the catalyst lowers the activation energy required for both the forward and reverse reactions. Therefore, a catalyst increases the rate of each reaction and achieves an equilibrium condition faster.

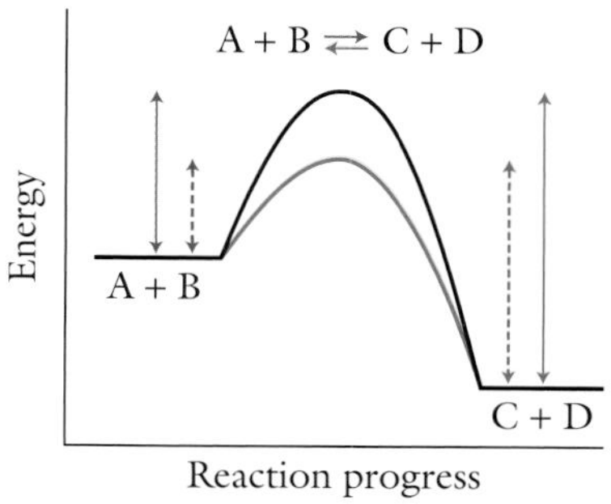

Figure 13.5 Catalyzed reaction pathway. A catalyst increases the rate of a reaction because the activation energy barrier is decreased, which allows a greater number of collisions to be successful. Notice that the catalyzed reaction path (red) lowers the activation energy barrier for both the forward and reverse reactions.

13.3 Values of the Equilibrium Constant

Learning Objective

Determine concentrations of reactants and products using the equilibrium constant.

When a poison enters the circulatory system, it encounters albumin, the main plasma protein. As stated earlier in this chapter, many compounds will react with and bind to albumin molecules. The amount of time a poisonous compound resides in the body depends on how strongly it is bound to albumin. A compound that is weakly bound will be better able to enter cells and do damage, but it is also more quickly eliminated from the body by the kidneys, liver, or digestive tract. If the poison is more strongly bound to albumin, it will have a longer half-life in the body.

Consider the binding of a toxic metal (M) to albumin (A) to form the metal-albumin product (MA) by the reaction:

$$\mathrm{M + A \rightleftarrows MA}$$

When the reaction has reached equilibrium, the rate at which the metal combines with albumin to form the metal-albumin product is equal to the rate at which the metal-albumin product (MA) dissociates to form the free metal (M) and albumin (A). But what are the concentrations of the substances when equilibrium is reached? If the binding forces between the metal and albumin are strong, the metal-albumin product will be the main component at equilibrium and very little free metal will be present. If, however, the binding forces are weak, there will be much more free metal. The **equilibrium constant (*K*)** is the concentration of the products raised to their stoichiometric coefficients, divided by the concentration of reactants raised to their stoichiometric coefficients at equilibrium. For the metal-albumin example, the equilibrium constant is

$$K = \frac{[\mathrm{MA}]^1}{[\mathrm{M}]^1[\mathrm{A}]^1} = \frac{[\mathrm{MA}]}{[\mathrm{M}][\mathrm{A}]}$$

The numerical value of the equilibrium constant for a chemical reaction is always the same, provided the temperature is held constant. The actual concentration of reactants and products may vary under different conditions, but their ratio is always the same. The equilibrium constant is also called the *stability constant,* the *association constant,* or the *binding constant,* especially when discussing toxins.

The magnitude of the equilibrium constant provides information about the relative amounts of reactants and products found in solution at equilibrium. Consider the following symbolic equation for the equilibrium constant:

$$K = \frac{[\text{Product C}]\,[\text{Product D}]}{[\text{Reactant A}]\,[\text{Reactant B}]}$$

If you have a high concentration of products and a low concentration of reactants,

$$\frac{[\text{large number}]}{[\text{small number}]} = \text{very large number} \Rightarrow \text{mostly products in solution}$$

If you have a low concentration of products and a high concentration of reactants,

$$\frac{[\text{small number}]}{[\text{large number}]} = \text{very small number} \Rightarrow \text{mostly reactants in solution}$$

Therefore, a large equilibrium constant indicates a solution consisting primarily of products. A small equilibrium constant indicates that mostly reactants are present in solution.

Table 13.1 lists the binding constants for the mercury(II) ion (Hg^{2+}), a toxin that can be fatal if ingested. The mercury(II) ion binds to the amino acids methionine and cysteine that are found in protein molecules such as albumin. Hg^{2+} forms a strong bond to the sulfur atom located in the methionine molecule. The equilibrium constant of 3.2×10^6 indicates that the bound form of the ion predominates at equilibrium. The bonding of Hg^{2+} to the sulfur atom in cysteine is even stronger, as reflected by an equilibrium constant of 1.6×10^{14} that is 50 million times

Table 13.1 Equilibrium Constants for Hg^{2+} Binding

Methionine	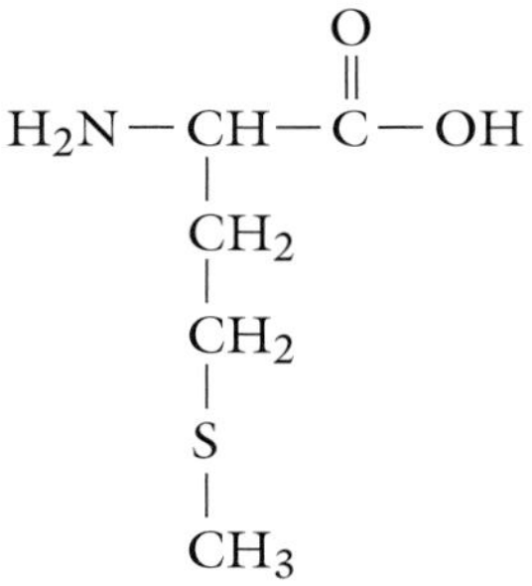	3.2×10^6
Cysteine	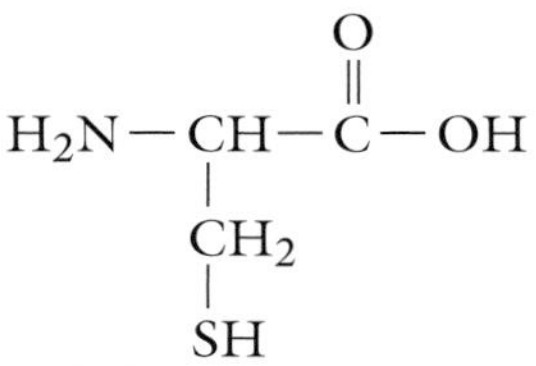	1.6×10^{14}
Dimercaprol	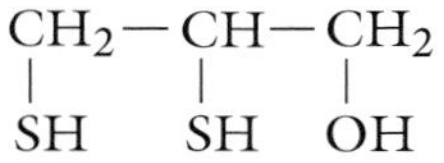	5.0×10^{25}

larger! The structures for cysteine and methionine are provided in Table 13.1, and it is evident that the mercury(II) ion has a stronger bonding preference for the R—SH group of cysteine rather than the R—S—R group of methionine.

The last compound listed in Table 13.1, dimercaprol, is the antidote for mercury poisoning because it exploits the mercury(II) ion's affinity for the —SH functional group. Dimercaprol can also be used to treat poisoning by heavy metals other than mercury. In fact, dimercaprol is often referred to as BAL, which stands for British Anti-Lewisite, because it was the antidote developed in World War I to combat the chemical warfare agent Lewisite, an arsenic-based compound. The binding constant for the mercury(II) ion with the —SH group of cysteine is extremely large, indicating that at equilibrium there is very little free mercury(II) ion. The antidote dimercaprol takes advantage of the strong bond between toxic metals and the —SH functional group by having two —SH groups within the same molecule. Both —SH groups can interact with a single metallic ion, yielding an extremely large binding constant.

Worked Example 3

Weak acids are in equilibrium with a hydrogen ion and the corresponding anion according to the reaction:

$$HA \rightleftarrows H^+ + A^-$$

Which of the following weak acid compounds will have the most acidic pH value?

Weak Acid	Formula	Equilibrium Constant
Hydrofluoric acid	HF	7.1×10^{-4}
Acetic acid	$HC_2H_3O_2$	1.8×10^{-5}

SOLUTION The most acidic pH value will correspond to the compound with an equilibrium constant that most favors the product side of the reaction ($HA \rightleftarrows H^+ + A^-$). Because hydrofluoric acid has a higher equilibrium constant than acetic acid, hydrofluoric acid will have the most acidic pH value.

Practice 13.3

An aqueous solution of silver nitrate can be sprayed on a fingerprint to develop it by reacting with the chloride ions in the fingerprint, thus forming solid AgCl. The precipitate-coated fingerprint is then exposed to ultraviolet radiation to decompose the AgCl to elemental silver, which appears as a black powder. Using the following information, explain why silver is a good choice for this application.

$$Ag^+ + Cl^- \rightleftarrows AgCl(s) \qquad (K = 5.6 \times 10^9)$$

ANSWER

Silver ion will react with chloride ion to precipitate silver chloride. The very large equilibrium constant value indicates that at equilibrium there is mostly product and very little reactants. This is beneficial for the development of fingerprints because it would be of little use to apply a reagent that did not form a stable compound.

13.4 Le Chatelier's Principle

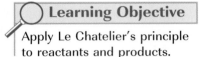

Learning Objective

Apply Le Chatelier's principle to reactants and products.

Henry Louis Le Chatelier (1850–1936). (SPL/Photo Researchers, Inc.)

Carbon monoxide, a tasteless, colorless, and odorless gas, is created by the incomplete combustion of fuels. It is associated with automobile exhaust systems, malfunctioning furnaces or oil heaters, and poorly ventilated fires inside a building. Carbon monoxide is also a deadly poison when inhaled. But if carbon monoxide poisoning is treated in time, the carbon monoxide can be removed from the body.

To understand the mechanism of carbon monoxide poisoning, we must first know a few facts about the chemical basis of respiration and how the body's equilibrium systems can be influenced by outside factors.

The protein hemoglobin (abbreviated Hb), found in red blood cells, is responsible for transporting oxygen gas from the lungs to tissues throughout the body. The equilibrium reaction established between hemoglobin, oxygen gas, and oxyhemoglobin (HbO_2) is

$$Hb + O_2 \rightleftarrows HbO_2$$

The respiratory system is illustrated in Figure 13.6, with emphasis placed on the regions where the Hb-HbO_2 equilibrium is altered. In the lungs, the hemoglobin reacts with oxygen gas to form HbO_2 in the alveoli—microscopic air sacs in which gases (carbon dioxide and oxygen) are exchanged. The blood is then pumped by the heart throughout the body to the extremities. As the blood circulates, a small amount of free O_2 is in equilibrium with the bound HbO_2. This free oxygen is available for use by the cells and is removed from the blood as needed.

How does the removal of free oxygen from the blood affect the overall equilibrium of hemoglobin, oxygen, and oxyhemoglobin? When oxygen is removed, the system is no longer at equilibrium because the concentrations have changed. The response to this change is summarized by a general rule called **Le Chatelier's principle,** which states:

> When a system at equilibrium is disturbed, the system will shift in such a manner as to counteract the disturbance and reestablish an equilibrium state.

A summary of the effects of changing the concentration of reactant(s) or product(s) is given in Table 13.2. In the hemoglobin–oxygen system discussed above, the response to the removal of O_2 is to produce more O_2 to replace the amount that was consumed by the cells. The hemoglobin–oxygen system replaces the oxygen that is removed by the cells by increasing the rate of the reverse reaction. The production of O_2 in the reverse reaction will continue until equilibrium conditions are reestablished.

The process of supplying more oxygen to the blood by adjusting the rates of the reactions repeats itself as the newly freed O_2 is taken in by more tissue cells. Eventually the hemoglobin can no longer supply O_2, and the deoxygenated hemoglobin returns to the lungs. Upon reaching

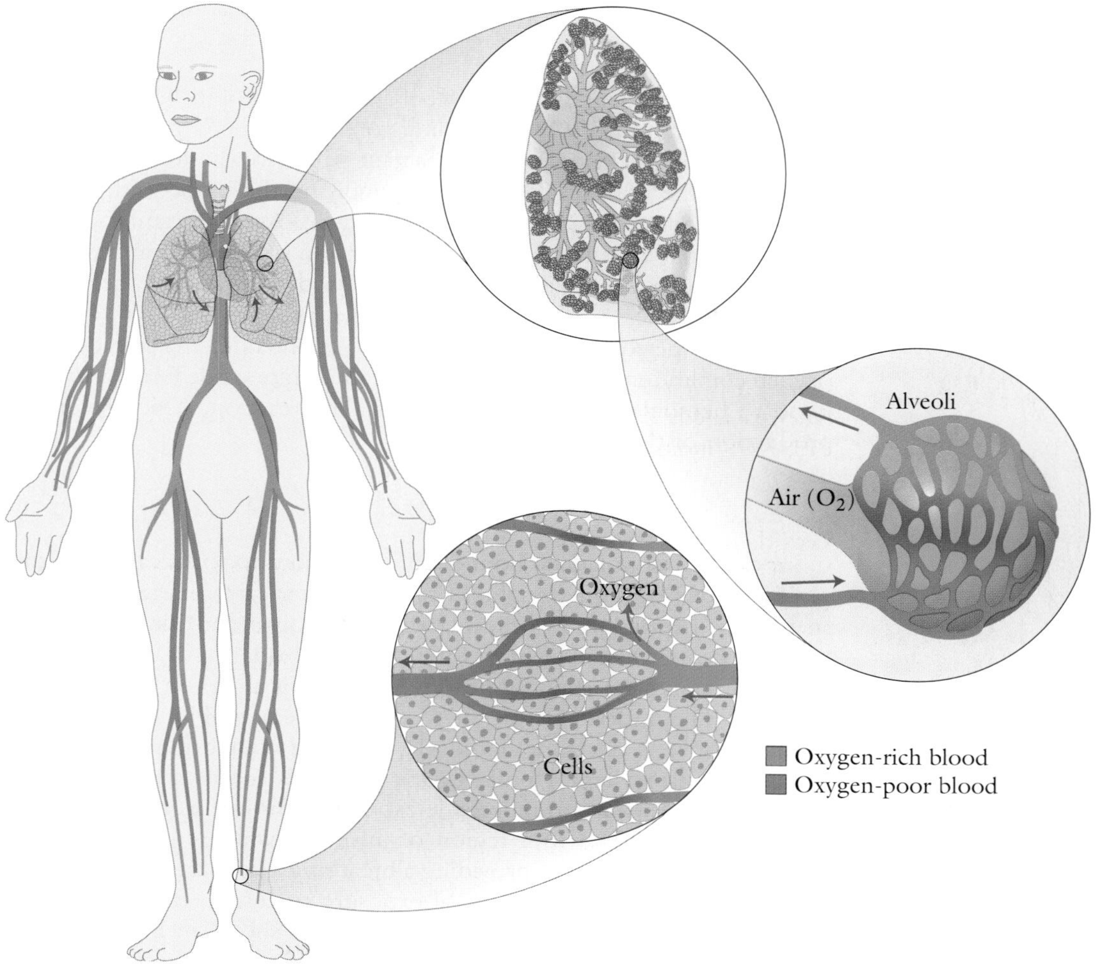

Figure 13.6 The respiratory system. The transportation of oxygen through the body is regulated by the $Hb + O_2 \rightleftarrows HbO_2$ equilibrium system. Hemoglobin in the blood enters the lung and passes through the alveoli, which are small air sacs in the lungs that provide a large surface area for oxygen to pass into the bloodstream and become bound with hemoglobin. The high concentration of O_2 shifts the equilibrium to produce a large amount of HbO_2. As the oxygenated hemoglobin passes into the body, the unbound oxygen is removed by cells. When a reactant is removed, the equilibrium shifts to counteract this change, releasing additional oxygen gas. When the hemoglobin is depleted of oxygen, it returns to the lungs to be replenished.

Table 13.2 Le Chatelier's Principle Applied

Action	Response to Action	Change in Equilibrium Constant
Increase reactant concentration	Shifts equilibrium to the right: Consume reactants, produce products	None
Decrease reactant concentration	Shifts equilibrium to the left: Consume products, produce reactants	None
Increase product concentration	Shifts equilibrium to the left: Consume products, produce reactants	None
Decrease product concentration	Shifts equilibrium to the right: Consume reactants, produce products	None

the capillaries surrounding the alveoli, the hemoglobin is exposed to a high concentration of O_2. According to Le Chatelier's principle, exposure to a high concentration of a reactant will drive the forward reaction to produce the bound HbO_2, thus resupplying the blood with a source of oxygen and starting the process over again. An important fact to remember is that even though the concentration of reactants and products has changed when equilibrium is reestablished, the ratio of the concentration of products to reactants has not changed.

When a person is exposed to carbon monoxide, the hemoglobin–oxygen equilibrium is disturbed because carbon monoxide forms a stronger bond to hemoglobin than oxygen does. The reaction, which favors the production of the products, is

$$HbO_2 + CO \rightleftarrows HbCO + O_2$$

Once carbon monoxide is bound to hemoglobin to form carboxyhemoglobin (HbCO), the hemoglobin can no longer carry oxygen to the tissues. Also, hemoglobin does not release carbon monoxide through the lungs to be exhaled, as it does with carbon dioxide.

Treatment for carbon monoxide poisoning is based on Le Chatelier's principle. The undesirable product is carboxyhemoglobin, and the desirable substance is oxyhemoglobin. To shift the equilibrium to favor the production of oxyhemoglobin, excess oxygen is supplied to the system. In cases of minor carbon monoxide exposure, this can be achieved by escorting the individual to a region of fresh air, which contains approximately 21% oxygen gas, or providing a breathing mask attached to a source of pure oxygen. In severe cases of exposure to carbon monoxide, a hyperbaric oxygen chamber is commonly used, as shown in Figure 13.7. A hyperbaric chamber is a pressurized chamber filled with pure oxygen gas. The high oxygen pressure forces the equilibrium to shift to the left side, favoring the production of oxyhemoglobin and minimizing the production of carboxyhemoglobin.

Figure 13.7 Hyperbaric oxygen chamber. A hyperbaric oxygen chamber is used to provide oxygen gas under high pressure to victims of carbon monoxide poisoning. The large amount of oxygen forces the hemoglobin bound with carbon monoxide to release the carbon monoxide and become bound to the oxygen. (Kike Calvo/V & W/The Image Works)

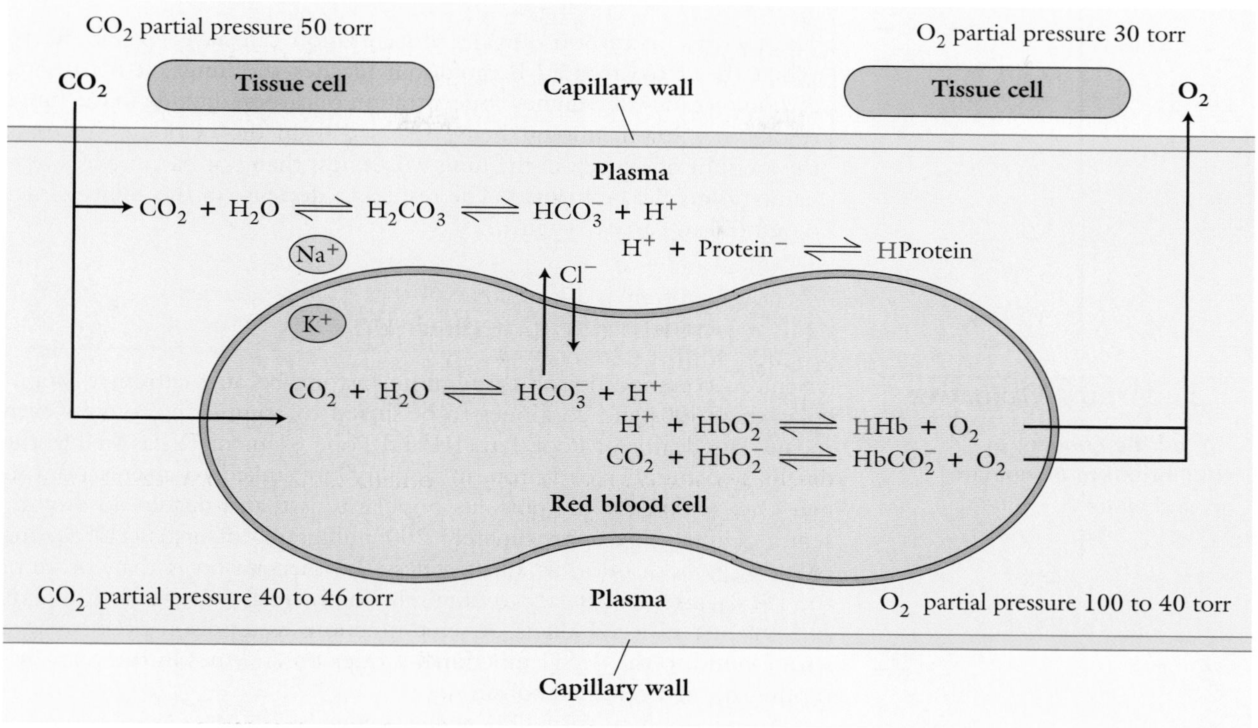

Figure 13.8 The equilibrium systems involved in respiration.

The above discussion of hemoglobin–oxygen equilibrium was greatly simplified by omitting the role of the transport of carbon dioxide through the body. When hemoglobin transports oxygen to the cells, it also carries away carbon dioxide, a waste product of the cells, back to the lungs to be exhaled. The carbon dioxide–hemoglobin equilibrium is influenced by another system involving carbonic acid and the bicarbonate ion. Figure 13.8 illustrates all of the equilibrium processes involved in normal respiration.

Worked Example 4

What will happen to the pH of blood if a person is exposed to elevated carbon dioxide levels?

SOLUTION According to the equilibrium reactions listed in Figure 13.8, increasing carbon dioxide (CO_2) will push the equilibrium position to favor the production of carbonic acid (H_2CO_3). Increasing the amount of carbonic acid shifts a second equilibrium, the dissociation of carbonic acid, to the product side, increasing the concentration of hydrogen ions (H^+). The increase in the concentration of H^+ corresponds to a decrease in pH.

Practice 13.4

What effect will breathing in a higher concentration of carbon dioxide have on the ability of hemoglobin to carry oxygen?

ANSWER

The increase in carbon dioxide concentration can have several effects. When the deoxygenated hemoglobin reaches the lungs, it is carrying carbon dioxide. The higher concentration of carbon dioxide in the lungs results in a lower amount being released from the hemoglobin. Also, the amount of oxygen in the lung will be less than normal, and less oxyhemoglobin will be formed. The result is a decrease in the ability of hemoglobin to carry oxygen.

13.5 Solubility Equilibrium

Learning Objective

Extend the concept of equilibrium to arsenic in ground water.

Arsenic, a common poison throughout history, became extremely popular in the 1500s and continues to be linked to criminal poisonings even today. The deaths of Pope Pius III and Pope Clement IV, as well as the deaths of a number of European nobility, are linked to arsenic poisoning. One reason for the poison's popularity is that a deadly dose of arsenic, delivered by approximately 200 milligrams of arsenic(III) oxide (As_2O_3), is both odorless and tasteless; another reason is that, up until the 1960s, arsenic was easy to obtain in commercial poisons used on ants and rats and in weed killers. Arsenic functions as a poison by forming a strong bond to the —SH functional groups on enzymes in the cells, disrupting normal cell metabolism.

Arsenic occurs naturally in the minerals realgar (AsS), orpiment (As_2S_3), and arsenopyrite (FeAsS), to name a few. These natural sources of arsenic have been a cause for concern as a possible source of unintended poisoning. The United States Geological Survey has been monitoring the amount of arsenic found in ground waters across the United States. The arsenic concentrations shown in Figure 13.9 are for samples taken directly from wells without prior treatment.

The EPA-recommended level for arsenic in water is not to exceed 10 μg/L. People living in a region with elevated arsenic levels in ground water are most likely not drinking untreated water. Arsenic would be removed in water treatment facilities to a safe level. It is also important to realize that even well water in an area with low reported arsenic levels is not necessarily safe to drink. Only analysis of a water sample, conducted by a certified laboratory, can provide information about the quality of ground water that is being directly used without treatment.

One possible source of accidental arsenic poisoning is the presence of naturally occurring arsenic minerals in contact with waters used for human consumption. Pictured here is the mineral realgar (AsS). (Maurice Nimmo/Frank Lane Picture Agency/Corbis)

How does arsenic get into ground water? The arsenic-containing minerals are in equilibrium with their dissolved ions. Using the mineral orpiment (As_2S_3) as an example, the equilibrium reaction is shown by the following equation:

$$As_2S_3(s) \rightleftarrows 2As^{3+}(aq) + 3S^{2-}(aq)$$

At equilibrium, the arsenic(III) and sulfide ions will re-form the solid orpiment mineral at the same rate at which the orpiment dissolves. The magnitude of the equilibrium constant is extremely small, so there is very little arsenic dissolved by the ground waters. Yet the dangers of arsenic are great enough that the amount is worth monitoring. If a solution contains a high level of free arsenic(III) ions, the addition of sulfide ion will

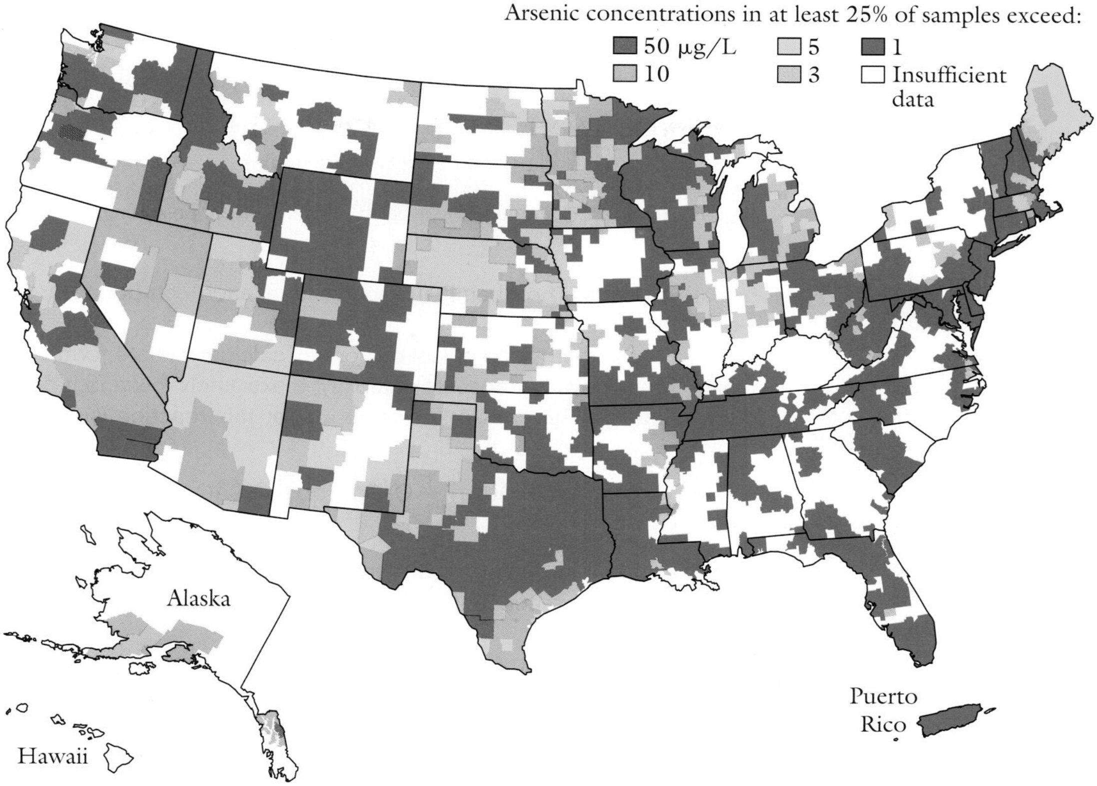

Figure 13.9 United States Geological Survey map of arsenic found in ground water. (water.usgs.gov/nawqa/trace/arsenic)

trigger the precipitation of the insoluble compound As_2S_3. The equilibrium reactions of arsenic minerals are actually more complicated than written above because other dissolved ions present in solution have a direct effect on the solubility of the mineral.

13.6 Case Study Finale: A Mother-in-Law's Justice

Nita Thompson and Kathy Turner determined that the similarities in their two sons' deaths were highly suspicious, so they went to the police. When Dr. Brian Frist, the medical examiner, learned of the similarity in the men's deaths, he reexamined the results of the 1995 autopsy of Glenn Turner and saw something that had been missed earlier: signs of a poison. He also found evidence of poison in Randy Thompson's exhumed body.

The discovery Frist made was the presence of small, insoluble crystals of calcium oxalate (CaC_2O_4) in the kidneys of both men. These crystals form when the poison ethylene glycol (1,2-ethanediol), the main ingredient in antifreeze, is consumed. The body metabolizes ethylene glycol to form oxalic acid, which then reacts with calcium ions in the kidneys to form calcium

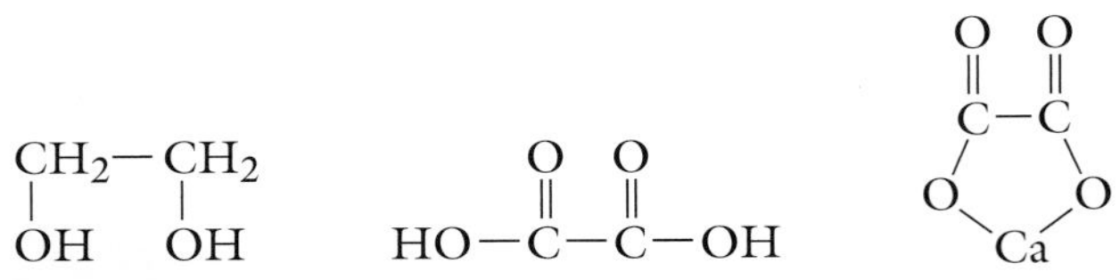

Figure 13.10 Compounds involved in ethylene glycol poisonings.

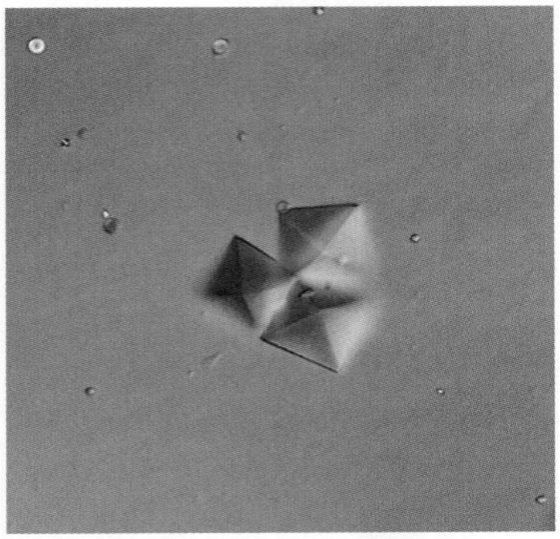

Calcium oxalate crystals in urine are formed when a person ingests ethylene glycol, the active ingredient in antifreeze. (Dr. Frederick Skvara/ Visuals Unlimited)

oxalate crystals. The chemical structures of each compound are given in Figure 13.10.

In a case of ethylene glycol poisoning, calcium oxalate forms solid crystals in the kidneys because its solubility is extremely low and it precipitates out of solution. The equilibrium equation for this reaction is

$$Ca^{2+}(aq) + C_2O_4^{2-}(aq) \rightleftarrows CaC_2O_4(s)$$

The equilibrium constant for the solubility of calcium oxalate is 2.3×10^{-9}, indicating that the concentration of the products (ions in solution) is very small.

Treatment for ethylene glycol poisoning consists of administering relatively large amounts of ethanol (C_2H_6O) to the individual who has ingested the poison. The justification for this procedure is that the oxidation of ethanol occurs preferentially before the oxidation of ethylene glycol. When excess alcohol is present, the body no longer produces oxalic acid and, therefore, no longer produces the calcium oxalate crystals. This delay allows the body more time to excrete and eliminate the ethylene glycol intact through the urinary system.

The prosecutor in the murder trial of Julia Turner would claim that Julia had slipped the antifreeze into the food of Glenn Turner. The calcium oxalate evidence found by Dr. Frist bolstered the case against Julia. On May 14, 2001, a jury found her guilty of the murder of Glenn Turner. She received a sentence of life in prison. While evidence from the death of Randy Thompson was used in the trial, Julia was never charged with his murder.

CHAPTER SUMMARY

- While all compounds can be toxic to living organisms, the term *poison* is usually reserved for those that cause harm in relatively small doses. A toxin is a poisonous compound that originates from either a plant or an animal.
- The ability of a poison to enter into and disrupt the normal function of cells is dependent on the chemical equilibrium of the poison with biological compounds.
- A chemical system achieves equilibrium when the rate of a forward reaction is equal to the rate of the reverse reaction. In an equilibrium state, reactants form products at the same rate that products regenerate reactants.
- A reaction in equilibrium appears to have stopped, but in reality, the reaction continues with the concentrations of products and reactants remaining constant.
- The ratio of the concentrations of products to the concentrations of reactants in a system at equilibrium is indicated by the equilibrium constant. A

very large value for the equilibrium constant means the concentration of products is much greater than the concentration of reactants at equilibrium. If the equilibrium constant has a small value, the concentrations of reactants are much greater than the concentrations of products at equilibrium.

- Although the ratio of product concentrations to reactant concentrations at equilibrium is constant for a particular reaction, the concentrations of individual species can vary dramatically from one equilibrium situation to another.
- Equilibrium can be lost by the addition or removal of either reactants or products because changes in concentration levels disrupt equilibrium.
- According to Le Chatelier's principle, when a system at equilibrium is disturbed, the system will shift in such a manner as to counteract the disturbance and reestablish an equilibrium state. If a reactant is removed, the reverse reaction attempts to restore the equilibrium status by preferentially producing the reactant until the equilibrium ratio is once again achieved.

KEY TERMS

poison, p. 385
toxic compound, p. 385
toxin, p. 385
chemical equilibrium, p. 387
reverse reaction, p. 388
equilibrium constant (K), p. 391
Le Chatelier's principle, p. 394

CONTINUING THE INVESTIGATION Additional Readings, Resources, and References

For additional information and updates on the trial of Julia Turner: www.courttv.com/trials/turner/

May, Meredith. "Fraternity Pledge Died of Water Poisoning. Forced Drinking Can Disastrously Dilute Blood's Salt Content," *San Francisco Chronicle,* February 4, 2005.

Smith, R. M., Martell, A. E., and Motekaitis, R. J. NIST Critically Selected Stability Constants of Metal Complexes, *NIST Standard Reference Database,* no. 46, version 3.0, Gaithersberg: U.S. Department of Commerce, 1997.

Trestrail, John Harris. *Criminal Poisoning: Investigational Guide for Law Enforcement, Toxicologists, Forensic Scientists, and Attorneys,* Totowa, NJ: Humana Press Inc., 2000.

For more on the fraternity pledge and water poisoning: www.cnn.com/2003/US/Southwest/11/17/fraternity.pledge.ap/

For additional information on arsenic in ground water: water.usgs.gov/nawqa/trace/arsenic/

REVIEW QUESTIONS AND PROBLEMS

Questions

1. What is the difference between a toxin and a poison?
2. Are all poisons toxins? Are all toxins poisons?
3. Explain the quote by Paracelsus on the nature of poisons.
4. All substances, even water, can be poisonous. But what property is shared by most substances that we consider to be "poisons"?
5. What are the conditions needed for the reverse reaction of a system to occur?
6. What is meant by a dynamic equilibrium?
7. Why does the rate of the forward reaction decrease as a function of reaction time?
8. Why does the rate of the reverse reaction increase as a function of reaction time?
9. Why does it *appear* that a reaction has stopped when it reaches equilibrium?
10. Will the concentrations of reactants and products always be the same at equilibrium?
11. How does a catalyst influence the equilibrium process?

12. What does the magnitude of the equilibrium constant represent?

13. When an equilibrium constant has a large value, is it the reactants or products that predominate in a solution? How about when an equilibrium constant has a small value?

14. In a style similar to Figure 13.2c, sketch (on a molecular scale) what the equilibrium system would look like for a system with a very small equilibrium constant.

15. In a style similar to Figure 13.2c, sketch (on a molecular scale) what the equilibrium system would look like for a system with a very large equilibrium constant.

16. The containers shown in parts (a) and (b) of the following figure represent snapshots of the same chemical solution taken at different times. Has the solution reached equilibrium? Explain your answer.

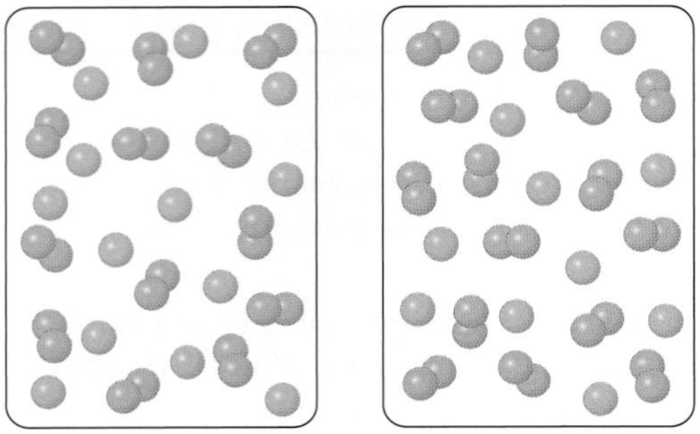

(a) (b)

17. Using the Web site www.nlm.nih.gov/medlineplus/encyclopedia.html as a resource, investigate the medical condition known as *acidosis*. What are some of the possible causes for acidosis? How does this condition affect the ability of the body to transport oxygen gas?

18. Using the Web site www.nlm.nih.gov/medlineplus/encyclopedia.html as a resource, investigate the medical condition known as *alkalosis*. What are some of the possible causes for alkalosis? How does this condition affect the ability of the body to transport oxygen gas?

19. Summarize the four types of stress that can be applied to a system in equilibrium, and the response to each, according to Le Chatelier's principle.

20. When a chemical reaction is carried out in an industrial setting, it is very common to have one of the reactants present in an excess amount above the stoichiometric quantity that would be needed for the reaction. Why?

21. Describe how oxyhemoglobin delivers oxygen gas for use by cells.

22. Describe how hemoglobin is converted to oxyhemoglobin in the lungs.

23. How does carbon monoxide interfere with the mechanism of hemoglobin?

24. What is a hyperbaric oxygen chamber, and why is it used to treat individuals with carbon monoxide poisoning?

25. If a sample of the arsenic-containing mineral realgar is in equilibrium with a solution, what could be done to decrease the amount of dissolved arsenic?

26. If a sample of the arsenic-containing mineral realgar is in equilibrium with a solution, what could be done to increase the amount of dissolved arsenic?

Forensic Chemistry Problems

27. Explain why dimercaprol is an effective antidote for mercury poisoning.

28. Dimercaptosuccinic acid (DMSA) is used to treat lead poisoning. Draw the structure of DMSA, given that it is an organic compound with a 4-C chain, two —SH groups, and two carboxylic acid groups.

29. When you exhale into a breathalyzer, the air from your lungs provides an accurate measure of the alcohol in your blood. Explain why, based on equilibrium principles.

30. The Chapter 4 case study, "A Killer Headache," focused on the death of several members of the Janus family by cyanide poisoning. Cyanide functions by binding very strongly with the Fe^{3+} ion in hemoglobin within the cell, which prevents the cells from utilizing the oxygen and results in suffocation. What antidotes are there for cyanide poisoning and what do they do chemically to counteract the effect of cyanide? Is oxygen present in the blood despite the suffocation? Consult the *Antidotes & Other Treatments* and *Laboratory Tests* sections at www.atsdr.cdc.gov/MHMI/mmg8.html for further information in constructing your answer.

Case Study Problems

31. When a firefighter discovers a body at the scene of a fire, a complete investigation into the cause and manner of death of the victim will follow. Recall from Section 9.2 that carbon monoxide is often produced as a result of incomplete combustion. What information would analysis of the hemoglobin for carbon monoxide provide investigators? Could it be the sole piece of evidence to determine whether the death was suicide, homicide, or accidental? Explain the limits of the information that could be provided from the lab.

32. In the Chapter 7 case study, "The Experts Agreed," the death of Ryan Stallings had initially been attributed to poisoning by 1,2-ethandiol from antifreeze. Calcium oxalate crystals had been found in his kidney and brain during an autopsy, supporting the theory that he was poisoned. However, Dr. Piero Rinaldo, the expert who testified on Patricia Stallings's behalf, instead blamed the use of ethanol in the mistreatment of Ryan and his genetic disorder of methylmalonic acidemia (MMA). According to Dr. Rinaldo, one of the metabolic products of ethanol oxidation in the body can be oxalic acid. Explain how a patient being administered a continual ethanol drip could quickly form calcium oxalate crystals. What would happen to the calcium ion levels in the blood if oxalic acid were being produced in the kidneys?

CHAPTER 14

Identification of Victims: DNA Analysis

This chapter will highlight the efforts of the Office of the Chief Medical Examiner of the City of New York to identify the victims of the September 11, 2001, attacks on the World Trade Center. (Corbis)

CASE STUDY: Remembering 9-11

Scenes of the chaos, death, and destruction from September 11, 2001, are ingrained in our memories and need not be shown here to retell the story of that day. The deaths of thousands of innocent citizens shocked the world. The horror was magnified as both World Trade Center towers collapsed, killing hundreds of firefighters, police officers, and port authority officers who had risked their lives rushing into the burning towers to save the occupants. Within a mere 90 minutes, the majestic World Trade Center towers were reduced to three billion pounds of twisted metal, concrete, and office debris. Also intermingled within this debris were the remains of 2749 victims known to have perished at the World Trade Center, remains that had to be recovered, preserved, and identified for their families.

Normally, investigators turn to dental X-rays, fingerprints, and visual identification for determining the identity of victims. However, those methods were of limited use because only 300 victims' bodies were found. All material from the site was transferred by truck to the Fresh Kills Landfill on Staten Island. Upon arrival at the landfill, large pieces of debris were removed, and small pieces were spread out over a field marked with grid lines. Officers working in pairs sifted through the material within each grid section to identify human remains and personal belongings that could be returned to the victim's families. Work continued at this site 24 hours a day for 10 months to search all three billion pounds of debris for human remains.

In all, 19,916 sets of human remains were found and sent to the Office of the Chief Medical Examiner of the City of New York for identification by DNA analysis. Samples were quickly sealed and preserved in a refrigerated storage facility to prevent further deterioration. The medical examiner's office set out to obtain DNA samples from personal items of the victims, such as hairbrushes or toothbrushes brought to them by family members. They also took DNA samples from relatives to add to the database.

Families were notified as samples were identified, and the remains, often just the smallest pieces of bone or tissue, were turned over to the family for burial. This process led to the identification of the remains of 1585 of the 2749 victims—leaving 1164 victims yet to be identified, largely because 9726 sets of these human remains cannot even be analyzed. The New York City medical examiner announced in February 2005 that all work on analyzing the human remains would cease for an indefinite length of time. The unidentified remains of these victims will be preserved and stored at the memorial being constructed at ground zero.

Why has the medical examiner stopped trying to identify the remains when the families of over a thousand victims still wait? . . .

14.1 Lipids: Fats, Waxes, and Oils

Learning Objective

Identify fats, oils, and waxes by their molecular structures.

When I was a youth, I knew an old Frenchman who had been a prison-keeper for thirty years, and he told me that there was one thing about a person which never changed, from the cradle to the grave—the lines in the ball of the thumb; and he said that these lines were never exactly alike in the thumbs of any two human beings.

—*Life on the Mississippi*, by Mark Twain (Samuel Clemens), 1883

Mark Twain's 1883 description of the uniqueness of fingerprints was remarkable, considering that the first criminal case based on fingerprint identification did not occur for another eight years. Because the technique was not common knowledge in the 1880s, Twain must have kept up-to-date with the scientific literature of the time, which discussed the use of fingerprinting as a method of identification. He also featured fingerprints in a later story, *Pudd'nhead Wilson,* written in 1894. But the widespread use of fingerprinting in the United States did not occur until the early 1900s, when the New York State prison system and the federal penitentiary in Leavenworth, Kansas, started keeping fingerprint records of prisoners.

Fingerprinting revolutionized the field of forensic science because it is a method of absolute identification connecting a person to an object such as a murder weapon or a doorknob at the location of a crime. A person guilty of a crime could no longer simply deny his or her presence at a crime scene without weaving a web of lies and multiple inconsistencies in a story. Fingerprints were used to identify some of the victims from the World Trade Center, but the vast majority of victims would require DNA analysis for identification.

(Astrid & Hanns-Frieder Michler/Photo Researchers, Inc.)

What makes up a fingerprint? Understanding how fingerprint evidence is gathered and preserved requires knowledge of the chemicals that are present. A fingerprint consists of many different compounds that can vary from one individual to another, from the young to the elderly—even from one time of day to another! However, there are several classes of compounds commonly found in fingerprints: inorganic ions, organic acids, water, and lipids.

Inorganic ions originate from the loss of the body's electrolytes through the sweat glands and consist of Na^+, Ca^{2+}, K^+, Cl^-, and Mg^{2+}. The organic acids may consist of various amino acids, acetic acid, or lactic acid. Water present on the fingerprint is produced by sweat glands in the skin. The final class of compounds, lipids, is one we have not discussed previously. **Lipids** include a variety of compounds such as oils, fats, and waxes that are defined by a shared common physical property: They are not soluble in water but are soluble in a nonpolar solvent. The deposition of lipids in fingerprints plays an important role in the forensic use of fingerprints.

You are already familiar with the physical differences between waxes (hard solids), fats (soft semisolids), and oils (liquids). But what are the *chemical* differences between these subclasses of lipid molecules? **Waxes** are made from a **fatty acid,** which is a long-chain carboxylic acid molecule, and a long-chain alcohol, as shown in Figure 14.1. Waxes are hard solids at room temperature because the large molecules have strong dispersion forces between them. Recall that a high degree of intermolecular forces is reflected in high melting and boiling points.

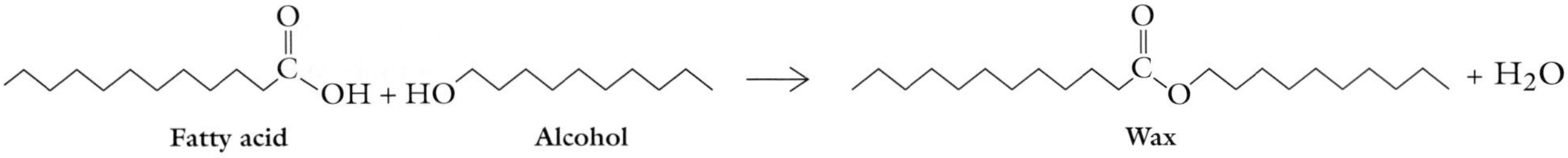

Figure 14.1 The structure of a wax.

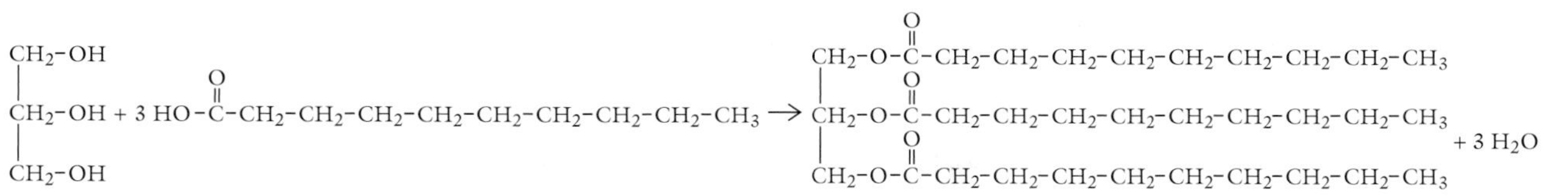

Figure 14.2 Formation of triglycerides.

Fats and oils are examples of **triglycerides,** compounds that are made by combining a glycerol molecule with three fatty acid molecules, as shown in Figure 14.2. The distinction between fats and oils lies in the nature of the fatty acid molecules composing the triglycerides. **Fats,** which are semi-solids at room temperature, have fatty acids whose carbon chains contain only single bonds between carbon atoms. **Oils,** which are liquids at room temperature, have fatty acids whose carbon chains contain one or more carbon-carbon double bonds.

Why does the presence or absence of double bonds have an effect on whether a compound is solid or liquid at room temperature? Consider Figure 14.3, which shows the structure of several fatty acids. When the fatty acid constituents consist of only single bonds, as in stearic acid, the molecule is linear. Oleic acid, on the other hand, has a double bond that puts a kink in the chain of atoms. The linoleic acid molecule with two double bonds has a bent shape with a 90° corner in the molecule. Each of the molecules in Figure 14.3 contains the same number of carbon atoms.

How does the shape of the fatty acid molecule, caused by the presence or absence of C=C atoms, influence the physical properties of the compounds? Table 14.1 lists the chain length, the number of C=C groups in the molecule, and the melting points of the compounds. The first three fatty acids listed in the table all have 18 carbon atoms in the chain. Yet the presence of a single C=C lowers the melting point from 69°C to 13°C, and the presence of two double bonds further lowers the melting point to −5°C. Stearic acid, with its linear shape, has the

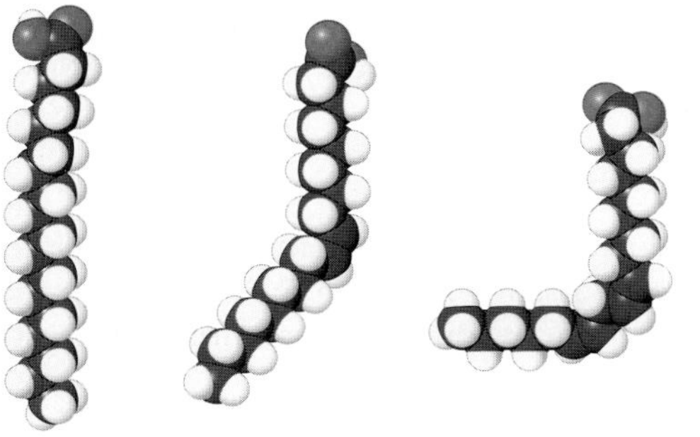

Figure 14.3 Effect of single and double bonds on the molecular shape of a fatty acid. Each molecule has the same number of carbon atoms.

Table 14.1 Properties of Saturated and Unsaturated Fatty Acids

Name	Length of Carbon Chain	Number of Carbon-Carbon Double Bonds (C=C)	Melting Point (°C)
Stearic acid	18	0	69
Oleic acid	18	1	13
Linoleic acid	18	2	−5
Arachidic acid	20	0	77
Arachidonic acid	20	4	−49

strongest intermolecular forces; linoleic acid, with a significantly bent shape, has the weakest intermolecular forces. The linear hydrocarbon portions of stearic acid molecules are better able to align so that dispersion forces have greater effect. The substantial bend in the linoleic acid molecule decreases the contact area between molecules and, therefore, decreases the melting point.

Fats and oils are often qualified with the terms *saturated* or *unsaturated*. The term **saturated** refers to lipid molecules in which only single bonds exist between carbon atoms. An **unsaturated** lipid molecule has one or more double bonds between adjacent carbon atoms.

The fats, oils, and waxes produced by a person's body and deposited on a surface as a fingerprint form the basis of many methods of fingerprint identification. The water quickly evaporates, but the nonvolatile lipids remain behind. Dusting for fingerprints by applying a fine powder to a surface reveals the fingerprints, as the powder clings to the lipid compounds present.

Another method for identifying fingerprints, especially on paper documents, involves exposing the fingerprint to iodine fumes. The nonpolar iodine vapor dissolves in the nonpolar lipids to form a dark brown image of the fingerprint. However, because the color will fade after a few minutes, iodine fuming is a temporary method of exposing fingerprints, so evidence must be photographed immediately. Fingerprints developed with iodine fuming can be sprayed with a starch solution to create a longer-lasting blue starch-iodine complex.

Worked Example 1

The forensic analysis of lipstick from a suspect's clothing may implicate the suspect in an assault. Based on the chemical properties of waxes and oils, discuss why these compounds are found in lipstick.

SOLUTION Waxes are solids at room temperature, an important feature for the lipstick to maintain its shape in the tube. The water-insoluble characteristic of waxes helps keep the lipstick coating on the lips instead of being dissolved by beverages. Oils are present to slightly soften the mixture because a hard wax would not spread well. The oil content also enhances shine and glossiness.

Practice 10.1

Margarine is produced from vegetable oil in a process called *hydrogenation*. Based on the different properties of fats and oils, what must happen to the vegetable oil during hydrogenation?

ANSWER

Vegetable oil is a liquid due to the double bonds in the fatty acid molecules. For oil to become a semisolid, the C=C bonds must be converted to C—C bonds through hydrogenation.

14.2 Carbohydrates

Learning Objective

Describe the structure of carbohydrates.

Locard's exchange principle states that whenever two objects come into contact, there is inevitably a bidirectional transfer of material from one object to the other. When a violent struggle has occurred between an assailant and a victim, there is always a transfer of material between the two individuals. Investigators try to obtain the clothing of both individuals as soon as possible to examine them for trace evidence. Hair and fibers are the most commonly sought-after materials that link two individuals together, but other trace materials such as body glitter and cosmetics can also be transferred.

The exchange of such trace evidence can be a *primary* transfer, which results from direct contact between individuals during a struggle. A *secondary* transfer occurs when fibers follow an indirect transfer route between two individuals. For example, if a person who owns a pet dog sits in a chair, hair from the pet can be transferred to the chair from the person's clothing. When a second person uses the same chair, the pet hair can be transferred to that individual as well.

The key to using this evidence is discovering unique fibers or hairs seldom found in the surroundings, because unique fibers specifically tie the individuals together. For instance, cotton fibers might have little investigative use because cotton is such a common fabric. But bright pink angora fibers would be unusual in most settings. Cotton fibers are derived from the cotton plant and consist primarily of cellulose, a carbohydrate. Angora fibers come from the hair of an angora rabbit or angora goat and are made of protein. The chemistry of carbohydrates is explored further in this section, and the chemistry of proteins will be discussed in the following section.

Carbohydrates, also known as *saccharides,* are literally the hydrates of carbon and have a carbon-hydrogen-oxygen ratio of 1:2:1. A **monosaccharide** is a small carbohydrate molecule, also referred to as a *simple sugar*. Examples include glucose, fructose, galactose, ribose, and deoxyribose. (Deoxyribose will be discussed further in Section 14.4.) Two monosaccharides can bond together to form a **disaccharide,** or double sugar. For instance, lactose (milk sugar) is made up of galactose bonded to glucose, sucrose (table sugar) is glucose bonded to fructose, and maltose (malt sugar) is two glucose molecules bonded together. The common term **sugar** is used to refer to any of the monosaccharides or disaccharides, all of which have a sweet taste.

The term **oligosaccharide** is sometimes used to describe a carbohydrate containing a small number of monosaccharides—between three and

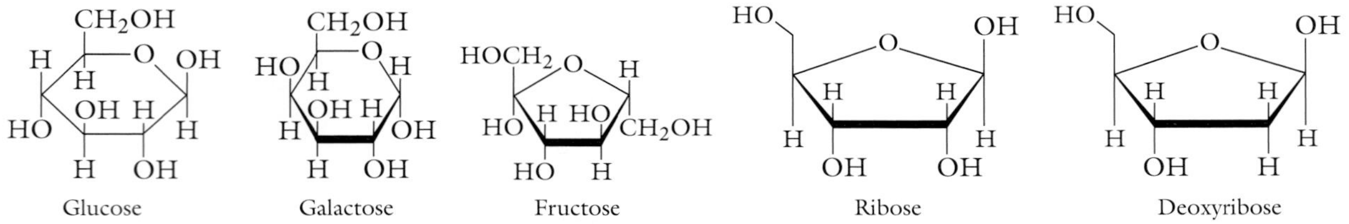

Figure 14.4 Chemical structures of several common monosaccharides.

ten. (The prefix *oligo-* means "a few.") Oligosaccharides played an important role in forensic science before DNA analysis because of their significance in the ABO blood typing system. For type A and type B blood (but not type O blood), red cells have different oligosaccharide units on the outer cell wall that enable the cell to differentiate between foreign and non-foreign material. Blood typing was based on the reactions of these oligosaccharides and was the primary way investigators attempted to match to a specific person the blood found at a crime scene. DNA analysis of blood, though, can now provide much more specific identification.

Carbohydrates can exist as very large molecules made by linking together many individual sugar molecules. The term **polysaccharide** is used to characterize large carbohydrates that can range in size from hundreds to thousands of individual sugar units. In this sense, polysaccharides are a type of polymer.

We usually think of plastics when we hear the term **polymer**—a long-chain molecule made by linking together smaller molecules called **monomers.** A carbohydrate is a polymer made up of sugar monomers. The sugar glucose ($C_6H_{12}O_6$), shown in Figure 14.4, is the basic monomer found in cotton fibers. The glucose monomers link up to form **cellulose,** a polysaccharide of glucose that is used in plants to make up the structural material of the cell walls. Cotton fibers are 92% cellulose.

Cellulose is also a component of foods derived from plants. Because humans are unable to digest cellulose, it is often referred to as **dietary fiber.** Dietary fiber supplements containing cellulose usually state on the label that sufficient amounts of water should be consumed when taking them. The reason is that cellulose is capable of absorbing a substantial quantity of water through hydrogen bonding. Consider the number of hydroxyl groups (—OH) on a glucose unit in Figure 14.4, all of which are capable of forming hydrogen bonds with water. The cellulose molecule consists of hundreds of glucose units, so the amount of water that can be associated with cellulose is considerable.

Green plants produce another polysaccharide of glucose called **starch,** a substance that serves as a source of energy for the body. Potatoes and rice are common sources, although all plants that undergo photosynthesis produce starch. Enzymes in the human digestive system are able to cleave off the individual glucose monomers for use in metabolism.

What is it about starch that makes it digestible by humans whereas cellulose is not? Both are polymers of glucose, but the key to understanding the difference in their digestibility lies in how the glucose units are bonded together. Figure 14.5 shows the bonding of glucose units in cellulose and starch. Notice how the glucose units in cellulose link in a way

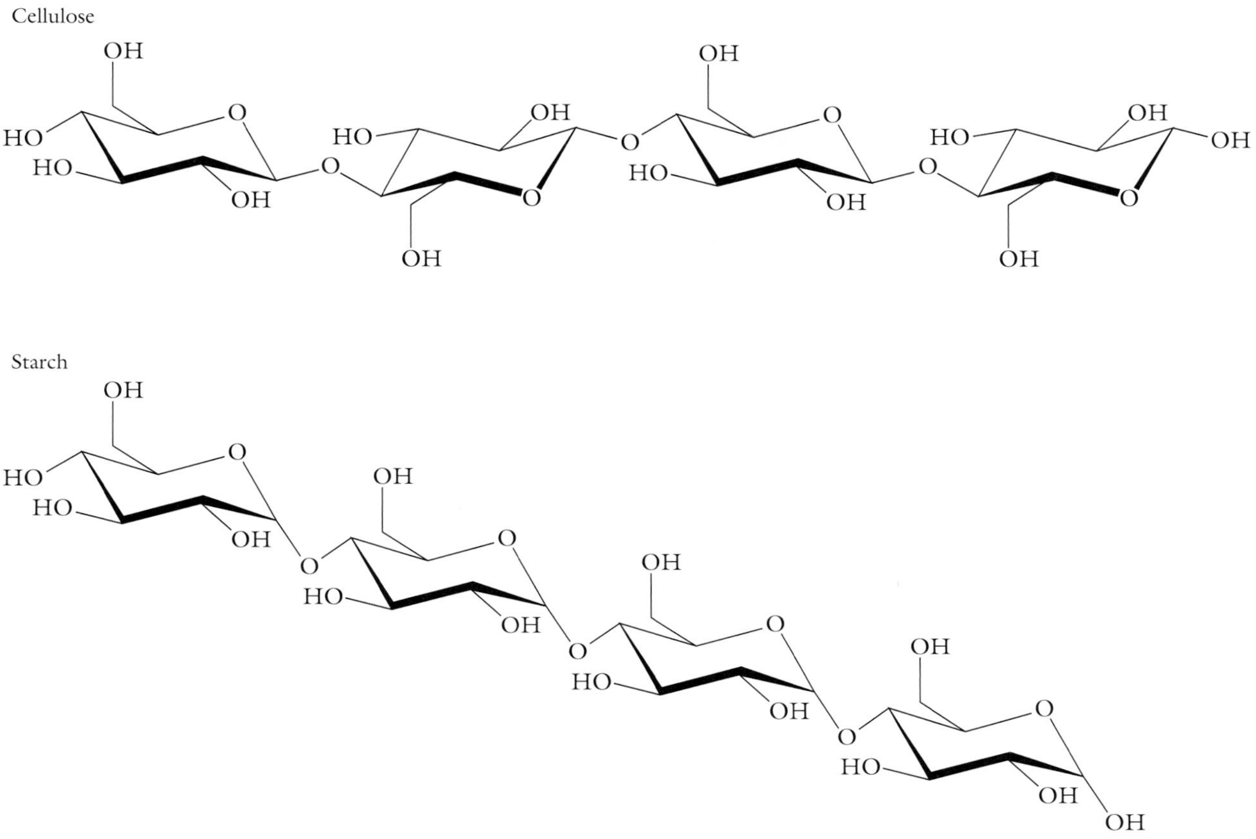

Figure 14.5 Cellulose and starch polysaccharides.

that results in a linear structure for the overall molecule. The linkages between glucose molecules in starch are oriented such that an angular molecule results.

The enzymes in the body are designed to cleave starch molecules apart, but the shape of the enzymes will not allow cellulose to be cleaved. Some animals such as cattle contain cellulose-digesting bacteria in their digestive tracts and are thus able to use cellulose as an energy source.

14.3 Proteins

Learning Objective
Describe the role of amino acids in the structure of proteins.

The fibers that compose our clothing belong to three classes: vegetable sources such as cotton, animal sources such as wool, and synthetic polymer fibers such as nylon. In addition to wool, other natural animal-based fibers include camel hair, alpaca, mohair, cashmere, silk, and angora. Each of these animal-based examples is a type of **protein,** a long polymer made of individual amino acid monomers. There are approximately 100,000 different proteins in the human body, each designed for a specific purpose—whether it is to form the structural component of a hair strand, to transport oxygen from the lungs throughout the body, or to enable digestion of starches into glucose for energy.

The key to the function of a protein is its shape, which is dictated by the sequence of its amino acids. Each protein has a specific shape that will

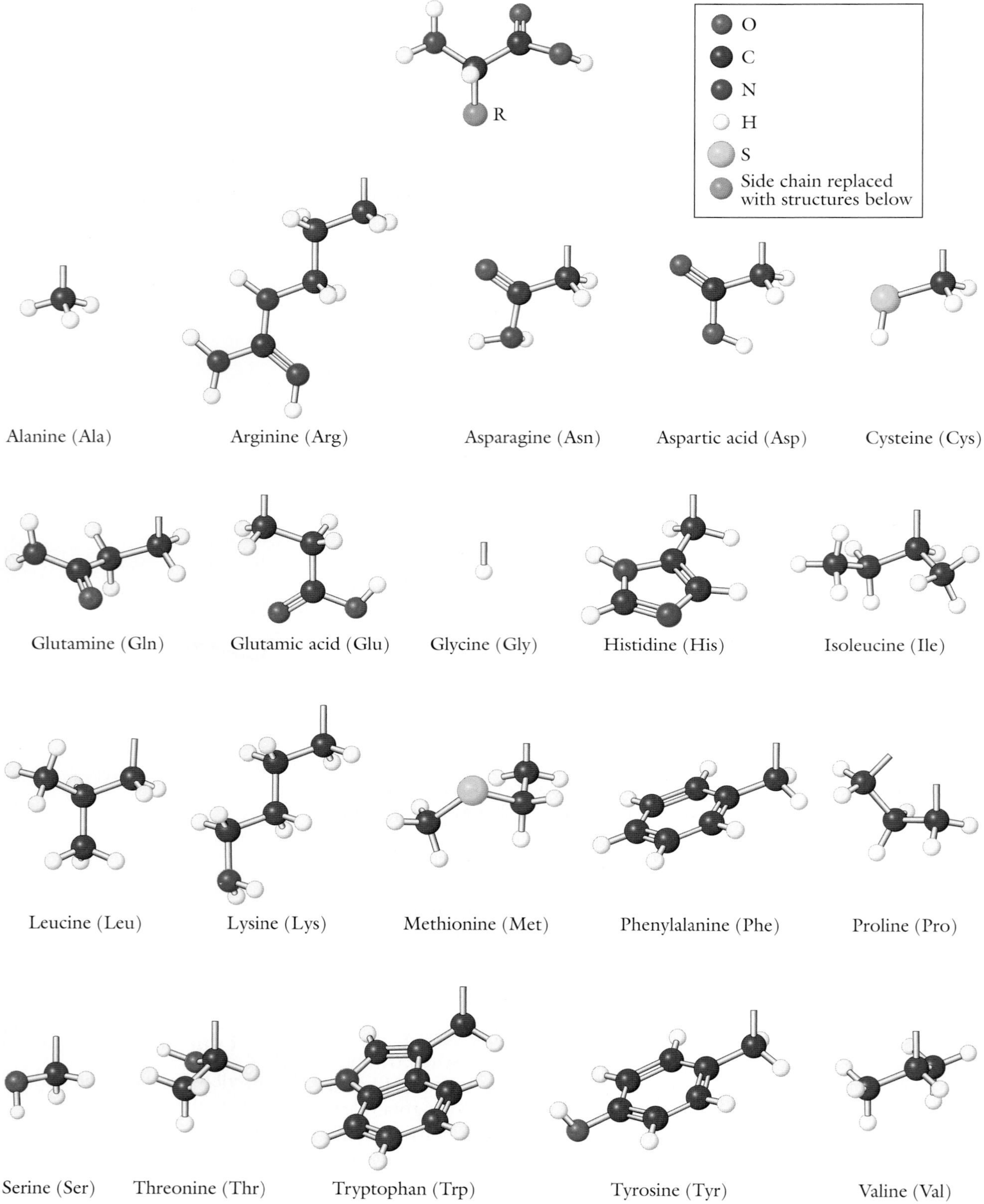

Figure 14.6 The 20 amino acids. The top structure is the backbone, with the purple sphere depicting the —R group. Replace the purple sphere on the amino acid backbone with the group listed beneath each name for the structure of each amino acid. It should be noted that the proline side chain actually bonds back to the nitrogen atom of the backbone.

allow it to interact with only its target compounds and no others. The body is able to produce needed proteins by storing within each cell a blueprint of every protein molecule in the body. That blueprint is called *DNA*.

The **amino acids** are a group of 20 naturally occurring compounds that are similar in structure. Each molecule contains an amino functional group ($-NH_2$) and a carboxylic acid functional group ($-COOH$), separated by a carbon atom. An organic group ($-R$) bonded to the middle carbon atom is what distinguishes one amino acid from another. Figure 14.6 shows the 20 naturally occurring amino acids and the standard abbreviations used when writing out the constituents of protein molecules. The top structure in Figure 14.6 is the generic amino acid backbone with the purple-colored sphere depicting the $-R$ group. For the specific amino acid structure, replace the purple sphere on the amino acid backbone with the group listed beneath each name.

Groups of 10 to 20 amino acids bonded together are called **oligopeptides.** If the chain ranges from 20 to 50 amino acids in length, it is commonly referred to as a **polypeptide.** The term *protein* is usually reserved for chains longer than 50 amino acids. Proteins are manufactured in the cells as needed, and the blueprint for each protein is encoded on an individual's DNA. The mechanism of DNA encoding for proteins is discussed in a later section.

The sequence of amino acids within a protein gives the protein its **primary structure.** For a silk protein fiber, the primary structure is [Gly-Ala-Gly-Ala-Gly-Ser]$_n$. This notation includes abbreviations for three amino acids—glycine (Gly), alanine (Ala), and serine (Ser)—and shows how they link together in a unit that repeats *n* times. Strands of silk, which can reach over a kilometer in length when spun from the silkworm, are composed of this protein.

Beyond the primary structure, proteins assume different shapes, called the **secondary structure,** based on how the amino acid chains interact. The secondary structure of silk is known as a **pleated sheet,** shown in Figure 14.7a. (Hydrogen atoms have been removed from the structure to simplify the image.) The distinctive shape is created by hydrogen bonding that occurs between two adjacent chains. Figure 14.7b is a top-down view in which the hydrogen bonding (green lines) can be seen.

Animal hair consists of a protein called *keratin* that does not have a fixed amino acid sequence. The primary structure is simply the order of

Silkworm cocoons that will be unraveled. The individual strands of silk, up to 1 km long, will be made into silk cloth. (blickwinkel/Alamy)

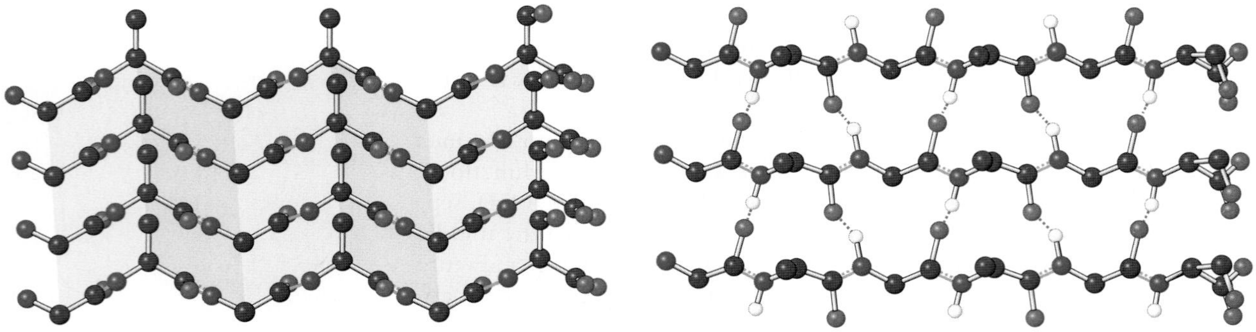

(a) Pleated sheet structure (b) Top-down view

Figure 14.7 (a) The pleated sheet secondary structure of silk. Shading has been added to help illustrate the pattern. (b) The individual strands of protein are held in place by hydrogen bonding (green lines).

amino acids that happened to be used in the creation of the molecule. The secondary structure of keratin is quite different from that of silk. Keratin assumes an **alpha helix** form, much like a spring. The coils of the spring are held in place by hydrogen bonds that form between every fourth amino acid. Figure 14.8a illustrates an alpha helix structure. The ribbon has been added to aid in visualization. Figure 14.8b provides a close-up view of the hydrogen bonding (green lines) present in the alpha helix structure.

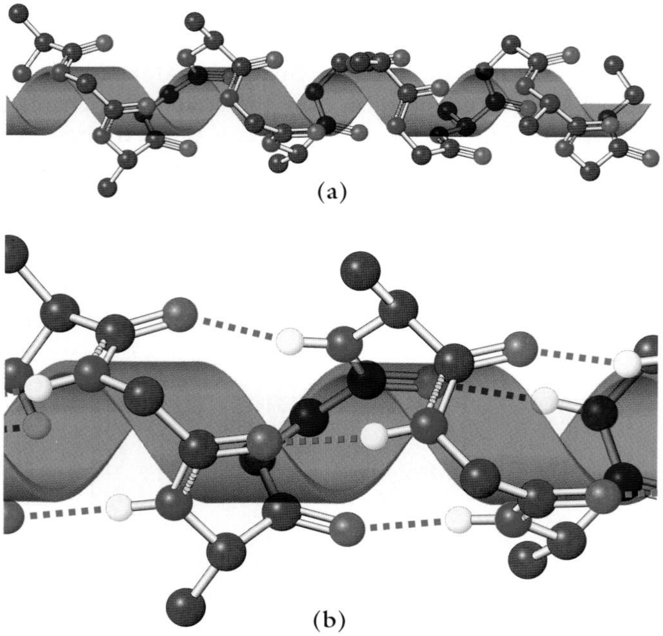

(a)

(b)

Figure 14.8 The structure of hair. (a) An alpha helix structure. The ribbon has been added to highlight the structure. (b) A close-up view of the hydrogen bonding (green lines) in the alpha helix structure.

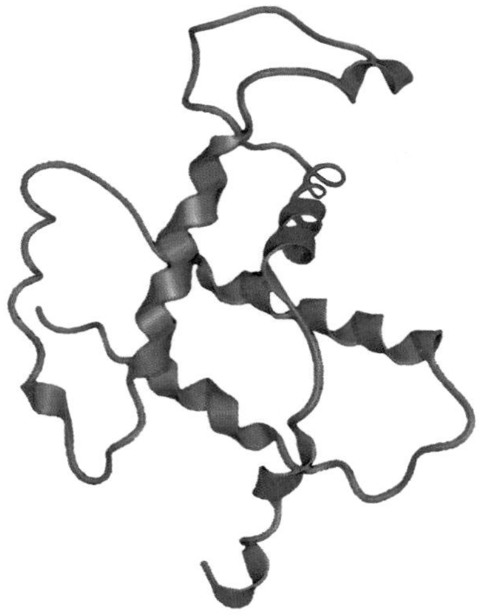

Figure 14.9 A single protein molecule from hemoglobin.

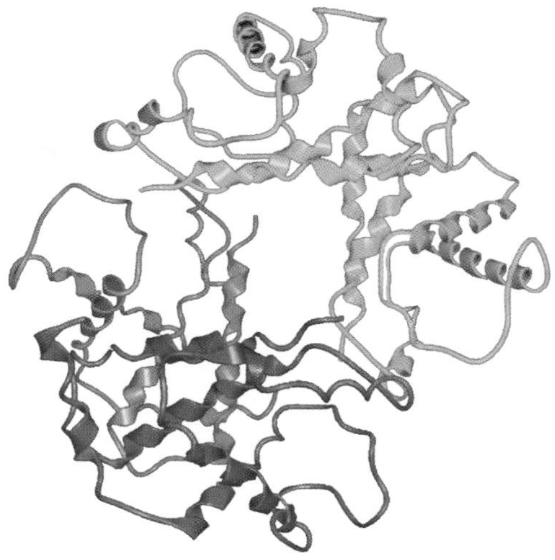

Figure 14.10 Protein molecules of hemoglobin.

One example of an important protein with an alpha helix structure is hemoglobin, the protein in blood that transports oxygen from the lungs to the rest of the body. However, proteins such as hemoglobin are much more complex than a simple alpha helix. The helix bends and twists upon itself, creating a **tertiary structure** that forms a larger three-dimensional shape. Figure 14.9 illustrates the protein structure present in hemoglobin.

Protein molecules can also join with one or more different protein molecules to form what is known as a **quaternary structure.** The full structure of hemoglobin consists of four separate protein molecules, as illustrated in Figure 14.10.

Protein molecules maintain their three-dimensional shape as a result of attractive forces created between the amino acid side-chain functional groups. Hydrogen bonding between peptide groups is the dominant force in the creation of the secondary structure of protein molecules. For example, curly hair is due to hydrogen bonding within the protein molecules of hair. When curly hair is dampened with water, it will straighten out because the water molecules form hydrogen bonds to the protein molecules. As hair dries out, the curls return as hydrogen bonds re-form between the peptide units of the protein molecule itself.

Also responsible for the curliness of a person's hair is a second type of bond due to the amino acid cysteine, which has a terminal —SH group. Cysteine in the hair can bond to another —SH group farther down the protein chain to form a **disulfide bond,** —S—S—. If a person has straight hair, the existing disulfide bonds are such that the protein molecules can lie flat. If the person receives a permanent, the disulfide bonds are first reduced to form the —SH bonds, the hair is then wrapped around curlers, and the —SH is ultimately oxidized back to form disulfide bonds that are now angled, producing a curl in the hair.

Attractions between large nonpolar functional groups on the amino acids are called **hydrophobic interactions.** Nonpolar functional groups tend to form in the tertiary and quaternary structures of a molecule. Ionic forces also influence the structure of proteins through an **ionic bridge**

Curls in hair come from a combination of hydrogen bonding and disulfide bonds that form along the protein fibers. (M. Thompsen/zefa/Corbis)

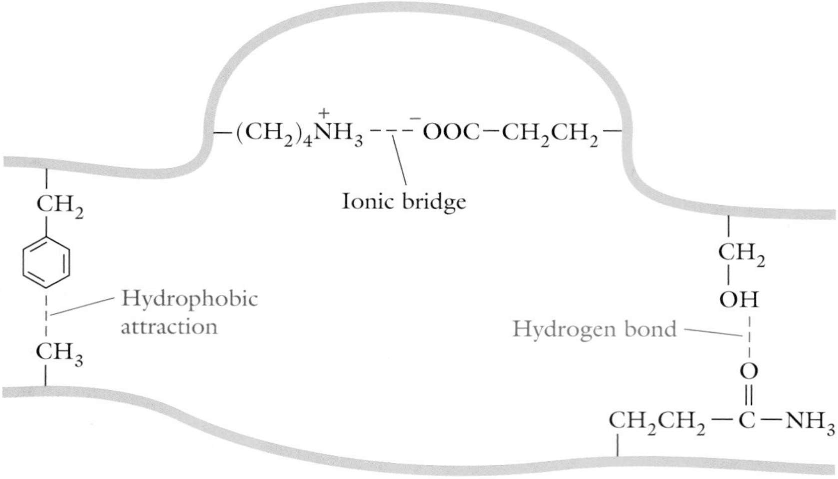

Figure 14.11 Protein molecule linkages.

that forms between an amine functional group and a carboxylic acid functional group. At normal physiological pH, the amine group gains a hydrogen ion to form $-NH_3^+$, and the carboxylic acid group loses a hydrogen ion to form $-COO^-$. These interactions are illustrated in Figure 14.11.

14.4 DNA Basics

Learning Objective

Describe the way the information in a DNA molecule is coded.

DNA is the abbreviation for **deoxyribonucleic acid.** It is the molecule responsible for transmitting all hereditary information from one generation to the next. The forensic value of DNA lies in the fact that each individual has DNA unlike that of any other person, and samples of DNA can be obtained from common types of evidence such as hair, skin, sweat, blood, and saliva. Even identical twins are believed to have slightly different DNA due to small variations and mutations that inevitably occur over time.

DNA analysis is a relatively new technique that was used for the first time in a criminal case in 1986. Since that time, the technology has improved. The ability to detect one billionth of a gram of DNA is now routine. There are still limits to what can be done with DNA, however. Almost 10,000 samples of human remains from the World Trade Center are being preserved for the future when new methods may be developed to identify the victims. To understand the problems associated with the identification of the World Trade Center victims, we must understand the molecular structure of DNA.

DNA has a double-stranded alpha helix shape in which each strand of DNA is a polymer made up of subunits, or building blocks, called **nucleotides.** A nucleotide consists of a phosphoric acid molecule bonded to a deoxyribose sugar and to one of four possible nitrogen-containing compounds: cytosine, thymine, adenine, or guanine. The six substances

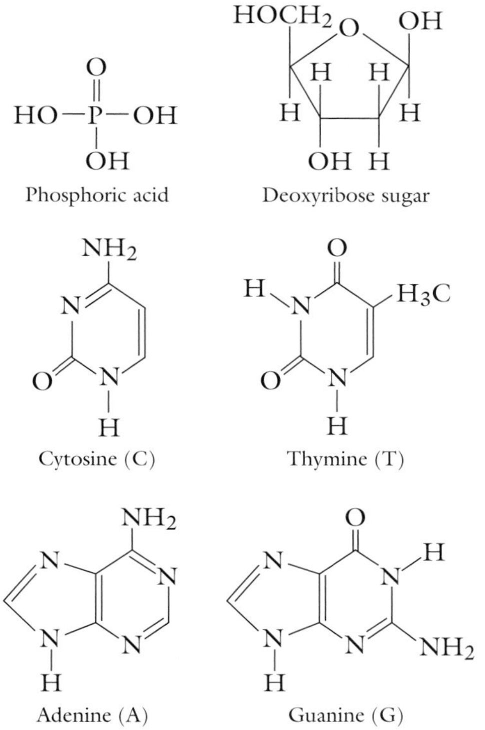

Figure 14.12 DNA components.

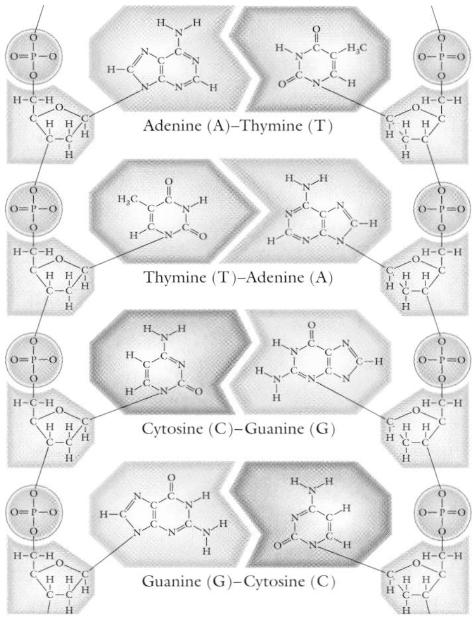

Figure 14.13 DNA polymer chain. The backbone of the DNA molecule is made of phosphoric acid and deoxyribose sugar. The side branches that link the two strands together consist of the nitrogen-containing compounds guanine (G), cytosine (C), thymine (T), and adenine (A). Due to their shapes and the hydrogen bonding that occurs between the side branches, C always binds to G, and A always binds to T.

that combine to create nucleotide building blocks of DNA are shown in Figure 14.12.

Phosphoric acid and deoxyribose sugar form the backbone of the DNA polymer, with the nitrogen base group extending across toward the nitrogen base on the complementary strand of DNA. The DNA molecule is held together in this configuration by hydrogen bonding between the nitrogen bases. The basic structure of a single DNA strand is shown in Figure 14.13.

The interaction between nitrogen bases is quite specific: Cytosine always bonds to guanine (C-G), and adenine always bonds to thymine (A-T), to form the classic double helix structure pictured at the start of this section. The combination of C-G or A-T is referred to as a **base pair** and is often used to refer to the length of a DNA molecule or portion of DNA. If the entire DNA material were extracted from one cell and extended full length, it would be one meter long and would consist of three billion base pairs.

Worked Example 2

What is the complementary strand of DNA for the strand given below?

A A A C G G T C T

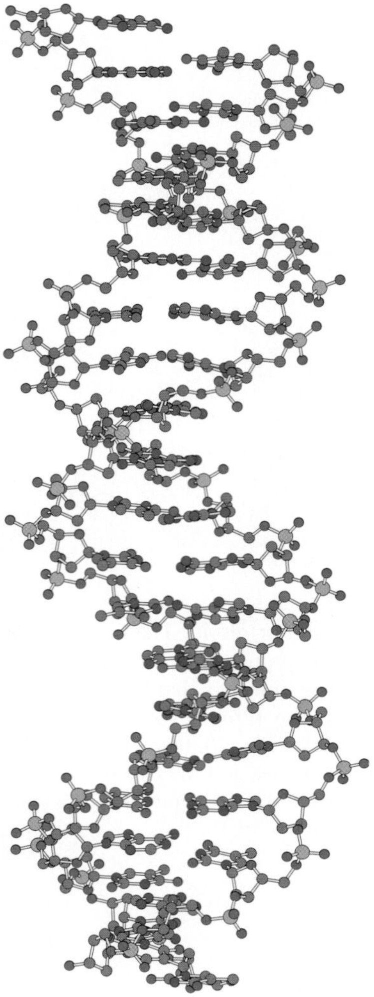

DNA forms a double helix shape, held in place by hydrogen bonding.

SOLUTION The nucleotide adenine always associates opposite to thymine (A-T or T-A), and cytosine to guanine (C-G or G-C). Therefore, the second complementary strand of the DNA is

T T T G C C A G A
A A A C G G T C T

Practice 10.2

What is the complementary strand of DNA for the strand given below?

C T A C C A G T C

ANSWER

G A T G G T C A G

DNA provides hereditary information by supplying the blueprints for the creation of the estimated 100,000 different protein structures found within the human body. The blueprint for each protein is referred to as a **gene** and provides instructions for the order in which the amino acids bind together in the protein. The order of each amino acid in the protein is stored by a series of three nucleotides called a **codon.** If a protein molecule has 100 amino acids, the gene would be $100 \times 3 = 300$ nucleotides long.

In the process by which a protein is created in a cell, the DNA strands untwist near the gene, and a copy of the blueprint is made by creating the complementary strand to the gene. The blueprint is then sent to the ribosome, which synthesizes protein molecules. The job of the ribosome is to assemble the individual amino acids into a protein in the order specified by the gene. If the nucleotide sequence AAA is detected at the ribosome, it is the specific code for the amino acid lysine. If the next amino acid in the protein is glutamate, the nucleotide pattern is GAA. Table 14.2 is a partial list of the nucleotide sequence used by the ribosome to assemble amino acids into proteins.

Table 14.2 Nucleotide Code for Selected Amino Acids

Amino Acid	Nucleotide Sequence
Alanine (Ala)	GCC
Arginine (Arg)	CGC
Asparagine (Asn)	AAC
Aspartate (Asp)	GAC
Glutamine (Gln)	CAA
Glutamate (Glu)	GAA
Glycine (Gly)	GGC
Histidine (His)	CAC
Lysine (Lys)	AAA
Proline (Pro)	CCC
Threonine (Thr)	ACC

Worked Example 3

What is the amino acid sequence indicated by the sequence of codons below?

G A A A A A G C C

SOLUTION The first codon is GAA, the unique pattern for the amino acid Glu, given in Table 14.2. Similarly, the second codon is AAA, which codes for Lys. The final codon, GCC, codes for Ala.

Practice 14.3

What is the nucleotide sequence for the section of a protein consisting of His-Gln-Arg-Asp?

ANSWER

CACCAACGCGAC

14.5 DNA Analysis

Learning Objective

Describe the techniques used to take a sample of DNA and analyze its genetic information.

DNA is packaged in a unit called a **chromosome,** found inside body cells. Humans have 23 pairs of chromosomes, totaling 46 chromosomes. The dominant means of DNA analysis today is a method called **short tandem repeat (STR) analysis.** This method makes use of the fact that 95% of the DNA in a chromosome is noncoding information—long stretches of DNA that don't code for any particular gene. Within these unused portions of DNA, patterns of nucleotides that repeat one after another can be identified.

For example, on chromosome number 2, a section of noncoding DNA called TPOX has the short tandem repeat sequence [AATG], which occurs anywhere from eight to twelve times. Each individual has two copies of chromosome number 2, one from each parent. By studying a large sample of the population, it has been determined that 28.5% of humans have eight repeats of [AATG] on each chromosome, but only 0.24% of humans have 10 repeats of [AATG] on each chromosome. The population frequency for each possible repeat combination of [AATG] for the TPOX location of chromosome number 2 is provided in Table 14.3.

Obviously, if an individual who committed a crime had left DNA at the crime scene, having the 12,12 (0.24%) repeat sequence would limit

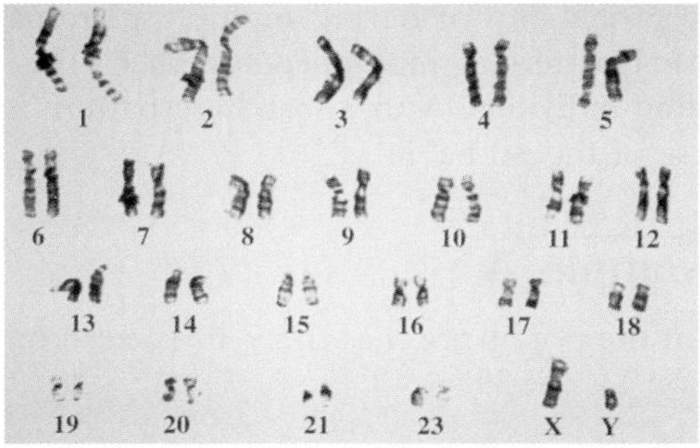

Humans have 23 pairs of chromosomes stored in each cell. One chromosome is inherited from each parent. The chromosome structure consists of protein and DNA molecules. (Scott Camazine & Sue Trainor/Photo Researchers, Inc.)

The short tandem repeat (STR) sections of DNA appear as dark lines during analysis. Each line represents an STR of a different length. (Mauro Fermariello/Photo Researchers, Inc.)

Table 14.3 STR Probabilities for Two of the 13 FBI CODIS Loci

Chromosome #2 TPOX STR	$[AATG]_n$ Repeats	Probability of Matching (%)	Chromosome #16: D16S539 STR	$[GATA]_n$ Repeats	Probability of Matching (%)	$[GATA]_n$ Repeats	Probability of Matching (%)
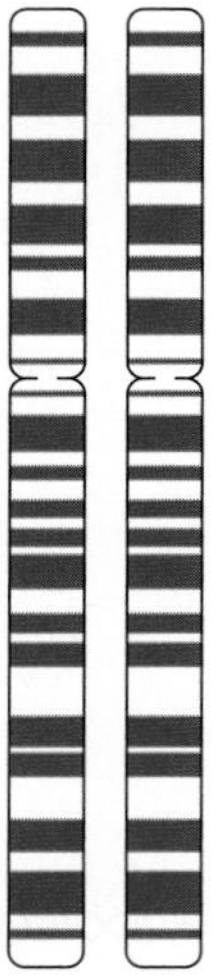	8,8	28.5		8,8	0.023	10,13	1.94
	8,9	12.7		8,9	0.38	10,14	0.29
	8,10	4.91		8,10	0.18	10,15	0.037
	8,11	2.70		8,11	0.93	11,11	9.67
	8,12	5.23		8,12	0.90	11,12	18.6
	9,9	1.41	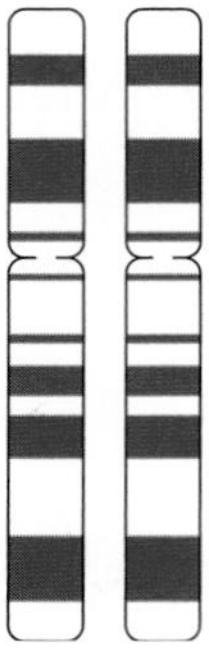	8,13	0.48	11,13	9.89
	9,10	1.09		8,14	0.072	11,14	1.49
	9,11	6.02		8,15	0.009	11,15	0.19
	9,12	1.17		9,9	1.64	12,12	8.94
	10,10	0.212		9,10	1.56	12,13	9.51
	10,11	2.33		9,11	7.96	12,14	1.44
	10,12	0.451		9,12	7.65	12,15	0.180
	11,11	6.40		9,13	4.07	13,13	2.53
	11,12	2.48		9,14	0.61	13,14	7.63
	12,12	0.24		9,15	0.077	13,15	0.095
				10,10	0.37	14,14	0.058
				10,11	3.79	14,15	0.014
				10,12	3.65	15,15	0.0009

The FBI has chosen 13 STR locations on the various chromosomes for use in STR analysis. While 28.5% of people have the TPOX 8,8 pattern, only one out of 15,500 people will have the TPOX 8,8 pattern *and* the D16S539 8,8 pattern. The chance that two individuals will have a matching pattern of STRs at all 13 locations is about one in a billion.

the possible number of suspects with matching DNA better than the 8,8 (28.5%) repeat sequence. The FBI has chosen a standard set of 13 different regions, or *loci,* of noncoding DNA in which to analyze DNA. After comparing the statistics of a person's genetic makeup at all 13 loci, the odds of two people sharing the same pattern usually becomes one in a billion or more. It is important to understand that even if two individuals are found to match at all 13 loci, both individuals still have unique DNA but just happen to share the same exact pattern of repeats at each of the 13 loci. Table 14.3 also shows data for the D16S539 region of chromosome number 16. At this location, the sequence [GATA] can repeat itself eight to 15 times.

To find the probability of having multiple patterns of repeats, the percentage of each pattern is multiplied together. The population size required to find an individual with a matching pattern is determined by taking the inverse of the probability.

Worked Example 4

What percent of the population has the 9,11 pattern on TPOX and the 10,13 pattern on D16S539?

SOLUTION From Table 14.3, we see that:

1. The TPOX 9,11 pattern is present in 6.02% of the population.

2. The D16S539 10,13 pattern is present in 1.94% of the population.

The percent of the general population that has both patterns is

$$0.0602 \times 0.0194 = 0.00117 \text{ or } 0.117\%$$

To find out how many people are needed to find one person with this pattern, take the inverse of 0.00117:

$$\frac{1}{0.00117} = 855$$

Practice 14.4

How large a sample of the human population would be required to find one individual with the TPOX pattern 11,11 and the D16S539 pattern 11,12?

ANSWER
84

14.6 Mitochondrial DNA

Learning Objective

Describe how mitochondrial DNA can be used to identify samples.

When attempting to identify victims, investigators use another form of DNA found within the body, called **mitochondrial DNA (mtDNA).** The mtDNA is found within the mitochondria, a cell structure that serves as the source of power for the cell. Chromosomal DNA is found in the nucleus of a cell, a cell structure that controls the cell's activities.

The differences between mitochondrial DNA and nuclear DNA are significant, one being that mitochondrial DNA has a total of approximately 16,569 base pairs compared with the three billion base pairs found in nuclear DNA. This difference in size is related to the number of genes for which each molecule codes. Nuclear DNA maintains the code for some 100,000 proteins whereas the mtDNA codes for only 37 proteins. There are only two copies of nuclear DNA corresponding to the two chromosome pairs stored in the nucleus of the cell. The number of mtDNA copies can range from the hundreds to thousands for a single cell and are located outside the nucleus of the cell. Another difference is that mtDNA comes only from the mother's side of the family, whereas DNA in the nucleus is contained in 23 chromosomes from the maternal side and 23 chromosomes from the paternal side.

The differences between mitochondrial DNA and nuclear DNA present some useful and not-so-useful properties in forensic science. The sheer number of mtDNA copies in each cell makes recovery of a usable sample much easier than the relatively scarce amount of nuclear DNA. And the mtDNA molecule does not decompose as easily as nuclear DNA molecules, which increases the odds of obtaining a usable sample. However, the direct maternal inheritance of the mtDNA results in an inability to distinguish between relatives on the mother's side of a family. This limits forensic usefulness of mtDNA to situations such as attempting to identify the remains of a missing person. If a set of human remains is located, but there is no source of DNA for a missing individual, the mtDNA of a sibling, mother, maternal aunt, or child of a maternal aunt can be compared with the mtDNA of the remains to establish whether there is a family connection. However, if several family members are among the missing, the identity of the remains cannot be established using mtDNA.

Architectural rendering of the site of the future World Trade Center memorial. (Sipa Press/Sipa)

14.7 Case Study Finale: Remembering 9-11

The guidelines for storing evidence provided by crime laboratories state that any blood or fluid suspected of containing DNA should be dried and then stored in an evidence bag made out of paper. When the crime laboratory receives the DNA evidence, it is kept in refrigerated storage, a process that dramatically slows the fragmentation of DNA molecules. The rate of DNA decomposition is substantially slower in the solid state than in an aqueous solution. The opaque paper bags prevent ultraviolet rays (from light sources) from reaching the sample, and refrigeration further decreases the decomposition rate of the DNA by slowing the kinetics.

Now consider that the landfill in Staten Island was searched for samples for more than eight months. The samples had been exposed to wind, rain, snow, and insects. Samples were also mixed with a multitude of debris and exposed to fires that smoldered in the rubble at ground zero for 99 days. To say that the conditions for optimal recovery of DNA were not present is an understatement. The success rate that was ultimately achieved was due to the heroic efforts of a collaborative group of academic researchers, commercial manufacturers, and state and federal laboratories known as the World Trade Center Kinship and Data Analysis Panel. This group not only used current technology to the fullest but also fast-tracked new technology and wrote software capable of comparing statistics from a mass casualty event that was unlike anything anyone could have imagined.

DNA analysis requires millions of copies of the DNA section containing the STR pattern of interest. This is accomplished by carefully cutting out the STR sections of DNA and replicating them. As the DNA was exposed to weather, fire, chemicals, and insects prior to collection, however, many of the DNA molecules started to break into smaller pieces. If the degradation of DNA occurs at a position near the STR band, amplification will not occur even if the repeat section is intact.

In response to this problem, several researchers were asked to accelerate their efforts to develop the miniSTR analysis, which functions by using a smaller section of DNA with the STR pattern. For example, in STR analysis, the TPOX band is usually amplified within a segment of noncoding DNA up to 249 base pairs long, even though the maximum length of the repeat unit is only 64 base pairs long. The new miniSTR primers produced DNA sections from amplification within a segment of DNA of 101 base pairs in length—a dramatic improvement. The smaller size of fragments allows for more degraded samples to produce a pattern. Of the samples from the World Trade Center, 70% were missing at least one of the 13 STR bands and 50% were missing three or more of the STR bands.

Mitochondrial DNA is an alternative to nuclear DNA for the analysis of missing individuals and seemed like a natural choice to use. An advantage to using mtDNA is that it is a more robust molecule than normal DNA and is present at a much higher concentration per cell. However, the fact that the samples were subjected to such extreme conditions prior to collection led to widespread degradation of even the more robust mtDNA. Another problem arose in that 25% of the victims did not have a sample submitted by a maternally related family member to the database for comparison.

CHAPTER SUMMARY

- Lipids are a large class of organic compounds that are soluble in nonpolar solvents but not in water. The main subclasses of lipids are fats, oils, and waxes. Fats and oils are triglycerides made up of a glycerol molecule bonded to three fatty acid molecules. A wax consists of a long-chain fatty acid bonded to a long-chain alcohol.
- Fatty acids, which are long-chain carboxylic acids, contain double bonds (unsaturated) between adjacent carbon atoms in oils, whereas fats contain only single bonds (saturated).
- Carbohydrates are compounds such as sugars, cellulose, and starch that provide both an energy source and the structural material of cell walls. Cellulose and starch are polysaccharides of glucose. Due to a difference in the bonding of the glucose units in cellulose as compared with starch, only starch is useable by humans as a food source. Cellulose is indigestible and is referred to as dietary fiber.
- Proteins are polymers of amino acids. Several types of fibers—such as wool, cashmere, angora, and silk—are made of proteins. The primary structure of a protein is simply the order in which the amino acids appear in the molecule. The chains of amino acids may interact to form a pleated sheet structure, which is typical of silk. Wool, cashmere, and angora are animal-hair-based proteins found in an alpha helix shape that resembles a coiled spring held in place by hydrogen bonds.
- Protein molecules may have tertiary structures in which the long chains fold back on one another, forming a larger three-dimensional shape. Several protein molecules may join together to form a superstructure referred to as a quaternary structure.
- DNA molecules are long polymer chains of nucleotides that serve as the blueprint for all proteins. The basic nucleotide consists of a phosphoric acid molecule and a deoxyribose sugar unit that together form the backbone of the polymer, and one of four nitrogen base compounds. The cross-linking interaction between chains of DNA occurs because of hydrogen bonding between the bases adenine and thymine (A-T or T-A) or between cytosine and guanine (C-G or G-C).
- The pattern of base pairs serves as the blueprint for DNA. Three base pairs, called a codon, provide the code for one of the 20 amino acids. The group of codons that produces a specific protein is referred to as a gene.
- Mitochondrial DNA is used in another form of DNA analysis. However, such analysis is limited in that all maternal relatives will share identical mitochondrial DNA.

KEY TERMS

lipid, p. 406
wax, p. 406
fatty acid, p. 406
triglyceride, p. 407
fat, p. 407
oil, p. 407
saturated, p. 408
unsaturated, p. 408
carbohydrate, p. 409
monosaccharide, p. 409
disaccharide, p. 409
sugar, p. 409
oligosaccharide, p. 409
polysaccharide, p. 410
polymer, p. 410
monomer, p. 410
cellulose, p. 410
dietary fiber, p. 410
starch, p. 410
protein, p. 411
amino acid, p. 413
oligopeptide, p. 413
polypeptide, p. 413
primary structure, p. 413
secondary structure, p. 413
pleated sheet, p. 413
alpha helix, p. 414
tertiary structure, p. 415
quaternary structure, p. 415
disulfide bond, p. 415
hydrophobic interactions, p. 415
ionic bridge, p. 415
deoxyribonucleic acid (DNA), p. 416
nucleotide, p. 416
base pair, p. 417
gene, p. 418
codon, p. 418
chromosome, p. 419
short tandem repeat (STR) analysis, p. 419
mitochondrial DNA (mtDNA), p. 421

CONTINUING THE INVESTIGATION Additional Readings, Resources, and References

Marchi, Elaine. "Methods Developed to Identify Victims of the World Trade Center Disaster," *American Laboratory*, March 2004, pp. 30–36.

For a Web site that spotlights children's fingerprints:

www.ornl.gov/info/ornlreview/rev28-4/text/tech.htm

For information on the Web about identification techniques:

archives.cnn.com/2002/US/07/13/wtc.identification/index.html

For more information about the use of short tandem repeats:

www.cstl.nist.gov/div831/strbase/

REVIEW QUESTIONS AND PROBLEMS

Questions

1. Explain what properties lipid molecules share and how a lipid molecule is defined.
2. Is there a restriction on what types of organic molecules can be considered a lipid?
3. What is the difference between a fatty acid and a wax?
4. What are the components of a triglyceride? What are the two classes of triglycerides?
5. What properties determine whether a triglyceride is a fat or an oil?
6. Is a saturated triglyceride most likely to be a fat or an oil? An unsaturated triglyceride?
7. If a fatty acid has a 24-carbon chain that is completely saturated, is it most likely a solid or liquid at room temperature?
8. What is a polymer? A monomer?
9. What is the monomer in a starch polymer? In a cellulose polymer?
10. The term *sugar* commonly refers to sucrose. However, in chemistry what type of compounds can be given the generic title of sugar?
11. Draw a sketch of both cellulose and starch to highlight the differences between the two molecules.
12. Why are proteins classified as polymers? What are the monomers?
13. What are the three distinct regions of an amino acid? What distinguishes one amino acid from another?
14. What is the difference between an oligopeptide and a polypeptide?
15. What is the primary structure of a protein molecule? How does that differ from the secondary structure of a protein?
16. What are the two secondary protein structures? Sketch each.
17. How are the secondary structures held together?
18. Discuss the importance of the tertiary and quaternary structures of proteins. Do all proteins have primary, secondary, tertiary, and quaternary structures?
19. What are the four types of bonds that hold the tertiary and quaternary structures together?
20. Why is DNA classified as a polymer? What are the monomers?
21. What force holds together the strands of DNA? Why would covalent bonding between strands be problematic?
22. What is a gene? How does it function? What is it responsible for creating?
23. List the order of nucleotides in the DNA strand complementary to the section with the sequence [TTAGACGAC]. How many codons does this section represent?
24. What is the sequence of amino acids represented by the nucleotides [GAAGCCAACAAACCCCAC]?
25. What is the sequence of amino acids represented by the nucleotides [AACCGCCGCGAACACAACAAC]?
26. What is the nucleotide sequence for the section of a protein consisting of [Gln-Arg-Arg-Ala-Lys-Pro-Lys]?

27. What is the nucleotide sequence for the section of a protein consisting of [Arg-Glu-His-Pro-Asp-Gly-Asp]?
28. Explain how the STR sections in the noncoding portion of DNA allow for identification of a person.
29. Explain the advantages and disadvantages of using mtDNA for identification purposes.

Forensic Chemistry Problems

30. In terms of the STR, miniSTR, and mtDNA analysis, explain why there are nearly 10,000 unidentified remains recovered from victims of the World Trade Center disaster.
31. Why is it important to know what kind of fibers have been transferred during a crime? Are all fibers that get transferred useful for investigative purposes?
32. One of the raw ingredients for the production of alcohol is glucose. Discuss why blood alcohol levels in postmortem samples can be skewed by the presence of starches in the person's system, whereas the presence of cellulose does not affect blood alcohol levels.
33. What is the sample of people statistically needed to find a person with both the TPOX 10,10 repeat pattern and the D16S539 8,15 repeat pattern?
34. It is estimated that only 5% of DNA actually constitutes genes, the rest being noncoding sections of nucleotides. Could DNA be used for identification purposes if it contained only genes?
35. Two cousins are implicated in an assault case where mtDNA has been recovered. However, only one of them is believed to have committed the crime. Could mtDNA be used to identify the guilty person?

Case Study Problems

36. It is known that the fingerprints of children do not last on objects as long as those of adults. Considering the nature of fingerprints, discuss why this might be. For further information, go to www.ornl.gov/info/ornlreview/rev28-4/text/tech.htm.
37. Can two identical twins be distinguished by the current method of DNA analysis, given that there are probably a small number of mutations found in their DNA?

Appendix

Conversion Factors

To convert from the number of . . .	into the number of . . .	multiply by . . .
Mass:		
pounds	kilograms	0.454
kilograms	pounds	2.20
tons	tonnes	0.907
tonnes	tons	1.102
kilograms	tonnes	0.001
tonnes	kilograms	1000
Volume:		
quart	liter	0.946
liter	quart	1.057
cm^3 or mL	liters	0.001
liters	cm^3 or mL	1000
m^3	liters	1000
liters	m^3	0.001
Length:		
inches	centimeters	2.54
centimeters	inches	0.394
yards	meters	0.914
meters	yards	1.094
miles	kilometers	1.61
kilometers	miles	0.621

Temperature

To convert temperatures in °F to those in °C, subtract 32 and then divide the result by 1.80.

To convert temperatures in °C to those in °F, multiply by 1.8 and then add 32 to the result.

To convert temperatures in °C to Kelvin scale, add 273.

To convert Kelvin temperature to °C, subtract 273.

Answers to Odd-Numbered Review Questions and Problems

Chapter 1

1. Biology, geology, physics, psychology, and especially chemistry
3. If the components are distributed evenly, it is a homogeneous mixture. Otherwise, it is a heterogeneous mixture.
5. The components of a mixture can be separated by physical changes.
7. Chromatography, distillation, filtration, centrifugation, or other techniques that use physical changes
9. Science is based on observations.
11. Some atomic symbols are based on the Latin name for the elements.
13. Most elements are reactive under the conditions found on Earth.
15. **(a)** Pure substance **(b)** Mixture **(c)** Mixture **(d)** Mixture
17. **(a)** Compound **(b)** Element **(c)** Element **(d)** Compound
19. **(a)** Hydrogen **(b)** Nitrogen **(c)** Oxygen **(d)** Sulfur
21. **(a)** Arsenic **(b)** Neon **(c)** Calcium **(d)** Platinum
23. **(a)** Mg **(b)** Fe **(c)** Ag **(d)** Hg
25. **(a)** I **(b)** Si **(c)** Cr **(d)** Ti
27. **(a)** Sulfur **(b)** Aluminum **(c)** Mn **(d)** Correct
29. **(a)** Metal **(b)** Metal **(c)** Nonmetal **(d)** Metalloid
31. **(a)** Metal **(b)** Nonmetal **(c)** Nonmetal **(d)** Metal
33. **(b)** Na_3PO_4
35. **(a)** Heterogeneous **(b)** Homogeneous **(c)** Homogeneous **(d)** Homogeneous
37. The center-to-center distance between adjacent letters should be the same throughout. The document used a more advanced printing press that compressed the spaces between adjacent letters by having different sized blocks for each of the letters.

Chapter 2

1. Physical changes alter only the physical properties of the substance, not its chemical identity.
3. Physical properties can be measured without altering the chemical identity of the substance.
5. Determine the nature of the problem: Who committed a crime and how? Collect and analyze all relevant data: Consider all physical evidence, witness statements, alibis, etc. Form a hypothesis: Make an educated guess about what happened. Test the hypothesis: Verify alibis, reconstruct the crime scene, etc. If the hypothesis holds up, you are finished. If not, go back to the second step.
7. Mass is the amount of substance; weight is the force of gravity acting on an object.
9. Kilograms and meters
11. Mega (M), kilo (k), deci (d), centi (c), milli (m), micro (μ)
13. 2.54 cm = 1 in.
15. Estimating the last digit on a measurement gives a more precise and potentially more accurate value for the measurement.
17. **(a)** Physical **(b)** Chemical **(c)** Physical **(d)** Chemical
19. **(a)** Physical **(b)** Chemical **(c)** Chemical **(d)** Physical
21. **(d)** Eliminate conflicting data
23. **(a)** Weight **(b)** Mass **(c)** Mass **(d)** Weight
25. **(a)** μ, 0.000001 **(b)** k, 1000 **(c)** c, 0.01
27. **(a)** m **(b)** k **(c)** d
29. **(a)** 1 mg **(b)** 5.755 L **(c)** 5.9 dm **(d)** 7.50 m
31. **(a)** 12 ft. **(b)** 28.3 yd. **(c)** 1040 mm **(d)** 16.13 cm
33. **(a)** 4 **(b)** 3 **(c)** 4 **(d)** 1
35. **(a)** 1×10^3 **(b)** 1×10^{-5} **(c)** 4.3×10^{-2} **(d)** 5.12×10^6
37. **(a)** 3320 **(b)** 0.04110 **(c)** 900,000 **(d)** 0.006617
39. **(a)** 5.12 **(b)** 3.55 **(c)** 5.70 **(d)** 2.69
41. **(a)** 1.00×10^{-3} **(b)** 0.0200 **(c)** 0.00215 **(d)** 1.00×10^2
43. **(a)** 1.0×10^2 **(b)** 1.00×10^{-4} **(c)** 1.585×10^6 **(d)** 8.8×10^3
45. **(c)** 4.79, 4.68, 4.81, 4.83
47. **(a)** 7.4 **(b)** 373.5 **(c)** 17.8 **(d)** 30
49. **(a)** 280 **(b)** 0.024 **(c)** 2300 **(d)** 2.71
51. **(a)** 50 **(b)** 2.8 **(c)** −36.2 **(d)** −130

53. Alkali borosilicate, alkali zinc borosilicate, and borosilicate; alkali zinc borosilicate
55. **(a)** 1.45 g/cc **(b)** 3.56 g/mL **(c)** 1.45 g/cm^3 **(d)** 0.847 g/mL
57. **(a)** 14.7 mL **(b)** 3.10 mL **(c)** 3.51 cc **(d)** 4.02 cm^3
59. **(a)** 21.8 g **(b)** 17.7 g **(c)** 127 g **(d)** 12.6 g
61. Determine the nature of the problem: The investigators believed that Pitera was linked to the mass grave, but they needed evidence to prove that link. Collect and analyze all relevant data: The investigators collected soil samples from all over Staten Island for comparison. Form a hypothesis: Pitera's shovel would have soil matching the soil type at the mass grave to link him to the crime. Test the hypothesis: Bruce Hall examined the color, texture, and composition of the soil in each of the reference samples and in the sample taken from Pitera's shovel. The soil on the shovel matched only the burial site and none of the alibi sites proposed by the defense.
63. Soil on the blade of a shovel is easily brushed off or mixed with other soils as the shovel is used repeatedly. The soil in the rounded-over flange of the shovel has been compacted and doesn't mix with soil as the shovel is used at new sites.
65. 279 mL
67. 0.00025 g/(dL · min)
69. 36.76 mL
71. A positive or negative result simply means an accelerant was or was not found in the sample. If a sample comes back positive, it could mean the fire burned in a location that already contained an accelerant. If a sample comes back negative, it could mean the arsonist did not use an accelerant. How an investigator uses the test results will depend on the location and type of fire and the suspects they are considering.
73. The shard of glass might come from a bottle or jar that is made in a particular location or manufacturing plant, so it might be possible to link the shard to a specific location. It might also be possible that the suspect could own containers that use glass with a unique composition. The color, refractive index, thickness, and shape of the glass shard should be closely compared with reference samples to determine whether a link can be made.

Chapter 3

1. Atoms were thought to be small, hard, indivisible particles of various sizes, shapes, and weights. They were thought to be in constant motion and combined to make up the various forms of matter.
3. Aristotle believed that matter could be divided infinitely, contradicting the theory proposed by the atomists.
5. The law of conservation of mass says that in a chemical reaction, matter changes form but is neither created nor destroyed.
7. Lavoisier conducted experiments in closed systems, which enabled him to show mass was conserved.
9. The law of definite proportions helped scientists to understand that atoms of different elements can combine in different ratios to make different compounds.
11. In science, a theory is the best current explanation of a phenomenon. In popular use, it is an opinion that can be easily swayed by argument.
13. (1) All matter is made up of tiny, indivisible particles called atoms. (2) Atoms cannot be created, destroyed, or transformed into other atoms in a chemical reaction. (3) All atoms of a given element are identical. (4) Atoms combine in simple, whole-number ratios to form compounds.
15. The law of definite proportions is a restatement of one of the principles of Dalton's atomic theory.
17. 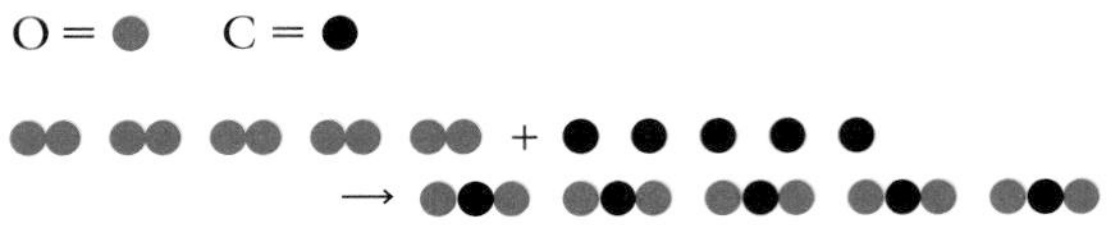

19. 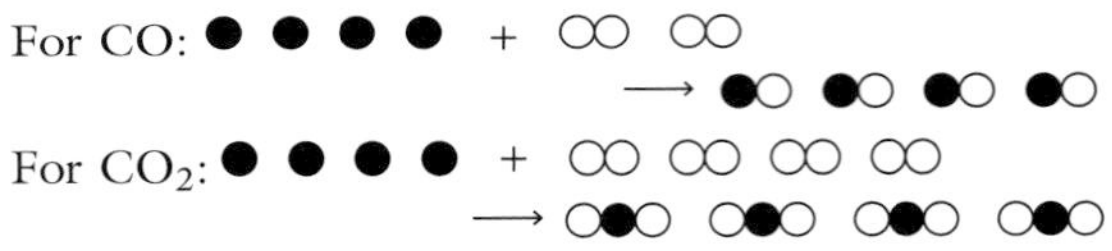

21. Electrons, protons, and neutrons
23. An atom consists of a small but dense positive nucleus with electrons rotating around the nucleus.
25. Isotopes of the same element have different numbers of neutrons.
27. An incandescent light bulb causes the emission of light at all wavelengths in the visible spectrum. A high voltage discharge tube containing an element in the gas phase produces a line spectrum, as do flame emissions of many inorganic salts.
29. A cesium atom contains more excited electrons than does a lithium atom.
31. According to the Heisenberg uncertainty principle, if the energy of an electron is known precisely, then its location cannot be known.
33. See Figures 3.15 and 3.16.
35. **(b)**, **(a)**, **(c)**, **(d)**
37. **(c)** The chemical reactions could exchange mass with the surroundings.
39. **(a)** 31.0 g **(b)** 40.0 g **(c)** 18.5 g **(d)** 4.4 g
41. **(b)** The majority of the atom must consist of empty space.

43.

Particle	Charge	Mass (amu)	Symbol
Electron	−1	0.0005444	e^-
Proton	+1	1	p^+ or H^+
Neutron	0	1	n

45.

Protons	Neutrons	Electrons	Isotope Symbol
27	32	27	$^{59}_{27}Co$
76	124	76	$^{200}_{76}Os$
6	8	6	$^{14}_{6}C$
21	24	21	$^{45}_{21}Sc$

47. **(a)** ^{25}Mg **(b)** ^{44}Ca **(c)** ^{75}As **(d)** ^{177}Hf

49. **(a)** Mg-24

51. **(b)** Electrons are stable in excited states.

53. **(a)** 3.33×10^{14} Hz **(b)** 4.62×10^{14} Hz **(c)** 7.50×10^{14} Hz **(d)** 9.09×10^{14} Hz

55. **(a)** 240 nm **(b)** 226 nm **(c)** 165 nm **(d)** 136 nm

57. **(a)** 2.21×10^{-19} J **(b)** 3.06×10^{-19} J **(c)** 4.97×10^{-19} J **(d)** 6.02×10^{-19} J

59. **(a)** 8.28×10^{-19} J **(b)** 8.81×10^{-19} J **(c)** 1.21×10^{-18} J **(d)** 1.45×10^{-18} J

61. **(a)** $1s^22s^22p^63s^23p^64s^23d^{10}4p^65s^24d^8$ **(b)** $1s^22s^22p^63s^23p^4$ **(c)** $1s^22s^22p^63s^23p^64s^23d^{10}4p^5$ **(d)** $1s^22s^22p^63s^23p^64s^13d^5$

63. **(a)** $[Ne]3s^23p^5$ **(b)** $[Ar]4s^23d^{10}4p^4$ **(c)** $[He]2s^22p^1$ **(d)** $[Xe]6s^1$

65. **(a)** Mo **(b)** Al **(c)** Ti **(d)** Co

67. **(a)** Si **(b)** I **(c)** Mg **(d)** Sb

69. **(a)** $[Kr]5s^24d^1$ **(b)** $[Ar]4s^23d^1$ **(c)** $1s^22s^22p^63s^23p^64s^23d^6$ **(d)** $1s^22s^22p^5$

71. All have 38 electrons and 38 protons. The number of neutrons ranges from 46 to 50.

73. $1s^22s^22p^63s^23p^64s^23d^{10}4p^65s^24d^{10}5p^3$, $E = 9.61 \times 10^{-19}$ J, $\nu = 1.45 \times 10^{15}$ Hz

75. Depending on the Internet source you find, your answer should contain lead (Pb), barium (Ba), and antimony (Sb) as the major primer elements. It may also include aluminum (Al), sulfur (S), tin (Sn), calcium (Ca), potassium (K), chlorine (Cl), silicon (Si), or strontium (Sr).

77. The evidence that the burn was caused by fireworks that match those used by the neighbors is consistent with the version of events as claimed by the burn victim. However, the evidence would also be consistent with a situation in which the burn victim had accidentally burned himself while lighting his own fireworks.

Chapter 4

1. See Figure 4.1.

3. Elements within the same group tend to have similar chemical and physical properties and undergo similar reactions. In addition, the charges of ions for some groups of elements follow a predictable pattern.

5. A monatomic ion is formed from a single atom when it gains or loses electrons. Some examples are H^+, Na^+, Ca^{2+}, Al^{3+}, Zn^{2+}, O^{2-}, and Br^-.

7. The melting point increases with the magnitude of the charge on the ion because of the attractive forces between oppositely charged ions in the crystal lattice.

9. Two atoms share electrons when their orbitals overlap.

11. Electronegativity is a measure of how much an element tends to pull electrons within a bond closer to itself. The electronegativities of the two elements in a covalent bond determine whether the bond is polar or nonpolar.

13. Balanced chemical equations are important because they reflect the fact that all chemical reactions obey the law of conservation of mass.

15. The mole relates the mass of a substance with the number of atoms present in it.

17. The limiting reactant determines the maximum theoretical amount of product formed. If the amount is unknown, it is impossible to predict how much product is formed.

19. The higher the concentration of the dye, the less light passes through the solution, so the solution appears darker.

21. **(a)** Cl, I **(b)** Br **(c)** None **(d)** F

23. **(a)** Group 7, transition metal, Period 4 **(b)** Group 6, transition metal, Period 4 **(c)** Group 15, Period 4 **(d)** Group 14, Period 6

25. **(a)** Mercury **(b)** Chlorine **(c)** Neon **(d)** Selenium

27. **(a)** +2 **(b)** −3 **(c)** −1 **(d)** Multiple

29. **(a)** $MgCl_2$ **(b)** LiF **(c)** Al_2O_3 **(d)** ZnI_2

31. **(a)** NO_3^-, −1 **(b)** OH^-, −1 **(c)** CN^-, −1 **(d)** PO_4^{3-}, −3

33. **(a)** $NH_4C_2H_3O_2$ **(b)** Na_2CO_3 **(c)** $Mg_3(PO_4)_2$ **(d)** $LiMnO_4$

35. **(a)** +3 **(b)** +2 **(c)** +2 **(d)** +2

37. **(a)** Calcium sulfide **(b)** Magnesium nitride **(c)** Lithium fluoride **(d)** Aluminum hydroxide

39. **(a)** Iron(III) chloride **(b)** Cobalt(II) sulfate **(c)** Copper(II) fluoride **(d)** Chromium(II) sulfide

41. (a) Sulfur hexafluoride (b) Carbon tetrachloride (c) Nitrogen trifluoride (d) Phosphorus pentachloride

43. (a) S_2Cl_2 (b) N_2S_5 (c) SF_4 (d) SO_3

45. $3NaCN + Fe(NO_3)_3 \rightarrow Fe(CN)_3 + 3NaNO_3$

47. (a) $3BaCl_2 + 2Na_3PO_4 \rightarrow Ba_3(PO_4)_2 + 6NaCl$
(b) $Na_2S + Fe(NO_3)_2 \rightarrow 2NaNO_3 + FeS$
(c) $C_3H_8 + 5O_2 \rightarrow 3CO_2 + 4H_2O$
(d) $Ca(C_2H_3O_2)_2 + 2KOH \rightarrow 2KC_2H_3O_2 + Ca(OH)_2$

49. (a) 0.0883 mol (b) 0.00320 mol (c) 0.5560 mol (d) 0.0236 mol

51. (a) 152 g (b) 53.2 g (c) 89.1 g (d) 251 g

53. 11.2 g

55. 42.32 g

57. 74.86 g

59. (a) Redox (b) Combustion (c) Precipitation (d) Acid-base

61. KOH; 15.9 g

63. Iodine is a halogen (Group 17) in Period 5. Phosphorus is in Group 15 and Period 3. Lithium metal is an alkali metal (Group 1) in Period 2.

65. 23.13 g

67. 310.3 g

69. 150 mg/L

71. $TlCl(aq) + AgNO_3(aq) \rightarrow TlNO_3(aq) + AgCl(s)$; TlCl, 0.3226 g

Chapter 5

1. A solvent is a substance into which a solute can dissolve. The most common solvent on earth is water.

3. The potassium ion (K^+) would be surrounded by water molecules such that the oxygen atom of water is preferentially oriented toward the cation. The solution would contain twice as many potassium ions as sulfide ions because of the 2:1 ratio in the formula unit. Each sulfide ion (S^{2-}) would be surrounded by water molecules such that the hydrogen of atoms are oriented toward the ion. See the *Student Solutions Manual* for the full diagram.

5. Ionic compounds dissociate when they dissolve in water, while molecular compounds such as sugar do not dissociate.

7. Electrolytes are substances that conduct electricity when dissolved in solution. Nonelectrolytes do not conduct electricity when they are dissolved. NaCl is an electrolyte, and sugar ($C_{12}H_{22}O_{11}$) is a nonelectrolyte.

9. A saturated solution is one in which the dissolved solute is in equilibrium with the undissolved solute. In an unsaturated solution, all the solute is completely dissolved. The unsaturated solution is uniform throughout and does not have any undissolved solute present.

11. Heat the solvent above room temperature to create a saturated solution. Allowing the solution to cool back to room temperature will cause a supersaturated solution to form.

13. Spectator ions balance the charges of the ions that participate and are left behind in solution as the solid precipitate forms. A precipitation reaction can occur without spectators, but only if all the ions in solution become part of the precipitate (quite an unusual situation).

15. Compounds (b) and (c) dissolve in compound 2. Compounds 2 and (a) dissolve in compound 1.

17. The net ionic equation shows only the ions that participate in the net reaction. It is useful because it leaves out the spectator ions that make the reaction look more complicated than it really is.

19. Increasing surface area, temperature, and rate of stirring

21. A strong acid completely ionizes in solution, but a weak acid does not. Likewise, a strong base completely dissociates and a weak base does not.

23. Neutralization reactions and reduction-oxidation reactions

25. The concept of pH was developed as a way of expressing the acidity of a solution.

27. A buffer is a solution that resists changes in pH.

29. (a) Electrolyte (b) Nonelectrolyte (c) Electrolyte (d) Nonelectrolyte

31. (a) Strong (b) Strong (c) Strong (d) Weak

33. (a) Saturated (b) Supersaturated (c) Saturated (d) Saturated

35. (a) 750 g (b) 1.38 g (c) 125 g (d) 250 g

37. KNO_3 and KBr are unsaturated, KCl is saturated, and NaCl is supersaturated.

39. (a) 32 g/100 mL (b) 42 g/100 mL (c) 64 g/100 mL (d) 52 g/100 mL

41. (a) 0.359 M (b) 0.341 M (c) 0.00347 M (d) 0.104 M

43. (a) 0.0249 mol (b) 0.025 mol (c) 0.263 mol (d) 4.28 mol

45. (a) 2.48 g (b) 2.5 g (c) 26.3 g (d) 428 g

47. (a) 4.84 M (b) 1.21 M (c) 0.242 M (d) 1.61 M

49. (a) 41.3 mL (b) 496 mL (c) 496 mL (d) 0.826 mL

51. (a) Soluble (b) Insoluble (c) Insoluble (d) Soluble

53. $Na_2SO_4(aq) + MgCl_2(aq) \rightarrow 2NaCl(aq) + MgSO_4(s)$
$2Na^+(aq) + SO_4^{2-}(aq) + Mg^{2+}(aq) + 2Cl^-(aq) \rightarrow 2Na^2(aq) + 2Cl^-(aq) + MgSO_4(s)$

$Mg^{2+}(aq) + SO_4^{2-}(aq) \rightarrow MgSO_4(s)$

55. $2Fe(C_2H_3O_2)_3(aq) + 3Na_2S(aq) \rightarrow Fe_2S_3(s) + 6NaC_2H_3O_2(aq)$
$2Fe^{3+}(aq) + 6C_2H_3O_2^-(aq) + 6Na^+(aq) + 3S^{2-}(aq) \rightarrow Fe_2S_3(s) + 6Na^+(aq) + 6C_2H_3O_2^-(aq)$
$2Fe^{3+}(aq) + 3S^{2-}(aq) \rightarrow Fe_2S_3(s)$

57. **(a)** Carbonic, weak **(b)** Nitric, strong **(c)** Phosphoric, weak **(d)** Sulfuric, strong

59. **(a)** Acid, acetic acid, $HC_2H_3O_2$, weak electrolyte **(b)** Base, sodium hypochlorite, NaClO, strong electrolyte **(c)** Acid, sulfuric acid, H_2SO_4, strong electrolyte **(d)** Acid, carbonic and phosphoric acids, H_2CO_3 and H_3PO_4, weak electrolytes

61. **(a)** 1.5 **(b)** 2.3 **(c)** 0.82 **(d)** 3.7

63. 3.3

65. 7.7 M (KCN), 9.8 M (NaCN), 0 M $[Cu(CN)_2]$, 0.10 M $[Cd(CN)_2]$

67. 0.00300 g/mL

69. $CaO(s) + H_2O(l) \rightarrow Ca(OH)_2(aq)$
The poem says that the man was chained "with fetters on each foot, Wrapped in a sheet of flame!" While quicklime does burn the skin, it doesn't do so with a sheet of flame. It slowly dissolves away the flesh from a person's body.

71. The first statement says that thallium nitrate was added; the second statement says thallium nitrate may have been added.

Chapter 6

1. Dipole-dipole forces and London dispersion forces (Van der Waals forces)

3. Ion-dipole forces

5. Dipole-induced dipole force

7. London dispersion forces are the weakest type of intermolecular forces because the induced dipole is formed only temporarily. Because it exists only for a limited amount of time, the force is weaker.

9. Hydrogen bonds form between molecules that have hydrogen bonded to nitrogen, oxygen, or fluorine atoms within them.

11. A system is in equilibrium when there is no net change in the number of molecules in any of the phases present within the system.

13. The boiling point of a solution is the temperature at which the vapor pressure of the solution is equal to atmospheric pressure.

15. The solute particles block solvent molecules from moving into the vapor phase by decreasing the available surface area.

17. If saltwater has a freezing point of −4°C, then lowering the temperature to −5°C will require that the entire sample become a solid. All the liquid water molecules line up around the dissolved salt ions and form a crystal that eventually spreads through the entire sample to make a solid block of frozen saltwater. See the *Student Solutions Manual* for the full diagram.

19. Calcium chloride has three ions formed for every 110.986 g of $CaCl_2$ while there are only two ions formed for every 58.44277 g of NaCl. This means that more ions per gram are formed from sodium chloride than from calcium chloride, making the sodium chloride more efficient.

21. High salt concentrations cause high osmotic pressures within living cells. Any bacteria or mold cells that attempt to survive in this environment will become extremely dehydrated and are unlikely to grow.

23. Intermolecular forces determine how strongly the molecules of a mixture are attracted to the stationary and mobile phases.

25. 5.7 cm^2, 19%

27. B < Al < Ga < In

29. **(a)** London dispersion forces, dipole-dipole forces, and hydrogen bonding **(b)** London dispersion forces, dipole-dipole forces, and hydrogen bonding **(c)** London dispersion forces

31. **(a)** Dipole-induced dipole forces, London dispersion forces **(b)** Hydrogen bonding, dipole-dipole forces, dipole-induced dipole forces, and London dispersion forces **(c)** Ion-dipole forces, hydrogen bonding, dipole-dipole forces, dipole-induced dipole forces, and London dispersion forces

33. HF has a higher boiling point than HBr because HF is capable of hydrogen bonding whereas HBr is not.

35. C > A > B

37. Water boils at 100°C only when it is completely pure and the atmospheric pressure is the same as that at sea level.

39. NaCl > Na_2S > $C_6H_{12}O_6$

41. 100.1°C (NaCl), 100.08°C (Na_2S), 100.04°C ($C_6H_{12}O_6$)

43. **(a)** 0.52°C **(b)** 3.40°C **(c)** 4.74°C

45. −0.37°C (NaCl), −0.28°C (Na_2S), −0.15°C ($C_6H_{12}O_6$)

47. **(a)** −7.56°C **(b)** −34.17°C **(c)** −70.81°C

49. **(a)** 1.83 m **(b)** 6.45 m **(c)** 4.88 m

51. **(a)** 1.83 m, 330 g **(b)** 6.44 m, 1160 g **(c)** 4.88 m, 879 g

53. See Figure 6.10.

55. No, because there was enough time for it to be absorbed through the stomach and small intestine.

57. Because the boiling point and density can change with the amount of dissolved solvent, it is helpful to know both.

59. Trace levels of the compounds will be found in the young man's lungs even if they were the cause of death because the petroleum-based cleaners will gradually diffuse out of his lungs.

Chapter 7

1. Organic compounds were those isolated from living or once-living organisms. Inorganic compounds were considered to be composed of the remaining elements (chiefly non-carbon based) of the periodic table.

3. C_nH_{2n+2}, solvents and fuel sources

5. C_nH_{2n} (alkenes), C_nH_{2n-2} (alkynes)

7. Each line represents a bond, and each corner (that has not been labeled otherwise) represents a carbon atom. Hydrogen atoms are implied and are bonded to each carbon atom to bring its total number of bonds to four.

9. By rearranging the atoms in a molecule, we change the shape of the molecule. This impacts polarity, which in turn affects solubility, melting/freezing point, boiling/melting point, vapor pressure, and many other properties.

11. It means that there is more than one way to arrange the electrons in the molecule and still have the same structure.

13. The amine functional group contains a nitrogen bonded to the molecule (—NH_2 or similar). A primary amine has only one carbon bonded to the nitrogen (CNH_2), while secondary and tertiary amines have, respectively, two (C_2NH) and three (C_3N) carbons bonded to the nitrogen.

15. Congeners are compounds that have the same functional groups. Methanol, ethanol, 1-propanol, 2-propanol, and other alcohols are congeners.

17. Ketones, ethers, and esters all contain oxygen atoms. Ketones have a C=O group near the middle of the molecule. Ethers have an O atom within the molecule: C—O—C. Esters have a C=O group in the middle of the molecule that is bonded to an O atom, which is in turn bonded to another carbon atom.

19. A Brønsted-Lowry base is a proton acceptor. An Arrhenius base is a compound that produces OH^- in aqueous solution. NaOH is an Arrhenius base and NH_3 is a Brønsted-Lowry base.

21. **(a)** Methane **(b)** Pentane **(c)** Propane **(d)** Hexane

23. **(a)** C_5H_{12} **(b)** C_3H_8 **(c)** C_8H_{18} **(d)** CH_4

25. **(a)** CH_4 **(b)** $CH_3CH_2CH_2CH_2CH_3$
(c) $CH_3CH_2CH_3$
(d) $CH_3CH_2CH_2CH_2CH_2CH_3$

27. **(a)**  **(b)** 
(c) 
(d) 

29. **(a)** C_4H_6, $HCCCH_2CH_3$, 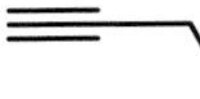
(b) C_5H_{10}, $CH_3CHCHCH_2CH_3$, 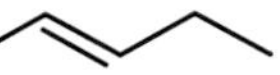
(c) C_6H_{12}, $CH_2CHCH_2CH_2CH_2CH_3$,

(d) C_8H_{14}, $CH_3CH_2CCCH_2CH_2CH_2CH_3$,
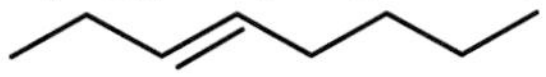

31. **(a)** Two **(b)** Three **(c)** Two **(d)** Four

33. **(a)** 2-methylpentane **(b)** 3-methylhexane
(c) 3-methylhexane **(d)** 2-methylbutane

35. **(a)** 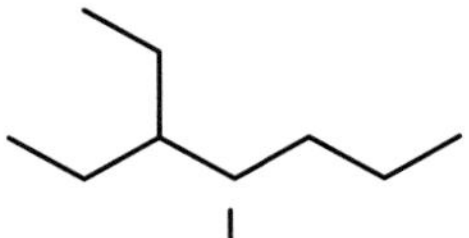
(b) 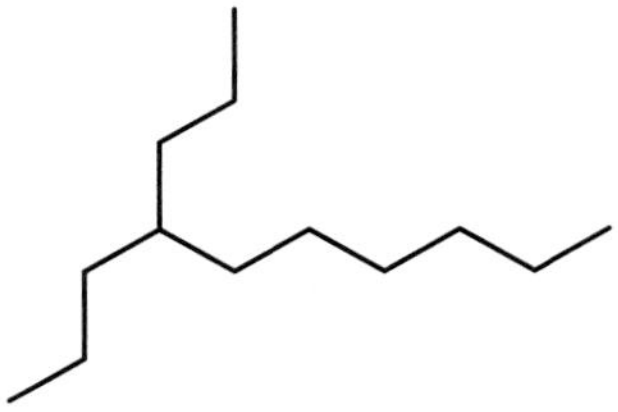
(c) 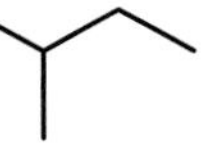
(d) 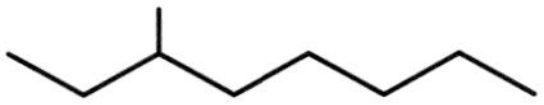

37. **(a)** $(CH_2)_3$ **(b)** $(CH)_6$ **(c)** $(CH_2)_6$
(d) $CH_3C(CH)_5$

39. **(a)** $CH_3C(O)CH_3$
(b) $CH_3CH_2OCH_2CH_2CH_3$
(c) $CH_3OC(O)CH_3$ **(d)** $CH_3C(O)CH_2CH_3$

41. **(a)** 3-hexanone **(b)** Ethoxybutane **(c)** Ethyl pentanoate **(d)** 3-hexanone

43. **(a)** $CH_3CH_2CH_2CH_2NH_2$
(b) $CH_3CH_2NH(CH_3)$ **(c)** $(CH_3CH_2)_3N$
(d) $CH_3CH_2CH_2CH_2N(CH_3)CH_2CH_3$

45. **(a)** 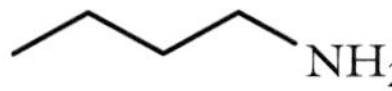

(b) 

(c) 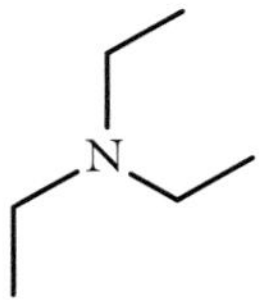

(d) 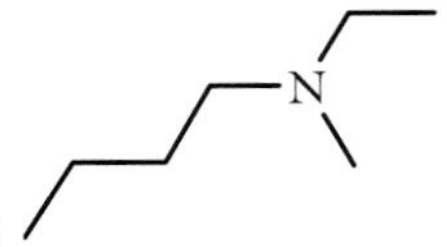

47. **(a)** $CH_3CH(OH)CH_2CH_3$
(b) $HOCH_2CH_2CH_3$
(c) $CH_3CH_2CH(OH)CH_2CH_3$
(d) $CH_3CH(OH)CH_2CH_2CH_2CH_3$

49. **(a)** Pentanal **(b)** Hexanoic acid **(c)** 2-pentanol
(d) Hexanal

51. (a) $CH_3CH_2CH_2CH_2CO(OH)$
(b) $CH_3CH_2CH_2CH_2CH_2CH_2CH_2CHO$
(c) $CH_3CH_2CH_2CH_2CH_2CO(OH)$
(d) $CH_3CH_2CH_2CH_2CH_2CH_2CHO$

53. $CH_3CH_2CH_2CH_2CH_2CH_2OH$, 1-hexanol; $CH_3CH_2CH_2CH_2CH(OH)CH_3$, 2-hexanol; $CH_3CH_2CH(OH)CH_2CH_2CH_3$, 3-hexanol

55. (a) 3-methylpropane is really butane.
(b) 2-propylethane is really pentane.
(c) 4-ethylheptane is named correctly.
(d) 2-methylethane is really propane.

57. (a) 3-propanoic acid is an incorrect name because the carboxylic acid group should receive the lowest possible position number. The correct name is 1-propanoic acid, or simply propanoic acid.
(b) 2-propanal does not exist because an aldehyde group cannot exist on the second carbon in a propane molecule. If there is a C=O group on the second carbon, it would be called 2-propanone, or simply propanone. (c) 1-butanone does not exist because a ketone group cannot exist on a terminal carbon in a chain. If the C=O group is on the terminal carbon of a butane molecule, it would be called butanal. (d) 3-butanol is more correctly named 2-butanol. The alcohol group should get the lowest possible position number, and the third carbon becomes the second carbon when the butane chain is numbered from the opposite group.

59. (a) Neutral compound (b) Acid (c) Acid
(d) Base

61. Marijuana, GHB, 1,4-butanediol, lorazepam, and morphine

63. Ketamine, OxyContin, and Vicodin

65. Heroin, marijuana, MDMA, MDA, OxyContin, Vicodin, codeine, and morphine all contain ether groups.

67. Methamphetamine, cocaine, heroin, marijuana, MDMA, MDA, ketamine, LSD, psilocybin, OxyContin, Vicodin, Ritalin, Xanax, lorazepam, clonazepam, and morphine

69. CH_3OH (methanol) → CH_2O (methanal) → $CHO(OH)$ (methanoic acid) and CH_3CH_2OH (ethanol) → CH_3CHO (ethanal) → $CH_3CO(OH)$ (ethanoic acid)

71. $NH_2CH_2CH_2CH_2CH_2CH_2NH_2$ (cadaverine) and $NH_2CH_2CH_2CH_2CH_2NH_2$ (putrescine)

73. The functional group region of the FTIR evidence shows only the *type* of functional groups (structures like alcohols, aldehydes, ketones, amines, etc.) but gives little data in terms of *how many* of these structures are present in the molecule. Therefore, it would be very difficult to differentiate 1-butanol (one C—OH group) from 1,4-butanediol (two C—OH groups) by examining only this region. However, the fingerprint region of the FTIR evidence should be unique, and a visible difference is present between the two compounds in question.

Chapter 8

1. Most single atoms do not have a full shell of valence electrons. To obtain a full shell, they can share electrons to form a bond:
$H\cdot + H\cdot \rightarrow H\cdot\cdot H \rightarrow H—H$

3. When atoms exchange or share electrons, they achieve a full valence shell. This is a stable configuration that minimizes the energy of the atoms present.

5. Atoms form covalent bonds when they share electrons equally. Atoms form ionic bonds when they exchange (donate or accept) electrons.

7. A double bond will form between two atoms when both atoms need the second bond to achieve a full valence shell. For example, oxygen is two electrons short of an octet. If each oxygen shares two electrons with the other, a double bond is formed.

9. Resonance structures show equivalent possible structures based on the placement of electrons in multiple bonds or lone pairs. They do not represent accurate bonding within the molecule because the actual bonding is a combination of the resonance structures.

11. According to the VSEPR theory, electron regions around an atom will move until they are as far apart as possible. The number of electron regions determines the angle required between regions to allow maximum separation. The angle describes the electron geometry of the molecule.

13. The nature of the electron region does not affect the overall electron geometry of the molecule. However, the "size" of an electron region does depend on its nature (single bond, double bond, triple bond, lone pair). Lone pairs take up more space than bonds do, so they cause a distortion of the bond angles to make more space for themselves. This slightly alters the electron geometry.

15. As described in Problem 13, the nature of electron regions slightly alters electron geometry. Because electron geometry determines the molecular geometry, the molecular geometry is also affected (albeit slightly) by the nature of the electron regions in the molecule.

17. Stereoisomers occur when two compounds share the same chemical formula and the same connections between atoms but exhibit differences in the way their atoms are arranged three-dimensionally. This is extremely significant in drug chemistry—different stereoisomers can have very different properties.

19. The molecular shape of neurotransmitters is the key function that controls their uptake into the uptake channel of the neuron. The tunnel walls are strands of protein molecules arranged so that neurotransmitters of a particular shape and size may pass through.

21. Cocaine and other illegal drugs cause a permanent change in brain function by destroying uptake channels. The resulting damage to neurons often leads to depression and other symptoms.

23. Immunoassays are a screening method because they do not provide definite identification of a chemical compound. Because molecular geometry is an important part of how the immunoassay functions, it is possible for a molecule that has a similar shape to an antigen (or its antibody) to cause the same reaction in the immunoassay as the target molecule.

25. H—$\ddot{\underset{..}{I}}$:

27. (a) ·$\ddot{\underset{..}{As}}$:
    (b) ·Be·
    (c) ·$\ddot{N}$:
    (d) ·Al·

29. (a) $[K]^{1+}[:\overset{..}{\underset{..}{P}}:]^{3-}[K]^{1+}$
    $$[K]^{1+}$$
    (b) $[:\ddot{\underset{..}{I}}:]^{1-}[Ba]^{2+}[:\ddot{\underset{..}{I}}:]^{1-}$
    (c) $[Ca]^{2+}[:\ddot{\underset{..}{O}}:]^{2-}$
    (d) $[Na]^{1+}[:\ddot{N}:]^{3-}[Na]^{1+}$
    $$[Na]^{1+}$$

31. (a) $[:N\equiv C:]^{1-}$
    (b)
    $$\left[\begin{array}{c} :\ddot{O}: \\ | \\ :\ddot{\underset{..}{O}}-P-\ddot{\underset{..}{O}}: \\ | \\ :\ddot{\underset{..}{O}}: \end{array}\right]^{3-}$$
    (c)
    $$\left[\begin{array}{c} :\ddot{O}: \\ | \\ \overset{C}{:\ddot{\underset{..}{O}}\quad\ddot{\underset{..}{O}}:} \end{array}\right]^{2-} \leftrightarrow \left[\begin{array}{c} :O: \\ || \\ \overset{C}{:\ddot{\underset{..}{O}}\quad\ddot{\underset{..}{O}}:} \end{array}\right]^{2-} \leftrightarrow \left[\begin{array}{c} :\ddot{O}: \\ | \\ \overset{C}{:\ddot{\underset{..}{O}}\quad\ddot{\underset{.}{O}}} \end{array}\right]^{2-}$$
    (d) $[:\ddot{\underset{..}{O}}-H]^{1-}$

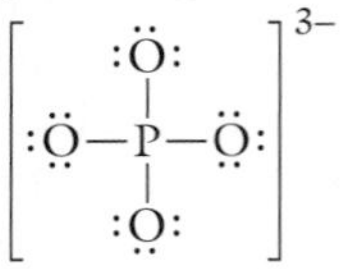

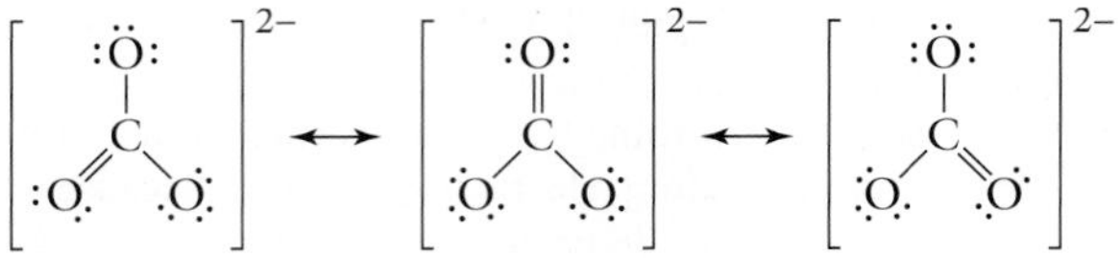

33. (a)
    $$\begin{array}{c} H \\ | \\ P: \\ H^{\diagup}\quad^{\diagdown}H \end{array}$$
    (b) $:\ddot{\underset{..}{I}}-\ddot{\underset{..}{I}}:$

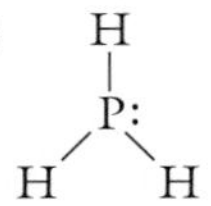

    (c)
    $$\begin{array}{c} :\ddot{O}: \\ :\ddot{\underset{.}{F}}\quad\ddot{\underset{.}{F}}: \end{array}$$

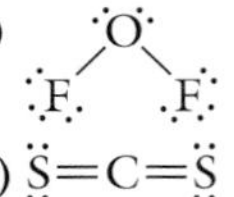

    (d) $\ddot{\underset{..}{S}}=C=\ddot{\underset{..}{S}}$

35. (a) Linear (b) Tetrahedral (c) Trigonal planar (d) Tetrahedral

37. (a) Tetrahedral (b) Tetrahedral (c) Tetrahedral (d) Linear

39. (a) Linear (b) Tetrahedral (c) Trigonal planar (d) Linear

41. (a) Trigonal pyramidal (b) Linear (c) Bent (d) Linear

43. (a) Nonpolar covalent (b) Polar covalent (→) (c) Polar covalent (→) (d) Nonpolar covalent

45. (a) Polar (b) Nonpolar (c) Polar (d) Nonpolar

47. (b) < (a) < (c) < (d)

49. (d) < (c) < (a) < (b)

51. The oxygen at the top of the ring is tetrahedral, and the oxygen at the right side of the ring is trigonal planar.

53. A typical blood sample does not have as many complicating factors as the sample taken in the vigilante Jell-O case. The body had been embalmed (which means that most of the blood has been removed from the body and replaced with methanal), and ground water had seeped into the coffin. These factors decrease the amount of LSD present in the body, making the analysis less reliable. A normal blood sample is pure, in a living body, and will have more reliable results.

Chapter 9

1. Energy is transferred as the atoms "bump" into each other, until the energy is distributed evenly throughout the sample. When the masses are equal, the equilibrium temperature is 275°C. When the mass of the hot copper is twice the mass of the cold copper, the equilibrium temperature is 350°C.

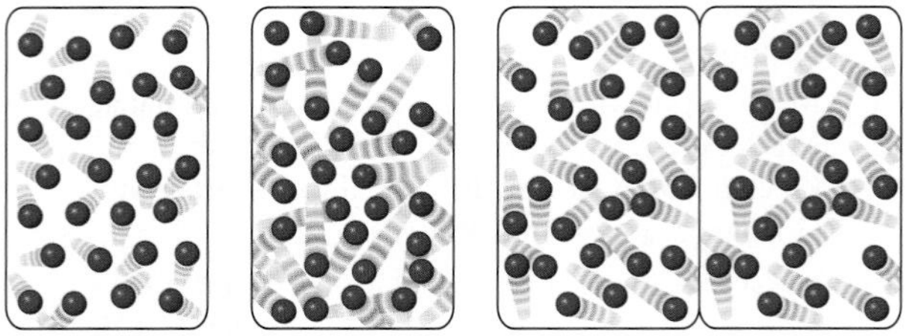

Cu at 50°C   Cu at 500°C   Cu at thermal equilibrium (275°C)

3. Heat is a form of energy that is transferred from a hot object in contact with a cold object. Temperature is a measurement of the average kinetic energy of the particles in a system.

5. Heat will flow either from the solution into the thermometer or from the thermometer into the solution, changing the temperature.

7. A backdraft forms when an excess of fuel is heated in the absence of oxygen. It can be prevented by ventilating the room.

9. $CH_4(g) + 3O_2(g) \rightarrow CO_2(g) + 2H_2O(g)$ (organic compound); $2Ca(s) + O_2(g) \rightarrow 2CaO(s)$ (metal); $S_8(s) + 12O_2(g) \rightarrow 8SO_3(g)$ (nonmetal)

11. Compare the oxidation number of the element on each side of the chemical equation. Oxidation is a loss of electrons, and reduction is a gain of electrons. Electrons have a negative charge, so oxidation results in a more positive oxidation number; reduction results in a less positive oxidation number.

13. If the products have more energy than the reactants, the system must absorb energy from the surroundings (endothermic). If the products have less energy than the reactants, the system must release energy into the surroundings (exothermic).

15. Flashover takes place when heat from a fire in a small area rises to the ceiling and then spreads out across the room. The heated layer of smoke and gases increases its temperature and thickness and radiates heat downward until the materials reach the autoignition temperature, overcoming the activation energy of the combustion reaction. Flashover conditions can be prevented by allowing the heated layer of smoke to escape.

17. Compounds with strong intermolecular forces have larger heat capacities than those with weak intermolecular forces. Compounds with hydrogen bonding (like water) have large heat capacities, and nonpolar compounds (like hexane) have small heat capacities.

19. Tile floor feels colder than carpeting because of the flow of heat from your bare foot into the material. Tile floor conducts heat more quickly than carpeting and has a greater relative heat capacity than carpeting.

21. Accelerants are often not completely consumed in a fire because the vaporization of the liquid fuel is an endothermic process. This lowers the temperature of the region surrounding the fuel to the point that there is insufficient heat for complete combustion to take place.

23. By heating the crude oil, compounds boil out of the mixture in quantities grouped by boiling point. Gasoline has components with boiling points from approximately 40°C to 220°C. Kerosene has components with boiling points ranging from 175°C to 270°C. Gasoline has more compounds of low molecular weight, and kerosene has more compounds of high molecular weight.

25. **(a)** 105 J **(b)** 460,200 J **(c)** 50,000 J **(d)** 308,000 J

27. **(a)** 46,000 J/g **(b)** 48,100 J/g **(c)** 41,800 J/g **(d)** 41,500 J/g

29. **(a)** 43,300 J/g **(b)** 43,500 J/g **(c)** 43,700 J/g **(d)** 44,000 J/g

31. **(a)** $CH_4 + 2O_2 \rightarrow CO_2 + 2H_2O$
(b) $C_5H_{12} + 8O_2 \rightarrow 5CO_2 + 6H_2O$
(c) $C_6H_{14}O + 9O_2 \rightarrow 6CO_2 + 7H_2O$
(d) $2C_8H_{16}O + 23O_2 \rightarrow 16CO_2 + 16H_2O$

33. **(a)** $C_3H_4 + 4O_2 \rightarrow 3CO_2 + 2H_2O$
(b) $C_4H_8 + 6O_2 \rightarrow 4CO_2 + 4H_2O$
(c) $2C_{10}H_{22} + 31O_2 \rightarrow 20CO_2 + 22H_2O$
(d) $C_8H_{16} + 12O_2 \rightarrow 8CO_2 + 8H_2O$

35. **(a)** +2 **(b)** O: −2, Cu: +2 **(c)** O: −2, S: +4 **(d)** O: −2, H: +1, N: −3

37. **(a)** CO_2 (reduced) and H_2O (oxidized)
(b) CH_4 (oxidized) and O_2 (reduced)

39. **(a)** CO (oxidized) and O_2 (reduced)
(b) Na (oxidized) and H_2O (reduced)

41. 0.899 J/g°C

43. 14 kJ

45. 4.5 g

47. 8.547×10^8 J

49. 213 mL

51. 46,000 J

53.

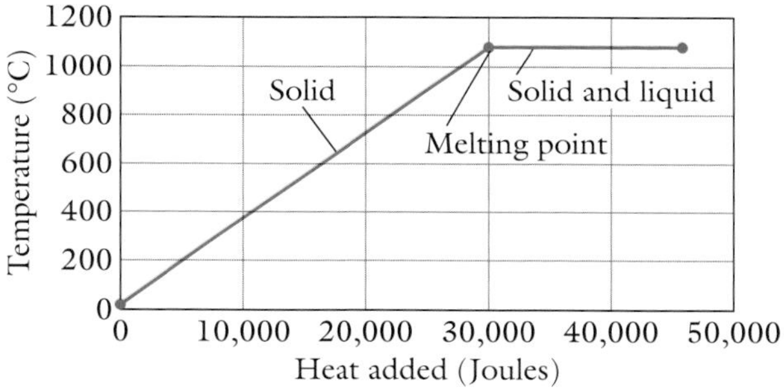

55. 2.38 g

57. 9200 J/g

59. A solid plug should have a melting temperature that allows it to melt at temperatures typically reached in a fire but not during normal building operation. The liquid in a vial should have a boiling temperature with the same characteristics.

61. Only if there is significantly more benzene in the suspected sample than in the blank could that be used to determine whether an accelerant was used to ignite the fire.

63. There also must be sufficient evidence to show that the electrical box was the source of the fire.

65. If the temperature of the smoke is high enough, the exposure of the smoke to the oxygen-enriched atmosphere outside the structure could allow the particles that make up the smoke to burst into flames once they reach open air.

67. If a large amount of heated smoke is trapped near the ceiling, it is entirely possible that a plastic fire detector might melt during a fire even if it had never been tampered with.

69. Charring between deck boards can easily result if the deck boards are sufficiently close together to limit the amount of oxygen.

Chapter 10

1. It must be capable of releasing a large amount of energy, react instantaneously, and release substantial amounts of gaseous products.
3. Most low explosives serve as propellants for guns and military artillery.
5. It is important for explosive molecules to contain C, H, and O because they have gaseous combustion products (carbon dioxide and water), and the compound already contains some (if not all) of the oxygen necessary to produce those products.
7. Gas particles are extremely small and have relatively large distances between them. Gas particles act independently of one another; there are no attractive or repulsive forces between gas particles. Gas particles are continuously moving in random, straight-line motion as they collide with one another and with the container walls. The average kinetic energy of gas particles is proportional to the temperature of the gas.
9. Gases can be compressed to a much greater extent than liquids and solids can because there is much more space between particles. See Figure 10.2.
11. Above the UEL (upper explosive limit), the fuel-air mixture is too fuel-rich for the fuel to react explosively because the reaction is limited by the amount of oxygen present in the mixture.
13. The volume is directly proportional to the number of gas particles. See Figure 10.2.
15. Gay-Lussac's law says that pressure is directly proportional to temperature. See Figure 10.4.
17. These are the conditions of 1 atm pressure and 0°C.
19. Dalton's law of partial pressures states that the total pressure of a gas mixture is the sum of the pressures of each component in the mixture. This requires that each gas behave independently of all other gases in the mixture and that each gas continue to undergo collisions in the same exact manner. This is consistent with the four premises of the kinetic-molecular theory of gases. If particles of one gas act independently of other particles of the same gas, it is a natural extension to say that particles of one gas act independently of particles of a second gas.
21. Chemical markers are high vapor pressure compounds that are included in plastic explosives by their manufacturers. The high vapor pressure compound forms a vapor surrounding the plastic explosive that is more easily detected than the explosive itself.
23. Taggants, markers, and isomers help security officials during all phases of explosive detection and identification. Chemical markers enable security officials to detect explosives before they are used. Taggants and isomers allow the identification of an explosive either before or after detonation, which helps security officials track the explosive from factory through sale to use and identify any illegal activity along the way.
25. **(a)** 3.5 mol **(b)** 6.5 mol **(c)** 3 mol **(d)** 3 mol
27. **(a)** 2 mol **(b)** 5 mol **(c)** 5.5 mol **(d)** 7.5 mol
29. **(a)** Increases **(b)** Increases **(c)** Decreases **(d)** Decreases
31. **(a)** 0.159 mol **(b)** 0.600 L **(c)** 2.59 L **(d)** 1.07 mol
33. **(a)** 3.93 atm **(b)** 0.0294 L **(c)** 1.30 L **(d)** 1.30 atm
35. **(a)** 2400 K **(b)** 2.18 atm **(c)** 1540 K **(d)** 16.8 atm
37. **(a)** 1470 K **(b)** 2.70 L **(c)** 1.20 L **(d)** 505 K
39. **(a)** 172 L **(b)** 741 L **(c)** 3.32 L **(d)** 22.3 L
41. **(a)** 4.61 L **(b)** 13.7 L **(c)** 62.5 L **(d)** 15.4 L
43. **(a)** 0.716 g/L **(b)** 0.143 g/L **(c)** 1.25 g/L **(d)** 1.97 g/L
45. 8.01 L
47. 151 L
49. 37.2 atm
51. 22.09 L
53. 6710 K
55. 36.7 atm
57. Negative
59. Explosives with a positive oxygen balance will produce white smoke because the major component will be steam condensing on the small particles produced by the explosion. Explosives with a negative oxygen balance will produce black, sooty smoke because there will be partially combusted particles composed primarily of leftover carbon.
61. Gases are in constant, random motion and collide with other gas molecules. An explosion requires that the fuel, oxygen, and ignition source are present simultaneously. If the partial pressure of fuel is below the LEL, an explosion will not occur; if it is above the UEL, there is insufficient oxygen for the reaction to continue. Because gaseous molecules of fuels are in constant random motion, a fuel initially present at a partial pressure above the UEL may soon become explosive as the fuel mixes with oxygen gas from the environment, thus lowering its partial pressure.
63. For proper breathing, a person's lungs must be able to expand and contract (via diaphragm movements)

to create a pressure difference with respect to the outside atmosphere. To inhale, the lungs expand (increase *V*), thereby reducing the pressure (decrease *P*) of the air in the lungs, drawing in outside air. In order to exhale, the lungs contract (decrease *V*), increasing the pressure (increase *P*) of the air in the lungs and forcing some outside of the body. If a puncture wound is suffered, then the lungs have great difficulty creating a pressure difference, and it is very hard to breathe.

65.

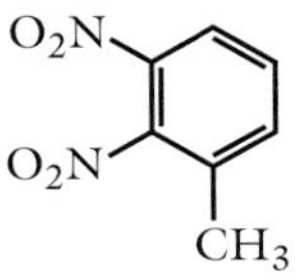

2,3-dinitrotoluene

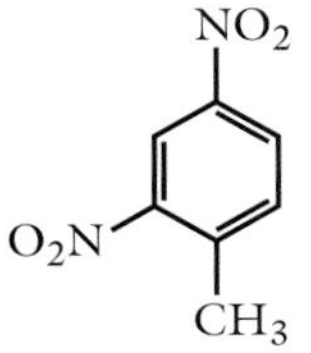

2,4-dinitrotoluene

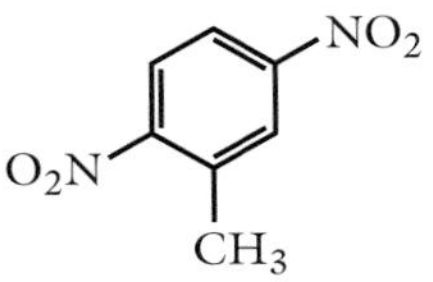

2,5-dinitrotoluene

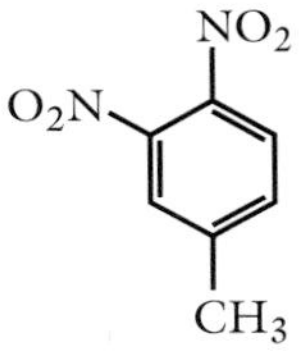

3,4-dinitrotoluene

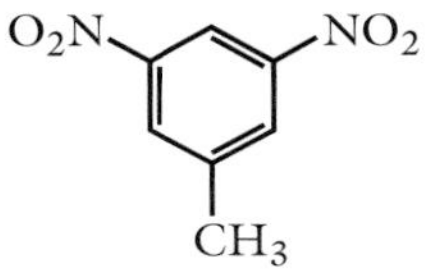

3,5-dinitrotoluene

67. Because explosives are used so frequently and in such large quantities, taggants will be spread to such an extent that it would be very difficult to differentiate taggants already in the environment from taggants that might indicate the presence of an explosive device at a crime scene.

Chapter 11

1. The rates of chemical reactions are very important in determining time of death.

3. The reactants must collide for a reaction to take place. Collisions must have high energy. The colliding particles must be properly oriented for a reaction to occur.

5. The initial concentration of the reactants determines the number of collisions that can potentially take place between reactant molecules.

7. If the clothes are dry, the biological evidence degrades more slowly. Thus, more information can be obtained by the forensic scientists.

9. No, it just allows the reaction to occur more quickly.

11. Adding a catalyst provides an alternative pathway (mechanism) for the reaction to take place—one that is lower in energy than the pathway for the uncatalyzed reaction.

13. The greater the surface area of a heterogeneous catalyst, the more reaction sites on the surface of the catalyst will be available.

15. See Figure 11.2. The collisions must also have the correct orientation for a reaction to take place. As a result, only a fraction of molecules in the shaded area of the bell curve will result in a reaction.

17. Heterogeneous

19. The method for this analysis relies on knowing the rate at which ethanol oxidizes into carbon dioxide in the body.

21. The rate of an enzyme-catalyzed reaction will double because the number of collisions will double. At some point, the enzyme molecules will impede the mixing and collision of reactant molecules, so the rate will eventually decrease.

23. This information can be used to estimate the blood serum level of those drugs at some earlier time (such as when the drugs were originally taken into the body). This can help predict the level of intoxication and determine whether a drug could have been the cause of death for a victim.

25. 200 minutes

27. If the headlights were on when broken, the filament would oxidize very quickly when the inert gases are replaced by air. If the filament is heavily oxidized, the headlights must have been on.

29. The delay of insect activity could be because of the lower average temperature in northern California in late winter/early spring.

Chapter 12

1. By observing what happened when he put the mineral crystal and photographic plate in a dark place, Henri Becquerel disproved the hypotheses that phosphorescence was related to X-rays from the sun.

3. He would have found that phosphorescent materials do not emit X-rays and would have tried to find another way to explain that phenomenon.

5. Neutrons exert attractive forces on protons and other neurons. This overcomes the repulsive forces between the positively charged protons.

7. Alpha decay decreases the number of neutrons by 2 and the number of protons by 2. Beta decay decreases the number of neutrons and increases the number of protons by 1. Gamma rays do not change the number of protons and neutrons.

9. Alpha particles have the greatest ionization power; gamma particles have the least ionization power.

11. Alpha particles, because they have the greatest ionization power. They also have the least penetrating power, so they remain in the body until they have impacted a large amount of tissue.

13. The total number of protons and neutrons on each side of the equation is conserved, not the identity of the elements.

15. Nondestructive analysis of samples is desirable because it leaves open the possibility of analyzing the sample using another technique at a later date.

17. See Figure 12.4. The reactor core system (orange-colored) generates heat through nuclear fission. The steam generation system (blue-colored) transfers the heat to steam, which drives the turbines. The cooling tower system (gray-colored) cools the steam so that no harm is done to the environment.

19. Fission reactions involve the fragmentation of a nuclear isotope into smaller particles. Fusion reactions involve the combination of small nuclear isotopes into a larger isotope. Fusion is more desirable for generating electricity because it releases much more energy and does not produce radioactive waste. Unfortunately, the temperature of fusion is so high that no known materials can contain the reaction and harvest the energy produced.

21. A nuclear chain reaction occurs when a fission reaction generates multiple small particles (often neutrons) that in turn initiate more fission reactions. Control rods are used in a nuclear power plant to absorb excess neutrons to limit the number of fission reactions. See Figure 12.5.

23. Radioisotopes used for medical diagnosis should have an affinity for a particular organ or target a specific system within the body that is being analyzed. They should also be short-lived within the body and have minimal ionization power. Radioisotopes used for cancer treatment should also be short-lived and should have minimal penetration power, to limit the damage done to organs or systems not intended to receive the treatment.

25. There is less chance that radioactive isotopes would be vaporized and dispersed into the atmosphere.

27. (a) $^{131}I \rightarrow {}^{131}Xe + {}_{-1}e$
(b) $^{220}Fr \rightarrow {}^{216}At + {}^{4}He$
(c) $^{60}Co \rightarrow {}^{60}Ni + {}_{-1}e$ (d) $^{226}Ra \rightarrow {}^{222}Rn + {}^{4}He$

29. (a) ${}_{-1}e$ (b) ^{235}U (c) ^{3}He (d) ^{231}Th

31. (a) ^{57}Fe (b) ^{58}Co (c) ${}_{+1}e$ (d) ^{36}Ar

33. Hair grows slowly and is made from waste protein and other compounds from within the body. Drug metabolites are incorporated as part of the hair. If the rate of hair growth were known (or could be estimated), NAA could be used to detect drug metabolites in the hair. How far above the follicle the metabolites are located could determine how long ago the person had ingested the drugs.

35. 724.5 days

Chapter 13

1. A toxin is a naturally occurring poisonous substance produced by a living organism. A poison is any compound that injures or harms a living organism.

3. Virtually any substance can be lethal if enough is present. Even water, which is necessary for life, can act as a poison by seriously disrupting the body's electrolyte balance, if ingested in excess.

5. The reverse reaction of a system can occur as long as product molecules collide with sufficient energy to overcome the activation barrier and the correct orientation to produce the reagents.

7. It decreases because the amount of reagent decreases as a function of time.

9. There is no net change in concentration of reactants or products with time because the rate of forward reaction is equal to the rate of reverse reaction.

11. A catalyst has no effect on the equilibrium constant or on the concentrations of the reactants or products at equilibrium. It simply speeds up the initial rates of the forward reaction, allowing it to reach equilibrium more quickly.

13. If the equilibrium constant is large, products predominate. If the equilibrium constant is small, reactants predominate.

15. The equilibrium system would have numerous C and D molecules and fewer A and B molecules.

17. Various types of lung disease and malfunction cause acidosis, which results from the body's failure to remove carbon dioxide from the blood. This makes the blood and other fluids more acidic than normal.

19. Adding reactant produces more product; removing reactant produces more reactant. Adding product produces more reactant; removing product produces more product.

21. Oxyhemoglobin (HbO_2) is transported to the cells in the blood where the oxygen is released to produce hemoglobin (Hb), which is then circulated back to the lungs to gather more oxygen.

23. Carbon monoxide binds to hemoglobin more strongly than oxygen does. This competitive binding makes carbon monoxide toxic because it prevents the delivery of oxygen to cells.

25. Adding a sulfide, such as Na_2S, to solution would cause the solubility equilibrium to shift toward production of more solid, minimizing the amount of dissolved arsenic.

27. Mercury poisoning requires the mercury(II) to be a free ion in solution. Dimercaprol binds to mercury because of the extremely large binding constant between the mercury(II) ion and the −SH group of the cysteine, thus hindering mercury's toxic effects.

29. The alcohol in your blood is in equilibrium with the alcohol in air in your lungs. This means the ratio of alcohol in the blood to alcohol in the air in your lungs remains constant.

31. Analyzing for carbon monoxide could not be the sole piece of evidence to determine whether the

death was suicide, homicide, or accident. Carbon monoxide could be the cause of death in any one of these instances.

Chapter 14

1. Lipid molecules include a variety of compounds such as oils, fats, and waxes. Their shared physical property is that they are not soluble in water but are soluble in a nonpolar solvent.
3. Waxes are made from a fatty acid, a long carbon chain carboxylic acid molecule, and a long-chain alcohol to form an ester. Both molecules (waxes and fatty acids) are long-chain carboxylic compounds, but the waxes are esters and the fatty acids are carboxylic acids.
5. Fats and oils are both examples of triglycerides. The difference between them is that fats are semi-solids at room temperature and oils are liquids at room temperature.
7. Solid
9. Glucose is the monomer in starch and cellulose.
11. Cellulose:

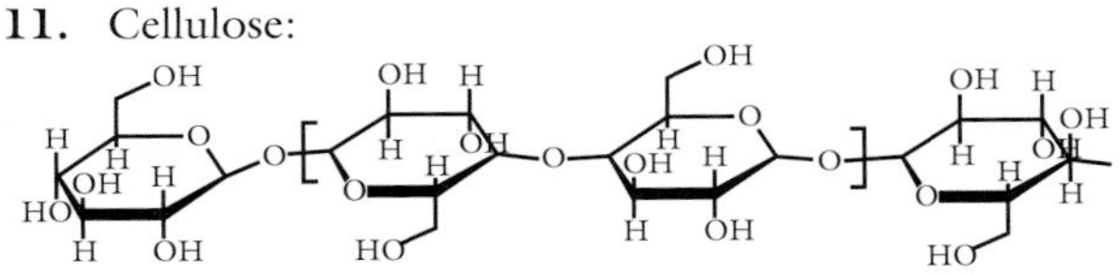

Starch:

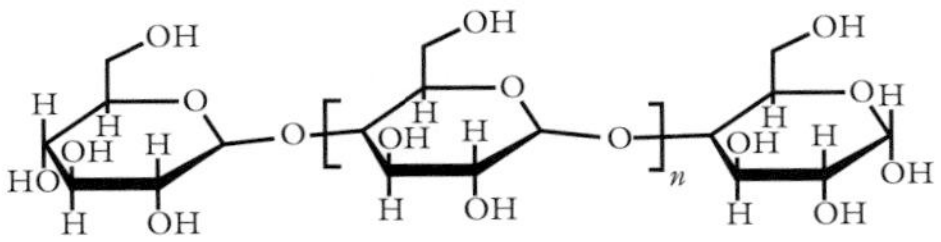

13. Amino acids contain an amino functional group, a carboxylic acid functional group, and an organic group bonded to the middle carbon atom. The group attached to the middle carbon atom distinguishes one amino acid from another.
15. The primary structure of a protein is the polypeptide chain that links amino acid monomer units. The secondary structure results from the interaction of amino acid chains to fold into various superstructures.
17. Secondary structures are held together through hydrogen bonding and disulfide bonds.
19. Tertiary and quaternary structures of proteins are held together by hydrophobic interactions, ionic bridges, disulfide bonds, and hydrogen bonds.
21. Strands of DNA are held together through hydrogen bonding of the DNA base pairs. Covalent bonding between strands would be problematic because DNA replication and other functions require the "unzipping" of the DNA strands. Covalent bonds would not allow this.
23. AATCTGCTG; three codons
25. Asn-Arg-Arg-Glu-His-Asn-Asn
27. CGCGAACGCCCCGACGGCGAC
29. Mitochondrial DNA (mtDNA) comes only from the maternal side of the family, which makes it impossible to distinguish among relatives on the mother's side of the family. However, mtDNA does not decompose as easily as nuclear DNA, and there are multiple copies of mtDNA within a single cell, making recovery of a usable sample very simple.
31. A certain carpet fiber, a certain automobile fabric fiber, and a certain animal hair could get caught on an individual's clothing or shoes en route to a crime scene and be deposited there. Any one of those fibers might not be useful by itself, but the combination could establish beyond reasonable doubt that a certain suspect was present on the crime scene.
33. 524
35. Not if their mothers were siblings.
37. Identical twins are not currently distinguishable using the known techniques.

Index

Note: Page numbers followed by f indicate figures; those followed by t indicate tables. Boldface entries indicate definitions.

B

F

H

N

O

T